Advanced Statistics in Research

Advanced Statistics in Research

Reading, Understanding, and Writing Up Data Analysis Results

Larry Hatcher
Saginaw Valley State University

Shadow Finch Media LLC

Advanced Statistics in Research: Reading, Understanding, and Writing Up Data Analysis Results

Published by
Shadow Finch Media
Saginaw, MI 48602.

Copyright © 2013 Larry Hatcher. All rights reserved. Except as permitted under U.S. Copyright Law, no part of this book may be reproduced or utilized in any form or by any means, electronic or mechanical, including but not limited to scanning, digitization, photocopying, recording, or by any information storage or retrieval system, without prior written permission from the copyright holder.

First printing 2013.

To request permission to use materials from this work, please submit written requests to:

Larry Hatcher
ShadowFinchMedia@gmail.com

Notice of Liability: The information in this book is distributed on an "as is" basic without warranty. While every precaution has been taken in the preparation of this book, the author and the publisher make no representations or warranties with respect to the accuracy or completeness of this book's contents. Neither the author nor the publisher shall have any liability to any person or entity with respect to any loss or damage caused or alleged to be caused directly or indirectly by the instructions or information contained in this book or by the products described in it.

The characters and incidents portrayed and the names herein are fictitious and any similarity to the name, character, or history of any actual person living or dead is entirely coincidental and unintentional.

Microsoft®, Excel®, Internet Explorer®, PowerPoint®, and Windows® are registered trademarks of Microsoft Corporation. PsycINFO® is a registered trademark of the American Psychological Association. SAS®, JMP®, and all other SAS Institute Inc. product names or service names are registered trademarks or trademarks of SAS Institute Inc. in the USA and other countries. SPSS® is a registered trademark of SPSS Inc. for its proprietary computer software. ® indicates USA registration.

General notice: Other brand names and product names used herein are used for identification purposes only and may be registered trademarks, trademarks, trade names, or service marks of their respective owners.

Includes bibliographical references and index

ISBN-13: 978-0-9858670-0-3 (Pbk)

ISBN-10: 0985867000

Library of Congress catalog card number 20-12919538

Printed in the United States of America

To the memory of Mary C. Miller:
When I grow up, I want to be like you.

Table of Contents

1 WHAT THIS BOOK WILL DO — 3
- Why Ruin a Good Story with Research? — 3
- Why This Book is Important — 4
- The Purpose of This Book — 6
- Book Contents — 8
- Chapter Conclusion — 12

2 BASIC CONCEPTS IN RESEARCH AND STATISTICS — 13
- Do You Speak *Statistics*? — 13
- Variables, Values and Related Concepts — 14
- Populations Versus Samples — 16
- Nonexperimental Versus Experimental Research — 17
- Tests of Differences Versus Tests of Association — 27
- Systems for Classifying Variables — 31
- The Generic Analysis Model — 41
- Assumptions Underlying Statistical Procedures — 44
- Chapter Conclusion — 48

3 CENTRAL TENDENCY, VARIABILITY, AND DESCRIPTIVE STATISTICS — 49
- This Chapter's *Terra Incognita* — 49
- The Basics of these Procedures — 50
- Study Providing Data for this Chapter's Statistics — 51
- Frequency Tables — 52
- Histograms and the Shapes of Distributions — 56
- Measures of Central Tendency — 63
- Measures of Variability — 69
- The Coefficient Alpha Reliability Estimate — 87
- Using Boxplots to Understand Distributions — 89
- Chapter Conclusion — 98

4 Z SCORES AND AREA UNDER THE NORMAL CURVE — 99
- This Chapter's *Terra Incognita* — 99
- Introduction to the Normal Curve — 100
- The Standard Normal Distribution — 103
- How Researchers Determine Area Under Curve — 108
- Area Captured by Whole-Number z-Score Intervals — 111
- Capturing the Central 95%, 99%, and 99.9% — 115
- Final Word: Is Any of this Really True? — 118

5 THE BIG-THREE RESULTS IN RESEARCH ARTICLES — 121
- This Chapter's *Terra Incognita* — 121
- Introduction to the Big-Three Results — 122
- Big-Three Results, Part 1: Statistical Significance — 125
- Big-Three Results, Part 2: Confidence Intervals — 147
- Big-Three Results, Part 3: Effect Size — 161
- Chapter Conclusion — 169

6 BIVARIATE CORRELATION — 171
- THIS CHAPTER'S *TERRA INCOGNITA* — 171
- THE BASICS OF THIS PROCEDURE — 172
- DETAILS, DETAILS, DETAILS — 175
- RESULTS FROM THE INVESTIGATION — 179
- ADDITIONAL ISSUES RELATED TO THIS PROCEDURE — 187

7 BIVARIATE REGRESSION — 195
- THIS CHAPTER'S *TERRA INCOGNITA* — 195
- THE BASICS OF THIS PROCEDURE — 196
- DETAILS, DETAILS, DETAILS — 198
- SUMMARIZING RESULTS IN A PUBLISHED ARTICLE — 212
- ASSUMPTIONS UNDERLYING LINEAR BIVARIATE REGRESSION — 220

8 PARTIAL CORRELATION AND STATISTICAL CONTROL — 223
- THIS CHAPTER'S *TERRA INCOGNITA* — 223
- THE BASICS OF THIS PROCEDURE — 224
- DETAILS, DETAILS, DETAILS — 227
- COMPUTING PARTIAL CORRELATIONS FROM RESIDUALS — 231
- SEMIPARTIAL (PART) CORRELATION — 238
- PRESENTING THE RESULTS IN JOURNAL ARTICLES — 243
- ADDITIONAL ISSUES RELATED TO THESE PROCEDURES — 244

9 MULTIPLE REGRESSION I: BASIC CONCEPTS — 249
- THIS CHAPTER'S *TERRA INCOGNITA* — 249
- THE BASICS OF THIS PROCEDURE — 250
- RESULTS FROM THE INVESTIGATION — 254
- ADDITIONAL ISSUES RELATED TO THIS PROCEDURE — 268
- ASSUMPTIONS UNDERLYING THIS PROCEDURE — 270

10 MULTIPLE REGRESSION II: ADVANCED CONCEPTS — 273
- THIS CHAPTER'S *TERRA INCOGNITA* — 273
- THE BASICS OF THIS PROCEDURE — 274
- PART 1: CONTINUOUS CONTROL VARIABLES — 280
- PART 2: CONTINUOUS AND CATEGORICAL CONTROL VARIABLES — 287
- ASSUMPTIONS UNDERLYING THIS PROCEDURE — 293

11 DISCRIMINANT ANALYSIS — 295
- THIS CHAPTER'S *TERRA INCOGNITA* — 295
- THE BASICS OF THIS PROCEDURE — 296
- DETAILS, DETAILS, DETAILS — 299
- RESULTS FROM THE INVESTIGATION — 301
- ADDITIONAL ISSUES RELATED TO THIS PROCEDURE — 312

12 LOGISTIC REGRESSION — 315
- THIS CHAPTER'S *TERRA INCOGNITA* — 315
- THE BASICS OF THIS PROCEDURE — 316
- DETAILS, DETAILS, DETAILS — 319
- EXAMPLE 1: DIRECT LOGISTIC REGRESSION — 324
- EXAMPLE 2: SEQUENTIAL LOGISTIC REGRESSION — 336
- ADDITIONAL ISSUES RELATED TO THIS PROCEDURE — 340

13 MANOVA AND ANOVA — 343
- THIS CHAPTER'S *TERRA INCOGNITA* — 343
- THE BASICS OF THIS PROCEDURE — 344
- DETAILS, DETAILS, DETAILS — 348
- RESULTS FROM THE INVESTIGATION — 356
- ADDITIONAL ISSUES RELATED TO THIS PROCEDURE — 364

14 ANALYSIS OF COVARIANCE (ANCOVA) — 373
- THIS CHAPTER'S *TERRA INCOGNITA* — 373
- THE BASICS OF THIS PROCEDURE — 374
- DETAILS, DETAILS, DETAILS — 382
- RESULTS FROM THE ANCOVA — 386
- ADDITIONAL ISSUES RELATED TO THIS PROCEDURE — 398

15 EXPLORATORY FACTOR ANALYSIS — 405
- THIS CHAPTER'S *TERRA INCOGNITA* — 405
- THE BASICS OF THIS PROCEDURE — 406
- DETAILS, DETAILS, DETAILS — 409
- RESULTS FROM THE INVESTIGATION — 412
- ADDITIONAL ISSUES RELATED TO THIS PROCEDURE — 438

16 SEM I: PATH ANALYSIS WITH MANIFEST VARIABLES — 441
- THIS CHAPTER'S *TERRA INCOGNITA* — 441
- THE BASICS OF THIS STATISTICAL PROCEDURE — 442
- DETAILS, DETAILS, DETAILS — 445
- RESULTS FROM THE INVESTIGATION — 452
- ADDITIONAL ISSUES RELATED TO THIS PROCEDURE — 473

17 SEM II: CONFIRMATORY FACTOR ANALYSIS — 477
- THIS CHAPTER'S *TERRA INCOGNITA* — 477
- THE BASICS OF THIS PROCEDURE — 478
- DETAILS, DETAILS, DETAILS — 482
- RESULTS FROM THE INVESTIGATION — 484
- ADDITIONAL ISSUES RELATED TO THIS PROCEDURE — 492

18 SEM III: PATH ANALYSIS WITH LATENT FACTORS — 495
- THIS CHAPTER'S *TERRA INCOGNITA* — 495
- THE BASICS OF THIS PROCEDURE — 496
- ILLUSTRATIVE INVESTIGATION — 498
- DETAILS, DETAILS, DETAILS — 504
- RESULTS FROM THE INVESTIGATION: OVERVIEW — 507
- STEP 1: DEVELOPING THE MEASUREMENT MODEL — 508
- STEP 2: COMPARING THREE LATENT-FACTOR PATH MODELS — 510
- ADDITIONAL ISSUES RELATED TO THIS PROCEDURE — 523

19 META-ANALYSIS — 529
- THIS CHAPTER'S *TERRA INCOGNITA* — 529
- THE BASICS OF THIS PROCEDURE — 530
- ILLUSTRATIVE INVESTIGATION — 532
- DETAILS, DETAILS, DETAILS — 533

  Fictitious Meta-Analysis on Video Game Violence ---------------------------------- 548
  Investigating Publication Bias in Meta-Analysis ---------------------------------- 554
  Psychometric Meta-Analysis -- 559
  Additional Issues Related to this Procedure -------------------------------- 569

20 LEARNING MORE ABOUT STATISTICS --------------------------------- 573
  I Want *Too* Much --- 573
  It's Raining Statistics Books -- 574
  There's This Thing Called *the Internet* ------------------------------------- 580
  There's Also This Thing Called *the Library* --------------------------------- 582
  You May Now Begin --- 583

APPENDIX A: QUESTIONNAIRE ON EATING AND EXERCISING --------------------------- 585

APPENDIX B: BASICS OF APA FORMAT (AND HOW THIS BOOK SOMETIMES DEVIATES) 595
  Introduction -- 595
  The Basics of APA Format -- 596
  Fonts (Typefaces) Used in the Manuscript -------------------------------------- 600
  Italicizing Symbols for Statistics -- 601
  Preparing Tables -- 603
  Conclusion -- 606

APPENDIX C: STATISTICAL TABLES --- 609
  Table C.1: Critical values of the χ^2 (chi-square) distribution
  Table C.2: Critical values of the t distribution
  Table C.3: Critical values of the F distribution (alpha = .05)
  Table C.4: Critical values of the F distribution (alpha = .01)

REFERENCES -- 615

INDEX --- 629

Acknowledgements

Much of this book was written while I was on a sabbatical granted by Saginaw Valley State University (SVSU) in the fall of 2010. I am grateful to Dr. Andrew Swihart, Dr. Mary Hedberg, Dr. Donald Bachand, and the SVSU Professional Practices Committee for their support related to this sabbatical.

My friends and colleagues in the SVSU Psychology Department have been a great source of encouragement and ideas. Many thanks to Dr. Margaret Borkowski, Dr. Marie Cassar, Dr. Louis Cohen, Dr. Eric De Vos, Dr. Ranjana Dutta, Dr. Julie Lynch, Dr. Jeanne Malmberg, Dr. Matthew Margres, Cheryl Michalski, Dr. Merlyn Mondol, Dr. Sandra Nagel, Dr. Gerald Peterson, Dr. Janet Robinson, and Dr. Andrew Swihart.

Thanks to Brandy Foster for editing the book. Her attention to detail, expertise on APA format, and willingness to go the extra mile helped me produce the book that I had envisioned.

Delores Gale is a dear friend who spent countless hours doing the final proofreading just prior to printing. I would not have proofed a book like this for less than $100,000, but Delores was insulted when I offered any payment at all. A friend indeed.

Finally, I owe the greatest debt to my wife Ellen. Her commitment, sacrifice, and long hours at the keyboard turned a set of computer files into a published book. In life's journey, some hope for true love, some hope for a faithful spouse, and some would settle for just a decent web-site designer. But to have all three in one person?

I mean, what are the odds?

Advanced Statistics in Research

Chapter 1

What This Book Will Do

Why Ruin a Good Story with Research?

A news correspondent is interviewing a college professor—an expert on the subject of video games. The correspondent notes that a lot of these games are violent. Does playing them cause people to be violent in real life?

The professor has heard this one a million times. He says:

> No. There have been tons of studies, and none of them show that. Playing video games makes people more excited. . . it makes their senses more excited. But it doesn't have any effect on real aggression.

The expert's comments were logical and reassuring. They were pretty much what you would have expected.

Unless you had read the research.

Consider an article by Anderson et al. (2010) in the prestigious journal *Psychological Bulletin*. The article describes a *meta-analysis*—a statistical synthesis of over 100 separate studies that investigated the effects of violent video games. The studies employed a variety of different research methods, measured multiple outcome variables, and included over 130,000 participants (in all) from the United States, Japan, and other nations.

The meta-analysis painted a much less reassuring picture. It concluded that playing violent video games was associated with significant increases in aggressive behavior, aggressive thoughts, and aggressive emotions. It showed that gaming was associated with significant *decreases* in empathy for victims as well as decreases in prosocial behavior (i.e., the willingness to help others).

The size of the effects were not large; depending on the dependent variable and research method used, they were generally somewhere between *small* and *medium* effects, according to Cohen's (1988) widely used criteria. But the findings were robust across cultures and across research designs— investigators obtained pretty much the same outcomes with cross-sectional studies, longitudinal studies, correlational studies, and true experiments.

An article like this is *gold*. It summarizes decades of research and provides information that could be of great use to the media, to lawmakers, and to harried parents trying to decide which video games to allow. But it comes with a catch:

Somebody has to read it.

Why This Book is Important

What would you call a society that awards millions of tax dollars for research on topics related to its health, prosperity, and general welfare, and then pretty much ignores that research when making decisions about how to govern and manage itself?

You would call it *our* society.

An Enigma Wrapped Inside a Bonferroni Adjustment

Most people avoid reading empirical research articles. They avoid journals like *Psychological Science, Health Psychology, The Journal of Counseling Psychology, The Journal of Educational Psychology, The Academy of Management Journal, Nature, Sport, Exercise, and Performance Psychology,* and *The New England Journal of Medicine.* They don't do this because they are lazy or stupid or apathetic. They do it—in large part—because of the statistics that the articles contain.

Just consider the terms that appear in a typical journal article: *heteroscedasticity, standard error of the estimate, Bonferroni adjustment, semi-interquartile range, failure to reject the null hypothesis, sum of squares,* and *Type II error.* And don't forget the symbols and abbreviations: ω^2, 95% CI, σ, p_{rep}, η^2, *MSE*, α, $r_{Y(1.2)}$, χ^2, and ΔR^2.

What We Could Learn by Reading Journal Articles

This collective aversion to statistics is unfortunate, because so much really good research is being done. All of us *should be* reading about this research. And any numerically literate person (college graduate or not) should be able to read about this research at the source—in the peer-reviewed academic journals where most important research first appears. These journals are bursting with articles that address important research questions—issues that affect each of us in one way or another.

Here are just a few examples of real-world questions that most of us would find interesting, along with a specific journal article that tackled this question (complete references are provided in the *References* section at the end of this book):

- Does maternal employment have a negative effect on child development? (Lucas-Thompson, Goldberg, & Prause, 2010)
- Does placing infants on their backs at bedtime decrease the likelihood of sudden infant death syndrome? (Gilbert et al., 2005)
- Does exposure to thin models in magazines contribute to eating disorders among young women? (Grabe, Ward, & Hyde, 2008)
- Are young people more alienated today than they were in 1960? (Twenge, Zhang, & Im, 2004)
- Does talking on a cell phone increase the likelihood of automobile collisions? (Horrey & Wickens, 2006)
- What are the differences between the people who do volunteer work at a homeless shelter versus those who do not? (Harrison, 1995)
- What are the differences between employees who report organizational wrongdoing versus those who see the wrongdoing and do not report it? (Miceli & Near, 1984)

And, yes, even the topic introduced at the beginning of this chapter is compelling:

- Does exposure to violent video games cause people to become more aggressive? (Anderson et al., 2010; Ferguson, 2007)

Reading the Middle of the Article

I have been teaching college courses since sometime around 1980. At the end of each semester, my students evaluate me using whatever instructor-course evaluation form is being used at the time. Of the thousands of evaluations I have received, the most memorable one came from a graduate student in an advanced statistics course who said something like this:

> Finally, I can read the middle of the article.

The student was referring to the fact that the middle of the typical research article is where the *Results* section is found—that part of the article that presents the statistical analyses. A lot of college students read the beginning of the article (the *Abstract* and *Introduction* sections), along with the end of the article (the *Discussion* section). From these disconnected parts they then attempt to piece together what was probably presented in the *Results* section. As illogical as this approach might sound, many students find it preferable to slugging their way through the *Results* section itself. The *Results* section is filled with Greek letters, p values, post-hoc tests, and dense tables with lots of notes at the bottom. They probably wonder, *what's the point*?

After taking my advanced statistics course, however, this student (the one with the memorable course evaluation) had cracked the code. The middle of the article was no longer *terra incognita*.

And if one student can do it, all students can do it. That's why I wrote this book.

The Purpose of This Book

The market offers many excellent textbooks that show how to use statistical applications (such as SAS® or SPSS® or Minitab®) to perform advanced statistical analyses. This is not one of those texts.

What This Book Will Do

This book has three interrelated goals:

1) To refresh your memory on basic concepts in research and elementary statistics.

2) To improve your ability to read, understand, and evaluate advanced statistical procedures that appear in empirical research articles.

3) To improve your ability to write about advanced statistics by summarizing results in tables, text, and figures according to standard conventions.

REFRESHING YOUR MEMORY ON THE BASICS. Reading that the F statistic for a multiple regression equation is "statistically significant" will not make much sense if you have forgotten the meaning of "statistical significance." Therefore, the first few chapters will review these basics.

I suspect that many individuals using this book have never completed an elementary statistics course (or may have completed it so long ago that those memories are no longer retrievable). If you are in this category, you will need an introduction (or a re-introduction) to the basic, foundational concepts in statistics. The first few chapters of this book provide just such an introduction. They cover concepts such as:

- The differences between experimental versus non-experimental research
- Systems for classifying variables
- Measures of central tendency and variability
- z Scores and area under the normal curve
- The meaning of statistical significance
- The interpretation of confidence intervals
- The interpretation of effect size

IMPROVING YOUR ABILITY TO READ AND UNDERSTAND ADVANCED STATISTICS. The majority of this book—the last two-thirds or so—shows how to read and understand the advanced statistical procedures that are commonly reported in empirical research journals in the social and behavioral sciences. It explains these procedures in relatively simple, non-mathematical terms, so that readers with a minimal background will be able to understand and critically evaluate these statistics when they are presented in the *Results* sections of articles. These procedures include multiple regression, discriminant analysis, exploratory factor analysis, structural equation modeling, meta-analysis, and others. For most procedures, you will learn:

- The research questions can be investigated with the procedure
- The assumptions must be met for the results to be valid
- The null hypothesis that is typically tested using the procedure
- The index of effect size that is produced
- How the results are typically presented in the text, tables, and figures of an article

IMPROVING YOUR ABILITY TO WRITE ABOUT ADVANCED STATISTICS. In addition to teaching how to read and understand advanced statistics, this book also shows how to write about advanced statistics. It illustrates the standard conventions that you should follow when summarizing the results of your own analyses in research articles.

The results presented here are prepared (for the most part) according to the *Publication Manual of the American Psychological Association 6th ed.* (2010): the format required by most journals in the social and behavioral sciences. To keep things concise, these guidelines will be referred to collectively as *APA format*.

Most of the individual chapters in this text are devoted to a specific statistical procedure. Within a given chapter, you will see:

- The symbols (e.g., t, F, η^2, M, R^2) that are widely used to represent the specific statistics produced by that procedure
- The standard conventions that are used when summarizing results from that procedure within the text (body) of the article
- The formats and approaches that are typically used when summarizing complex results in tables
- The formats and approaches that are typically used when illustrating results graphically in bar charts, path diagrams, and other types of figures

Students and instructors tell me that they need lots of examples of how to write up specific statistical procedures such as multiple regression or factor analysis. Most chapters in this book describe some empirical investigation and then provide "excerpts" from fictitious research articles that summarize the results obtained. In most cases, the chapter provides examples of how the results could be summarized in text, in tables, and in figures. Readers can use these excerpts as models to guide them as they write up the results from their own investigation.

Appendix B of this book is titled *Basics of APA Format (and How This Book Sometimes Deviates)*. The appendix touches on a few of the more important aspects of APA format: margins, preferred font, how to arrange headings, how to prepare tables, and so forth. It is a short chapter, so it will not even mention many aspects of APA format (e.g., citations within text, preparing references, and about 60,000 other things). But it will refresh the reader's memory on the basics of APA format, so it is a good place to start for those who need to write up research results.

The title includes the words ...*and How This Book Sometimes Deviates*. This is to emphasize the fact that some aspects of the research article excerpts that appear in this book are *not* consistent with APA format. For example, the tables presented here are typed using a font called *Phoenica Std Mono*, although APA Format recommends *Times New Roman*. This appendix summarizes deviations such as this in one location, so that readers will not be misled when using this book's fictitious article excerpts as models for their own manuscripts.

What This Book Will Not Do

As was noted earlier, this book is not intended to show you how to actually *perform* statistical analyses. It will not show you how to use statistical applications such as SAS® or SPSS®, and it will not show you how to use matrix algebra to perform the analyses by hand (yes, my readers—there was once a time when researchers had to do this by hand). Remember—this book is about *reading and understanding* statistics; it is not about computing them.

Intended Audience

Three groups can benefit from this book. They include:

- Undergraduate students in upper-level courses dealing with statistics and research
- Graduate students enrolled in graduate-level courses on the same topics
- Individual students, researchers, or laypeople who want to read and understand the advanced statistical procedures that often appear in scientific articles.

Do You Need To Be Good at Math?

Math plays a fairly minor role in this book. Although you will encounter formulas from time to time, they are included as supplements to your understanding, not as a focus. The concepts in this course are described in relatively simple verbal terms, not as purely mathematical concepts.

Advanced statistics are like a foreign language. If you are like most readers, you will find this book to be more like a language text than a math text.

Book Contents

This section provides a preview of things to come: (a) the topics to be covered in the five sections that constitute the book, along with (b) a general outline of what you will find in most chapters.

Topics Covered

This book is divided into five sections, with two to six chapters per section. Here is what you will find in the five sections:

SECTION I: BASIC CONCEPTS AND BASIC STATISTICS. This is the refresher section described above—the foundation chapters designed to get everyone up to speed. The chapters in this section review basic concepts in research and statistics: measures of central tendency, measures of variability, descriptive statistics, *z* scores, area under the normal curve, and the *big-three* results reported in most research articles—significance, confidence intervals, and effect size.

SECTION II: CORRELATION, REGRESSION, AND PARTIAL CORRELATION. Many of the advanced statistics covered in this book are essentially correlational statistics, so the chapters in this section are provided to refresh your memory regarding bivariate correlation and regression (*bivariate* means *two variables*). This section also includes a chapter on partial correlation and statistical control to prepare you for the more advanced procedures to follow.

SECTION III: ADVANCED STATISTICAL PROCEDURES. This section constitutes the heart of the book. It covers an array of data-analysis procedures that allow researchers to investigate the relationship between (a) one or more predictor variables and (b) one or more criterion variables. Most of the statistics in this section are the more sophisticated procedures that are seldom covered in introductory statistics courses. The section begins with chapters on basic multiple regression and hierarchical (sequential) multiple regression. From there, it moves to various members of multiple-regression's extended family: discriminant analysis, logistic regression, MANOVA, ANOVA, and ANCOVA. Each chapter shows how researchers use a specific procedure to (a) determine whether there is a significant relationship between the variables, (b) understand the nature of the relationship (as revealed by regression coefficients and similar statistics), and (c) evaluate the strength of the relationship.

SECTION IV: FACTOR ANALYSIS AND STRUCTURAL EQUATION MODELING. This section begins with a dimension-reduction procedure called *exploratory factor analysis* (*EFA*). EFA is widely used to discover the factor structure that underlies responses to questionnaire rating items (as well as other sets of variables). From EFA, the section moves to three popular applications of *structural equation modeling* (SEM). SEM is a family of statistical procedures that includes path analysis with manifest variables, confirmatory factor analysis, and path analysis with latent factors. Path analysis with manifest variables allows researchers to investigate complex causal models by analyzing correlational data. Confirmatory factor analysis (CFA) allows researchers to test hypotheses about the number and nature of the latent factors that underlie a dataset (think of it as EFA on steroids). Finally, path analysis with latent factors allows researchers to investigate complex models that propose causal relationships between latent factors (similar to the factors that are investigated with CFA). These analyses are sometimes called *LISREL analyses*, because LISREL® was the first widely-available statistical application which made them possible.

SECTION V: META ANALYSIS AND BEYOND. The first chapter in this section discusses *meta-analysis:* the statistical synthesis of results from multiple empirical studies. For many research questions, a large number of studies have already been conducted by teams of researchers working independently across the globe. Meta-analysis allows researchers to combine these results quantitatively in order to see what position is supported by the "bulk of the evidence." This chapter discusses the various indices of effect size that are reported in a meta-analyses, how researchers test for moderator variables, how they investigate possible publication bias, and the differences between the most popular approaches to meta-analysis. The section ends with suggestions for further reading.

Contents of a Typical Chapter

Different chapters in this book are organized in different ways, depending on the topic being covered. However, once you get to the main part of the book (Sections III through VI), most chapters follow an outline that goes something like this:

THIS CHAPTER'S *TERRA INCOGNITA*. This section provides a short excerpt from the *Results* section of a fictitious article. Its purpose is to illustrate the cryptic, obtuse language that researchers sometimes use when reporting results.

What, exactly, is **terra incognita?** It is the *unknown territory*. Centuries ago, map makers drew the outlines of newly-discovered continents as they were surveyed and described by explorers. The inland (and unknown) parts of the continents were sometimes labeled *terra incognita* and illustrated with fearsome monsters.

At the beginning of most chapters, you may feel that you are lost in terra incognita. But don't be afraid—by the end of the chapter, all symbols, terms, and conventions shall be revealed, and you will be the master of this statistic, not its victim.

THE BASICS OF THIS PROCEDURE. Each chapter begins at the beginning: What types of variables may be analyzed with this statistic? What research questions may we investigate? To make things more concrete, the section typically describes a specific investigation that will be used to illustrate all aspects of the procedure.

DETAILS, DETAILS, DETAILS. Some statistics require more explaining than others. For the really nasty procedures, this section introduces the basic terms and concepts that will help you make sense of things.

RESULTS FROM THE INVESTIGATION. This section constitutes the largest part of most chapters. It provides results from the illustrative investigation, and shows how the results might be presented in journal articles. You will see how researchers determine whether there is a statistically significant relationship between the variables, how they investigate trend, and how they evaluate the strength of the relationship. A major emphasis is placed on reporting practices recommended by the *Publication Manual of the American Psychological Association 6th ed.* (2010). For example, you will learn about the importance of reporting confidence intervals and effect size. Where possible, the chapter describes *best practices* for data analysis and write-up as recommended by well-regarded sources such as

Field (2009a), Howell (2002), Huck (2009; 2012), Osborne (2008), Pedhazur (1982), Stevens (1996), Tabachnick and Fidell (2007), Warner (2008), and Wilkinson and the Task Force on Statistical Inference (1999).

ADDITIONAL ISSUES RELATED TO THIS PROCEDURE. This is another *Details, Details, Details* section. It reviews the assumptions underlying the procedure, alternative data-analysis procedures, and related topics.

Is Dr. O'Day Fictitious? Is Her Research Fictitious?

Most of this book's statistical procedures are illustrated by a researcher named *Dr. O'Day*. In a typical chapter, you will see how Dr. O'Day refines a research question, gathers data, and uses the statistical procedure of interest to answer her question. You will see the problems that she encounters along the way, and how she solves those problems.

Dr. O'Day is a work of fiction. Her arch-enemy, Dr. Grey, is also a work of fiction. They are intended to add an element of drama to the book (assuming we define "drama" very, very loosely).

Most chapters in this text report an empirical investigation that produces data that must be analyzed. In some cases these investigations are completely fictitious, and others they are only semi-fictitious.

First, the investigations reported in *Chapter 5* and *Chapter 19* are entirely fictitious. For those chapters, I created fictitious data sets from scratch in order to obtain specific results that would allow me illustrate specific concepts. In those chapters, I clearly indicate that the data were contrived.

However, most of the results reported in the remaining chapters are only *semi-fictitious*. The results presented in those sections are based on the analysis of real-world data sets. This applies to most of the findings related to the so-called *Questionnaire on Eating and Exercising*, which appears in *Appendix A*. For what it is worth:

- The *Questionnaire on Eating and Exercising* is a real questionnaire.
- It was, in fact, administered to about 260 college students.
- Most of the results reported in this book are the actual statistical results that I obtained when I analyzed the resulting data.

Despite this, you must not treat these results as if they were legitimate research findings appearing in a research journal. All of these analyses were performed merely to illustrate specific statistical procedures such as multiple regression or factor analysis. In most cases, I selected specific variables for a given analysis because those variables allowed me to illustrate some specific concept (such as "a negative multiple regression coefficient"). In most cases, my analyses were not performed as part of an actual program of scientific research. Although the investigation was approved by my university's Internal Review Board, none of the results reported here have been published in any peer-reviewed journal.

After all, I had a more important audience in mind: *you*.

Chapter Conclusion

Enough with the previews—it's time to get started.

The next chapter begins at the beginning, with elementary concepts in research and statistics. Think of it as a crash course for absolute beginners. Or think of it as the refresher course for seasoned learners. Either way, this chapter lays down the foundation: the key terms and core concepts that will allow you to make sense of the advanced and multivariate procedures to follow. So take a deep breath, gather all your courage, and turn the page.

The *terra incognita* awaits.

Chapter 2

Basic Concepts in Research and Statistics

Do You Speak *Statistics*?

It has been said that learning about statistics is like learning a foreign language. This is good news and bad news.

- First the good news: Mathematically speaking, a course on statistics will not be as difficult as that course on *Partial Differential Equations* that you dropped.

- Now the bad: Those three semesters of *Spanish* were no piece of cake either.

If you have taken a course on a foreign language, you know that it is standard to begin with the basics—a few simple nouns, verbs, and expressions that provide a foundation for the more difficult material to follow. In the same way, this chapter introduces the basic terms, definitions, and concepts related to advanced statistics. It provides the conceptual frameworks that will help you understand:

- the distinction between variables, values, and observations;

- the systems that are used to classify variables;

- the *big-four assumptions* underlying most statistical procedures; and

- other concepts which are central to an understanding of statistics.

This chapter moves quickly, offering a short explanation for each concept before moving on. Its purpose is to provide just enough information so that you will be able to make sense of the more complex ideas to follow.

If you plan to venture into the distant and exotic land of statistics, you had better learn the language. Think of this chapter as your pocket dictionary.

Variables, Values and Related Concepts

A ***data set*** is a collection of records that indicates how the participants scored on one or more variables. This section provides an example of a data set, and discusses its constituent parts: observations, variables, and values.

Illustrative Data Set

Assume that you obtain information on five different variables from a small number of college students. The fictitious data are displayed in Table 2.1.

Table 2.1

College Major and Achievement Motivation Scores for 10 College Students

Observation	Name	College major[a]	Achievement motivation[b]
1	Holly	B	65
2	Brandon	E	61
3	Jonathan	P	55
4	Jerrod	P	26
5	Alysha	B	89
6	Sachiko	E	51
7	Carlos	B	77
8	Erin	P	92
9	Andre	P	83
10	Jason	E	60

Note. $N=10$.
[a] B = Business major; E = Education major; P = Psychology major. [b] Scores could range from 1-100, with higher scores indicating greater achievement motivation.

Observations

An ***observation*** is an individual case that provides the scores that you will analyze. Observations are sometimes referred to as *observational units* or *records*.

In Table 2.1, individual people serve as the observations: The participant named Holly serves as Observation 1, the participant named Brandon serves as

Observation 2, and so forth. A total of 10 observations (i.e., 10 participants) appear in the table.

In most research in the social and behavioral sciences, individual people serve as the observations. There are exceptions, however. For example, in psychology, non-human animals may serve as observations; in educational research, individual schools may serve as observations; and in business research, individual business organizations may serve as observations.

Variables

A *variable* is some characteristic of an observation which may display two or more different values (i.e., two or more different scores) in a data set. For example, one of the variables displayed in Table 2.1 is *college major*. You can see that one of the columns in the table is headed "College major," and that this column contains the letters "B," "E," and "P." A note at the bottom of this table explains that the letter "B" is used to identify business majors, "E" is used to identify education majors, and "P" is used to identify psychology majors.

College major is an example of a *classification variable*. A **classification variable** does not measure quantity; instead, it simply indicates the category to which the participant belongs. You have already seen that this variable tells us whether a given subject is in the "business major" group, the "education major" group, or the "psychology major" group. Classification variables are sometimes called **categorical variables**, **qualitative variables**, or **nominal-scale variables**.

In contrast, some of the variables in Table 2.1 are **quantitative variables:** variables that indicate *amount*. For example, the column headed "Achievement motivation" provides scores on a scale that measures achievement motivation (the desire to succeed and achieve goals). Scores on this measure may range from 1 to 100, with higher scores indicating greater amounts (i.e., higher levels) of the construct being measured.

Values

A *value* is a score or category that a variable may assume. If the variable is a classification variable, then the values simply identify the category to which a given participant belongs. For example, the classification variable *college major* could assume the values of "B," "E," or "P." These values tell us whether a given participant was in the "business major" category, the "education major" category, and so forth.

On the other hand, if the variable was a quantitative variable, then the values are almost always numbers, and these numbers reflect some actual amount or quantity. For example, Table 2.1 shows that Participant 8 (Erin) has a score of 92 on the achievement motivation variable, and the value indicates that she scores relatively high on this construct (the highest possible score was 100).

Populations Versus Samples

This section discusses the differences between populations versus samples, parameters versus statistics, and descriptive statistics versus inferential statistics. These are basic concepts that reflect the fact that researchers typically perform data analyses on *samples* of data, and then attempt to make inferences about the likely characteristics of larger *populations*.

Populations and Parameters

A ***population*** refers to the complete set of observations that constitute some specified, well-defined group. For example, a researcher may be interested in studying *students enrolled at all four-year universities in the United States*. If it were possible to obtain data from every single student enrolled at all such universities, then the researcher would be conducting research with a population.

A ***parameter*** is a value that describes some characteristic of a population. For example, imagine that a certain researcher is able to determine the age of every single student in the population of college students in the United States. Imagine that this researcher then goes on to compute the mean of these ages, and finds that the mean age is 23.52 years. If the researcher was, in fact, able to do all of this, then the resulting mean of 23.52 would be a parameter. It would be a parameter because it is a value that describes a characteristic of an entire population.

Researchers typically use lower-case Greek letters (such as μ, ω or σ) to represent parameters. For example, the standard symbol for the mean of a population is the Greek letter μ (pronounced *mu*). This means that, if the researcher actually computed the mean age for all students in the population of college students, she might report it as "$\mu = 23.52$."

Samples and Statistics

A ***sample*** is a subset of observations drawn from a population. Because it is a subset, a sample (by definition) must be smaller than the population from which it was drawn. When you think population, think *large*, but when you think sample, think *small* (smaller than the population, anyway).

Samples are important because, with most research, it is not feasible to obtain data from every single member of the population of interest. Therefore, the researcher instead collects data from a smaller sample that represents the population of interest, and analyzes data from that sample. The researcher then uses the characteristics of the sample to estimate the likely characteristics of the larger population.

For example, consider the researcher who is interested in studying the population of college students in the United States. She realizes that she will never be able to study every single member of this population, so she instead draws a sample. From the U.S. Department of Education, she obtains a list of every college student in the United States (remember that this is a fictitious example—there is no such list). She randomly selects 1,500 names from the list, and this pool of 1,500 names now constitutes her sample. She obtains data from the 1,500 participants in this sample and performs her data analyses on this subset.

Which brings us to the concept of a *statistic*. A **statistic** is a value that describes some characteristic of a sample. A statistic is used (a) to describe the sample itself, and/or (b) to make inferences about the likely characteristics of the larger population from which the sample was drawn.

Remember how an earlier paragraph made the point that Greek letters (such as μ) are typically used to represent population parameters? The same convention dictates that Latin letters (the letters of the standard alphabet used in the United States, such as *A*, *B*, *C*, and *D*) are typically used to represent sample statistics. For example, the Greek letter μ is used to represent the mean of a population, whereas the italicized Latin letter *M* is often used to represent the mean of a sample.

Descriptive Statistics Versus Inferential Statistics

A ***descriptive statistic*** is a value that is computed by analyzing sample data and is used merely to describe some characteristic of that sample (without making any inferences about the larger population from which the sample was drawn). For example, consider the researcher described above. She obtains a representative sample of 1,500 college students drawn from the much larger population of college students in the United States. She computes the mean age for these 1,500 students and finds that it is 22.17 years. If she uses this mean to simply describe the mean age of her sample of 1,500 students, then she is using it as a descriptive statistic.

In contrast, an ***inferential statistic*** is a value that is (a) computed by analyzing sample data, but is then used to (b) estimate some likely characteristic of a larger population. For this reason, inferential statistics are sometimes referred to as ***parameter estimates*** or simply as ***estimators***.

The following quote illustrates how the researcher might have used the same sample mean as an inferential statistic:

> The mean age for the representative sample of 1,500 college students was 22.17 years. Therefore, the best estimate for the mean age of an entire population of college students is also 22.17 years.

You may recall that the actual mean for the entire population was reported earlier as being μ = 23.52 years. This indicates that the researcher's "best estimate" based on the sample (M = 22.17) was somewhat inaccurate. This is to be expected when using sample data to make inferences about the larger population. The problem is called *sampling error*, and you will learn more about it in a later chapter.

Nonexperimental Versus Experimental Research

Many different research methods are used in the social and behavioral sciences. These include qualitative research, case studies, survey research, laboratory research, true experiments, quasi-experiments, and others. To simplify things, this chapter will divide most of these approaches into two very broad categories: experimental research versus nonexperimental research. It will show how the two approaches could be used to investigate a research hypothesis involving a proposed causal relationship between two variables.

Using Research to Investigate Causal Relationships

Scientists conduct research for a variety of purposes. One of their most important goals is to indentify and understand the *causal relationships* that exist between between variables.

A **causal relationship** between Variable *A* and Variable *B* exists when the values of Variable *A* bring about the values of Variable *B*. Causal relationships are also called **cause-and-effect relationships**, and are often represented using figures that include directional arrows, such as the one in Figure 2.1.

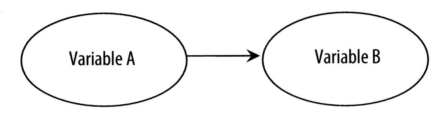

Figure 2.1. Causal relationship between Variable A and Variable B.

Sometimes researchers discuss causal relationships between *events*, rather than causal relationships between *variables*. For example, a researcher might predict that Event A causes Event B. In this scenario, Event A would be the **antecedent event** (the event that occurs earlier), and Event B would be the **consequent event** (the event that occurs later, because it was brought about by Event A).

To make things more concrete, imagine that you believe that there is a causal relationship between the consumption of caffeine and class attendance among college students. Your belief is summarized by the following research hypothesis:

> There is a positive causal relationship between the consumption of caffeine and class attendance: When students consume greater amounts of caffeine, it causes them to display higher levels of class attendance than they would display if they consumed smaller amounts of caffeine.

The nature of your hypothesis is illustrated in Figure 2.2. With this hypothesis, you define *consumption of caffeine* as the number of milligrams of caffeine (a stimulant) that are consumed per day in a sample of college students. You define *class attendance* as the number of class meetings that the students actually attend over the course of a semester for a specific class.

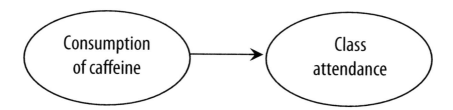

Figure 2.2. Hypothesized causal relationship between consumption of caffeine and class attendance.

If you wanted to investigate a possible causal relationship (such as the relationship described above), you would find that some research methods are better-suited for this purpose than others. The section that follows the current section shows how you could design a nonexperimental investigation to explore this hypothesis. There, you will learn that—although nonexperimental research has an important role in science—it typically does not provide terribly strong evidence of cause and effect. A later section then shows how you could instead design a different type of study—a true experiment—to investigate the same hypothesis. There, you will see why true experiments tend to provide stronger evidence of cause and effect.

Nonexperimental Research (Correlational Studies)

In **nonexperimental research**, researchers measure naturally occurring variables and investigate the nature of relationships between these variables. Nonexperimental research is sometimes called **correlational research**.

The **naturally occurring variables** described above are variables that are not manipulated or controlled by the researcher—they are simply measured as they normally exit. Typically, naturally occurring variables are **subject variables:** traits already possessed by the research participants at the time that they are observed or measured. These might be cognitive variables (such as general mental ability or achievement motivation), behavioral variables (such as helping behaviors or aggressive acts), demographic variables (such as gender or ethnic group) or other variables. Whatever they are called, the point is this: In nonexperimental research, investigators do not *manipulate* variables; instead, they simply *measure* variables as they naturally exist.

CRITERION VARIABLES AND PREDICTOR VARIABLES. In most of this book's chapters, variables are referred to as being either *criterion variables* or *predictor variables*. This section explains the difference.

In nonexperimental research, the **criterion variable** is the study's outcome variable—the variable that is the main focus of the investigation. For example, in the current investigation, the criterion variable is class attendance. You know this because class attendance is an important construct—an outcome that is of great interest to some individuals (such as college teachers). A criterion variable is sometimes alternatively referred to as the ***response variable***, the ***outcome variable***, or the ***Y variable***. In addition, some researchers refer to a criterion variable as being a ***dependent variable***. This is fine if the study is a true experiment (to be discussed shortly), but is questionable if it is a nonexperimental investigation. The reasons for this will be explained later in this chapter.

In nonexperimental research, a ***predictor variable*** is a variable that may be correlated with the criterion variable. Typically, the researcher wishes to determine whether values on the predictor variable may be used to predict scores on the criterion (hence the name *predictor variable*). In the current study, the predictor variable was consumption of caffeine. You know this because you hope to determine whether consumption of caffeine is correlated with class attendance—you hope to determine whether scores on caffeine consumption can be used to predict scores on class attendance. A predictor variable is sometimes alternatively referred to as the ***X variable***. It is also sometimes called the ***independent variable***. Calling it an independent variable is fine if the study is a true experiment, but is questionable if the study is a nonexperimental investigation. Again, the reasons for this will be covered later.

Table 2.2

Class Attendance and Amount of Caffeine Ingested for 10 College Students

Observation	Name	Class attendance[a]	Caffeine ingested/day (mg)
1	Holly	30	705.00
2	Brandon	25	678.12
3	Jonathan	15	151.32
4	Jerrod	24	421.98
5	Alysha	29	112.30
6	Sachiko	23	90.15
7	Carlos	28	0.00
8	Erin	21	237.88
9	Andre	27	249.99
10	Jason	19	74.87

Note. $N = 10$.
[a]Number of class meetings attended out of 30.

EXAMPLE OF A CORRELATIONAL STUDY. As an illustration, imagine that you conduct a simple correlational study to determine whether there is a positive association between the consumption of caffeine and class attendance, as predicted by the preceding research hypothesis. You identify a random sample of 10 college students enrolled in a General Psychology course and ask each student to report the amount of caffeine that they ingest in a typical day. From their General Psychology instructors, you also determine the number of class meetings they have attended this semester (out of a possible 30 class meetings). The results are reported in Table 2.2.

In Table 2.2, the first student is Holly. You can see that her score on the class attendance variable is "30," and her score on the caffeine variable is "705.00" mg of caffeine ingested per day. Scores for the remaining nine students may be interpreted in the same way.

In your study, the criterion variable is *class attendance*. You know class attendance is the criterion because it is the outcome variable which is the main focus of your study. In the investigation, the predictor variable is *amount of caffeine ingested*. You know that this is the predictor variable because you hope to determine whether its values can be used to predict scores on the main outcome variable, class attendance.

You analyze your data and find that there is a statistically significant positive correlation between caffeine consumption and class attendance. You will learn what is meant by *statistically significant* in *Chapter 5*, and you will learn how to interpret correlation coefficients in *Chapter 6*. For the moment, however, we will assume that the correlation is in the direction predicted by the research hypothesis, is statistically significant, and is indicative of a relatively strong relationship between the two variables.

At first glance, you might think that these results have proven that your research hypothesis is true. After all, your research has shown that there is a correlation between the two variables. But don't open the champagne just yet.

THE PROBLEM OF ALTERNATIVE EXPLANATIONS. Most researchers would see these results as constituting only weak evidence that there is a causal relationship between caffeine consumption and class attendance. In fact, many researchers see the results as providing no evidence at all. They would argue this because it is relatively easy to generate *alternative explanations* for your research findings. In this context, an **alternative explanation** is an explanation that:

- accepts that there might be a correlation between the variables, but…
- offers a plausible explanation for the observed correlation that differs from the research hypothesis.

A skeptic's criticism might go like this:

> I don't doubt that there is a correlation between caffeine consumption and class attendance, but that doesn't prove that caffeine consumption has an effect on class attendance. Maybe the two variables are correlated because of causation in

the opposite direction. In other words, maybe class attendance causes caffeine consumption: When student go to class, it causes them to ingest more caffeine so that they can stay awake.

The above is one alternative explanation for the observed correlation. A different critic might have generated an entirely different explanation. The larger point is this:

> Correlational research typically provides only weak evidence of causal relationships.

When nonexperimental research reveals a correlation between two variables, there may be dozens or even hundreds of alternative explanations that can plausibly account for the observed relationship. When a researcher implies that an observed correlation *proves* causation, it is your job—as the reader—to be skeptical. It is your job to consider alternative explanations that could reasonably account for the findings.

When researchers want stronger evidence of cause and effect, there is a better tool in the tool chest: *the true experiment.*

Experimental Research (True Experiments)

We will stay with the previous research hypothesis involving caffeine and class attendance. This section shows how a different type of investigation—an experiment—might have been designed to investigate this proposed relationship. It will also show why true experiments tend to provide stronger evidence of cause and effect.

THREE CHARACTERISTICS OF A TRUE EXPERIMENT. There are different ways of defining "true experiment," and these definitions vary according to the researcher's discipline, the topic being studied, and other factors. However, in the social and behavioral sciences, the typical ***true experiment*** displays the following three characteristics:

- *Characteristic 1:* The researcher randomly assigns participants to treatment conditions.
- *Characteristic 2:* The researcher actively manipulates an independent variable.
- *Characteristic 3:* The researcher maintains a high degree of experimental control over environmental conditions and other potential confounding variables.

Scientists value experiments because they tend to provide stronger evidence regarding causal relationships, compared to nonexperimental studies. This is because a well-designed experiment typically produces results that are less subject to the problem of alternative explanations (as described above).

EXAMPLE OF A TRUE EXPERIMENT. Imagine that you design a new investigation, this time beginning with a pool of 200 potential participants enrolled in various sections of General Psychology at a university. You randomly assign half of the participants to be in the *high-caffeine* condition. Over the course of the semester, these subjects each drink multiple cups of coffee containing a total 1,000 mg of caffeine each day (this is enough to give most of us the jitters). You randomly assign the other half of the participants to be in the *no-caffeine* condition. They also drink multiple cups of coffee, but in their case it is decaffeinated coffee containing 0 mg of caffeine. None of the subjects are told whether they were getting regular or decaffeinated coffee.

At the end of the semester, you compute the mean class attendance scores for the two groups. You determine that that the participants in the high-caffeine condition display a mean score on class attendance that is significantly higher than the mean score displayed by the no-caffeine condition. The index of effect size shows that the difference represents a large effect (you will learn about effect size in *Chapter 5*).

INDEPENDENT VARIABLES AND DEPENDENT VARIABLES. Because the current study was a true experiment rather than a correlational study, we will use different terms when describing it. Most notably, we will now refer to the variables as being independent variables and dependent variables, rather than predictor variables and criterion variables.

An ***independent variable*** is a variable that is manipulated by the researcher to determine whether it has an effect on the dependent variable. In the current study, you manipulated the amount of caffeine consumed by each participant. This means that the independent variable in the current study was the *amount of caffeine consumed*.

In contrast, a ***dependent variable*** is an outcome variable—it is some aspect of the participant's behavior or performance which may be influenced by the independent variable. In the current study, the dependent variable was *class attendance*.

In general, an independent variable will consist of two or more **treatment conditions** (also called **levels** or just **conditions**). A treatment condition is an environmental circumstance or physiological state to which the individual research participant is exposed. For example, the current caffeine study includes two treatment conditions: the high-caffeine condition and the no-caffeine condition.

Treatment conditions are sometimes discussed in terms of the *experimental condition* versus the *control condition*. In general, the **experimental condition** is the condition in which the participant receives the treatment of interest, and the **control condition** is the condition in which the participant does not receive the treatment of interest. In the current study, the treatment of interest was "caffeine." This means that the experimental condition was the high-caffeine condition, and the control condition was the no-caffeine condition.

WHY EXPERIMENTS PROVIDE STRONGER EVIDENCE. In general, experiments tend to provide relatively strong evidence of cause and effect, whereas nonexperimental studies tend to provide relatively weak evidence of cause and effect. This is because it is usually easy to generate alternative explanations for the results obtained in a nonexperimental study, but it is usually more difficult to generate alternative explanations for the results obtained in an experiment.

To understand why this is so, consider the true experiment dealing with caffeine and class attendance, as described above. Imagine that, at the end of the study, the mean class-attendance score displayed by the high-caffeine condition was significantly higher than the mean displayed by the no-caffeine condition. As the ecstatically self-satisfied researcher, you offer this explanation for the results:

> My research hypothesis predicted that, when college students consume caffeine, it causes them to display higher scores on class attendance. The current results are consistent with this hypothesis.

That is *your* explanation for the results. Can anyone generate a plausible *alternative* explanation?

It doesn't seem likely that causation could be operating in the opposite direction. After all, you were in control of the events, and you saw to it that the amount of caffeine ingested by the students (the independent variable) was manipulated at one point in time, while the number of classes that they attended (the dependent variable) occurred at a later point in time.

Some skeptic might suggest that maybe the students in high-caffeine condition just happened to be higher on some subject variable (such as *academic motivation*) before the study even began, and that is the real reason that they scored higher on class attendance. But this criticism is not terribly persuasive either. After all, you randomly assigned subjects to conditions, and the conditions were pretty large (with 100 people in each group). So it seems unlikely that the two groups could have differed substantially on academic motivation or any other variable at the very beginning of the study.

In short, it is difficult to generate any plausible explanation for the study's results other than *your* explanation—the explanation which says that consuming caffeine had a positive effect on class attendance. The larger point is this: *Experimental research typically provides stronger evidence of causal relationships, compared to correlational research.* No research method is capable of proving a hypothesis with 100 percent certainty, but when researchers need to investigate cause-and-effect relationships, experimental research is the gold standard.

BETWEEN-SUBJECTS VERSUS WITHIN-SUBJECTS RESEARCH DESIGNS. As long as we are discussing true experiments, this would be a good place to introduce some additional terms related to research design. Here, you will learn about *between-subjects* research designs versus *within-subjects* research designs.

With a ***between-subjects design*** (also called an ***independent-samples design*** or a ***between-groups design***), scores obtained under one treatment condition are independent of scores obtained under a different treatment condition. With a between-subjects research design, a given participant is assigned to just one treatment condition under the independent variable—no participant is assigned to more than one condition.

The caffeine study described above is an example of a between-subjects design. You began with a pool of 200 participants. A given participant was assigned to either the high-caffeine condition or to the no-caffeine condition. No participant was assigned to both conditions. It is called a between-subjects design because you are making comparisons *between* groups of subjects: Did the high-caffeine subjects score higher on class attendance compared to the no-caffeine subjects?

A different approach is used with the **within-subjects design** (also called the **related-samples design**, the **correlated-samples design**, the **paired-samples design**, the **repeated-measures design**, or the **within-groups design**). In a within-subjects design, scores obtained under one treatment condition tend to be correlated with scores obtained under a different treatment condition. Researchers typically create a within-subjects research design by using either (a) a repeated-measures procedure, or (b) a matched-subjects control design. The repeated-measures approach is by far the more popular, and so it will be illustrated with reference to the caffeine study.

Assume that you redesign your caffeine study so that each participant is eventually exposed to both treatment conditions under the independent variable and provides a separate score on the dependent variable under each condition. To make this realistic, we will have to assume that the General Psychology course was actually taught over a two-semester sequence (i.e., the one course was spread out over two semesters). For the sake of consistency, we will assume that the total number of participants is still 200.

Assume that, for one of these two semesters, a given participant (Subject 1) is assigned to the high-caffeine condition. During this semester, she is made to ingest 1,000 mg of caffeine each day. At the end of the semester, you record the number of class meetings that she attends out of 30. Then, during the following semester, the same participant (Subject 1, still) is now assigned to the no-caffeine condition. During this semester, she is now made to ingest 0 mg of caffeine each day and, at the end of the semester, you again record the number of class meetings that she attends out of 30. You follow a similar procedure with each of the remaining subjects.

When researchers use a repeated-measures research design, it is possible that the results may be contaminated by order effects. **Order effects** occur when scores on the dependent variable are influenced by the sequence in which subjects experience the treatment conditions. Order effects are a bad thing.

To control for possible order effects, you use counterbalancing. **Counterbalancing** is an experimental procedure in which (a) subjects are divided into subgroups, and (b) different subgroups experience the treatment conditions in different sequences in an effort to balance out the potential order effects. In this case, counterbalancing means that half of your subjects are in the high-caffeine condition during Semester 1 and are in the no-caffeine condition during Semester 2. For the other half of your subjects, the order is reversed: They are in the no-caffeine condition during Semester 1 and are in the high-caffeine condition during Semester 2.

At the end of the study, you analyze the data and find that the mean class attendance score obtained under the high-caffeine condition is significantly higher than the mean score obtained under the no-caffeine condition. An index of effect size shows that the difference represents a large effect.

You can see why this second research design is called a *within-subjects design*: Comparisons are essentially made *within* individual participants. For a given subject, was the class attendance score higher under the high-caffeine condition, or under the no-caffeine condition? Rather than making comparisons between groups, you are instead making comparisons within individuals.

Why Conduct Nonexperimental Research?

Much of this chapter has been devoted to bad-mouthing correlational research—emphasizing that correlational research tends to provide relatively weak evidence of cause-and-effect (compared to experimental research, anyway). But does this mean that nonexperimental research is worthless and should be abandoned?

Not at all. Despite its weaknesses, nonexperimental research remains an essential tool for doing good science.

Here is the bigger picture—the more *nuanced* picture—regarding the role played by correlational research methods: In science, there are many areas in which it is either physically or ethically impossible to conduct a true experiment. For these topics, nonexperimental research remains a valuable tool.

For example, imagine that you wish to investigate the following hypothesis:

> When individuals experience extreme physical abuse as children, it increases the likelihood that they will display psychological disorders such as depression when they become adults.

Having read the preceding sections, you now know that the "correct" way to investigate this hypothesis is to (a) randomly assign 100 infants to the experimental condition and 100 infants to the control condition, (b) instruct the parents in the experimental condition to physically abuse their children for the next 20 years, and (c) look up kids 20 years later and see how things turned out. (By the way, if you were not horrified by the preceding sentence, you have no business being in a statistics course—please put down this book immediately and see what offerings your university has in the way of *professional ethics*.)

As the preceding paragraph makes clear, there are some topics that just cannot be investigated using true experiments. For these situations, nonexperimental research helps us to understand the relationships as they naturally exist. And this understanding—limited as it might be—is still good for science. Nonexperimental research increases the body of knowledge related to a research topic and can lead to better models and theories.

Independent, Dependent, Predictor, and Criterion Variables

It is time to clarify our use of certain terms. Some sections of this chapter have referred to constructs as being independent and dependent variables, while other sections have referred to the constructs as being predictor and criterion variables. But this does not mean that the two sets of terms are completely interchangeable. This section provides some guidelines for appropriate use.

INDEPENDENT AND DEPENDENT VARIABLE: EXPERIMENTAL RESEARCH ONLY. Technically, the terms *independent variable* and *dependent variable* should be used only with investigations that are true experiments. In other words, it is best to use these terms only with studies in which an independent variable is actively manipulated by the researcher and a high degree of control is exercised over most aspects of the study. Technically, it is bad form to use these terms with a nonexperimental study (although you will see this done in some articles and even in textbooks).

PREDICTOR AND CRITERION VARIABLE: BOTH NONEXPERIMENTAL AND EXPERIMENTAL RESEARCH. On the other hand, when the investigation is a nonexperimental investigation, it is best to use the terms *predictor variable* and *criterion variable*. In a nonexperimental study, the researcher has not actually manipulated anything, so it is best to use the term predictor variable (rather than independent variable). Similarly, in a nonexperimental study we cannot be terribly confident that the outcome variable was actually caused by any manipulated variable, so it is best to use the term criterion variable (rather than dependent variable).

But the news gets even better. To keep things simple, most chapters in this book use the terms predictor variable and criterion variable exclusively and avoid the terms independent variable and dependent variable entirely. This is because the terms predictor variable and criterion variable are *general-purpose* terms—terms that may be used regardless of whether the study was a true experiment or a nonexperimental study. In other words, the terms predictor and criterion variable are *play-it-safe terms* that may be used with any type of research design.

Tests of Differences Versus Tests of Association

If you read much research, you will encounter many different statistics and data-analysis procedures. To make sense of these procedures, you will need a classification system that sorts them into categories. You will need multiple systems, in fact.

One such system sorts statistical procedures into *tests of differences* versus *tests of association*. This section defines the two categories and provides examples for each. It shows how either type of test could be used to investigate the previous research hypothesis involving caffeine consumption and class attendance.

This section also shows how these two categories are not really so different after all. It shows that they are merely two specific instances of a broader (and very important) conceptual framework call the *General Linear Model*.

Tests of Differences

Before defining *tests of differences*, we must first understand *measures of central tendency*. A **measure of central tendency** is a value which reflects the most typical score in a distribution. Measures of central tendency include the mean, the median, and the mode.

A statistical procedure is a **test of differences** if (a) your data are divided into two or more conditions, (b) you plan to compute some measure of central tendency (such as the mean) for each condition, and (c) you wish to determine whether there is a significant difference between these conditions with respect to the measure of central tendency. A test of differences answers questions such as "Is there a difference between the means displayed by Condition 1 versus Condition 2?"

As suggested above, most tests of differences are tests of the differences between *means*. For example, consider the "true experiment" that investigated the effects of caffeine on class attendance presented earlier. In that study, you randomly assigned 100 participants to a high-caffeine condition and assigned another 100 participants to a no-caffeine condition. At the end of the study, you determined whether the mean class attendance score for the first group was significantly higher than the mean for the second. If the right assumptions were met, you would probably analyze your data by performing an *independent-samples t test*, and this test would tell you whether there is a significant difference between the two means. The independent-samples t test is a classic example of a test of differences and will be described in greater detail in *Chapter 5*.

Tests of Association

In contrast, a procedure is a **test of association** if you are studying just one group of participants (i.e., the subjects have not been divided into sub-groups). With a test of association, you will determine whether there is a relationship between a predictor variable and a criterion variable within that group.

For example, consider the correlational study on caffeine and class attendance described in the previous section on nonexperimental research. In that study, you assessed two naturally-occurring variables within a single sample of college students. For each student, you measured the amount of caffeine that they normally consumed each day, along with the number of class meetings they attended over the course of the semester. If your data met certain statistical assumptions, it would be possible to compute a Pearson correlation coefficient (symbol: r) to determine the nature of the relationship between these two variables. Among other things, this correlation coefficient could tell you whether the students who consumed more caffeine did, in fact, display better class attendance, and whether the correlation was statistically significant. The Pearson correlation coefficient is a prototypical example of a test of association, and will be covered in *Chapter 6*.

Although the definition for *test of association* provided in this section refers to just one predictor variable and one criterion variable, there are also tests of association that allow for multiple predictor variables and/or multiple criterion variables. A classic example is multiple regression, which is covered in *Chapter 9* and *Chapter 10* of this book.

The General Linear Model

The previous sections implied that there are two types of statistical procedures. It implied that a given procedure is either a test of differences or it is a test of association, and that no procedure can be both. But this is not exactly true.

DEFINITION FOR THE GLM. Most statistical procedures that are used to analyze data from empirical studies (such as the *t* test and the Pearson correlation coefficient, *r*) are merely special cases of something called the *General Linear Model* (abbreviation: GLM). The general linear model is a framework that allows us to see the similarity between different statistical procedures that might otherwise appear to have nothing in common. Technically, the ***general linear model*** can be defined as a broad mathematical framework that views a given subject's score on a criterion variable as being the sum of a constant, one or more optimally-weighted predictor variables, and an error term (Aron, Aron, & Coups, 2006).

If the preceding sentence made no sense at all, don't panic—you will learn about "constants," "optimally-weighted predictors" and all that when you get to the chapters in this book that deal with regression. For the moment, focus instead on Thompson's (2008) approach to understanding the GLM. He describes the general linear model in terms of the following three concepts:

- All statistical analyses are correlational in nature.
- All statistical analyses produce indices of effect size analogous to the r^2 statistic (indices of effect size such as r^2 will be discussed in *Chapter 5*).
- All statistical analyses apply weights to measured variables (observed variables) so as to estimate scores on composite variables (or latent variables).

Think about the first concept listed above: "All statistical analyses are correlational in nature." It seems clear that the test of association described earlier (the Pearson correlation coefficient, *r*) is consistent with this idea, but how can a test of differences (such as the independent-samples *t* test) also be a test of association?

The GLM assumes that, when we perform an independent-samples *t* test (a test of differences), we are actually computing the correlation between (a) a criterion variable that is quantitative and continuous in nature, and (b) a predictor variable that is categorical in nature. And yes, we can compute correlations even if the predictor variable is a categorical variable. This may be easiest to understand if we see how the data could be set up for an independent-samples *t* test.

EXAMPLE OF A DATA SET. Imagine that you scaled back your caffeine experiment so that you just had 10 participants in it (instead of the 200 that had been described earlier). Table 2.3 shows how you might arrange the data for analysis once it had been gathered.

In Table 2.3, the column headed "Class attendance" indicates the number of class meetings attended by each participant by the end of the semester. The column headed "Caffeine treatment condition" is used to identify the group to which each subject had been assigned: The value of "0" is used to identify the participants who had been assigned to the no-caffeine condition (the first five participants), and the value of "1" is used to identify the participants who have been assigned to the high-caffeine condition (the last five participants). Using 0s and 1s to code group membership in this way is called ***dummy coding***.

Table 2.3

Class Attendance and Caffeine Treatment Condition for 10 College Students

Observation	Name	Class attendance[a]	Caffeine treatment condition[b]
1	Heather	15	0
2	Fred	12	0
3	Paul	16	0
4	Sheila	20	0
5	Alysha	19	0
6	Sachiko	24	1
7	Carlos	28	1
8	Andrea	26	1
9	Pierre	27	1
10	Jerome	29	1

Note. $N = 10$.
[a]Number of class meetings attended out of 30. [b]Coded 0 = no-caffeine condition, 1 = high-caffeine condition.

STATISTICAL SIGNIFICANCE. When researchers analyze data from an investigation such as this, they typically want to know (among other things) whether the results are statistically significant. When results from a sample of observations are ***statistically significant***, it means that the results observed in the sample would be unlikely if the null hypothesis were true. The ***null hypothesis*** typically states that, in the population, there is no relationship between the predictor variable and the criterion variable. You will learn all about statistical significance in

Chapter 5; for the moment, just remember that researchers typically hope that their results will be statistically significant.

DIFFERENT STATISTICAL PROCEDURES LEADING TO THE SAME CONCLUSION. If you asked a group of researchers how to analyze the data set presented above, they would probably suggest a test of group differences, such as the independent-samples *t* test. They would want to determine whether there is a significant difference between the mean score on class attendance displayed by the no-caffeine group versus the mean displayed by the high-caffeine group. To do this, they would determine whether the resulting *t* statistic is statistically significant.

But the GLM assumes that this same analysis can also be viewed as a test of association (i.e., a test of correlation). Since the predictor variable (caffeine treatment condition) has been dummy-coded, it is possible to compute the Pearson correlation (*r*) between this caffeine-treatment-condition variable and the class-attendance variable. As part of the analysis, it is fairly straightforward to determine whether the Pearson *r* statistic is statistically significant.

Here is the important part: Both data-analysis procedures will lead to exactly the same conclusion regarding the "statistical significance" of the results. If the test of differences (the independent-samples *t* test) indicates that the results are statistically significant, then the test of association (the Pearson *r* statistic) will also indicate that the results are statistically significant. Researchers will arrive at the same conclusion regardless of which type of test is used (Huck, 2009).

This is just one example of the first of Thompson's (2008) concepts related to the GLM: the idea that all statistical analyses are correlational in nature. Although an independent-samples *t* test appears to be a test of differences, it can also be viewed as a test of association. The point is this: Although the various statistical procedures covered in this book might appear to be very different from one another, the differences are more superficial than real. The majority of the procedures covered in this book are correlational in nature, even when they are called tests of differences.

Systems for Classifying Variables

When examining a research report, one of the first things a reader must evaluate is this: Did the researcher use the correct data-analysis procedure? The answer will be determined, in part, by the number and nature of the variables being analyzed. Were the variables discrete or continuous? Were they measured on an interval scale of measurement? How many values did they display?

All of these questions beg for classification schemes. The following sections introduce a number of simple systems that that can be used to clarify the nature of the variables included in a data set so that you can determine whether it has been analyzed with the appropriate statistical procedure.

Steven's Four Scales of Measurement

Almost all statistics textbooks cover Steven's (1946) *four scales of measurement*: the widely-used system in which variables are classified as being either nominal-scale variables, ordinal-scale variables, interval-scale variables, or ratio-scale variables (Stevens, 1946). The scales differ with respect to the amount and quality of information that they provide: The nominal scale provides the least information, followed by the ordinal, interval, and ratio scales (in that order).

This classification system is important because the statistical procedure that is appropriate for a given situation is determined, to some extent, by the scale of measurement demonstrated by the variables included in the analysis. The scale of measurement used with the criterion variable is particularly important. Some procedures (such as chi-square test of independence) allow the criterion variable to be assessed on a nominal scale, whereas other procedures (such as multiple regression) require that the criterion variable must be assessed on an interval or ratio scale. In this section, you will see examples of variables at each of the four scales of measurement.

NOMINAL SCALE. A *nominal-scale variable* merely identifies the group to which a participant belongs, and does not measure quantity or amount. It is essentially just a *classification variable*: a variable that identifies group membership without measuring quantity. Examples of nominal-scale variables include *subject sex* (male versus female), *political party membership* (Democrat versus Republican versus Independent) and *academic major* (English versus Business versus Secondary Education). Notice that each of the preceding variables simply identifies the group to which a person belongs; none of these variables reflect quantity.

ORDINAL SCALE. With an *ordinal-scale variable*, subjects are placed in categories, and the categories are ordered according to amount or quantity of the construct being measured. When you think of an ordinal-scale variable, think "ordered categories." Ranking variables are typically on an ordinal scale (as we shall see below).

Ordinal-scale variables do *not* display the characteristic of equal intervals. A variable displays *equal intervals* when differences between consecutive numbers represent equal differences in the underlying construct being measured.

For example, imagine that a college instructor wants to measure the construct of *social outgoingness* in each of the 30 students in her class. To do this, she rank-orders the students from 1 to 30 according to how outgoing each student is. The most outgoing student is given a value of 1, the second-most outgoing student is given a value of 2, and so forth through the last student, who is given a value of 30.

Ordinal-scale variables provide more information than would be obtained with a nominal-scale variable, but not much. This is because ordinal variables do not display the characteristic of equal intervals, defined above. For example, consider the way that the teacher measured the construct of social outgoingness in her students. Assuming that her rankings were valid, we can assume that the student ranked 1 is more outgoing than the student ranked 2, but how much more? With respect to the underlying construct of "outgoingness," the difference between student 1 and student 2 may not necessarily be equal to the difference between student 2 and student 3. As measured here, the construct of social outgoingness does not display equal intervals.

In the social and behavioral sciences, ordinal scales are often produced in two situations: (a) when researchers use *ranking variables*, and (b) when researchers use *single-item rating scales*. First, **ranking variables** are created when people, objects, or other things are placed in categories, and the categories are ordered according to some criteria. You saw an example of a ranking variable when the teacher rank-ordered her students according to social outgoingness.

In contrast, a ***single-item rating scale*** is a single statement that subjects respond to by rating themselves (or something else) on some construct of interest. For example, instead of rank-ordering her students, the teacher in the preceding example might have instead asked each student to rate himself or herself using the following single rating item:

> 1) To what extent are you socially outgoing? (Circle your response)
>
> 5 = To a Very Great Extent
> 4 = To a Great Extent
> 3 = To a Moderate Extent
> 2 = To a Small Extent
> 1 = To a Very Small Extent
> 0 = To No Extent

A given subject's score on this variable may range from 0 (indicating a relatively low level of outgoingness) to 5 (indicating a relatively high level of outgoingness).

Many (but not all) researchers would argue that scores from a single-item rating scale such as this do not display the characteristic of equal intervals. They would argue that this is because the differences between the consecutive numbers (0, 1, 2, 3, 4, and 5) do not necessarily represent equal differences in the underlying construct being measured (social outgoingness, in this case). With respect to this construct, the difference between a rating of 1 and a rating of 2 may not necessarily be equal to the difference between a rating of 2 and a rating of 3. This is why scores from single-item rating scales are often viewed as being on an ordinal scale of measurement.

For the sake of completeness, it should be noted that some researchers believe that variables assessed using single-item rating scales may, in some cases, be viewed as being on an interval scale rather than an ordinal scale. This is a controversial issue, and Warner (2008) provides an excellent review of the competing perspectives.

An ordinal scale is a very primitive measurement scale, not much more sophisticated than a nominal scale. Although there are many statistical procedures that can analyze criterion variables assessed on an interval or ratio scale (to be described below), there are relatively few procedures that can handle a criterion variable assessed on an ordinal scale. Because of this, experienced researchers typically avoid measuring variables (especially criterion variables) on an ordinal scale.

INTERVAL SCALE. An *interval-scale variable* is a quantitative variable that possesses the property of equal intervals, but does not possess a true zero point (more on this concept of a "true zero point" later). When a variable is assessed on an interval scale of measurement, the differences between consecutive numbers on the scale represent equal differences on the construct being measured. This is the characteristic that makes the interval scale more sophisticated—and more useful—than the ordinal scale.

Scores that are obtained from multiple-item summated rating scales are usually interpreted as being on an interval scale. A *multiple-item summated rating scale* (also called a *summative scale*) is a group of questionnaire items, all designed to measure the same construct. Participants respond to each individual item, often by indicating the extent to which they agree with that item. Their responses are then summed or averaged to arrive at a single score, and this score represents where the participant stands on the construct.

> For each of the following items, please circle the number which indicates the extent to which you agree or disagree with that item. Use the following response format in making your responses:
>
> 5 = Agree Strongly
> 4 = Agree
> 3 = Agree Slightly
> 2 = Disagree Slightly
> 1 = Disagree
> 0 = Disagree Strongly
>
> Circle your response
> _____
>
> 0 1 2 3 4 5   1) I am an outgoing person.
>
> 0 1 2 3 4 5   2) I like to spend time with my friends.
>
> 0 1 2 3 4 5   3) I am very friendly toward other people.
>
> 0 1 2 3 4 5   4) I enjoy socializing with others.

For example, consider the case in which the college instructor wanted to assess the construct of "social outgoingness" in her class of 30 students. Instead of ranking the students (which would have produced scores on an ordinal scale), she might instead have asked each student to complete a multiple-item summated rating scale (which would produce scores on an interval scale). Above is an example of such a multiple-item summated rating scale.

Assume that each of the 30 students in the class rates himself or herself on each of the items in this scale. For a given student, the instructor then:

- Adds together the response numbers that had been circled by the student, and
- divides this sum by the number of items (4, in this case).

For a given student, this would result in a single score that represents where the student stands on the construct of social outgoingness. Scores on this new variable could range from 0.00 (if the student *Disagreed Strongly* with each item) to 5.00 (if the student *Agreed Strongly* with each item). Scores with decimal values such as "3.25 or 2.50" would also be possible. Given the way that the items were worded, higher scores would represent higher levels of social outgoingness.

An earlier section indicated that variables on an interval scale display two characteristics: (a) equal intervals and (b) the absence of a true zero point. These characteristics will be discussed with respect to the multiple-item scale just developed.

First, as you have already learned, when a variable displays the characteristic of *equal intervals*, it means that the differences between consecutive scores on the variable represent equal differences on the underlying construct being measured. For example, consider the construct of *social outgoingness* being measured by the multiple-item scale just described: Most researchers would assume that the difference between a score of 2.00 and 3.00 is pretty much equal to the difference between a score of 3.00 and 4.00. If this assumption is reasonable, then the rating scale displays the characteristic of equal intervals.

The second defining characteristic of an interval scale is the fact that it does *not* display a true zero point. When a scale has a **true zero point**, this means that a score of zero on the scale indicates that the subject has absolutely no amount of the construct being measured. With multiple-item summated rating scales, a score of zero is often possible, but these zeros are typically not true zero points.

For example, with the rating scale measuring outgoingness (presented above), it is possible for a student to receive a score of zero if the student circled "0" (which represents *Disagree Strongly*) for each of the four items. But does this mean that the student has absolutely no amount of social outgoingness? Probably not—it probably just means that the student rates himself or herself relatively low on the construct.

Many variables that are studied in the social and behavioral sciences are assessed on an interval scale because the discipline makes such heavy use of paper-and-pencil instruments. Some examples:

- Multiple-item summated rating scales that assess personality traits such as social outgoingness, pleasantness, and emotional stability.

- Multiple-item summated rating scales that assess political attitudes such as liberal versus conservative attitudes toward national defense, or liberal versus conservative attitudes toward government-sponsored health care.
- Multiple-choice tests that assess mental abilities such as verbal skills, analytical skills, and math skills.

Notice that each of the instruments described above included *multiple* items (as opposed to *single* items). An earlier section of this chapter indicated that when a single questionnaire item is used to assess some construct, most researchers assume that the construct is being measured on an ordinal scale of measurement. However, when *multiple* items are used to measure the construct, most researchers instead assume that the construct is being measured on an interval scale.

This is probably the logical place to address the concept of a *Likert scale*. Research articles often indicate that some variable was measured using a "Likert scale" or a "Likert-type scale." In these instances, they are referring to the scale format credited to the legendary organizational psychologist, Rensis Likert (1903-1981). The format is described here because scores obtained using a multiple-item Likert scale are typically assumed to be on an interval scale of measurement.

It is important to distinguish between a *Likert item* versus a *Likert scale*. First, the prototypical **Likert item** is a single questionnaire item that uses a 5-point response format in which 1 = *Strongly disagree*, 2 = *Disagree*, 3 =*Neither agree nor disagree*, 4 = *Agree*, and 5 = *Strongly agree*. In contrast, a **Likert-scale** or a **Likert-type scale** is a multiple-item summated rating scale consisting of a group of Likert items, with all items measuring the same construct. And, yes, in most cases it is better to measure the construct using a multiple-item Likert scale rather than a single Likert item, for the reasons offered in this section.

RATIO SCALE. One level above the interval scale is the ratio scale of measurement. A *ratio-scale variable* is a quantitative variable that does both of the following:

- Displays all of the desirable characteristics of the interval scale (i.e., equal intervals)
- Displays a true zero point

Because ratio-scale variables possess a true zero point, a score of zero on a ratio-scale variable means that the participant has absolutely no amount of the construct being assessed. So you can think of a ratio-scale variable as being an interval-scale variable for which zero really means zero.

Two types of variables that are typically viewed as being on a ratio scale are (a) most *physical measurements*, and (b) *count variables*. These will be discussed in turn.

First, ***physical measurements*** are measures of the objective characteristics of physical objects or substances. Examples include measuring the length of some object in centimeters, measuring the weight of an animal in grams, and measuring the number of seconds required for a child to complete a task. Notice that, with the variables assessed in each of these examples:

- there are equal intervals between consecutive numbers (i.e, the variable displays the property of *equal intervals*), and
- a score of zero indicates that the subject has no amount of the construct being measured (i.e., the variable displays a *true zero point*).

Second, a **count variable** is a variable that reflects the number of instances that some specified event has occurred. For example, consider the instructor who wants to measure social outgoingness among her students. She might invite observers to her class, and these observers might keep track of the number of times that specific students engage in verbal interactions with other students. The rationale behind this approach is the idea that verbal interaction is one aspect of social outgoingness. The *verbal interaction* variable created by this procedure is a *count variable*: A student with a score of zero has not engaged in any interactions at all; a student with a score of 3 has engaged in 3 interactions, and so forth.

Most researchers assume that a count variable displays equal intervals and a true zero point. Therefore, most would view a count variable as being on a ratio scale of measurement.

Scale of Measurement

Characteristic of Scale	Nominal	Ordinal	Interval	Ratio
Applies names or numbers to categories?	Yes	Yes	Yes	Yes
Orders categories according to quantity?		Yes	Yes	Yes
Displays equal intervals between consecutive numbers?			Yes	Yes
Displays a "true zero point?"				Yes

Figure 2.3. Characteristics of Steven's four scales of measurement.

SUMMARY: THE FOUR SCALES OF MEASUREMENT. Figure 2.3 lists Steven's (1946) four scales of measurement and summarizes the characteristics associated with each. The left side of the figure lists characteristics that a given scale may or may not display. The figure's vertical columns represent the four scales of measurement. Moving from left to right, you see they are organized from the most primitive scale that provides the least information (i.e., the nominal scale) to the most sophisticated scale that provides the most information (i.e., the ratio scale).

Researchers in the social and behavioral sciences typically prefer to measure their criterion (or dependent) variables on an interval or ratio scale of measurement. This is because there are so many statistical procedures that may be used for criterion variables assessed on an interval or ratio scale (e.g., the Pearson correlation, multiple regression, ANOVA). By comparison, there are relatively few statistical procedures that may be used if the criterion variable is assessed on a nominal or ordinal scale.

Although Steven's (1946) conceptualization of the four scales of measurement is important in helping researchers identify the correct statistical procedure for a specific investigation, it is not the only issue that researchers must consider. In the sections that follow, you will learn about some additional classification systems that help investigators select the correct data analysis procedure for a given situation.

Discrete Versus Continuous Variables

A *discrete variable* is a variable that can assume only whole-number values. Whole numbers are values such as 5, 8, 17, 120, and so forth. Whole numbers are numbers that do *not* display decimal places (e.g., 3.12, 15.765) or fractions (1½, 7¾).

Below are some examples of discrete variables. With each, notice that there are no decimal places and no fractions.

- The number of pets that you own is a discrete variable (e.g., 0 pets versus 1 pet versus 4 pets).
- The number of times you have been married is a discrete variable (e.g., 0 times versus 1 time versus 9 times).
- The number of children that you have is a discrete variable (e.g., 0 children versus 3 children versus 7 children).

In contrast, a *continuous variable* is a variable that is capable of displaying decimal places or fractions. So a continuous variable may assume values such as 0.173, 120.3, or 9 ½. With continuous variables, the number of different values that are possible is potentially infinite (it does not have to be infinite with the specific device that you are using, it just has to be *potentially* infinite). Examples of continuous variables include:

- Body temperature (e.g., 98.6°; 99.123°).
- Length measured in inches (e.g., 9⅛ inches, 2⅞ inches).
- Time measured in seconds (e.g., 2.13 seconds, 10.876 seconds).

Some variables are viewed as being *theoretically continuous*, even though they display only whole-number values in practice. For example, the variable *subject age* might appear on the surface to be a discrete variable, as most people give only whole-number values when asked their age (e.g., "I am 54 years old"). However, most researchers view age as being a theoretically continuous variable, as it would be easy to give values with decimals or fractions if we liked (e.g., it would be appropriate to give your age as 54.5 if it had been 54 years and 6 months since you were born).

Classifying Variables According to the Number of Values

Traditionally, statistics textbooks have advised readers to rely heavily on the two classification systems covered up to this point when selecting the correct statistical procedure for a given situation. My experience in working with students, however, has taught me that these two classification systems by themselves are often inadequate. I have learned that it is also important that students carefully consider the *number of values* displayed by each of the variables to be included in an analysis. For example, does a given variable display just two values (in the way that that the variable *subject sex* displays just the values of *male* versus *female*) or does it display a great many values (in that the *Scholastic Aptitude Test* might display a value of 200, a value of 800, or any of the dozens of values in-between: 320, 480, 740, and so forth).

I have developed a simple system for classifying variables according to the number of values that they display and summarized in one of my previous books (Hatcher, 2003). This system divides variables into just three groups: dichotomous variables, limited-value variables, and multi-value variables. These categories are described below.

Some of the terms used in this system are widely-used by other textbooks, whereas others were developed specifically for the current classification system. For example, the term *dichotomous variable* is a standard term that has been used by researchers for a very long time. On the other hand, the terms *limited-value variable* and *multi-value variable* (as defined here) are unique to the current classification system. In other words, do not ask your professors about *limited-value variables* because they will have no idea what you are talking about.

Table 2.4 provides a data set that will be used to illustrate this classification system. An earlier section of this chapter described a fictitious study that investigated caffeine consumption, class attendance, and other variables in a sample of college students. Table 2.4 is an expanded version of that table.

DICHOTOMOUS VARIABLES. First, a ***dichotomous variable*** is a variable that displays just two values in the sample being studied. In Table 2.4, *subject sex* is a dichotomous variable, because it displays only the values of male versus female. Similarly, the variable *campus status* is also a dichotomous variable: Under the heading "Campus status," the value "ON" is used to identify students who live on-campus and the value "OFF" is used to identify those who live off-campus.

It is important to remember that the variables are classified not according to the number of values that are theoretically possible, but instead according to the number of values that are *actually displayed* in the sample being currently investigated. For example, if all of the participants in Table 2.4 were male, then *subject sex* would not be a dichotomous variable, because it would no longer be displaying two values.

Table 2.4

Caffeine Consumption, Class Attendance, and Related Variables in a Sample of 10 College Students

Observation	Name	Sex[a]	Campus status[b]	Probation status	Class status	Class attendance[d]	Caffeine ingested/day (mg)
1	Holly	F	ON	No-Never	FR	30	705.00
2	Brandon	M	ON	Yes-Previous	FR	25	678.12
3	Jonathan	M	ON	Yes-Current	FR	15	151.32
4	Jerrod	M	ON	No-Never	FR	24	421.98
5	Alysha	F	ON	Yes-Previous	SO	29	112.30
6	Sachiko	F	ON	Yes-Current	SO	23	90.15
7	Carlos	M	OFF	No-Never	SO	28	0.00
8	Erin	F	OFF	No-Never	JU	21	237.88
9	Andre	M	OFF	No-Never	SE	27	249.99
10	Jason	M	OFF	Yes-Current	GR	19	74.87

Note. $N = 10$.
[a]F = Female; M = Male. [b]ON = Students who live on campus; OFF = Students who live off campus. [c]FR = Freshmen; SO = Sophomores; JU = Juniors; SE = Seniors; GR = Graduate students. [d]Number of class meetings attended out of 30.

LIMITED-VALUE VARIABLES. This book defines a ***limited-value variable*** as a variable that displays just two to six values in the sample. For example, in Table 2.4, one of the headings is "Probation status." Below this heading, the following values are used to identify three categories of students:

- "No-Never" is used to identify those students who have never been on academic probation.

- "Yes-Previous" is used to identify those students who were previously on probation but are not on probation at the current time.

- "Yes-Current" is used to identify those students who are currently on probation.

A review of the tables shows that five students are in the "No-Never" category, two students are in the "Yes-Previous" category, and three students are in the "Yes-Current" category. For this variable, three values (i.e., three categories) actually appear in the sample of 10 students. This means that probation status is a limited-value variable.

MULTI-VALUE VARIABLES. Finally, this book defines a ***multi-value variable*** as a variable that displays seven or more values in the sample. In Table 2.4, the column headed "Class attendance" indicates the number of class meetings that were attended by each of the 10 students (out of 30 possible class meetings). The values appearing in this column are presented below (re-ordered from low to high):

15, 19, 21, 23, 24, 25, 27, 28, 29, 30

In short, 10 different values were displayed for the class attendance variable (each student happened to have a unique score on this variable). Because this exceeds the "seven or more" criterion, we may classify class attendance as being a multi-value variable in this sample.

Classifying Variables: Summary

If you have learned nothing else from this chapter, you have learned that there are many different ways of classifying variables. And these classification schemes are important: When researchers use statistics that are not appropriate for the variables in question, it casts doubt on how well they handled other aspects of their investigation.

Most of the chapters in this book describe specific statistical procedures such as *multiple regression* or *analysis of covariance*. Most of these chapters begin with a quick description of the nature of the variables that are usually analyzed by that chapter's statistical procedure.

Having learned these classification schemes, you will now understand when one of these sections says something like "this procedure is typically used with a limited-value predictor variable and a multi-value criterion variable assessed on an interval or ratio scale of measurement." By learning these classification systems, you now have one more set of tools for determining whether researchers have used the appropriate statistical procedures in their investigations.

The Generic Analysis Model

A ***generic analysis model*** is a figure that graphically illustrates the nature of the variables that are typically analyzed with a specific statistical procedure. The generic analysis model for a given procedure classifies these variables according to the classification systems covered in this chapter. Think of these models as templates that concisely summarize the type of situation for which a given statistic might be appropriate.

The generic analysis model for a given statistical procedure consists of a path diagram. This path diagram illustrates (a) the number of predictor variables included in the typical analysis, (b) the number of criterion variables included in the typical analysis, and (c) the nature of these variables—the number of values that they usually display, their scale of measurement, and related information.

Examples

For example, Figure 2.4 presents a generic analysis model for a statistical procedure called the *independent-samples t test*. Earlier in this chapter, you learned that an independent-samples *t* test is often used when researchers wish to determine whether there is a significant difference between the means displayed by two groups of participants.

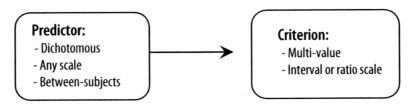

FIGURE *2.4.* Generic analysis model: Nature of the variables included in a typical independent-samples *t* test.

With any generic analysis model, the rounded rectangles on the left side of the figure represent the predictor variables in the investigation, and those on the right side represent the criterion variables. The information inside a rounded rectangle indicates the characteristics that are usually displayed by that variable.

For example, the rounded rectangle on the right side of Figure 2.4 shows that the criterion variable in an independent-samples *t* test typically (a) is a multi-value variable and (b) is assessed on an interval scale or ratio scale of measurement. The rounded rectangle on the left side of the figure shows that the predictor variable usually (a) is a dichotomous variable, (b) may be assessed on any scale of measurement (i.e., nominal, ordinal, interval, or ratio), and (c) results in a *between-subjects* research design (as opposed to a *within-subjects* research design).

When more than one rounded rectangle appears on the right side in a generic analysis model, it means that this statistical procedure usually involves multiple criterion variables. When more than one appears on the left side, it means that the procedure usually includes multiple predictors. The latter situation is illustrated in Figure 2.5.

Figure 2.5 presents the generic analysis model for *multiple regression*, a statistical procedure that is often used to analyze data from correlational studies. Notice that on the left, the predictor variables are identified as "Predictor 1" through "Predictor *p*." The three vertical dots on the left side indicate that the number of predictor variables is different in different studies.

Multiple regression is typically used in studies that include just one criterion variable. This is reflected by the fact that there is just one rounded rectangle on the right side of the figure.

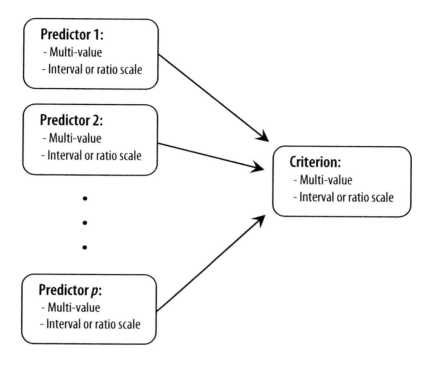

Figure 2.5. Generic analysis model: Nature of the variables included in a typical multiple regression analysis.

Caveats Regarding These Models

The generic analysis models contained in this book are merely used to describe—in shorthand form—the basic characteristics of the situations in which the statistical procedures are typically used. These models do not portray all of the situations for which a given statistic might be appropriate. Given the flexibility of many statistical procedures, it is not likely that this could be done with a single figure. Similarly, these generic analysis models do not list all of the statistical assumptions that must be met for a given data-analysis procedure to be appropriate (a ***statistical assumption*** is a characteristic that the data must display in order for the results to be correct). Most statistics have many assumptions—again, too many to include in a single figure.

The generic analysis models included in this book are meant to serve as a sort of introduction to the statistical procedures. Think of them as executive summaries of the situations in which a given procedure might be appropriate.

The system for drawing these figures—as well as the term *generic analysis model*—is once again unique to this book. So don't ask your professor "What's the generic analysis model for this procedure?" as they once again will have no idea what you are talking about.

Assumptions Underlying Statistical Procedures

Researchers have an obligation to ensure that the results of the statistical analyses that they report in research articles are correct. Among other things, the results of an analysis are "correct" when the null hypothesis significance test results in a correct decision and when the index of effect size is an accurate estimate of the actual strength of the relationship between the variables (you will learn about these concepts in greater detail in *Chapter 5*).

Researchers can increase the likelihood that their results will be correct by first verifying that they have satisfied the statistical assumptions underlying their data-analysis procedures. ***Statistical assumptions*** refer to characteristics that the data must display for the results of an analysis to be correct. For example, many data-analysis procedures are based on the statistical assumption that the criterion variable is measured on an interval or ratio scale of measurement. If the criterion variable in a specific study is actually measured on a lower scale of measurement (such as the ordinal scale), it is likely to cause the results to be incorrect. For example, the researcher might tell the world that the results are statistically significant, not knowing that she is wrong in doing so.

Most of the chapters in this book are each devoted to a specific statistical procedure. At the end of most chapters, you will find a section titled "Assumptions Underlying this Procedure," and this section will briefly list the conditions that should be met for the statistic to be valid. So as to avoid a lengthy description of the same assumptions in more than one chapter, detailed descriptions of some of the more common assumptions will be presented just once, in the current section.

This section is divided into two sub-sections. The first sub-section is titled "The Big-Four Assumptions." These are the conditions required by most parametric statistics. ***Parametric statistics*** are data-analysis procedures that have fairly strict assumptions involving population parameters. Earlier, you learned that ***parameters*** are characteristics of populations, such as the population's variance or mean. Examples of parametric statistical procedures include the independent-samples *t* test and one-way ANOVA. To give valid results, most of these parametric procedures require that the "Big-Four Assumptions" must be met.

The second sub-section is titled "Additional Assumptions." This section discusses assumptions that are more specialized, such as the assumptions underlying regression or some multivariate procedures such as MANOVA.

The Big-Four Statistical Assumptions

As explained above, many parametric statistical procedures require that the four assumptions described here should be satisfied. These assumptions are taken from Aron et al. (2006), Field (2009a), Gravetter and Wallnau (2000), Heiman (2006), Tabachnick and Fidell (2007), and Yaremko, Harari, Harrison, and Lynn (1982).

INTERVAL-SCALE OR RATIO-SCALE MEASUREMENT. Most parametric statistics require that the criterion variable should be assessed on an interval-scale or a ratio-scale of measurement. Variables that are typically viewed as being on an interval scale are constructs measured using multiple-item summated rating scales (such as political attitudes) and multiple-item paper-and-pencil tests of abilities (such as tests of math ability). Variables that are usually viewed as being on a ratio-scale include count variables (such as the number of times that a child displays some specific behavior) and most physical measurements (such as weight measured in grams).

NORMALLY DISTRIBUTED DATA. Many parametric statistics require that the sampling distribution of the statistic must be normally distributed (you will learn about sampling distributions in *Chapter 5*). The **normal distribution** is a symmetric, bell-shaped probability distribution whose exact shape is defined by a fairly complex formula. When a given variable is normally distributed in the sample, it is usually safe to assume that the sampling distribution will be normally distributed as well. Because of this, researchers typically review frequency histograms or other types of graphs to verify that their variables follow an approximately normal distribution. Even when the sample is not normally distributed, a mathematical principle called the *central limit theorem* tells us that the sampling distribution will still be normally distributed as long as the sample includes 30 or more observations.

Some statistics do not necessarily require that the observed variables must be normally distributed, but instead require that the errors in the model must be normally distributed. Multiple regression is an example of a statistical procedure that makes this assumption, and you will learn about the concept of "errors" when you read about this procedure in *Chapter 9*.

HOMOGENEITY OF VARIANCE. The way that we describe *homogeneity of variance* differs somewhat, depending on whether we are discussing a test of differences versus a test of association. When performing a test of group differences (such as an independent-samples *t* test), ***homogeneity of variance*** means that sample data in the various treatment conditions must come from populations with equal variances. When this assumption is met, we say that the variances are ***homogeneous*** (i.e., "similar to one another"). When this assumption is not met, we say that the variances are ***heterogeneous*** (i.e., "different from one another"). Researchers can perform a variety of statistical tests to determine whether this assumption has been violated. Examples include the *F*-max test and Levene's test for equality of variances. It is typically not required that each treatment condition must have exactly the same number of subjects, but when the sample sizes are not equal, the homogeneity assumption becomes more important.

When performing a test of association involving continuous quantitative variables (such as bivariate regression or multiple regression), this homogeneity assumption translates into a requirement that the variance of one variable should be the same at each level of the other variable. When this assumption is met, we say that the data display ***homoscedasticity;*** when it is not met, we say that they display ***heteroscedasticity.*** Researchers typically inspect plots of residuals to verify that this assumption is met.

INDEPENDENCE OF OBSERVATIONS. If there is no consistent, predictable relationship between one observation and a second observation, the two observations are said to be ***independent***. Many statistical procedures require that observations in the sample must be independent observations. The independence assumption is met if no observation or score on the criterion variable is influenced by or is predictable from any other observation or score on the criterion variable. In behavioral science research, the independence assumption is often satisfied by drawing random samples of participants who are not related to each other in any way (Gravetter & Wallnau, 2000).

Find the preceding just a wee-bit abstract? An easier (and more concrete) way to understand the independence assumption is to think about the ways that researchers sometimes violate it. Here are two examples:

1) *The researcher obtains repeated measures from the same subject.* When the researcher obtains repeated measures on the criterion variable from the same subject, the observations are no longer independent (the score that a given subject provides at one point in time will probably be predictive of the score that he or she will provide at a later point in time). This does not mean that researchers must never take repeated measures from the same subjects, because there are many statistical procedures that are designed to analyze repeated-measures data. For example, it may be okay for a researcher to obtain repeated measures from the same subjects as long as he analyzes the data using a paired-samples t test (because the paired-samples t test is designed for repeated-measures data). However, it is *not* okay for the researcher to obtain repeated-measures data and analyze them using the independent-samples t test (because this t test assumes that the observations are independent).

2) *The researcher collects scores from students who share the same classroom.* Since the students in the same class have the same teacher, it is likely that the score provided by one student will be systematically related to the score provided by another student—there is probably some similarity in their scores due to the fact that they share the same teacher (among other things). Because of this, the independence assumption has been violated. Again, this does not mean that researchers must never collect data from students in the same class, because there are statistical procedures that are designed to analyze data of this sort (one such procedure called *hierarchical linear modeling*). However, these data must not be analyzed using statistical procedures which assume independence of observations (such as linear multiple regression).

Additional Assumptions

The previous section covered the big-four assumptions—the assumptions that are common to most of the widely-used workhorse statistics such as t tests, regression, and ANOVA. As we branch out into more specialized statistics, however, we encounter more specialized assumptions. In no particular order, some of those assumptions are described here.

LINEARITY (LINEAR RELATIONSHIP). This assumption is common among some tests of association in which the researcher investigates the nature of the relationship between a quantitative predictor variable (i.e., an *X* variable), and a quantitative criterion variable (i.e, a *Y* variable). When two variables satisfy the assumption of **linearity**, it means that, with each increase of one unit on the predictor variable, there is a change in the criterion variable of a constant amount. The result is a scatterplot through which it is possible to fit a best-fitting straight line (a *scatterplot* is a type of graph that will be covered in *Chapter 6*).

The linearity assumption is violated when there is a *curvilinear* relationship (also called a *nonlinear* relationship) between the predictor variable and the criterion variable. A **curvilinear relationship** means that, as the *X* scores increase, the *Y* scores change, but they do not change in just one consistent manner. As the *X* scores change, the *Y* scores may change by differing amounts, or they may even change their *direction* of change (e.g., by creating a U-shaped distribution). As the name would suggest, curvilinear relationships can only be fit by a curved line. *Chapter 7* presents figures which graphically illustrate these relationships.

BIVARIATE NORMALITY. With statistical procedures based on correlation and regression, it is often assumed that the pairs of scores are sampled from a bivariate normal distribution. A distribution displays **bivariate normality** when, for any specific score on one variable, the scores on the second variable follow the shape of the normal distribution.

For example, assume that you are computing the correlation between IQ scores and college grade-point average (GPA). Your distribution would display bivariate normality if:

- At each score on the IQ variable, the scores of the GPA variable follow a normal distribution, and
- At each score on the GPA variable, the scores of the IQ variable also follow a normal distribution.

MULTIVARIATE NORMALITY. Statistical procedures that include more than one criterion variable are collectively referred to as **multivariate statistics**. With these procedures, the assumption of normally-distributed data (described above) often translates into an assumption of **multivariate normality**. This assumption requires that:

- Each variable is normally distributed.
- Each linear combination of variables is normally distributed.
- The residuals (errors) of the analysis are normally distributed and independent.

At this time, there are no simple and effective ways of assessing multivariate normality in a direct fashion. Instead, researchers typically inspect the normality, homoscedasticity, and linearity of the individual variables. When the individual variables fail to satisfy these basic assumptions, it casts doubt on likelihood that the data set as a whole satisfies the assumption of multivariate normality (Tabachnick & Fidell, 2007).

ABSENCE OF MULTICOLLINEARITY. The issue of multicollinearity is important for some statistical procedures that include multiple predictor variables (such as multiple regression). Predictor variables display ***multicollinearity*** when they are strongly correlated with one another. One rule of thumb says that predictor variables display multicollinearity when the Pearson correlation (r) between them is greater than ±.80 or perhaps ±.90 (Field, 2009a). Some statistics textbooks provide relatively sophisticated approached for investigating potential multicollinearity (see, for example, Field, 2009a, as well as Tabachnick & Fidell, 2007).

Chapter Conclusion

Earlier, this book warned you that learning statistics is like learning a foreign language. In this crowded chapter, you have learned the basic nouns and verbs of the dialect. More important, you have learned *schemas:* conceptual frameworks that you will use to organize and make sense of these terms.

This leaves you ready to begin where the *Results* section of most research articles begins: with the descriptive statistics. Most articles begin by telling us something about the nature of the participants and the variables included in the study: How old was the typical subject? Was there a lot of variability in the age of the subjects? How reliable were the scales used to measure the study's variables?

You cannot make sense of this information until you can make sense of measures of central tendency, measures of variability, and the other indices that fall under the general heading of *descriptive statistics*. These concepts await in the following chapter.

Chapter 3

Central Tendency, Variability, and Descriptive Statistics

This Chapter's *Terra Incognita*

You are reading a research article. It describes a study in which the investigator measured a variable called *health motivation* in a sample of 238 college students. In the following excerpt, she describes this variable:

> A multiple-item summated rating scale was used to assess health motivation. Scores on this variable could range from 1.00 to 7.00, with higher values indicating greater motivation. The mean score (M = 5.76) was lower than the median (Mdn = 6.00), indicating that the distribution might be negatively skewed. This was confirmed by a skewness statistic that was negative and statistically significant, skew = −1.16, SE = 0.16, z = −7.25, p < .001. A frequency boxplot revealed a long tail, outliers, and extreme values in the direction of lower scores. Despite this asymmetrical distribution, coefficient alpha reliability for the multiple-item scale was acceptable, α = .83.

Do not panic. All shall be revealed.

The Basics of these Procedures

This section introduces *descriptive statistics:* measures of central tendency, measures of variability, and other statistics that represent key characteristics of the study's variables. It shows how to interpret frequency tables, boxplots, and other exhibits that illustrate the shape of variable distributions. These statistics, tables, and figures quickly communicate the most basic characteristics of the variables being investigated:

- Was the frequency distribution for this variable symmetrical? Was it skewed? Were there outliers?
- What score was the most typical score?
- Did the participants display a lot of variability in their responses?

Why Descriptive Statistics Are Important

In some cases, the answers to the preceding questions are directly relevant to the study's main research questions. In other cases, they are not. But you, as the reader of the article, should give the *Descriptive Statistics* section of the article a close reading, as it will help you decide just how seriously you should take the investigation. Among other things, the descriptive statistics section reveals the following:

- How large was the sample? Was it large enough to produce reasonably precise estimates and confidence intervals?
- Is there evidence that the variables were skewed? If yes, did the researcher analyze data using a procedure that actually requires a symmetrical distribution?
- Did the researcher screen the data for outliers or other data points that could lead to misleading findings?
- Do the variables appear to satisfy the assumptions required by the statistical procedures used? Did the researchers discuss these assumptions?

Preview of Things To Come

This is a crowded chapter, because there are a wide variety of statistics, figures, and other exhibits that researchers use to describe their data. Here is a quick description of the topics to be covered:

FREQUENCY TABLES. The frequency table for a given variable typically presents the various values displayed by that variable, along with frequencies and percentages for each value. This chapter's section on frequency tables serves as a foundation for related concepts such as percentiles and quartiles.

FREQUENCY HISTOGRAMS. The frequency histogram for a given variable is a graph (similar to a bar chart) in which the length of the bars represent the frequencies of the values. Frequency histograms make it easy to understand the shape of a distribution, so this section leads to related topics such as the normal distribution, skewness, and kurtosis.

MEASURES OF CENTRAL TENDENCY. Measures of central tendency are statistics that reflect the most typical value in a distribution. Examples include the mean, the median, and the mode.

MEASURES OF VARIABILITY. Measures of variability are statistics that indicate the amount of spread that is displayed by a distribution of scores. Examples include the range, the interquartile range, the variance, and the standard deviation.

COEFFICIENT ALPHA. Coefficient alpha is a *reliability estimate*—a statistic that represents the consistency or stability displayed by a set of scores. Coefficient alpha is widely used to estimate the reliability of scores created from multiple-item summated rating scales.

BOXPLOTS. Boxplots are figures used to graphically illustrate the median, the interquartile range, and other basic characteristics of a frequency distribution. Boxplots are particularly useful for determining whether a variable displays outliers or extreme values.

A separate section is devoted to each of the preceding topics. Most sections provide a definition, a conceptual formula (when appropriate), and one or two examples.

Study Providing Data for this Chapter's Statistics

The various tables, figures, and descriptive statistics to be covered in this chapter come from a fictitious investigation conducted by a fictitious researcher. The researcher is named Dr. O'Day, and she is interested in attitudes and behaviors related to eating, exercising, and health.

The Questionnaire on Eating and Exercising

As a part of her program of research, Dr. O'Day developed an instrument called the *Questionnaire on Eating and Exercising*. A copy of the questionnaire appears in *Appendix A* of this book.

She administered the questionnaire to 260 students enrolled in a variety of psychology courses at a regional state university in the Midwest, and obtained usable responses from 238 participants. To understand the general nature of the variables measured by this questionnaire, she created frequency tables and graphs and computed the descriptive statistics that you will see in this chapter.

Multiple-Item Summated Rating Scales

Many of the variables analyzed by Dr. O'Day were measured using multiple-item summated rating scales. For example, one such variable was *health motivation*.

This variable was measured by items 27-29 on the questionnaire which appears in *Appendix A*.

Subjects responded to each of these items using a 7-point Likert-type scale whose values ranged from 1 (*extremely unlikely*) to 7 (*extremely likely*). For a given subject, Dr. O'Day computed the average of that subject's responses to items 27, 28, and 29. The resulting mean was that participant's score on the *health motivation* variable. Scores could range from 1.00 to 7.00, with higher scores reflecting higher levels of health motivation. When subjects scored high on health motivation, it indicated that it was important to them to stay healthy and avoid medical problems.

The questionnaire in *Appendix A* included a number of additional multiple-item summated rating scales designed to measure additional variables. For example, the variable *desire to be thin* was measured by items 31-33, the variable *appetite for high-fat foods* was measured by items 35-37, and so forth. In most cases, headings on the questionnaire identify the variable being assessed by a given group of questionnaire items.

Other Items

With the *Questionnaire on Eating and Exercising*, other types of items were used to measure other variables. For example, Item 60 allowed participants to check a category to indicate their sex, and Item 62 allowed them to write a number to indicate their age in years. Some of these items produced nominal-scale categorical variables, and others produced roughly continuous quantitative variables.

Data Analysis and Results

Dr. O'Day performed a variety of analyses to investigate the basic nature of the data obtained from this questionnaire. These analyses included the following:

- She created frequency tables to summarize the frequency of responses to specific questions.
- She computed means, standard deviations, and similar statistics to understand the central tendency and variability of subject responses.
- She created frequency histograms and boxplots to understand the shape of distributions and to identify potential outliers.

The remainder of this chapter presents Dr. O'Day's findings. When appropriate, it shows how Dr. O'Day might have prepared her findings to be included in a published article.

Frequency Tables

A *frequency table* presents the values (i.e., the scores or categories) that a variable displays in a sample, along with information about the number of participants who display each value. This information may take the form of simple frequencies, proportions, percentages, or other statistics.

Frequency tables are not a standard part of the descriptive statistics section of a published article. They do, however, provide an excellent starting place for the analyses in which researchers will first get acquainted with the data.

Example: Frequency Table for Age

In the *Questionnaire on Eating and Exercising*, Item 62 asked participants to indicate their age in years. Table 3.1 presents the frequency table which summarizes subjects' responses to this item. A note at the bottom of the table tells us $N = 238$, indicating that data from 238 participants are represented.

Table 3.1

Frequencies Table for Age

Value	Frequency	Relative frequency	Percent	Cumulative percent
57	1	.004	0.4	100.0
47	2	.008	0.8	99.6
44	1	.004	0.4	98.7
43	1	.004	0.4	98.3
39	1	.004	0.4	97.9
34	1	.004	0.4	97.5
31	1	.004	0.4	97.1
30	2	.008	0.8	96.6
29	1	.004	0.4	95.8
27	3	.013	1.3	95.4
26	2	.008	0.8	94.1
25	2	.008	0.8	93.3
24	2	.008	0.8	92.4
23	7	.029	2.9	91.6
22	11	.046	4.6	88.7
21	31	.130	13.0	84.0
20	33	.139	13.9	71.0
19	73	.307	30.7	57.1
18	63	.265	26.5	26.5

Note. $N = 238$.

VALUES. Table 3.1 consists of five columns. The column headed "*Value*" provides the scores or categories that the variable displayed in this sample. At the top of Table 3.1, this column displays the value "57," indicating that at least one participant in the study was 57 years old. At the bottom of the table, this column displays the value "18," meaning that at least one participant was 18 years old. In this table, the values are ordered from largest value (at the top) to smallest value (at the bottom), but frequency tables sometimes arrange values in the opposite order.

SIMPLE FREQUENCIES. The column headed "Frequency" presents *simple frequencies:* the raw number of participants who displayed a given value. The table shows that there was one subject at age 57, two subjects at age 47, and so forth—these are simple frequencies for those values. Simple frequencies are often represented by the symbol "*f*."

RELATIVE FREQUENCIES (PROPORTIONS). Simple frequencies can be difficult to interpret because they do not indicate the relative size of the group of participants who displayed a given score in relation to the total size of the sample. For example, the table shows that the simple frequency for people at the age of 18 is 63. This would be a relatively large number if there were only 70 people in the entire sample, but it would be a relatively small number if there were 1,000 people in the sample.

One solution to this problem is to report *relative frequencies,* (symbol: *rel. f*) also known as *proportions.* These appear in the third column of the table. For a given value, its relative frequency is computed by dividing the simple frequency for that value by the total sample size, N, as is illustrated in this formula:

$$rel.f = \frac{f}{N}$$

For example, the table shows that the simple frequency for participants at age 18 is $f = 63$. You will also recall that the total sample size is $N = 238$. The relative frequency is therefore computed by inserting these values into the formula:

$$rel.f = \frac{f}{N} = \frac{63}{238} = .2647 = .265$$

This means that the relative frequency (or proportion) of people at age 18 is .265. The relative frequencies for the remaining values in the table were computed in the same way.

Relative frequencies may range from a low of .000 to a high of 1.000 and may be reported to any number of decimal places, depending on the desired level of precision. They are useful because they are interpreted in the same way regardless of the sample size. For example, if the relative frequency for a given score is .25, it means that one-fourth of the subjects displayed that score, regardless of the total number of people in the sample.

PERCENTAGES. A *percentage* is a relative frequency multiplied by 100. Here is the formula:

$$percentage = \left(\frac{f}{N}\right) \times 100$$

Converting a relative frequency (proportion) for a given value into a percentage is simple: Take the relative frequency, move the decimal point two places to the right, and the result is the corresponding percentage. Table 3.1 shows that the relative frequency of participants at age 18 is .265; this means that the 26.5% of the participants are at age 18. From this, you can see that a percentage communicates essentially the same information as a relative frequency—the only difference lies in the placement of the decimal point.

The percentages in Table 3.1 provide a general sense of how the ages of the participants were distributed in Dr. O'Day's sample. A little over one-fourth of the subjects (26.5%) were at age 18, and no participant was younger than 18. This makes sense, given that the subjects were all college students. The 19-year-olds constituted the single largest age group, representing 30.7% of the sample. The percentages trail off after age 19: The 20-year-olds constituted 13.9% of the sample, the 21-year-olds constituted 13.0% of the sample, and so forth.

CUMULATIVE PERCENTAGES. The last column of Table 3.1 presents cumulative percentages (symbol: *cum. %* or *Σ%*). The ***cumulative percentage*** for a given value indicates the percent of participants who displayed that value or a lower value. If you begin at the top Table 3.1 (in the column headed "Cumulative percent") and read down one score at a time, you can see that:

- 100% of the subjects were at age 57 or younger.
- 99.6% of the subjects were at age 47 or younger.
- 98.7% of the subjects were at age 44 or younger.

If you skip down to the bottom of the table, you can see that:

- 57.1% of the subjects were at age 19 or younger.
- 26.5% of the subjects were at age 18 or younger.

Percentiles and Quartiles

Once researchers have constructed a frequency table such as Table 3.1, it is fairly straightforward to slice it into 100 little slivers (to determine percentiles) or slice it into four larger chunks (to determine quartiles). This section explains how this is done.

PERCENTILES. A ***percentile*** is one section of a frequency distribution that has been divided into 100 equal sections, with $1/100^{th}$ of the participants appearing in each section. The percentile for a given value indicates the percent of participants who displayed that value or a lower value. For example, if a given participant scored at the 81st percentile, it means that 81% of the participants displayed that score or a lower score. Sound familiar? Yes, it is very similar to the concept of *cumulative percentage*.

Consider the cumulative percentages in Table 3.1. If a student in this study were at age 20, we would say that he is at the 71st percentile with respect to the age variable. This is because 71% of the subjects were at age 20 or younger. If a student in this study were at age 21, we would say that she is at the 84th percentile (because 84% of the subjects were at age 21 or younger).

A common convention is to represent percentiles using the letter "*P*" with numeric subscript to indicate specific percentiles. For example, subjects in the current study who were 20 years old would be at P_{71}, and subjects who were 21 years old would be at at P_{84}.

QUARTILES. A *quartile* is one section of a frequency distribution that has been divided into four equal sections with 25% of the participants in each section. Quartiles are typically represented by the letter "*Q*" with numeric subscripts:

- If a given value is at the first quartile (symbol: Q_1), it means that 25% of the subjects displayed that score or a lower score. In other words, Q_1 is equivalent to the 25th percentile, P_{25}.
- If a given value is at the second quartile (symbol: Q_2), it means that 50% of the subjects displayed that score or a lower score. In other words, Q_2 is equivalent to P_{50}. The second quartile is therefore also equal to the *median*, an important measure of central tendency that will be discussed later in this chapter.
- If a given value is at the third quartile (symbol: Q_3), it means that 75% of the subjects displayed that score or a lower score. In other words, Q_3 is equivalent to P_{75}.

Understanding this concept of quartiles is essential to understanding *boxplots*: an important type of graph that is often used to investigate the shape of a distribution and to determine whether the distribution displays outliers. Boxplots are covered in a section near the end of this chapter.

Histograms and the Shapes of Distributions

Researchers use different types of graphs to understand the *shape* of a frequency distribution. These graphs help them determine whether a given variable displays an approximately normal distribution, a skewed distribution, or some other departure from normality.

One widely used graph is the *frequency histogram:* a graph that plots a variable's values on the horizontal axis and the frequencies for each value on the vertical axis. A histogram looks like a bar chart, except that the bars typically touch each other in a histogram, whereas they typically do not touch in a bar chart.

Figure 3.1 displays the frequency histogram for *healthy-diet behavior,* one of the variables measured by a multiple-item summated rating scale on the *Questionnaire on Eating and Exercising.* Scores on this variable could range from 1.00 to 7.00, and scores with decimal values such as 2.40 and 5.20 were possible. High scores indicated that the subject was eating a healthy diet (consisting largely of fruits and vegetables), and low scores indicated that the subject was eating an unhealthy diet (consisting largely of junk food).

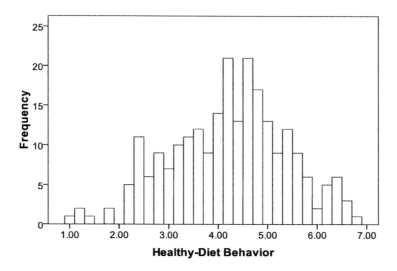

Figure 3.1. Frequency histogram for scores from the *healthy-diet behavior* scale; this variable displays an approximately normal distribution.

Points on the horizontal axis of the histogram represent the values that the variable could display. The figure shows that these values ranged from 1.00 to 7.00. Each vertical bar appearing above one of these values represents a simple frequency: the number of people who actually displayed that value in the sample. In Figure 3.1, you can see that most subjects scored somewhere between 3.00 and 5.00 on this 7-point scale—very few people displayed really low scores (such as 1.00 or 1.20), and very few displayed really high scores (such as 6.80 or 7.00).

Normal and Approximately Normal Distributions

One of the reasons that researchers create frequency histograms is to determine whether their variables follow an approximately normal distribution. This is important, because some statistical procedures are based on an assumption (i.e., a statistical requirement) that the data must be normally distributed. Such statistics will not provide correct results unless this assumption is satisfied.

THE PERFECT NORMAL DISTRIBUTION. The concept of an approximately normal distribution will not make sense unless we first understand the concept of the **perfect normal distribution**. The **theoretical normal distribution** (also called the *perfect normal distribution,* the *true normal distribution,* or the *Gaussian distribution*) is a bell-shaped, symmetrical probability distribution whose form is described by a precise mathematical formula. Figure 3.2 illustrates this theoretical normal distribution.

From Figure 3.2, you can see that the perfect normal distribution is a smoothed-out frequency histogram. The horizontal axis plots the values of some quantitative variable (from low scores on the left to high scores on the right), and the vertical axis plots frequencies for these values.

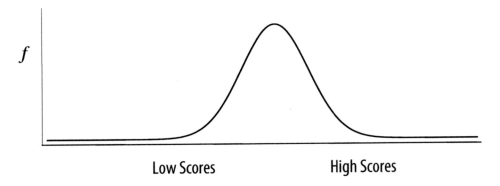

Figure 3.2. The theoretical normal distribution.

When conducting investigations in the real world, researchers just about never encounter variables that display a theoretical perfect normal distribution. It is, after all, a *theoretical* distribution. Among other things, the tail on the left side of the perfect normal distribution extends in the direction of negative infinity, and the tail on the right side extends in the direction of positive infinity. Think of this distribution as a model—an ideal that is never quite achieved.

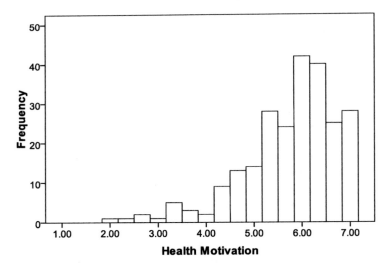

Figure 3.3. A distribution that displays a negative skew.

APPROXIMATELY NORMAL DISTRIBUTIONS. On the other hand, when conducting research in the real world, researchers often encounter ***approximate normal distributions***: variables that produce histograms that are symmetrical, bell-shaped, and follow the basic form of the theoretical normal distribution. Scores on the healthy-diet behavior scale (presented earlier as Figure 3.1) is an example of such a variable. As long as the variable is approximately normal, most investigators assume that they have satisfied the requirement of normally distributed data.

Skewness in Distributions

A distribution is said to be *symmetrical* if the left side of the distribution is a perfect mirror image of the right side. If a distribution is not symmetrical, it is said to be *skewed*. With a **skewed distribution**, one of the tails is longer than the other tail. Two types of skew are possible.

NEGATIVE SKEW. When a distribution displays a **negative skew**, it means that most of the observations are bunched up where the higher scores appear, and there is a long tail in the direction of the lower scores. Figure 3.3 presents a frequency histogram for the *health motivation*—a variable which displays a negative skew.

POSITIVE SKEW. When a distribution displays a **positive skew**, it means that most of the observations are bunched together where the lower scores appear, and there is a long tail in the direction of the higher scores. Figure 3.4 presents a frequency histogram for the variable *age*. Each bar in this histogram represents the simple frequencies for two age categories combined: The first bar gives the frequencies for 17-year-olds and 18-year-olds combined; the second bar gives the frequencies for 19-year-olds and 20-year-olds combined; and so forth. You can see that the age variable displays a positive skew.

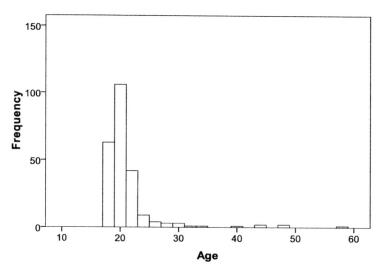

Figure 3.4. A distribution that displays a positive skew.

Here is a mnemonic device (i.e., a memory trick) to help you remember the difference between a negative skew versus a positive skew:

- If the distribution has one long tail that points in the direction of the lower (more negative) values, it has a negative skew.

- On the other hand, if the distribution has one long tail that points in the direction of the higher (more positive) values, it has a positive skew.

THE SKEWNESS STATISTIC. Researchers sometimes report a skewness statistic whose values indicate the nature and severity of the skew. With this statistic:

- a value of zero means that the distribution is not skewed at all,
- a negative value means that the distribution displays a negative skew, and
- a positive value means that the distribution displays a positive skew.

This skewness statistic is presented in Table 3.2. This table actually presents results for nine different variables from Dr. O'Day study (e.g., age, hours employed, healthy-diet norms, and others). For most variables, it presents results on seven different statistics (e.g., the mean, the median, the standard deviation, and others). For the moment, however, we will just focus on just the skewness statistic. This index appears in the next-to-last column of the table.

Table 3.2

Means, Standard Deviations, and Other Descriptive Statistics.

Variable	M	Mdn	SD	α	Range	Skewness (SE)	Kurtosis (SE)
Age	20.58	19.0	4.85	--	18-57	4.53(0.16)	24.18(0.31)
Hours employed	11.88	7.5	14.22	--	0-72	1.11(0.16)	0.98(0.31)
Healthy-diet norms	4.83	5.0	1.42	.87	1.00-7.00	-0.82(0.16)	0.18(0.31)
Healthy-diet attitudes	6.04	6.2	0.92	.88	1.60-7.00	-1.79(0.16)	4.46(0.31)
Healthy-diet behavior	4.22	4.2	1.17	.72	1.00-6.80	-0.19(0.16)	-0.30(0.31)
Health motivation	5.76	6.0	0.98	.83	2.00-7.00	-1.16(0.16)	1.55(0.31)
Desire to be thin	5.30	5.67	1.35	.85	1.00-7.00	-0.92(0.16)	0.42(0.31)
Appetite high-fat foods	4.25	4.33	1.49	.88	1.00-7.00	-0.07(0.16)	-0.76(0.31)
Current physical health	5.46	5.75	0.94	.80	1.50-7.00	-1.00(0.16)	1.32(0.31)

Note. N = 238. α = Coefficient alpha reliability.

The next-to-last column in Table 3.2 is headed "Skewness(*SE*)." This heading indicates that this column provides two entries for each variable. The first entry is the skewness statistic itself, and the second entry is the standard error (symbol: *SE*) for the skewness statistic. A ***standard error*** is the standard deviation of a sampling distribution of some specific statistic. You will learn more about standard errors in *Chapter 5* of this book.

For example: the variable appearing on the top row of Table 3.2 is *Age* (the participant's age in years). At the location where the row headed "Age" intersects with the column headed "Skewness(*SE*)," the table presents two numbers. The first number is 4.53, and this number is the skewness statistic for age. This index is a positive value, indicating that the age variable displays a positive skew. This is consistent with the shape of the frequency histogram for age that was presented in Figure 3.4—this frequency histogram revealed a positive skew for age.

At the location where the row headed "Age" intersects with the column headed "Skewness(*SE*)," the second number was 0.16. This was the standard error for the skewness statistic. You will learn how researchers often make use of this standard error in the section titled "Testing Skewness Statistics for Significance."

The fifth row of Table 3.2 presents information for the variable *healthy-diet behavior*. The skewness statistic for this variable is −0.19. This value is pretty close to zero (remember that zero means "no skew at all"). We should not be surprised that the skewness index for healthy-diet behavior is so close to zero, since the frequency histogram in Figure 3.1 revealed an approximately normal distribution.

In Table 3.2, the sixth row down presents information for the *health motivation* variable. The skewness statistic for health motivation is −1.16, and this negative value comes as no surprise, given the long tail in the negative direction displayed by the frequency histogram for health motivation in Figure 3.3.

TESTING SKEWNESS STATISTICS FOR SIGNIFICANCE. Researchers sometimes report analyses in which they test skewness statistics for significance. This is done by dividing the skewness statistic for a given variable by standard error for the skewness statistic. This converts the skewness statistic into a *z* statistic. If the *z* statistic computed in this way is greater than ±1.96 (in absolute value), then the skewness index is interpreted as being statistically significant. You will learn more about significance testing in *Chapter 5* of this book, but for the moment we will test a few of the skewness statistics in Table 3.2 for statistical significance.

For the variable *age*, Table 3.2 shows that the skewness statistic was 4.53 and the standard error was 0.16. To determine whether it is statistically significant, we divide the skewness statistic by its standard error, as demonstrated below:

$$z = \frac{skew}{SE} = \frac{4.53}{0.16} = 28.313 = 28.31$$

So the skewness statistic for participant age resulted in an obtained *z* statistic of *z* = 28.31. Any *z* statistic that is larger than ±1.96 (in absolute value) is statistically significant. This assumes that we have set alpha at $\alpha = .05$ and we are using a two-tailed test (you will learn about *alpha* and *two-tailed tests* in Chapter 5). For the age variable, the obtained *z* statistic (28.31) is larger than the critical value of *z* (±1.96). This indicates that the skew displayed by the age variable is statistically significant.

Table 3.2 shows that, for *healthy-diet behavior*, the skewness statistic was −0.19 and its standard error was 0.16. We can test this skewness statistic for statistical significance using the same formula presented above:

$$z = \frac{skew}{SE} = \frac{-0.19}{0.16} = -1.188 = -1.19$$

So the skewness statistic for participant age resulted in an obtained z statistic of $z = -1.19$. This obtained z statistic is smaller (in absolute value) than the critical value of ± 1.96. This tells us that the skew displayed by healthy-diet behavior is not statistically significant.

Having gone through all of these mathematical calculations, it is time for an important caveat: When the sample size is larger than 200, the z test described here becomes very sensitive—so sensitive that it is often statistically significant even when the distribution displays only a small skew. Therefore, with larger samples, researchers are generally advised to skip the significance test and stick with the visual inspection of the histogram or other graphs (Field, 2009a).

In the current analysis, the sample size was $N = 238$. This indicates that it was probably inappropriate to use the z tests described above.

Kurtosis in Distributions

Kurtosis refers to the peakedness of a frequency distribution (i.e., the extent to which the distribution is relatively peaked or relatively flat). If a distribution follows the shape of the perfect normal curve, then it is neither peaked nor flat—it is described as being ***mesokurtic***. Distributions which are not mesokurtic are either *leptokurtic* or *platykurtic*, as illustrated in Figure 3.5.

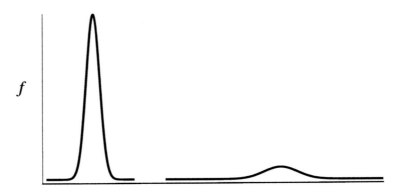

Figure 3.5. A leptokurtic distribution (left) and platykurtic distribution (right).

A ***leptokurtic distribution*** (on the left in Figure 3.5) is a pointy distribution with a relatively high peak. In contrast, a ***platykurtic distribution*** (on the right in Figure 3.5) is a relatively flat distribution.

THE KURTOSIS STATISTIC. Just as there was a statistic that reflects the skewness in a distribution, there is a different statistic that reflects kurtosis. With this statistic:

- a value of zero means that the distribution is mesokurtic (i.e., similar to the perfect normal distribution),

- a negative value means that the distribution is platykurtic, and
- a positive value means that the distribution is leptokurtic.

This kurtosis statistic appeared in Table 3.2, which had been presented earlier. This table shows that the kurtosis statistic for the variable labeled *healthy-diet behavior* was –0.30. This value was fairly close to 0.00, the value that would indicate no problem with kurtosis at all.

On the other hand, the kurtosis statistic for the variable *age* was 24.18. This index was far away from zero, and was positive in sign, indicating a leptokurtic distribution. This comes as no surprise, given that the histogram presented earlier in Figure 3.4 showed that the frequency distribution for age was tall and skinny with a narrow peak around the ages of 19-20.

TESTING KURTOSIS STATISTICS FOR SIGNIFICANCE. As was the case with skewness, it is possible to test the statistical significance of a kurtosis statistic by dividing the statistic itself by its standard error. This produces a *z* statistic, and you determine whether the *z* statistic is significant by using the same criteria that are used for the skewness statistic. In other words, if the *z* statistic is greater than ±1.96, then the kurtosis index is statistically significant at the $p < .05$ level.

As was the case with the skewness statistic, this *z* test for kurtosis tends to be overly sensitive in large samples, producing statistically significant results when the distribution displays only a slight departure from normality. Therefore, when samples include more than 200 subjects, researchers are advised to forgo the test and instead rely on visual inspection of the frequency histogram (Field, 2009a).

Measures of Central Tendency

A ***measure of central tendency*** is a value which reflects the most "typical" score in a distribution. The most widely used measures of central tendency are the mean, the median, and the mode.

Different measures of central tendency are appropriate in different situations. The "correct" measure of central tendency for a given situation is largely determined by (a) the scale of measurement used with the variable and (b) the shape of the variable's frequency distribution. These considerations are illustrated in the sections to follow.

The Mean

DEFINITION. The ***mean*** (also known as the ***average*** or ***arithmetic mean***; symbol: M or $\bar{X}$) is the mathematical center of a distribution of scores. It is computed by summing the scores and dividing by the number of scores. This is illustrated in the following formula:

$$\bar{X} = \frac{\Sigma X}{N}$$

Where

> X = each subject's score on the variable (the Σ symbol to the left of the symbol X tells us to add together these scores)
>
> N = the total number of subjects

For example, imagine that the ages of nine research participants are 19, 19, 20, 20, 21, 22, 22, 22, and 25. The mean age would be computed as follows:

$$\bar{X} = \frac{\Sigma X}{N} = \frac{19+19+20+20+21+22+22+22+25}{9} = \frac{190}{9} = 21.111 = 21.11$$

For this sample, the mean age is 21.11.

WHEN APPROPRIATE. It is appropriate to compute the mean when:

- The variable of interest is a quantitative variable assessed on an interval scale or ratio scale, and
- The distribution of scores is approximately normal (or at least is symmetrical).

The importance of the mean is reflected in the fact that it is so frequently used as a measure of central tendency in research articles as well as in the popular media. Its importance is further demonstrated by the fact that it is often used as one component in formulas that compute other statistics, such as the variance and the standard deviation.

Statisticians value the mean as a measure of central tendency for a number of reasons. First, unlike the median or the mode (to be described next), the mean is influenced by every score in a distribution, and therefore does a better job of representing the distribution as a whole. You saw this when we calculated the mean age in the sample of nine participants. Second, compared to the median or the mode, the mean tends to be more stable as a measure of central tendency across samples (Jaeger, 1993). For example, if we drew from a population of 10 samples with 17 participants in each sample and computed the mean, median, and mode for each group, we would find that mean would display less variability from sample to sample, compared to the other two measures. Statisticians like stable measures, so they like to use the mean.

The Median

Although the mean has a number of advantages as a measure of central tendency, there are situations in which its use is not appropriate. In those situations, it is sometimes more appropriate to compute the median.

DEFINITION. The ***median*** (symbol: Mdn, P_{50}, Q_2) is the score at or below which 50% of the scores in a distribution fall. For example, if the median annual income for a sample of individuals is $30,000, it means that 50% of the people in this sample make $30,000 or less each year.

The concept of "the median" is closely related to two other concepts already covered. First, the median is equivalent to the score at the 50th percentile (P_{50}), and therefore is also equivalent to the score at the second quartile (Q_2). In fact, some articles will use the symbol P_{50} or Q_2 in place of the symbol *Mdn*.

To compute the median in smaller samples (i.e., samples of the size typically encountered in research), the researcher re-orders the values from lowest to highest. If the number of observations in the sample is an odd number, the median is simply the value that appears in the very middle of this sequence of values. For example, imagine that the ages of nine research participants are as follows (below, these values have been ordered from the youngest to the oldest):

> 19
> 19
> 20
> 20
> 21
> 22
> 22
> 22
> 25

The preceding sequence contains nine values, which is an odd number. The value in the exact middle of the sequence is "21," as there were four values before it and four values after it. Therefore, the median for this distribution is 21.

In contrast, if the number of observations in the sample is an even number, the median is the mean of the two values that appear in the middle of the sequence. For example, imagine that we now have 10 different participants, and their ages have again been ordered from youngest to oldest as follows:

> 19
> 19
> 20
> 20
> 21
> 22
> 23
> 23
> 23
> 26

The preceding sequence contains 10 values, which is an even number. The two values in the middle of the sequence are "21" and "22." The mean of 21 and 22 is 21.5, so the median of this distribution is 21.5.

Notice that the median does not necessarily have to be a score that actually appears in the distribution. In the preceding distribution, no participant was exactly at age 21.5, and yet the median was still 21.5.

WHEN APPROPRIATE. There are two situations in which it may be appropriate to use the median as a measure of central tendency:

- *When the variable is assessed on an ordinal scale.* In *Chapter 2*, you learned that a ranking variable is a good example of a variable assessed on an ordinal scale. Therefore, imagine that you conducted a study in which each of 200 participants rank-ordered a set of 10 automobiles in terms of personal preference. With these rankings, a value of "1" represented *car I most prefer* and a ranking of "10" represented *car I least prefer*. In analyzing the data, you wanted to determine the typical ranking that was given to the Ford Mustang. Since your ranking variable was on an ordinal scale of measurement, you computed the median, and found that the median ranking given to the Mustang was a ranking of 3.

- *When the variable is assessed on an interval scale or ratio scale but displays a skewed distribution.* Earlier, you learned that a **skewed distribution** is a set of scores that is not symmetrical; it is a set of scores with a disproportionately long tail on one side. It turns out that, when a distribution is skewed, this greatly distorts the size of the mean, causing the mean to be a misleading estimate of the "most typical" score. When the long tail is on the right side of the distribution (where the higher scores appear), this causes the mean to be become a misleadingly high estimate of central tendency. Similarly, when the long tail is on the left side of the distribution (where the lower scores appear), it causes the mean to becomes a misleadingly low estimate of central tendency. Compared to the mean, however, the median is relatively unaffected when a distribution is skewed. The median always represents the score that is in the middle of the distribution, regardless of whether the distribution displays a skew, outliers, or similar problems. Because the median remains a pretty good estimate of the "most typical" score even when the distribution is not symmetrical, the median is generally preferred over the mean when variables are on an interval or ratio scale but are skewed.

The Mode

In the previous section, we saw that the median was a somewhat primitive statistic, compared to the mean. In this section, we will see that the mode is even more primitive than the median.

DEFINITION. The ***mode*** (symbol: *Mo*) is the most frequently-occurring value in a distribution. For example, consider the following distribution of the ages of nine participants:

19
19
20
20
21
22
23
23
23
26

The mode of the preceding distribution is 23. This is because the age "23" was the most frequently-occurring score (there were three people at age 23; no other age occurred this many times).

The preceding distribution is **unimodal**, which means that it has only one mode (the age of 23, in this case). Think of a unimodal distribution as being represented by a frequency histogram with just one "peak."

On the other hand, some distributions are **bimodal**, which means that they have two modes. Think of a bimodal distribution as being represented by a frequency histogram with two "peaks." The following set of scores has two modes: One mode is the group of three people at age 19, and the second mode is the group of three people at age 23.

18
19
19
19
21
23
23
23
26

WHEN APPROPRIATE. The mode is typically used with variables that are assessed on a nominal scale of measurement (i.e., categorical variables that do not convey quantitative information at all). Yes, that means that my preceding example involving the variable *age* was a pretty bad example.

Here's a better example: In a study dealing with international students, an article might indicate that "The participants represented 12 different nations, with the mode being Canada." In this case, the variable of interest was "nation of origin." This variable was assessed on a nominal scale, and so it was appropriate to use the mode as a measure of central tendency.

Comparing the Mean to the Median To Identify Skew

An earlier section of this chapter used frequency histograms to illustrate negatively-skewed, positively skewed, and symmetrical distributions. However, when you are reading an article that does not present histograms or other graphs, it is still possible to determine whether a distribution is skewed by referring to two measures of central tendency: the mean and the median.

This tactic is based on the fact that when a distribution is skewed, the skew usually has a greater effect on the mean than on the median. In the previous section titled "The Median," you learned that (a) when there is a negative skew, the mean is pulled down, making it lower than the median and (b) when there is a positive skew, the mean is pulled up, making it higher than the median.

Table 3.3

Means, Standard Deviations, and Other Descriptive Statistics

Variable	M	Mdn	SD	α	Range	Skewness (SE)	Kurtosis (SE)
Age	20.58	19.0	4.85	--	18-57	4.53(0.16)	24.18(0.31)
Hours employed	11.88	7.5	14.22	--	0-72	1.11(0.16)	0.98(0.31)
Healthy-diet norms	4.83	5.0	1.42	.87	1.00-7.00	-0.82(0.16)	0.18(0.31)
Healthy-diet attitudes	6.04	6.2	0.92	.88	1.60-7.00	-1.79(0.16)	4.46(0.31)
Healthy-diet behavior	4.22	4.2	1.17	.72	1.00-6.80	-.19(0.16)	-0.30(0.31)
Health motivation	5.76	6.0	0.98	.83	2.00-7.00	-1.16(0.16)	1.55(0.31)
Desire to be thin	5.30	5.67	1.35	.85	1.00-7.00	-0.92(0.16)	0.42(0.31)
Appetite high-fat foods	4.25	4.33	1.49	.88	1.00-7.00	-0.07(0.16)	-0.76(0.31)
Current physical health	5.46	5.75	0.94	.80	1.50-7.00	-1.00(0.16)	1.32(0.31)

Note. N = 238. α= Coefficient alpha reliability.

THE GUIDELINES. As a reader of research articles, you can take advantage of these known characteristics of means and medians to make educated guesses about the likely shape of a variable's distribution. No set of guidelines is infallible, but the following will be correct most of the time:

- When the mean is equal to the median, it is likely that the distribution is symmetrical (i.e., it is likely that the tail on the left side of the distribution is about the same length as the tail on the right side).

- When the mean is greater than the median, it is likely that the distribution displays a positive skew (i.e., it is likely that the distribution has a long tail in the direction of the higher scores).

- When the mean is less than the median, it is likely that the distribution displays a negative skew (i.e., it is likely that the distribution has a long tail in the direction of the lower scores).

EXAMPLE. As an illustration, consider the measures of central tendency that had previously been presented in Table 3.2. For convenience, that table is reproduced again at this point as Table 3.3.

In the table, the second variable is "Hours employed" (the number of hours per week that the participants are employed at wage-earning jobs). The table shows that, for this variable, the mean is higher than the median ($M = 11.88$; $Mdn = 7.5$). According to the guidelines presented above, this may indicate the presence of a positive skew—a long tail in the direction of the higher (more positive) scores. To determine whether this prediction is correct, the frequency histogram for this variable is presented in Figure 3.6.

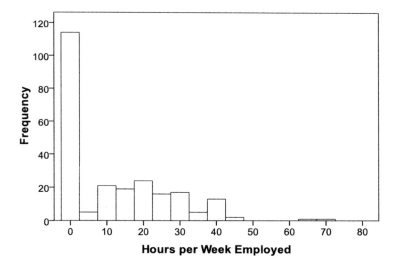

Figure 3.6. Frequency histogram indicating the number of hours per week that participants were employed at paying jobs.

As predicted, the frequency histogram in Figure 3.6 does in fact display a positive skew. Most subjects are bunched together where the lower scores appear, and a long tail stretches out in the direction of the higher scores.

Measures of Variability

A ***measure of variability*** is a statistic that indicates the amount of dispersion displayed by a set of scores. A measure of variability indicates the extent to which the scores are "spread out."

Statisticians and researchers have developed a number of different indices to communicate the amount of dispersion in their variables, and some are more useful than others. Put yourself in the role of one of these researchers, and think about the characteristics that would be displayed by the "perfect" measure of variability. The perfect statistic would be *simple*: If it were a small number, it would mean that the scores were not spread out very much; if it were a large number, it would mean that the scores were spread out a good deal. In addition, it would also be *meaningful*—it would assume values that are related to real-world metric of the variable being studied. Finally, it would not be misleading—it would provide an accurate picture concerning the dispersion in the distribution, even when the distribution is skewed or has outliers.

The current section provides a short inventory of measures of variability that are reported in empirical research articles. As we go through this list, see how each statistic fares when assessed against our characteristics of the "perfect" statistic (but don't be disappointed if no statistic scores 100%).

The Range

The ***range*** is the difference between the highest value and the lowest value in a distribution. In an article, the range is sometimes represented using two numbers (the highest value and the lowest value), or as a single number (the difference between those two values).

For example, Table 3.3 (presented earlier) provided descriptive statistics for several variables in Dr. O'Day's *Eating and Exercising* study. The sixth column in this table was headed "Range," and this column presented the highest and lowest value for each of the included variables. The first variable in the table was *age*, and the table indicated that the range for this variable was "18-57." This meant that the youngest person in the sample was 18 years old, and the oldest was 57 years old. If this table had presented the range as a single number, the range for age would have been 39 (because 57 − 18 = 39).

As a measure of variability, some researchers like the range because it is easy to compute and is easy for readers to understand. One of the negative features of the range is the fact that it is based entirely on just two values: the two most extreme scores in the distribution (the highest score and the lowest score). This can cause the range to be misleading as a measure of variability.

For example, in the current study, the range for age was computed as 39 (again, because 57 - 18 = 39). This connotes images of a very diverse sample—some subjects are quite young at age 18, many are a bit older at age 30 or so, and some are older still, at age 50 or so. But maybe this image is entirely misleading. It is possible that 237 of the 238 subjects in the sample were between the ages of 18 and 22, and just one person was at age 57. If this latter scenario were true, then the reported range of 18 to 57 would be very misleading. Because of this weakness, researchers seldom use the range as the sole measure of variability.

The Interquartile Range

The ***interquartile range*** (symbol: IQR) can be defined in two ways:

- It represents the difference between the score at the third quartile (Q_3) versus the score at the first quartile (Q_1).
- It consists of the two scores which capture the middle 50% of a distribution.

Earlier, you learned that a ***quartile*** is a section of a frequency distribution that has been divided into four equal sections, with 25% of the scores in each section. There, you learned that the symbols for the quartiles have the following interpretations:

- The score at the first quartile (Q_1) is equal to or greater than 25% of the scores in the distribution. This means that Q_1 is equal to the 25th percentile.
- The score at the second quartile (Q_2) is equal to or greater than 50% of the scores in the distribution. This means that Q_2 is equal to the 50th percentile, or the median.
- The score at the third quartile (Q_3) is equal to or greater than 75% of the scores in the distribution. This means that Q_3 is equal to the 75th percentile.

The interquartile range is the difference between the score at Q_3 and the score at Q_1. It is computed as follows:

$$IQR = Q_3 - Q_1$$

EXAMPLE. Imagine that a researcher analyzes the variable *age* in a very large sample. Imagine that the score at the third quartile (i.e., the 75th percentile) is 21, and the score at the first quartile (the 25th percentile) is 18. The interquartile range is therefore computed as:

$$IQR = Q_3 - Q_1$$
$$= 21 - 18$$
$$= 3$$

The interquartile range is similar to the classic range (presented earlier) in that a researcher may present it as two values (e.g.: "The interquartile range extended from 18 to 21"), or as a single value that represents the difference between Q_3 versus Q_1. (e.g., "The interquartile range was 3").

Compared to the classic range (i.e., the two most extreme scores), the IQR tends to be less misleading when used to indicate the amount of dispersion in a set of scores. This is especially likely to be the case when the distribution has outliers (extreme values) or is skewed. To compute the IQR, the researcher essentially "trims" the highest 25% of scores from the distribution, as well as the lowest 25% of scores. This "trimming" effectively eliminates outliers and long tails. Because of this, the interquartile range is a range of values which captures the middle 50% of the distribution—the heart of the distribution.

WHEN APPROPRIATE. To be meaningful, the interquartile range should be used with a quantitative variable that displays a good number of different values (the variable needs to display a fair number of values so that it is possible to divide the distribution into four sections with 25% in each section). In practice, the interquartile range is often used with the same types of variables for which we compute the median:

- when the variable is assessed on an ordinal scale, or
- when the variable is assessed on an interval or ratio scale, but displays a skew or outliers.

The Semi-Interquartile Range

The ***semi-interquartile range*** (symbol: $SIQR$) is the interquartile range divided by 2. Please do not ask me to defend the necessity of this statistic. I assume that the guy who created the IQR was told that he needed to create one more statistic in order to get tenure, and the result was the $SIQR$.

The preceding section described a fictitious study with a large number of participants. It indicated that, for the variable *age*, the score at the third quartile (Q_3) was 21, and the score at the first quartile (Q_1) was 18. We will insert these values in the following formula to compute the semi-interquartile range:

$$SIQR = \frac{Q_3 - Q_1}{2} = \frac{21 - 18}{2} = \frac{3}{2} = 1.5$$

The semi-interquartile range is typically used in the same situations in which a researcher might compute the interquartile range:

- when the variable is assessed on an ordinal scale, or
- when the variable is assessed on an interval or ratio scale, but displays a skew or outliers.

The Mean Average Deviation (MAD)

The ***mean average deviation*** (abbreviation: MAD), is the mean of the absolute values of the deviations from the mean. The mean average deviation is not terribly popular in published research, but it is first-cousin to a statistic that *is* terribly popular: the *standard deviation* (symbol: SD). We will therefore learn about the MAD, and that will ease our way toward understanding the SD.

The concept of a deviation is central to understanding both the mean average deviation as well as the standard deviation. A ***deviation*** is the difference between:

- the score of an individual participant versus
- the mean of the distribution in which the score appears.

EXAMPLE. This is illustrated in Table 3.4, which uses the symbol X to represent the variable *age* for five individuals. Under the heading "X," you can see that the age of Subject 1 is 18, the age of Subject 2 is 19, and so forth.

Table 3.4

Raw Age Scores (and Corresponding Deviations) for a Sample of Five Participants (Mean Age = 20).

Subject	X	$X - \bar{X}$	$\|X - \bar{X}\|$
1	18	-2	+2
2	19	-1	+1
3	20	0	0
4	21	+1	+1
5	22	+2	+2
Sum:	100	0	6

We will begin by computing the mean of the *age* variable, since this mean is one component in the formula for the *MAD*. The formula for the mean had been presented much earlier in this chapter:

$$\bar{X} = \frac{\Sigma X}{N}$$

In this formula, the term "ΣX" represents the sum of the X variable, which appears in the table at the location where the column headed "X" intersects with the row headed "Sum." In Table 3.4, this value is 100. In the formula, the term "N" represents the number of observations, which in this case is 5. Inserting these values into this formula produces the following:

$$\bar{X} = \frac{\Sigma X}{N} = \frac{100}{5} = 20$$

So the mean age is $\bar{X} = 20$ for these five subjects. We can now compute a deviation score for each subject by subtracting this mean from their age. Subject 1 is 18 years of age, so the deviation score is computed as:

$$\text{Deviation} = X - \bar{X}$$
$$= 18 - 20$$
$$= -2$$

Which means that, for Subject 1, the age deviation score is equal to -2. This is just another way of saying that the age for Subject 1 was 2 units below the sample mean. You can see that this deviation score of -2 appears in Table 3.4 at the location where the row for "Subject 1" intersects with the column for "$X - \bar{X}$." The deviation scores for the remaining participants also appear in the column headed "$X - \bar{X}$."

We would like to compute the mean average deviation (after all, that is the title of this section), and our first impulse is to simply add up these deviations and divide by five (the number of participants). But immediately we encounter a problem: For any variable, the sum of the deviations from the mean is equal to zero. In Table 3.4, at the place where the column headed "$X - \bar{X}$" intersects with the row headed "Sum," you can see the value of "0." This is no accident—the sum of deviations from the mean will always be equal to zero. This is a problem, because if the sum of the deviations is always equal to zero, the average of the deviations will also always be equal to zero (because zero divided by anything is always equal to zero). And a measure of variability that is always equal to zero is of no use to anyone.

One solution to this problem is to take the absolute value of the deviations prior to summing them. When we take the **_absolute value_** of a number, it means that we treat the number as if it were a positive number, regardless of whether it had started out as a positive or negative number. We use vertical bars to show that we are taking the absolute value of a number, and you can see that the last column of Table 3.4 is headed "$|X - \bar{X}|$," which means that this column reports the absolute value of each deviation. You can see that each value in the column headed "$|X - \bar{X}|$" is identical to the values in the column headed "$X - \bar{X}$," except that the negative signs have now been converted to positive signs.

COMPUTING THE MAD. Finally we are ready to compute the mean average deviation for this sample. At the bottom of Table 3.4, we can see that the sum of the absolute values of the deviations from the mean is 6. We compute the mean average deviation by dividing this sum by the number of observations (N), which is 5 in this case:

$$MAD = \frac{\sum |X - \bar{X}|}{N} = \frac{6}{5} = 1.2$$

So the mean average deviation is equal to 1.2. This tells us that the typical subject in the sample deviates 1.2 units from the mean. Notice that this value of 1.2 sort of makes sense when you look at the actual variability of scores in Table 3.4. The mean age in this table is 20. Some participants (like Subject 3) do not deviate from this mean at all, some participants (like Subjects 2 and 4) deviate just 1 unit from this mean, and other participants (like Subjects 1 and 5) deviate 2 units. So a mean deviation of 1.2 units feels just about right.

WHEN APPROPRIATE. The *MAD* is an appropriate measure of central tendency when (a) the variable is assessed on an interval scale or ratio scale and (b) the distribution of scores is fairly symmetrical. In other words, the *MAD* is appropriate under pretty much the same circumstances that the mean is appropriate.

Appropriate or not, the mean average deviation suffers from the shortcomings of being easy to compute and easy to understand, thus making it entirely unacceptable as a measure of variability. As a result, you will seldom see the *MAD* reported in a research article.

The shortcomings exhibited by the mean average deviation have led to the development of two statistics that are much more widely computed and reported: the variance and the standard deviation. We will learn about these statistics after first taking a brief detour to learn about the term that is at the very heart of their formulas: the *sum of the squares*.

The Sum of the Squares *(SS)*

In the preceding section, we saw that the sum of the deviations from the mean is always equal to zero, creating a problem if we wish to create a measure of variability based on deviations from the mean. In that section, we learned that one solution to this problem is to take the absolute values of the deviations. This creates a sum of the deviations that is not equal to zero—a sum which can then be used to create the mean average deviation.

There is, however, an alternative way of solving this problem. Instead of taking the absolute value of each deviation, we could instead take the *square* of each deviation. The squared deviations produced by this operation will always be positive, even if the original deviation had been negative (because squaring a negative number always results in a positive number). This second approach—squaring the deviations from the mean—is the approach followed when computing the variance and standard deviation. And since this concept of the *sum of the squares* is so important to the field of statistics, we will take a moment to show exactly how it is computed.

EXAMPLE. Table 3.5 presents the same data set that had been presented as Table 3.4. In the present case, however, the last column now represents the squared deviations from the mean, represented by: "$(X - \bar{X})^2$." In this term, the symbol "X" still represents each individual's score on the *age* variable, and $\bar{X}$ still represents the mean of the age variable (earlier, we computed mean as $\bar{X} = 20$).

Table 3.5

Raw Age Scores (Along with Deviations and Squared Deviations) for a Group of Five Participants (Mean Age = 20)

Subject	X	$X - \bar{X}$	$(X - \bar{X})^2$
1	18	-2	+4
2	19	-1	+1
3	20	0	0
4	21	+1	+1
5	22	+2	+4
Sum:	100	0	10

To compute a value in this last column, we subtract the mean (20) from a given subject's age, and then square the resulting deviation. For Subject 1, the deviation from the mean is –2. When we square this deviation, it produces a positive value: +4. The same procedure was used for each of the remaining participants, producing the values appearing in the last column of Table 3.5.

DEFINITION. In Table 3.5, at the location where the column headed "$(X - \bar{X})^2$" intersects with the row headed "Sum," we find the value "10." This value represents the sum of the squared deviations from the mean. This term is typically shortened to just the ***sum of the squares*** (symbol: *SS*), and it can be defined as the value that is obtained when we compute each score's deviation from the mean, square that deviation, and add together the squared deviations for all subjects. The formula is as follows:

$$SS = \Sigma(X - \bar{X})^2$$

WHY THE *SS* IS IMPORTANT. The concept of the sum of the squares is important because it is at the heart of many statistical procedures used to analyze data from empirical investigations: the *variance*, the *standard deviation*, *analysis of variance*, and *multiple regression*, just to name a few. If you understand what is being done mathematically when we compute the sum of the squares, you will understand what is being done when we compute many of our more advanced statistics.

The Variance

Having computed the sum of the squares, we are now ready to compute the variance. In general, the ***variance*** is a measure of variability computed by taking the average of the squared deviations from the mean. The variance is important because we must compute the variance in order to compute the standard deviation (our single most-importance measure of variability). The variance also plays a major role in a number of other statistical procedures, such as analysis of variance (ANOVA).

Complicating matters is the fact that there are three different types of variance: the *population variance*, the *sample variance*, and the *estimated population variance*. We shall discuss each in turn.

THE POPULATION VARIANCE. You will recall that a ***population*** consists of an entire set of observations that are of interest to researchers in some specific context. The ***population variance*** is the average of the squared deviations from the mean that is obtained when analyzing all observations in a population. It is appropriate to compute the population variance when you have scores for every single observation contained in the population, and you wish to compute a parameter that describes the variability of those scores. Remember that a ***parameter*** is a value that describes some characteristic of a population.

Table 3.6

Raw Age Scores (Along with Deviations and Squared Deviations) for a Population Consisting of Five Participants (Mean Age = 20)

Subject	X	$X - \mu$	$(X - \mu)^2$
1	18	-2	+4
2	19	-1	+1
3	20	0	0
4	21	+1	+1
5	22	+2	+4
Sum:	100	0	10

For example: Imagine that the Olympics has added a new sport to the summer games: *freeze tag* (this is the game in which the person who is "it" chases people and makes them freeze when caught). A sports psychologist is studying this new phenomenon and is curious about the ages of the people who typically win the event. He learns that the sport has only been offered in the last five summer Olympic games. The ages of these individuals are presented in Table 3.6

Table 3.6 should look familiar—it contains exactly the same values that appeared in Table 3.5. However, one of the symbols in the table has changed: The symbol for the sample mean ($\bar{X}$) has been replaced with the symbol for the population mean (μ). But this change of symbols does not have any effect on the size of the deviations or on the size of the squared deviations.

The symbol for the population variance is "σ^2." The formula for computing the population variance is as follows:

$$\sigma^2 = \frac{\sum(X - \mu)^2}{N}$$

Where:

σ^2 = the population variance

X = each participant's score on the variable of interest (age, in this case)

μ = the population mean for the variable of interest

N = the number of observations in the population

You may recall that Greek letters are typically used to represent population parameters. That is why σ^2 is used to represent the population variance, and μ is used to represent the population mean.

You may also recall that the quantity "$\sum(X - \mu)^2$" is our formula for sum of the squares (*SS*). We have already computed this value for the age scores appearing in Table 3.6, and found it to be *SS* = 10. This means that computing the population variance is simply a matter of dividing the sum of the squares (10) by the number of people in the population (5):

$$\sigma^2 = \frac{\sum(X - \mu)^2}{N} = \frac{10}{5} = 2$$

So the population variance for the age variable is σ^2 = 2. For the moment, we will not worry about why this number is so important to researchers. At the moment, we still have two other types of variance to compute: the sample variance and the estimated population variance.

THE SAMPLE VARIANCE. In this section, we will change some of the details of our thought experiment involving freeze tag. Let's imagine that we are no longer interested in the population of people who win gold medals in this sport. Instead, let's imagine that we are interested in a much larger population: The population of regular people who play freeze-tag each summer just for fun. Such a population would almost certainly consist of millions and millions of people world-wide.

Using a sophisticated computer algorithm, we have drawn a random sample of people from this population, and have recorded the ages for each of these individuals. Their ages are presented in Table 3.7.

Table 3.7

Raw Age Scores (along with Deviations and Squared Deviations) for a Sample of Five Participants (Mean Age = 20)

Subject	X	$X - \bar{X}$	$(X - \bar{X})^2$
1	18	-2	+4
2	19	-1	+1
3	20	0	0
4	21	+1	+1
5	22	+2	+4
Sum:	100	0	10

Notice that, in Table 3.7, the symbol for the mean is now $\bar{X}$ rather than μ. This is because the five people in the table now constitute a *sample*, not a population. You will recall that a sample is a subset of a larger population.

We now wish to compute the sample variance for these five scores on the age variable. It is appropriate to compute a **sample variance** (symbol: S^2) when you have scores for just a sample (e.g., a subset) of scores drawn from the population, and you wish to compute a *descriptive statistic* that describes the variability of scores in just that sample; you have no wish to estimate the variance in the larger population. You may recall from *Chapter 2* that a **descriptive statistic** is an index that describes some characteristic of a sample; it is not used to make inferences about the larger population. Notice that the symbol for the sample variance is an upper-case "S^2"; later, you will see that a lower-case "s^2" is used for a different type of variance.

The sample variance presented here is fine if you merely want to describe the variance in your sample, but it is not okay if you wish to analyze the sample data so as to estimate the variance in the population. If you use the formula presented here for that purpose, you will probably obtain a value that *underestimates* the actual variance in the population. For that reason, some textbooks refer to this type of variance as a ***biased estimate of population variance***.

With those caveats in mind, here is the formula for the sample variance (the biased estimate):

$$S^2 = \frac{\Sigma(X - \bar{X})^2}{N}$$

Where:

S^2 = the sample variance (biased estimate)

X = each participant's scores on the variable of interest (age, in this case)

$\bar{X}$ = the sample mean on the variable of interest

N = the number of observations in the sample

Since every single value (including the mean) in Table 3.7 is identical to the corresponding values in Table 3.6, and since the formula for the sample variance involves essentially the same operations that had been performed by the formula for the population variance, it only makes sense that the formula shown here will produce exactly the same result that had been produced by the formula for the population variance. In fact, that is exactly what happens:

$$S^2 = \frac{\Sigma(X - \bar{X})^2}{N} = \frac{10}{5} = 2$$

So the sample variance is $S^2 = 2$. This value is identical to the population variance ($\sigma^2 = 2$) that had been computed earlier.

Remember, researchers should use this formula only if they wish to *describe* the variance in the sample itself—they should not use this formula if they wish to compute an *unbiased* estimate of the population variance. The formula used for that purpose is presented next.

THE UNBIASED ESTIMATE OF POPULATION VARIANCE. We compute the ***unbiased estimate of population variance*** (i.e., the *estimated population variance*, the *unbiased variance estimate*) when:

- We have obtained data from a sample (a subset of observations drawn from a larger population), and
- we wish to analyze this sample in order to estimate the variance in the population.

In this situation, the variance is no longer a *descriptive statistic*; it is now an *inferential statistic*. Remember that an ***inferential statistic*** is a value computed by analyzing data from a sample for the purpose of estimating some characteristic of the larger population.

The symbol for the estimated population variance is s^2. Notice that this symbol makes use of a lower-case "s," whereas the symbol for the sample variance (presented earlier) made use of an upper case "S."

The good news is that the formula for the estimated population variance is identical to the formula for the sample variance, with just one twist: We will divide the sum of the squares by $N-1$, rather than N. First we will do the math, and then we will see why we use $N-1$ as the denominator rather than N.

These calculations will once again be based on the values appearing in Table 3.7. The estimated population variance is computed as follows:

$$s^2 = \frac{\Sigma(X - \bar{X})^2}{N - 1} = \frac{10}{5 - 1} = \frac{10}{4} = 2.5$$

So the unbiased estimate of the population variance is $s^2 = 2.5$. Notice that this estimate ($s^2 = 2.5$) is somewhat larger than the sample variance that had been computed in the preceding section ($S^2 = 2$). The sample variance was a *biased* estimate—it probably underestimated the actual variance in the population. The estimated population variance computed in this section was slightly larger, and was probably closer to the actual variance in the population (that is why we call it the *unbiased* estimate).

THE DEGREES OF FREEDOM. You have already seen that the formula for the biased variance estimate has N in the denominator, whereas the formula for the unbiased estimate has $N-1$ in the formula. This term $N-1$ has a special name: the *degrees of freedom* (symbol: *df*). The concept of *df* is important to the concept of variance, as well as to many other statistics to be covered in this book. So we will take a brief detour to introduce the concept here.

The ***degrees of freedom*** refer to the number of observations in a sample that are free to vary when estimating a population parameter such as the population variance. It is typically computed as:

- the number of scores in the sample…
- …minus the number of restrictions placed on the freedom of those scores to vary.

If you found the preceding definition a bit difficult to understand, join the crowd. The concept of degrees of freedom is very important to statistics, but is abstract and fairly challenging to grasp. So let's just cut to the three most important things to remember:

1) **THE *df* ALLOW US TO COMPUTE AN UNBIASED ESTIMATE OF POPULATION VARIANCE.** When analyzing sample data in order to estimate the variance in the population, we do not divide the sum of squares by N (the sample size); instead, we divide by $N - 1$ (the degrees of freedom). Dividing by the

degrees of freedom produces an unbiased estimate of the population variance.

2) **THE FORMULA FOR COMPUTING *df* IS DIFFERENT FOR DIFFERENT STATISTICAL PROCEDURES.** We have already established that the *df* are computed as $N - 1$ when estimating population variance from a sample, but the formula for computing the *df* are different for different statistics. For example, the *df* are computed as $N - 2$ when performing an independent-samples *t* test (to be covered in a later chapter). The *Publication Manual of the American Psychological Association 6th ed.* (2010) advises that the degrees of freedom should generally be reported for most statistical procedures, so this book will show how to compute the *df* for most of the statistical procedures that it covers.

3) **RESEARCHERS MUST COMPUTE *df* IN ORDER TO USE A TABLE OF CRITICAL VALUES.** To determine whether the results of an analysis are "statistically significant," researchers often compare their obtained statistics (such as an obtained *t* statistic) against the "critical value" of that statistic. They typically look up these critical values in a table in the back of a statistics textbook, and the *df* is one piece of information that they must have in order to use such a table. You will learn more about the concept of *statistical significance* in *Chapter 5* of this book.

In summary, the concept of degrees of freedom is important for several different reasons. To return to the topic at hand (variance), the *df* are important because they allow us to compute an *unbiased* estimate of the population variance. If we divide the sum of squares by *N*, the resulting value tends to underestimate the actual variance in the population. On the other hand, if we divide the sum of squares by the degrees of freedom ($N - 1$), the resulting value provides a more accurate estimate.

The current section has introduced three measures of variability: the population variance, the sample variance, and the estimated population variance. These measures are important, in part, because they can be used to compute what may be the most important measure of variability of all: the *standard deviation*.

The Standard Deviation: Introduction

The ***standard deviation*** is the square root of the variance. It is a measure of variability which communicates the approximate extent to which the typical score in a distribution deviates from the mean.

Notice that the preceding definition softened things up by saying that the standard deviation communicates the *approximate* extent to which a typical score deviates from the mean. If you have excellent recall, you will remember that it is another statistic—the *mean average deviation* (i.e., the *MAD*)—which represents the *exact* extent to which the average score deviates from the mean. The value of standard deviation typically comes pretty close to this average deviation, but is not exact.

WHY THE STANDARD DEVIATION IS PREFERRED AS A MEASURE OF CENTRAL TENDENCY. The standard deviation is generally the preferred measure of variability for quantitative variables used in research. When you read the section on descriptive statistics in a published research article, you will see that the standard deviation is reported more often than any other measure of variability.

The standard deviation is popular for two reasons. First, it is derived from the variance (and the variance is an *uber*-important measure of variability). Second, it is easier for readers to comprehend, compared to the variance. The variance is in *squared* units: It represents the average of the *squared* deviations from the mean. Because of this, the variance for a given variable is often a very large number—so large that it might seem unrelated to the variable being investigated.

For example: consider Dr. O'Day's *Eating and Exercising* investigation. Here, we are describing the large study in which the total number of subjects was $N = 238$, not the tiny study in which $N = 5$. In Dr. O'Day's large-sample investigation, the mean age for the 238 participants was $\bar{X} = 20.58$ years, and the estimated-population variance was $s^2 = 23.53$ years. The variance for this variable ($s^2 = 23.53$ years) is a large number—so large that the typical reader will have difficulty making sense of it. The typical reader will have a hard time deciding whether this value indicates that there was a lot of variability in age variable, very little variability, or somewhere in-between.

In contrast, the standard deviation for the same age variable was much smaller at $s = 4.85$ years. This standard deviation roughly communicates the idea that the typical participant deviated about 4.85 years from the mean of 20.58 years. Most readers will find this concept standard deviation easier to grasp—they will be able to imagine some of the subjects deviating about 4.85 years below the mean of 20.58 years, and some of the subjects deviating about 4.85 years above the mean.

In summary, most readers cannot relate to the variance, but can relate to the standard deviation. As a result, the standard deviation has become the most widely reported measure of variability for quantitative variables assessed on an interval or ratio scale.

WHEN APPROPRIATE. It is appropriate to compute the standard deviation under the same circumstances when we compute the variance:

- when the variable is assessed on an interval or ratio scale of measurement, and
- when the variable displays a fairly symmetrical distribution.

Given these requirements, it is clear that it was not a great idea to use either the variance or the standard deviation as a measure of variability for the age variable in Dr. O'Day's large-*N* study. In that investigation, age displayed a positive skew: Most participants were bunched together at the low end (with low scores on age, such as "18" and "19") and only a small number had really high scores of "45" and "50" and so forth. Because the distribution was skewed with a long tail in the direction of higher scores, the resulting variance and standard deviation were misleadingly large: Although the obtained standard deviation was $s = 4.85$ years, it would have been much smaller if the distribution had not displayed that long tail. The moral: The variance and standard deviation are most appropriate with *symmetrical* variables.

The Standard Deviation: Three Versions

There are three types of standard deviation corresponding to the three types of variance that you have already encountered: The *population standard deviation,* the *sample standard deviation,* and the *unbiased estimate of the population standard deviation.* In a research article, the typical symbol for the standard deviation is *SD.* In a statistics textbook, however, the symbol is typically:

σ for the population standard deviation,

S for the sample standard deviation (the biased estimate), and

s for the estimated population standard deviation (the unbiased estimate).

Notice that the preceding symbols correspond to the symbols used for the variance. In each case, the superscript "2" has been dropped to represent the fact that a standard deviation is the square root of a variance.

The preceding symbols will be used in this book. Be warned, however, that other books and articles may use different symbols. For example, some books use the symbol "*S*" to represent the unbiased estimate of the population standard deviation, and some do not cover the biased estimate at all.

THE POPULATION STANDARD DEVIATION. The population standard deviation (symbol: σ) is the square root of the population variance. Remember that a ***population*** is an entire set of observations, not a subset. This version of the standard deviation assumes that we have access to all observations in a population, and are analyzing all of these observations. The formula is as follows:

$$\sigma = \sqrt{\frac{\Sigma(X - \mu)^2}{N}}$$

The preceding formula is identical to the formula for the population variance (symbol: σ^2) provided earlier, except (a) to the left of the equals sign, the symbol for the variance (σ^2) has been replaced with the symbol for the standard deviation (σ), and (b) all of the operations to the right of the equals sign have been placed under the "square root" symbol. This tells us that the population standard deviation is merely the square root of the variance.

An earlier section of this chapter presented the *age* scores for a very small population in which $N = 5$ (you will recall that this was the entire population of people who had won the gold medal in the Olympic *freeze-tag* event). In that section, the population variance for the age variable was computed as $\sigma^2 = 2$. To compute the population standard deviation, we simply take the square root of that variance:

$$\sigma = \sqrt{2} = 1.414 = 1.41$$

THE SAMPLE STANDARD DEVIATION (BIASED VERSION). The sample standard deviation (symbol: S) is the square root of the sample variance (S^2). Remember that a *sample* is a subset of observations, drawn from the population of interest. Also remember that we are using this statistic as a ***descriptive statistic***—we are merely describing the standard deviation of the sample itself, and we are not analyzing the sample data in order to estimate the population standard deviation. That is why this statistic is called the *biased* estimate—if we attempted to use it as an estimate of the population standard deviation, it would be a bit too low.

The formula is as follows:

$$S = \sqrt{\frac{\sum(X - \bar{X})^2}{N}}$$

All of the above is identical to the formula for the sample variance (presented earlier) except (a) the symbol to the left of the equals sign has been changed from S^2 to S, and (b) the square root symbol has been drawn around the operations that appear to the right of the equals sign. All of this conveys the idea that the sample standard deviation is merely the square root of the sample variance.

In a previous section, we computed the sample variance as $S^2 = 2$. To compute the sample standard deviation, we merely take the square root:

$$S = \sqrt{2} = 1.414 = 1.41$$

THE UNBIASED ESTIMATE OF THE POPULATION STANDARD DEVIATION. As you would expect, this statistic is the "standard deviation" counterpart to the estimated population variance. It is computed when:

- We have obtained data from a sample (a subset of observations drawn from a larger population), and
- we wish to analyze this sample in order to estimate the standard deviation in the larger population.

The symbol for the estimated population standard deviation is "s." The formula is as follows:

$$s = \sqrt{\frac{\sum(X - \bar{X})^2}{N - 1}}$$

In one of the analyses presented earlier, we treated the five freeze-tag players as if they were a sample drawn from a larger population. With that example, we computed the unbiased estimate of the population variance as being $s^2 = 2.5$. If we now wish to compute the unbiased estimate of the population standard deviation, we need only take the square root of this variance estimate:

$$s = \sqrt{2.5} = 1.581 = 1.58$$

So the unbiased estimate of the population standard deviation is $s = 1.58$. Notice that this value is larger than the sample standard deviation, which earlier had been computed as $S = 1.41$. That section indicated that the sample standard deviation is a biased estimate, typically underestimating the actual standard deviation in the population. The unbiased estimate, on the other hand, is a bit larger ($s = 1.58$) and is probably closer to the actual population value.

The Standard Deviation: Summary

The preceding sections on variance and standard deviation have presented a lot of information. Figure 3.7 tries to simplify things by presenting symbols for the various statistics and parameters.

		Symbol or abbreviation	
Measure of central tendency	**Divisor in formula**	**Standard deviation**	**Variance**
Population standard deviation or variance	N	σ	σ^2
Sample standard deviation or variance (biased)	N	S	S^2
Estimated population standard deviation or variance (unbiased)	$N-1$	s	s^2

Figure 3.7. Differences between the three types of variances and standard deviations covered in this chapter.

The section headed "Divisor in formula" shows that, when computing the population variance or sample variance, the sum of the squares is divided by N. It shows that, in order to obtain an unbiased estimate of the population variance, we must instead divide by $N - 1$ (the degrees of freedom).

Remember that, in published articles, the standard deviation is often represented by the symbol *SD*, rather than σ, S, or s. Some researchers use the symbol *SD* regardless of whether the standard deviation is a population standard deviation, a sample standard deviation, or an unbiased estimate of the population standard deviation.

The Coefficient Alpha Reliability Estimate

Reliability refers to the consistency or stability displayed by a set of scores. The concept of reliability is discussed here because it is commonplace for research articles to report the reliability of the variables being analyzed as part of the descriptive statistics section. This is a good practice, because readers can use these reliability estimates to help them evaluate the overall quality of the investigation. Seasoned readers know that, in higher-quality investigations, researchers use reliable measures and report the evidence of that reliability in the article.

Key Concepts in Scale Reliability

There are different approaches to assessing reliability, including the test-retest method, the equivalent-forms method, and others. When a given variable has been measured using a multiple-item summated rating scale, researchers often assess the internal-consistency reliability of the resulting scores. A multiple-item scale displays high levels of ***internal-consistency reliability*** when scores from the individual items constituting that scale display strong correlations with one another or with the total scale score. When the items that constitute the scale are Likert-type rating items (as is the case with most of the scales on the *Questionnaire on Eating and Exercising*), the most popular internal-consistency reliability estimate is ***coefficient alpha*** (symbol: α; Cronbach, 1951).

Theoretically, coefficient alpha represents the lowest reliability estimate that can be expected for a multiple-item scale or test. The upper bound on coefficient alpha is +1.00. Values above +.70 are generally viewed as being acceptable (Nunnally, 1978), and values in the range of +.80 to +.90 are viewed as being ideal (Clark & Watson, 1995; DeVellis, 1991). With other factors held constant, coefficient alpha assumes larger values when (a) a large number of items are included in the multiple-item scale, and (b) the items display strong correlations with one another.

Example from Dr. O'Day's Study

Near the beginning of this chapter, Table 3.3 presented coefficient alpha reliability estimates for seven multiple-item rating scales included in Dr. O'Day's *Questionnaire on Eating and Exercising*. For convenience, that table is reproduced again here as Table 3.8.

In Table 3.8, coefficient alpha reliability estimates appear in the column headed with the symbol "α." In the table, you can see coefficient alpha is not reported for the first two variables: *age* and *hours employed*. Neither of these variables was measured using multiple-item rating scales, and it was therefore not possible to compute coefficient alpha for either variable.

Table 3.8

Means, Standard Deviations, and Other Descriptive Statistics

Variable	M	Mdn	SD	α	Range	Skewness (SE)	Kurtosis (SE)
Age	20.58	19.0	4.85	--	18-57	4.53(0.16)	24.18(0.31)
Hours employed	11.88	7.5	14.22	--	0-72	1.11(0.16)	0.98(0.31)
Healthy-diet norms	4.83	5.0	1.42	.87	1.00-7.00	-0.82(0.16)	0.18(0.31)
Healthy-diet attitudes	6.04	6.2	0.92	.88	1.60-7.00	-1.79(0.16)	4.46(0.31)
Healthy-diet behavior	4.22	4.2	1.17	.72	1.00-6.80	-.19(0.16)	-0.30(0.31)
Health motivation	5.76	6.0	0.98	.83	2.00-7.00	-1.16(0.16)	1.55(0.31)
Desire to be thin	5.30	5.67	1.35	.85	1.00-7.00	-0.92(0.16)	0.42(0.31)
Appetite high-fat foods	4.25	4.33	1.49	.88	1.00-7.00	-0.07(0.16)	-0.76(0.31)
Current physical health	5.46	5.75	0.94	.80	1.50-7.00	-1.00(0.16)	1.32(0.31)

Note. N = 238. α = Coefficient alpha reliability.

The remaining variables in Table 3.8 were all multiple-item summated rating scales, and coefficient alpha is reported for each. For example, the third variable in the table is *healthy-diet norms,* and for this scale coefficient alpha was found to be α = .87. This is well above the minimum criterion of .70 suggested by Nunnally (1978), and is in the "ideal" range of .80 to .90 suggested by Clark and Watson (1995) and DeVellis (1991). In fact, for six of the seven scales that report coefficient alpha, the obtained value is in this ideal range. The only exception is *healthy-diet behavior,* which displays a coefficient alpha of .72. This value is acceptable, but is not ideal.

How Coefficient Alpha is Typically Presented

Researchers sometimes present coefficient alpha within the text of their article, using the symbol "α" to represent the reliability estimate. Here's an example:

> The coefficient alpha reliability estimate for the healthy-diet norms variable was acceptable, α = .87.

As an alternative, researchers sometimes present coefficient alpha estimates in a table. This was done in Table 3.8, where the coefficient alpha estimates for a number of scales appeared under the heading, "α." In other articles, these estimates might be listed under the heading "r_{xx}," "Coefficient alpha" or just "Reliability."

Using Boxplots to Understand Distributions

Boxplots (also called box-and-whisker charts) are useful for exploratory data analysis. A ***boxplot*** is a figure that graphically illustrates several of the concepts taught in this chapter: the median, the interquartile range, the range, outliers, and others. Among other things, boxplots help researchers quickly determine whether a data set is symmetrical versus skewed.

Different statistical applications draw boxplots using somewhat different conventions. Most of the conventions presented here are based on boxplots created by the SPSS application (Field, 2009a), along with concepts described by Jaeger (1993) and Tukey (1977).

Example 1: An Approximately Normal Distribution

Figure 3.8 presents a frequency histogram for *healthy-diet behavior*, one of the variables assessed by Dr. O'Day's *Questionnaire on Eating and Exercising*. Scores on this variable could range from 1.00 to 7.00, with higher scores indicating healthier eating habits. The histogram shows that the distribution is fairly symmetrical, with the right side of the distribution coming pretty close to being a mirror image of the left side.

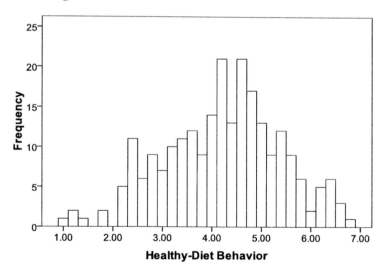

Figure 3.8. Frequency histogram for scores on the *healthy-diet behavior* scale (an approximately normal distribution).

Figure 3.9 presents a boxplot for the same variable, and this figure will be used to introduce some of the basic features of boxplots. In doing this, the current chapter will present some fairly precise numerical values, such as "the bottom of the box corresponds to a score of 3.40 on the vertical axis." This chapter is able to offer such precise values because I (the all-knowing author of this book) have access to frequency tables and additional information to which you (the lowly reader) do not have access. So do not be concerned if you are not able to arrive at such precise numerical values through a simple visual inspection of the boxplot. The purpose of this graph is to provide a simple drawing of a distribution, not to provide precise statistics.

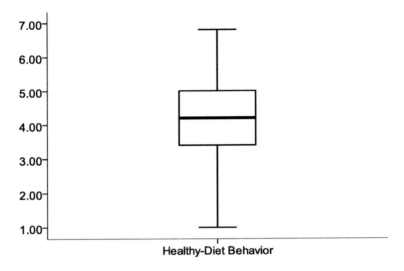

Figure 3.9. Boxplot for scores on the *healthy-diet behavior* scale (an approximately normal distribution).

The Box

The central part of the boxplot is the box itself. This part of the graph illustrates the 75th percentile, the 25th percentile, the interquartile range, and the median.

THE INTERQUARTILE RANGE. The top edge of the box corresponds to the score that is at the third quartile (Q_3). You will recall that the third quartile is equal to the 75th percentile (P_{75}). Figure 3.9 shows that the top of the box corresponds to a score of 5.00 on the vertical axis. This means that a subject with a score of 5.00 on the 7-point *healthy-diet behavior* scale has a score that is equal to or greater than 75% of the other subjects in this study.

The bottom edge of the box corresponds to the score that is at the first quartile (Q_1). Earlier, you learned that the first quartile is equal to the 25th percentile. Figure 3.9 shows that the bottom of the box corresponds to a score of 3.40 on the vertical axis. This means that a subject who has a score of 3.40 on healthy-diet behavior has a score that is equal to or greater than 25% of the other subjects.

The vertical length of the box (that is, the difference between the score at the top of the box versus the score at the bottom of the box) represents the interquartile range for this variable Earlier, you learned that **the interquartile range** (symbol: *IQR*) is the difference between the score at the 75th percentile versus the score at the 25th percentile.

In Figure 3.9, the top of the box was at a score of 5.00, and the bottom of the box was at a score of 3.40. This means that, if we wished to report the *IQR* as the two boundary values, we would describe it in this way:

> For healthy-diet behavior, the interquartile range extended from 3.40 to 5.00.

If we instead wished to report the interquartile range as a single number, it would be computed as follows:

$$IQR = Q_3 - Q_1 = 5.00 - 3.40 = 1.60$$

The single-value version of the *IQR* would be reported in this way:

> For healthy-diet behavior, the interquartile range was 1.60

If you visually inspect Figure 3.9, you can see that the vertical length of the box is indeed equal to about 1.60 units. This box length is an important concept, as it will be used to determine some of the other concepts related to a boxplot (such as the length of the "whiskers"). Since the ***box length*** represents the interquartile range, it follows that this box length "captures" the middle 50% of the distribution (i.e., the part that lies between the 25% percentile and the 75th percentile). For the current sample, this tells us that about 50% of the subjects received a score between 3.40 and 5.00.

Remember that, whenever this chapter refers to a box length, it is talking about the *vertical* box length (i.e., the height of the box, from top to bottom). The *horizontal* box width (going from the left side to the right side of the box) does not represent any statistical concept.

THE MEDIAN. The horizontal line that appears inside the box in a boxplot represents the ***median*** for the variable: the score at the 50th percentile. In Figure 3.9, the horizontal median line corresponds to a score of 4.20 on the vertical axis. This tells us that subjects who received a score of 4.20 on healthy-diet behavior have a score that is equal to or greater than 50% of the other subjects.

The Whiskers

In a boxplot, the ***whiskers*** are the vertical lines that extend above and below the box, ending in a short horizontal bar. The bar at the end of the whisker that appears above the box represents the highest observed value in the data set (that is, the highest observed value that is within 1.5 box lengths of the top edge of the box—more on this in a moment). In Figure 3.9, you can see that the highest observed value in the data set is 6.80.

The bar at the end of the whisker that appears below the box represents the lowest observed value in the data set (that is, the lowest observed value that is within 1.5 box lengths of the bottom edge of the box). In Figure 3.9, you can see that the lowest observed value is 1.00.

THE 1.5 RULE. The previous section indicated that the end of the top whisker represents the "highest observed value that is within 1.5 box lengths of the top edge of the box." To put this in the form of an equation, we can express the maximum length of a whisker in this way:

$$Maximum\ whisker\ length = (1.5)(Box\ length)$$

In other words, a whisker will be no more than 1.5 box lengths from the edge of the box (let's call this the "1.5 rule"). This is not to say that a whisker *must* be as long 1.5 box lengths—in some cases it will be shorter. The "1.5" value is merely the upper limit.

If observed values are more than 1.5 box lengths from the box, they are called "outliers" or "extreme values" and are represented with special symbols A later section will discuss outliers and extreme values in greater detail.

ASSESSING THE SYMMETRY OF THE DISTRIBUTION. The boxplot in Figure 3.9 represents an approximately symmetrical data set (a data set that it not skewed). You can tell that a variable is approximately symmetrical when its boxplot displays the following characteristics:

- The upper whisker is pretty much the same length as the lower whisker, and

- The median is pretty much in the middle of the box.

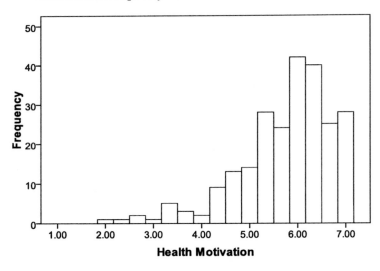

Figure 3.10. Frequency histogram for scores on the *health motivation* scale (a negatively-skewed distribution).

Example 2: A Negatively-Skewed Distribution

Figure 3.10 displays the frequency histogram for a different variable: *health motivation*. This histogram shows that this variable displays a negative skew. You know that it has a negative skew because:

- Most of the scores are bunched together where the higher scores appear, and
- there is a long tail going out in the direction of the lower (i.e., more negative) scores.

Figure 3.11 displays a boxplot for the same variable. We will use this boxplot to learn about outliers, extreme values, and other concepts relevant to skewed distributions.

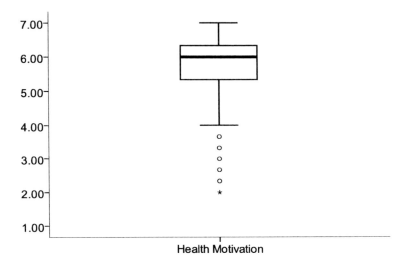

Figure 3.11. Boxplot for scores on the *health motivation* scale (a negatively-skewed distribution).

The top of the box in Figure 3.11 corresponds to a score of 6.33. Since the top of a box in a boxplot is at Q_3, this tells us that 75% of the participants displayed a health motivation score of 6.33 or lower. The bottom of the box corresponds to a score of 5.33. Since the bottom of the box in a boxplot is at Q_1, this tells us that 25% of the participants displayed a score of 5.33 or lower.

With a boxplot, the interquartile range (IQR) is equal to one vertical box length, and one vertical box length is computed by taking the difference between Q_3 and Q_1. For the variable health motivation, the length of one vertical box length is calculated as follows:

$$IQR = Q_3 - Q_1 = 6.33 - 5.33 = 1.00$$

The above shows that one vertical box length is equal to 1.00, indicating that the IQR is also equal to 1.00. Think about what it means to have such a small value for the interquartile range. Earlier, we established that the IQR captures the middle 50% of the distribution. The value of the current IQR tells us that 50% of the participants must have displayed scores within a very narrow interval—the interval from 5.33 to 6.33. This is what statisticians mean when they say that skewed distributions have many scores "bunched together" at one end.

The Ends of the Whiskers

The preceding section found that one box length was equal to 1.00 for the current variable. According to the 1.5 rule, the end of a whisker may be a maximum of 1.5 box lengths from the end of the box. This means that the end of a whisker for the current boxplot may be no more than 1.50 units from the edge of the box (because $1.5 \times 1.00 = 1.50$).

The bottom edge of the box in Figure 3.11 corresponded to a score of 5.33. This indicates that the end of the lower whisker should be no lower than a score of 3.83 (because $5.33 - 1.50 = 3.83$). As predicted, Figure 3.11 shows that the end of the lower whisker is in the general area of 3.83 (technically, the end of the lower whisker is at 4.00 because none of the subjects displayed a score of exactly 3.83, but two subjects displayed a score of 4.00).

The top edge of the box in Figure 3.11 corresponded to a score of 6.33. This means that the end of the top whisker should be no higher than a score of 7.83 (because $6.33 + 1.50 = 7.83$). But in Figure 3.11, the top of the whisker actually ends at a score of 7.00. This is because the multiple-item rating scale for health motivation was created in such a way that the scores could only range from 1.00 to 7.00.

Outliers

In general, an *outlier* is a score this is very different from most scores in a distribution, typically because it is either much higher or much lower than most of the other scores. One of the reasons that researchers create boxplots is to determine whether the variables they are about to analyze have any outliers.

WHY OUTLIERS ARE IMPORTANT. Outliers are important, in part, because they can distort the obtained value of some statistics, causing them to become misleadingly large, misleadingly small, or misleading in some other way. For example, in the previous section on "Measures of Central Tendency," you learned that an outlier can cause the mean to be a misleading index of the "most typical" score in a sample. Outliers have similar undesirable effects on many other statistics and statistical procedures, including t tests, ANOVA, multiple regression, and others. Particularly worrisome is the fact that when a researcher is determining whether specific results are "statistically significant," one type of outlier might lead to an incorrect conclusion that the results are significant, whereas a different type of outlier might lead to an incorrect conclusion that they are not (you will learn about the concept of statistical significance in *Chapter 5*).

In other words, outliers can be very bad news for researchers, particularly when researchers do not know that they are present. When reading articles published by others, you should always look for evidence that the authors screened their data for outliers. When conducting your own research, you should do the same.

IDENTIFYING OUTLIERS. Many procedures are available to researchers who wish to screen their data for possible outliers. This section shows how to do this by reviewing boxplots. The definitions used here (e.g., the definitions for "outlier" and "extreme value") are those used by the SPSS statistical application.

With this approach illustrated here, an **outlier** is defined as a data point that is located 1.5 to 3.0 box lengths from the edge of the box. You will recall that a whisker may extend up to 1.5 box lengths from the edge. If an observed score is more than 1.5 box lengths from the edge (but is less than 3.0 box lengths from the edge), it is labeled an *outlier*, and is represented by the symbol "o."

The boxplot in Figure 3.11 reveals five outliers for the health motivation variable—notice that the symbol "o" appears five times below the end of the lower whisker. These five scores were 3.67, 3.33, 3.00, 2.67, and 2.33. With the type of graph shown here, it is not possible to determine how many participants displayed each of these five scores, or who these participants were. With most statistical applications (including SPSS), however, it is possible to dig deeper and determine which specific subjects produced the outliers. Once this is done, the researcher may choose to transform the scores so that they are no longer outliers, or to deal with them in some other way.

Extreme Values

An extreme value is also an outlier, but it is a particularly *bad* outlier. With the conventions used here, an **extreme value** is defined as a score that is more than three box lengths from the edge of the box . In other words, an extreme value is a score that is more than three interquartile ranges from the edge of the box.

SPSS uses an asterisk ("*") to identify an extreme value in boxplots. In Figure 3.11, one extreme value appears near the bottom of the figure. The asterisk shows that this extreme value corresponds to score of 2.00 on the 7-point health motivation variable.

Example 3: A Positively Skewed Distribution

For the sake of completeness, this section ends with a positively skewed distribution. The graphs presented here illustrate a new variable: *dislike of vegetables*. This was measured using a multiple-item summated rating scale consisting of items 39-41 on Dr. O'Day's *Questionnaire on Eating and Exercising*. The scale consists of items such as "I *hate* to eat vegetables," and participants used a 7-point rating scale to indicate the extent to which they agreed or disagreed with each item. Scores on the summated scale could range from 1.00 to 7.00, with higher scores indicating a stronger dislike of eating vegetables. The frequency histogram for responses ($N = 238$) is presented as Figure 3.12.

The histogram in Figure 3.12 displays a strong positive skew: Most scores are bunched together at the low end of the scale, and a long tail stretches in the direction of the higher (more positive) scores. In fact, 115 of the 238 participants (almost 50%) displayed the lowest possible score: 1.00 on the 7-point scale. These are the people who indicated that they strongly disagreed with any assertion that they hate vegetables. Their mothers would be so *proud!*

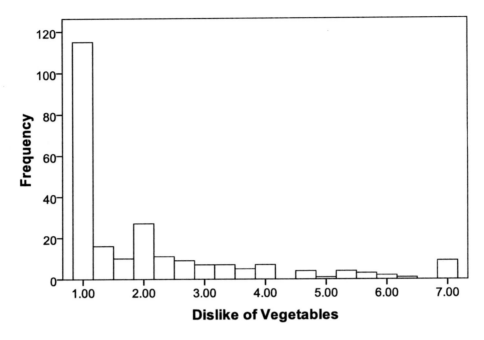

Figure 3.12. Frequency histogram for scores on the *dislike of vegetables* scale (a positively skewed distribution).

Figure 3.13 presents the boxplot for the same variable. The vertical axis (on the left side of the graph) represents scores on the scale, where 1 = *Extremely Disagree*, and 7 = *Extremely Agree*. Remember that this scale consisted of items such as "I *hate* to eat vegetables."

Notice that this boxplot does not have a whisker extending below the box, at the bottom of the figure. When a distribution displays a very strong positive skew, the scores may be so piled up on the bottom that there is no need for a whisker to represent the participants with the lower scores. That is what happened with the dislike of vegetables scale. For the same reason, when a distribution displays a strong negative skew, the boxplot may not have a whisker extending above the box, at the top of the figure.

Since these concepts were covered in previous sections, it is not necessary to review Figure 3.13 in great detail. Here is a summary of the most important results revealed by the boxplot:

- The heavy horizontal line inside the box shows that the median is equal to 1.33. This tells us that 50% of the participants displayed a score of 1.33 or lower.
- The top of the box corresponds to a score of 2.42, indicating that 75% of the participants displayed a score of 2.42 or lower.
- The length of the box is 1.42 units. We know this because the bottom of the box is at a score of 1.00, the top of the box is at a score of 2.42, and 2.42

− 1.00 = 1.42. Since the box length is equal to 1.42, we know that the interquartile range *(IQR)* is also equal to 1.42.

- The six scores represented by the symbol "0" are *outliers* (scores that fall between 1.5 and 3 box lengths from the top edge of the box).
- The one score represented by an asterisk ("*") is an extreme value—a score that lies more than 3 box lengths beyond the top edge of the box).

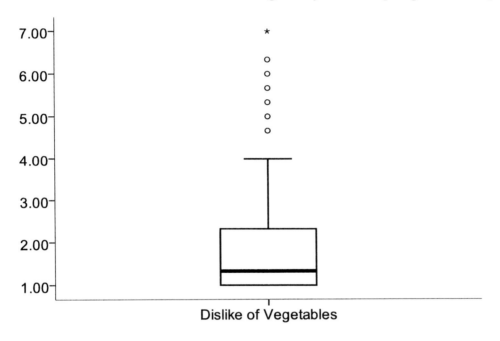

Figure 3.13. Boxplot for scores on the *dislike of vegetables* scale (a positively-skewed distribution).

Summary Regarding Boxplots

One boxplot contains a lot of information. With a glance, you can evaluate central tendency and variability by inspecting the variable's median, range, and interquartile range. You can also quickly determine whether the variable is skewed and whether it displays any outliers or extreme values.

This means that a boxplot is a valuable tool for researchers who want to quickly inspect and screen their data prior to performing statistical tests. It also means that, when included as a figure in a published article, a boxplot can help the audience quickly grasp the basic characteristics of a variable's distribution.

Chapter Conclusion

You have now covered the basics of ***descriptive statistics:*** statistical procedures that are used to explore and describe the fundamental characteristics of a distribution. When researchers describe a data set at the beginning of the *Results* section of an article, they typically stick to the basics: measures of central tendency (e.g., the mean, the median), measures of variability (e.g., the standard deviation, the interquartile range), estimates of reliability (e.g., coefficient alpha), along with a few other indices.

Researchers know that one picture is worth a dozen statistics, so they sometimes use graphs to communicate the nature of their variables. Having completed this chapter, you should be able to review the frequency histograms and boxplots that are sometimes published for this purpose.

One of the key concepts covered in this chapter was the nature of the normal distribution. This is because the theoretical normal distribution plays an important role in several procedures used to determine whether results are "statistically significant." Because of its importance, the normal distribution is given more detailed coverage in the following chapter. There, you will learn about the relationship between standard scores and "area under the curve." This will set the stage for the ultra-important concepts to follow: significance tests, confidence intervals, and effect size.

Chapter 4

z Scores and Area Under the Normal Curve

This Chapter's *Terra Incognita*

You are reading about a program of research in which the investigators have developed a new intelligence test (i.e., a new IQ test). An excerpt from the article reports the following:

> The final version of the test displayed a population mean of $\mu = 100$ and a population standard deviation of $\sigma = 15$. This indicated that 95% of test-takers could be expected to display scores between 70.6 and 129.4, whereas 99% could be expected to display scores between 61.3 and 138.7.

Do not fear. All shall be revealed.

Introduction to the Normal Curve

The ***normal distribution*** (also called the ***normal curve*** and the ***Gaussian distribution***) is a theoretical probability distribution whose shape is defined by a mathematical formula. Its basic form is illustrated in Figure 4.1.

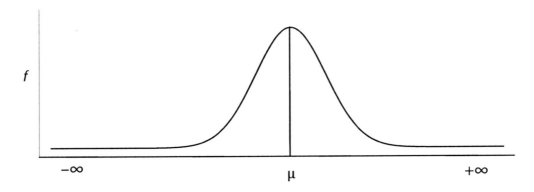

Figure 4.1. The normal distribution.

Think of the distribution in Figure 4.1 as being a frequency distribution for a continuous, quantitative variable assessed on an interval or ratio scale of measurement. The variable's scores appear on horizontal axis (the X axis) with the scores increasing in size as you move from the left to the right.

The vertical axis (the Y axis) plots the frequency for each value of this X variable. Technically, it actually plots the *probability density* for each value, but let's just think of it as frequencies for the moment. The horizontal line at the bottom of the figure represents a frequency of zero. You can see that the curve is highest in the middle, just above the symbol μ (which, you will recall, is our symbol for the population mean). This tells us that the most frequent score in a normal distribution is the mean.

Figure 4.1 shows that the normal distribution has the following properties:

- It is bell-shaped.
- It is *symmetrical* (this is just another way of saying that the right side of the distribution is a perfect mirror image of the left side).
- It is *unimodal* (i.e., it has just one mode).
- The most frequent value is the mean, and the mean is also equal to the median and the mode.
- Exactly 50% of the distribution lies below the mean, and exactly 50% lies above it.
- It is *asymptotic*, which (in this context) means that the curve stretches outward and downward as we move away from the mean and out into the

tails, but it never quite becomes a straight line and never actually touches the horizontal axis below it. You will recall that this horizontal axis represents a frequency of zero. In other words, the normal curve stretches to the left in the direction of negative infinity ($-\infty$), and stretches to the right in the direction of positive infinity ($+\infty$) and never quite touches a frequency of zero.

Some people mistakenly believe that any bell-shaped curve is the normal curve, but this is not the case. The true normal curve (i.e., the theoretical *perfect* normal curve) has a shape that is very precisely described by a mathematical formula. This formula will be presented later in the chapter.

Why the Normal Curve is Important

The concept of the normal curve is important (in part) because it makes possible the data-analysis practices of *interval estimation* and *significance testing*. This section explains.

INTERVAL ESTIMATION. When a variable is normally distributed (and the sample is large), researchers can estimate the percentage of individuals who will display a specific range of scores on that variable. For example, the *terra incognita* excerpt at the beginning of this chapter described a fictitious intelligence test with a mean of 100 and a standard deviation of 15. Because the mean and standard deviation were known values and because the distribution of test scores was normally distributed (and large), the researchers could estimate that 95% of individuals taking the test would display scores between 70.6 and 129.4. Later, we will see how researchers use established characteristics of the normal distribution to make such estimates.

SIGNIFICANCE TESTING. One of the most widely-used conventions in science involves determining whether the results from a study are *statistically significant*. Many (but not all) significance tests require that the distributions being analyzed must be normally distributed. In *Chapter 5*, you will see what is meant by "statistically significant," and you will learn why this practice has become controversial.

Approximately Normal Versus Normal Distributions

In published articles, researchers sometimes describe a variable as being "normally distributed." By this, they actually mean that the variable was "*approximately* normally distributed." This is because real-world variables just about never display a theoretical *perfect* normal distribution. The following sections explain why.

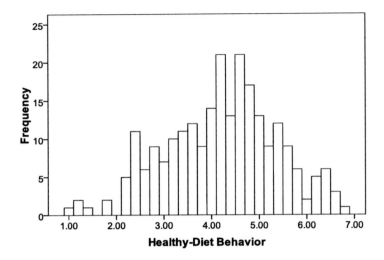

Figure 4.2. Frequency histogram for sample of scores on the *healthy-diet behavior* variable (an approximately-normal distribution).

AN APPROXIMATELY NORMAL DISTRIBUTION. *Chapter 3* described a fictitious study in which a researcher named Dr. O'Day obtained scores on a variable called *healthy-diet behavior*. This variable indicated the extent to which an individual eats nutritious foods and avoids unhealthy ones. Scores on this variable could range from 1.00 (indicating an unhealthy diet) to 7.00 (indicating a healthy diet). Figure 4.2 presents a frequency histogram for scores on this variable obtained from 238 participants.

The distribution of scores in this figure constitutes an ***approximately normal distribution***—a distribution that is bell-shaped, fairly symmetrical, unimodal, and more or less follows the basic shape of the theoretical normal distribution. But it does not constitute a *perfect* normal distribution.

A NORMAL DISTRIBUTION. The ***theoretical normal distribution*** (also called the ***perfect normal distribution*** or the ***Gaussian distribution***) is a mathematical concept—a mathematical ideal. If the population of scores on healthy-diet behavior displayed a mean of 4.22 and followed the shape of the theoretical perfect normal distribution, it would look like Figure 4.3.

Figure 4.3 shows why real-world variables can never follow the exact shape of the perfect normal distribution. Among other things, the perfect normal distribution extends toward negative infinity ($-\infty$) on the left, and toward positive infinity ($+\infty$) on the right, never quite touching the horizontal axis (i.e., never quite displaying a frequency of zero). The theoretical normal distribution is just an ideal—an ideal that researchers can hope for but never quite attain with real-world data. With real-world data, the best that they can attain is an approximately normal distribution.

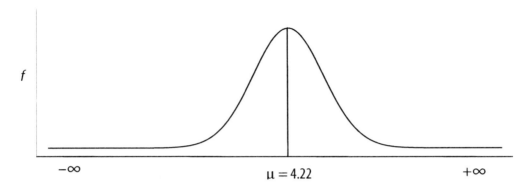

Figure 4.3. Frequency distribution for a population of scores on the *healthy-diet behavior* variable if it were distributed as a perfect normal distribution.

The Standard Normal Distribution

We have learned about the theoretical perfect normal distribution. We will now see an even more useful version of it: the *standard normal distribution*.

The **standard normal distribution** is a theoretical perfect normal distribution in which all of the scores are z scores. This section discusses the concept of z scores, and describes the advantages of working with a normal distribution consisting of z scores rather than raw scores.

Raw Scores Versus z Scores

A **raw score** is a participant's original score on some variable, before it has been transformed in any way. A **z score**, on the other hand, is a transformed version of this raw score. Once a participant's raw score has been converted into a z score, the resulting z score indicates the number of standard deviations that the raw score falls above (or below) the mean.

The process of converting all of the scores in a sample into z scores is called **standardizing** the variable. The process is called standardizing, in part, because it produces a variable that will have known values for the mean and standard deviation. Most researchers know that a **z-score variable** always has a mean of zero and a standard deviation of one.

FORMULA FOR CREATING z SCORES. To convert a raw score to a z score, the researcher subtracts the variable's mean from the raw score and divides the difference by the variable's standard deviation. When working with a sample of scores (as opposed to a population) the following formula is used to transform raw scores into z scores:

$$z_X = \frac{X - \bar{X}}{S}$$

Where:

> z_X = a given participant's score on the variable, after being converted to z-score form,
>
> X = the participant's score on the variable, while in its original, raw-score form,
>
> $\bar{X}$ = the sample mean for the variable, and
>
> S = the sample standard deviation for the variable (this must be the *biased* estimate of the standard deviation—the version whose formula uses N in the denominator rather than $N-1$).

On the other hand, when working with a population of scores (as opposed to a sample) the following formula is used:

$$z_X = \frac{X - \mu}{\sigma}$$

Where:

> z_X = a given participant's score on the variable, after being converted to z-score form,
>
> X = the participant's score on the variable, while in its original, raw-score form,
>
> μ = the population mean for the variable, and
>
> σ = the population standard deviation.

EXAMPLE. As an illustration, we will convert a few raw scores into z scores. Think about the healthy-diet behavior variable which was discussed earlier. Scores on this variable could range from 1.00 (indicating a diet of junk food) to 7.00 (indicating a diet of healthy and nutritious foods). In a sample of 238 participants, the mean for this variable was $\bar{X} = 4.22$, and the sample standard deviation was $S = 1.17$ (as was noted above, we must use the *biased* estimate of the sample standard deviation, not the *unbiased* estimate).

Let's assume that Subject 1 has a raw score of 5.00 on this healthy-diet behavior variable. Below, we convert his raw score into a z score:

$$z_X = \frac{X - \bar{X}}{S} = \frac{5.00 - 4.22}{1.17} = \frac{0.78}{1.17} = 0.667 = +0.67$$

So the raw score of 5.00 displayed by Subject 1 coverts into a corresponding z score of +0.67. The *sign* and *size* of this subject's z scores provides a great deal of information about where that person stood on the variable being analyzed:

- The *sign* of a z score (positive versus negative) tells us whether the corresponding raw score is above or below the mean. If the z score is positive (such as +0.53 or +2.18), we know that the raw score was above the mean. If the z score is negative (such as −0.21 or −1.99), we know that the raw score was below the mean.

- The *size* of a z score (in absolute value) tells us how many standard deviations the raw score differs from the mean. For example, if the z score is +0.53, it indicates that the raw score is 0.53 standard deviations above the mean. If the z score is −1.99, it indicates that the raw score is 1.99 standard deviations below the mean. If the z score is 0.00, the raw score is equal to the mean.

Since the z score for Subject 1 was +0.67, we know that his raw score on healthy-diet behavior was 0.67 standard deviations above the mean. If we can assume that scores on this variable are valid, such a z score would indicate that his diet is a bit healthier than the diet of the typical participant (he *is* above the mean, after all).

In this example, we converted just one subject's raw score into a z score. But what if we had converted all 238 raw scores from the study into z scores? The resulting distribution would display a mean of $M = 0$ and a standard deviation of $SD = 1$. This will be the case whenever we standardize a sample of raw scores by converting them into z scores.

Let's take things one step further. What if the distribution was not a sample with just a few hundred observations, but was instead a population with an infinite number of observations? And what if this distribution was a symmetrical, asymptotic, and perfectly normal distribution? We would call such a distribution a *standard normal distribution*, and it would prove useful to researchers in all sorts of ways.

The Standard Normal Distribution

The **standard normal distribution** is a theoretical perfect normal distribution consisting of z scores. It has all of the characteristics of the perfect normal distribution described earlier (e.g., is bell-shaped, follows the shape of the theoretical normal curve, etc.) but has one additional property: It has been standardized so that all scores are z scores. This tells us that the mean of this distribution will be $\mu = 0$ and its standard deviation will be $\sigma = 1$.

Figure 4.4 illustrates the standard normal curve. The values on the horizontal axis are z scores.

Notice that the mean of the distribution (at the very center of Figure 4.4) is represented by solid vertical line and a z score of $z = 0$. To the right of this mean, a dashed vertical line represents a z score of $z = +1$. This dashed line represents scores that are 1 standard deviation above the mean. The next dashed line represents a z score of $z = +2$ (i.e., scores that are 2 standard deviations above the mean).

To the left of the distribution's mean, a dashed vertical line represents a z score of $z = -1$. This dashed line represents scores that are 1 standard deviation below the mean. The remaining dashed lines may be interpreted in the same way.

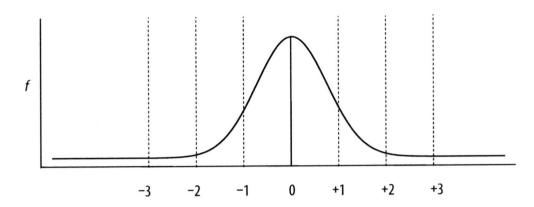

Figure 4.4. The standard normal distribution (values on horizontal axis are *z* scores).

Figure 4.4 shows that, in a standard normal curve, just about all scores lie somewhere between a *z* score of −3 and a *z* score of +3. Later in this chapter, we will learn about the specific percentages of the distribution that are "captured" between specific whole-number *z* scores.

Area Under the Curve for Whole-Number *z* Scores

The theoretical perfect normal distribution is useful in part because there is a precise relationship between specific *z* scores and "area under the curve." This section explains the meaning of this expression, and summarizes the area under the curve that is associated with some whole-number *z* scores.

AREA BETWEEN THE MEAN AND WHOLE-NUMBER *z* SCORES. When researchers talk about "area under the curve," they are typically referring to the percent (or, alternatively, the proportion) of scores that appear in some carefully delineated section of the theoretical perfect normal curve. In many cases, they are referring to the percent of scores that lie between two specific *z* scores (such as between $z = -1$ and $z = +1$). In other cases, they are referring to the percent of scores that lie between the mean of the distribution (which is always $z = 0$ if it is a standard normal curve) and some specific *z* score. The current section focuses on the latter situation. Figure 4.5 illustrates the percent of area under the curve that lies between the mean ($z = 0$), and three whole-number *z* scores: $z = +1$, $z = +2$, and $z = +3$.

Figure 4.5 shows that about 34% of the scores in a standard normal distribution lie between the mean ($z = 0$), and a *z* score of $z = +1$. To be more precise, the percent is closer to 34.13%, but we will round it to 34%. If we were expressing it as a proportion rather than a percentage, the proportion would be .3413.

Figure 4.5 shows that about 48% of the scores in the standard normal curve lie between the mean ($z = 0$), and a *z* score of $z = +2$. The actual percent is closer to 47.78%, and this value rounds to 48%. If we were expressing it as a proportion, it would be .4778.

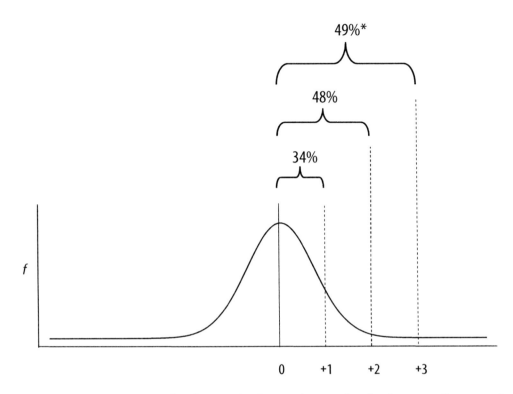

Figure 4.5. Area under the standard normal curve that lies between the mean ($z = 0$) and several whole-number z scores (*area between the mean and $z = +3$ is actually 49.87%).

Finally, the figure shows that about 49% of the scores in the standard normal curve lie between the mean ($z = 0$), and a z score of $z = +3$. In truth, this value of 49% is something of a fib, because the actual percentage is more like 49.87%, and 49.87% should actually round up to 50%, right? The reason that we fib is because we want students to remember that 50% of the curve lies between the mean of the distribution ($z = 0$) and a z score of positive infinity ($z = +\infty$). In other words, the entire right side of the distribution (which stretches out to infinity) constitutes 50% of all of the distribution's scores. To keep the two ideas separate, some books (including this one) say that 49% of the distribution lies between the mean and a z score of $+3$, and 50% lies between the mean and a z score of positive infinity.

AREA ASSOCIATED WITH POSITIVE VERSUS NEGATIVE z SCORES. Remember that the theoretical perfect normal curve is *symmetrical*—the tail on the left side of the distribution is a mirror image of the tail in the right side. This tells us that the percentages associated with positive z scores (as illustrated in Figure 4.5) will also apply to the corresponding negative z scores. For example, since 34% of the area under the curve lies between the mean and a z score of $+1$, we automatically know that 34% of the area under the curve must also lie between the mean and a z score of -1.

How Researchers Determine Area Under Curve

At this point, you should be wondering how we know that 34% of the area under the curve lies between the mean of the distribution and a z score of $z = +1$. How did researchers arrive at this percentage, or the other percentages described above?

In most cases, students and researchers use special tables to determine the area under the curve that is associated with specific z scores. The values in these tables, in turn, have been generated by a mathematical formula.

Formula for Area Under the Curve

Earlier, this chapter indicated that not just any bell-shaped distribution is a theoretical perfect normal distribution. It indicated that the shape of the theoretical perfect normal curve is quite specific and is described by a mathematical formula. That formula (from Hays, 1988) is as follows:

$$f(z) = \frac{1}{\sqrt{2\pi}} e^{-z^2/2}$$

Where:

$f(z)$ represents the height of the normal curve (i.e., the probability density) for a given value of z.

π is the standard mathematical constant equal to about 3.14.

e is the mathematical constant equal to about 2.72.

The above formula allows researchers to compute the height of the normal curve (i.e., the *probability density* of the curve) for any supplied value of z. If you were to insert a series of different z scores into this formula, you would notice the following trend:

- z scores that are close to zero (such as $z = 0.00$, $z = -0.02$, or $z = +0.04$) have relatively large probabilities. This is why the normal curve in Figure 4.5 was at its highest point at $z = 0$ (i.e., at the mean).

- z scores that are far away from zero (such as $z = -3.15$ or $z = +2.93$) have relatively small probabilities. This is why the normal curve in Figure 4.5 drops lower and lower as z scores move away from zero, going out into the tails. The tails contain the z scores that are higher in absolute value (such $z = -3$ or $z = +3$), and these z scores have small probabilities.

Tables Reporting Percent of Area Under the Curve

In truth, researchers who need to know the area under the curve that is associated with a specific z score are not likely to use the preceding formula to compute these values. Instead, they are more likely to refer to tables of values (tables which have been derived by using this formula). These tables typically appear as an appendix in the back of a statistics textbook, and have titles such as *The Normal Distribution, The Standard Normal Curve*, or *Area Under the Standard Normal Curve*. In most cases, the tables are quite detailed, covering (at a minimum) every possible z score from $z = 0.00$ to $z = 3.00$ in increments of .01 (e.g., $z = 0.00$, $z = 0.01$, $z = 0.02$ and so forth).

For each z score, the table indicates the percent of area between the mean and that z score. Many tables also provide the percent of area that lies beyond that z score, out in the tail. As an illustration, Table 4.1 provides a very small portion of one of these tables.

Table 4.1

Detail from a Table Reporting Percent of Area Under the Standard Normal Curve

z	% of area between mean and z	% of area beyond z in tail
.	.	.
.	.	.
.	.	.
2.00	47.72%	2.28%
2.01	47.78%	2.22%
2.02	47.83%	2.17%
.	.	.
.	.	.
.	.	.

Table 4.1 consists of three columns. The column headed "z" provides specific z scores. In most tables, these z scores range from $z = 0.00$ to $z = 3.00$, and in many tables the scores go to even higher values such as $z = 4.00$ or $z = 4.50$. Table 4.1 provides just three values: $z = 2.00$, $z = 2.01$ and $z = 2.02$ (the vertical dots represent the fact that the current table presents just a small portion of the information presented in an actual table of area under the curve).

The second column in Table 4.1 presents the percent of area in a standard normal curve that lies between the mean ($z = 0.00$) and the z score listed in the first column. The table shows that:

- 47.72% of the area of the standard normal curve lies between the mean and a z score of $z = 2.00$.
- 47.78% of the area of the standard normal curve lies between the mean and a z score of $z = 2.01$.
- 47.83% of the area of the standard normal curve lies between the mean and a z score of $z = 2.02$.

The third column in Table 4.1 presents the percent of area in a standard normal curve that lies *beyond* the z score listed in the first column. In other words, the third column presents the percent of area that lies past that z score, out in the tail. This column shows that:

- only 2.28% of the area of the standard normal curve lies beyond a z score of 2.00, out in the tail.
- only 2.22% of the area of the standard normal curve lies beyond a z score of 2.01, out in the tail.
- only 2.17% of the area of the standard normal curve lies beyond a z score of 2.02, out in the tail.

Table 4.1 indicates that, for a z score of 2.00, 47.72% of the area under the curve lies between the mean and this z score, and 2.28% lies beyond this z score, out in the tail. Below, we see that adding these two percentages together results in 50%:

$$47.72\% + 2.28\% = 50\%$$

This makes sense, because the tail on a given side of the perfect normal curve constitutes 50% of the complete distribution. For any given z score, if you add together (a) the percent of area that lies between the mean and that z score plus (b) the percent of area that lies beyond that z score out in the tail, it must sum to 50%. As an exercise, you can verify this by using the percentages for a z score of $z = 2.01$ from Table 4.1.

Tables Reporting Proportion of Area Under the Curve

Many statistics textbooks provide tables that report *proportions* of area under the curve, rather than *percentages* of area under the curve. You will recall from *Chapter 3* that a proportion provides the same information as a percentage, except that (in a proportion) the decimal point is moved two decimal places to the left.

Table 4.2 presents an example of a table that uses proportions instead of percentages. You can see that it provides essentially the same information that had been provided in Table 4.1, except that the decimal point (for the proportions) has now been moved two spaces to the left.

Table 4.2

Detail from a Table Reporting Proportion of Area Under the Standard Normal Curve

z	Proportion of area between mean and z	Proportion of area beyond z in tail
.	.	.
.	.	.
.	.	.
2.00	.4772	.0228
2.01	.4778	.0222
2.02	.4783	.0217
.	.	.
.	.	.
.	.	.

Area Captured by Whole-Number z-Score Intervals

A table of the area under the standard normal curve is useful for a variety of purposes. Among other things, it allows researchers to determine the percent of area under the curve that is "captured" by specific whole-number z scores.

Intervals Bounded by Whole-Number z Scores

This concept is illustrated graphically in Figure 4.6. This figure indicates the percent of area under the curve that is captured by (a) the interval that extends from $z = -1$ to $+1$, (b) the interval that extends from $z = -2$ to $+2$, and (c) the interval that extends from $z = -3$ to $+3$.

Figure 4.6 shows that, if a variable is distributed as the theoretical standard normal curve:

- About 68% of scores will fall within one standard deviation of the mean (i.e., between a z score of -1 and $+1$).

- About 96% of scores will fall within two standard deviations of the mean (i.e., between a z score of -2 and $+2$).

- About 99% of scores will fall within three standard deviations of the mean (i.e., between a z score of -3 and $+3$). Okay, this is another fib. The actual percentage is more like 99.74%, which otherwise would round up to 100%.

But many teachers want their students to remember that you have to go all the way out to negative infinity (−∞) and positive infinity (+∞) to capture 100% of the area under the curve. So we will pretend that 99.74% rounds to 99%, just to maintain this distinction.

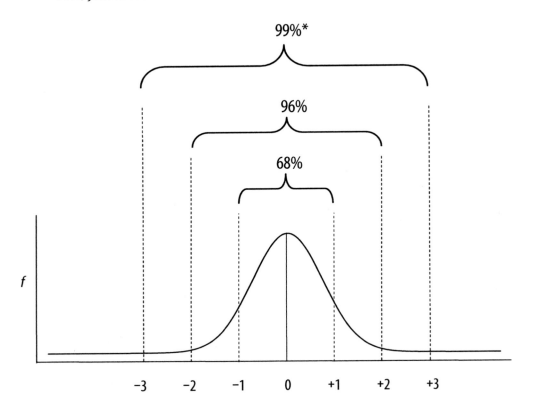

Figure 4.6. Area under the standard normal curve captured by whole-number z scores (*area captured by $z = -3$ and $z = +3$ is actually 99.74%).

Converting z-Score Intervals into Raw-Score Intervals

Now that we know the percentage of area under the curve that is captured by specific z scores, it is a simple matter to convert those z scores into *raw scores*. You will recall that **raw scores** are the original scores on the variable—the scores that have not been standardized or transformed. Once the z scores have been converted into raw scores, we can estimate the percent of participants who are likely to display a specific range of raw scores. For example, we could identify the two raw scores that capture the central 68% of the distribution. This practice is common, especially among test developers, as it identifies the scores that would likely be displayed by test-takers, as well as the scores that are not likely to be displayed.

As an illustration, imagine that we have developed a new intelligence test. With the scores on this test, the population mean is $\mu = 100$ and a population standard deviation is $\sigma = 15$.

Assume that we would like to determine the two specific raw scores on this test that capture the central 68% of the distribution. If the test scores were in z-score form, we know that the interval that extends from $z = -1$ to $z = +1$ would capture the central 68% of the distribution (this was illustrated in Figure 4.6). So our question becomes:

> What raw scores correspond to the z scores of $z = -1$ and $z = +1$?

CONVERSION FORMULA BASED ON POPULATION MEAN AND STANDARD DEVIATION. To answer this question, we will use a formula that converts z scores into raw scores. That formula is presented below (this is the version of the formula that uses the *population* mean and standard deviation, rather than the *sample* mean and standard deviation):

$$X = \mu + (z_X)(\sigma)$$

With the preceding formula:

X = The participant's score on the variable while in raw-score form.

μ = The population mean for the variable of interest.

z_X = The participant's score on the variable while in z-score form.

σ = The population standard deviation.

Remember that, with this IQ test, the population mean is $\mu = 100$ and the population standard deviation is $\sigma = 15$. Below, we compute the raw score that corresponds to a z score of $z = +1$:

$X = \mu + (z_X)(\sigma)$

$X = 100 + (+1)(15)$

$X = 100 + (15)$

$X = 115$

So a z score of $z = +1$ corresponds to a raw score of 115. Next, we compute the raw score that corresponds to a z score of $z = -1$:

$X = \mu + (z_X)(\sigma)$

$X = 100 + (-1)(15)$

$X = 100 - (15)$

$X = 85$

So a z score of $z = -1$ corresponds to a raw score of 85. Our original question had been "What raw scores correspond to the z scores of $z = -1$ and $z = +1$?" We now know that the raw scores are 85 and 115, respectively.

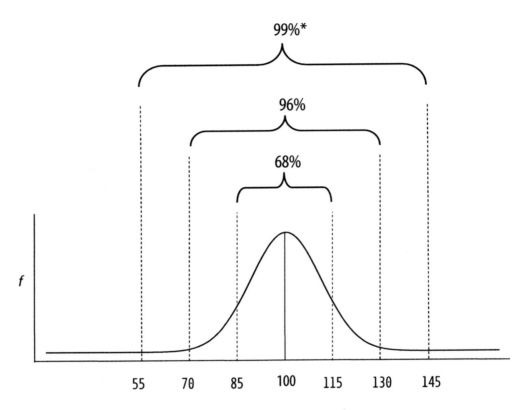

Figure 4.7. Area under the normal curve captured by raw scores on an IQ test whose mean is $\mu = 100$ and standard deviation is $\sigma = 15$ (*area captured by 55 and 145 is actually 99.74%).

Since 68% of the standard normal curve falls in the interval that extends from $z = -1$ and $z = +1$, it follows that 68% of distribution of IQ scores should fall in the interval that extends from 85 to 115. In other words, 68% of the people who take this test should obtain a score somewhere between 85 and 115. This is illustrated in Figure 4.7, under the bracket labeled "68%."

But why stop at $z = -1$ and $z = +1$? If we used the preceding formula to convert each whole-number z score (e.g., $z = -2$, $z = +2$, etc.) into its corresponding raw score, it would produce the other raw-score values appearing on the horizontal axis of Figure 4.7.

The figure shows that, when an IQ test has a mean of 100 and a standard deviation of 15, we can expect that:

- About 68% of individuals will display scores between 85 and 115 (this corresponds to the z-score interval that extends from −1 to +1).
- About 96% of individuals will display scores between 70 and 130 (this corresponds to the z-score interval that extends from −2 to +2).
- About 99% of individuals will display scores between 55 and 145 (this corresponds to the z-score interval that extends from −3 to +3).

CONVERSION FORMULA BASED ON SAMPLE MEAN AND STANDARD DEVIATION. The previous formula showed how to convert z scores into raw scores using the population mean (μ) and population standard deviation (σ). If we were instead working with a *sample* mean ($\bar{X}$) and *sample* standard deviation (S), the formula becomes:

$$X = \bar{X} + (z_X)(S)$$

Where:

X = The participant's score on the variable while in raw-score form.

$\bar{X}$ = The sample mean.

z_X = The participant's score on the variable while in z-score form.

S = The sample standard deviation.

Capturing the Central 95%, 99%, and 99.9%

One of the preceding figures showed that the z scores of $z = -2$ and $z = +2$ capture about 96% of the center of the standard normal distribution. But "about 96%" sounds so sloppy and unscientific. Wouldn't it be better if we knew the *exact z scores* that capture *exactly 95%* of the center of the distribution?

z Scores that Capture the Central 95%, 99%, and 99.9%

Figure 4.8 shows the same normal curve presented in earlier sections, but the dashed vertical lines are now in slightly different locations. The dashed vertical lines now represent the precise z values (out to two decimal places) that capture the central 95%, 99%, and 99.9% of the standard normal distribution.

Figure 4.8 shows that, if a variable is distributed as the standard normal distribution:

- 95% of scores will fall in the interval from $z = -1.96$ to $z = +1.96$.
- 99% of scores will fall in the interval from $z = -2.58$ to $z = +2.58$.
- 99.9% of scores will fall in the interval from $z = -3.29$ to $z = +3.29$.

The preceding z scores are actually quite important, especially the values of −1.96 and +1.96. This is because they are often used when performing significance tests and creating confidence intervals. Don't be surprised when you see them again in later chapters.

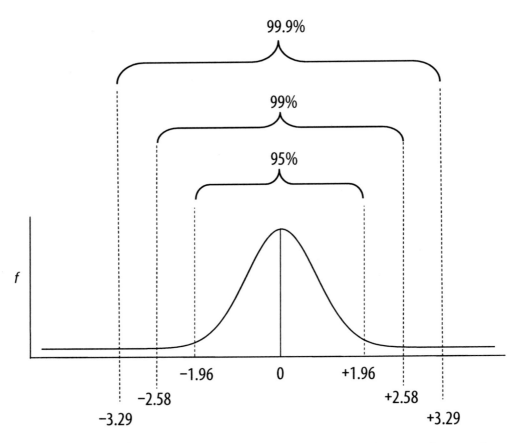

Figure 4.8. z Scores that capture the central 95%, 99%, and 99.9% of area under the standard normal curve.

Raw Scores that Capture the Central 95%, 99%, and 99.9%

Think about the IQ test with a mean of $\mu = 100$ and a standard deviation of $\sigma = 15$. With this test, what are the two raw scores that capture the central 95% of the distribution? What about the central 99%?

CAPTURING THE CENTRAL 95%. When a distribution of test scores is in z-score form, we have already established that 95% of the scores will fall between $z = -1.96$ and $z = +1.96$. To find the two raw scores that capture the central 95% of a distribution, it is necessary to convert these two z scores into raw scores.

We have already established that a z score can be converted into a raw score by using the following formula:

$$X = \mu + (z_X)(\sigma)$$

We will begin by finding the *upper bound:* the raw score that corresponds to a z score of $z = +1.96$. To do this, we insert the raw-score mean ($\mu = 100$), the desired z score ($z_X = +1.96$), and the raw-score standard deviation ($\sigma = 15$) into the formula:

$X = \mu + (z_X)(\sigma)$

$X = 100 + (+1.96)(15)$

$X = 100 + (29.4)$

$X = 129.4$

So the upper bound on this 95% interval is 129.4. To find the *lower bound*, we replace the z score of $+1.96$ with a z score of -1.96:

$X = \mu + (z_X)(\sigma)$

$X = 100 + (-1.96)(15)$

$X = 100 + (-29.4)$

$X = 100 - 29.4$

$X = 70.6$

So the lower bound for the interval is 70.6. Putting the two values together, we can see that 95% of the people who take this test should obtain a score somewhere between 70.6 and 129.4. This is illustrated by the inner interval in Figure 4.9 (the interval identified with the bracket labeled "95%").

CAPTURING THE CENTRAL 99%. If we wish to compute the two raw scores that capture the central 99% of the distribution (rather than the central 95%), we would repeat these calculations, but would replace the value "1.96" with "2.58." This is because (as we learned earlier), 99% of the standard normal distribution falls in the interval that extends from $z = -2.58$ to $z = +2.58$. The general-purpose formula for computing the 99% interval is:

$X = \mu \pm (2.58)(\sigma)$

Remember that the mean for the current test is $\mu = 100$ and the standard deviation is $\sigma = 15$. The formula therefore becomes:

$X = 100 \pm (2.58)(15)$

If we did the math, the preceding formula would provide a lower bound of 61.3 and an upper bound of 138.7, indicating that 99% of the people who take this test would be expected to obtain a score somewhere between 61.3 and 138.7. This is illustrated in Figure 4.9, under the bracket labeled "99%."

CAPTURING THE CENTRAL 99.9%. Finally, to compute the two raw scores that capture the central 99.9% of the distribution, we would repeat these calculations, replacing the value "2.58" with "3.29." The general-purpose formula for computing the 99.9% interval is:

$$X = \mu \pm (3.29)(\sigma)$$

For the current test, the formula becomes:

$$X = 100 \pm (3.29)(15)$$

This formula produces a lower bound of 50.65 and an upper bound of 149.35, indicating that 99.9% of the people who take this test (just about *everybody*) would be expected to obtain a score somewhere between 50.65 and 149.35. This is illustrated in Figure 4.9, under the bracket labeled "99.9%."

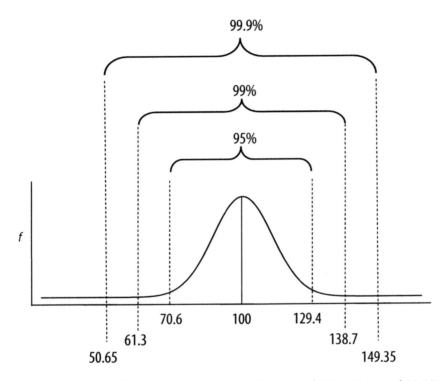

Figure 4.9. Raw scores that capture the central 95%, 99%, and 99.9% of area under the normal curve for an IQ test whose mean is $\mu = 100$ and standard deviation is $\sigma = 15$.

Final Word: Is Any of this Really True?

In this chapter, you have devoted a lot of time and effort to material dealing with the theoretical normal distribution. But was any of this material really true? Consider the following scenario:

> In an empirical investigation, you obtain scores from 17 subjects and then convert their raw scores into z scores. Is it true that 34% of these scores will fall somewhere between the mean and a z score of $z = +1$? Is it true that 95% will fall

somewhere between $z = -1.96$ and $z = +1.96$? Is it true that 99% will fall somewhere between $z = -2.58$ and $z = +2.58$?

The answer is yes and no. But mostly no. Let me explain.

The statistical "facts" presented in this chapter (e.g., the fact that says "34% of scores lie between the mean and $z = +1$") really are true for the theoretical standard normal distribution. This is the *perfect* normal distribution—the one whose shape is defined by the mathematical formula presented earlier.

The statistical "facts" may also be true for a real-world sample, as long as it displays two characteristics. First, the sample must contain a really large number of participants. Second, the sample must display a shape that is very similar to the shape of the theoretical perfect normal curve.

This means that if the fictitious researchers described in this chapter had obtained IQ test scores from thousands of participants and found that the shape of the distribution of test scores was very similar to the shape of the perfect normal distribution, it would be okay for them to make statements of the sort presented earlier in this chapter. For example, it would be reasonable for them to say that "34% of the test scores fall between the mean and $z = +1$."

However, the statistical facts presented here will not necessarily be true for real-world samples that (a) are small or (b) are skewed, platykurtic, leptokurtic, bimodal, or otherwise non-normal in shape. It is hard to say just how small is too small, but I will go out on a limb and assert that the sample of 17 subjects mentioned at the beginning of this section is probably too small for the percentages described in this chapter to hold true.

Why, then, have we spent so much time on this topic? Because the statistical facts described here do hold true for certain important *sampling distributions*—theoretical distributions that are used for significance testing, for the creation of confidence intervals, and for other procedures that are central to the statistical analysis of data.

So take heart—the concepts learned in this chapter are important, and are well worth the time and effort. They are core concepts, essential to understanding *the region of rejection, null-hypothesis significance testing, 95% confidence intervals, Cohen's d statistic*, and similar topics. You are introduced to these important concepts in the very next chapter.

Chapter 5

The Big-Three Results in Research Articles

This Chapter's *Terra Incognita*

You are reading a research article. It describes an experiment in which subjects in the experimental group consumed a large amount of caffeine, whereas those in the control group consumed none. The dependent variables were four different measures of subject irritability.

The *Results* section reports the following:

> Because the study included four dependent variables, the Bonferroni adjustment was used to maintain a familywise Type I error rate of .05. The resulting Bonferroni-adjusted alpha level was α_{ADJ} = .0125
>
> For self-rated irritability, the mean score for the high-caffeine condition (M = 575.00, SD = 35.49) was higher than the mean for the no-caffeine condition (M = 544.73, SD = 36.13). The observed difference between means was 30.27, 95% CI [18.71, 41.82], $t(148)$ = 5.18, p <.001. The point-biserial correlation between the independent variable and the dependent variable was r_{pb} = .39, and this fell just short of a large effect according to Cohen's (1988) criteria.

Have courage. All shall be revealed.

Introduction to the Big-Three Results

Research articles provide three types of statistical results that are used to make sense of the study's findings: significance tests, confidence intervals, and indices of effect size. This chapter explains the meaning of these results and shows how they are usually presented in the tables and text of an article.

The three results are discussed separately because each provides a somewhat different perspective regarding the investigation's findings. Specifically:

- The significance test communicates *probability*: It tells us how likely the current results would be if the study's null hypothesis were true.

- The confidence interval communicates *precision*: It provides a range of plausible values for the population parameter being estimated. Confidence intervals help us evaluate whether this range is relatively precise or relatively imprecise.

- The index of effect size communicates *strength*: It tells us about the strength of the relationship between the predictor variable and the criterion variable.

Illustrative Study: The Effects of Caffeine

Assume that a fictitious researcher named Dr. O'Day wants to determine whether consuming caffeine causes people to be more irritable (i.e., to be more easily upset or angered). To determine this, she conducts an experiment in which she manipulates the amount of caffeine consumed and observes how this manipulation affects scores on four measures of irritability.

MANIPULATING THE INDEPENDENT (PREDICTOR) VARIABLE. Dr. O'Day began with a pool of 150 potential participants. She randomly assigned half ($n = 75$) to the experimental group, and the other half ($n = 75$) to the control group. Regardless of condition, all participants were instructed to drink five cups of coffee each day. The coffee was provided by Dr. O'Day, and she saw to it that:

- Participants in the experimental group consumed regular coffee that contained a total of 1,000 mg of caffeine per day (we will call this the *high-caffeine* condition).

- Participants in the control group consumed decaffeinated coffee that contained a total of 0 mg of caffeine per day (this is the *no-caffeine* condition).

MEASURING THE DEPENDENT (CRITERION) VARIABLES. Dr. O'Day collected scores on the following four dependent variables:

- **SELF-RATED IRRITABILITY.** This variable indicated the extent to which the subject reported feeling irritable, upset, and out-of-sorts. Each day, the subject completed a brief multiple-item summated rating scale. Scores on this scale could range from 200 to 800 with higher scores indicating greater feelings of irritability. For a given subject, Dr. O'Day computed the mean of that subject's scores over the course of the study, and that mean served as the self-reported irritability score for that participant.

- **OTHER-RATED IRRITABILITY.** This variable indicated the extent to which the subject felt irritable, upset, and out-of-sorts, as rated by someone who saw the subject on a daily basis (such as a friend or relative). It was measured and scored in the same way that the self-rated irritability scale was scored.

- **SELF-REPORTED OUTBURSTS.** This variable indicated the number of anger-related outbursts (e.g., temper tantrums) that the participant displayed over the four-week course of the study. Scores on this variable were reported by the participant.

- **OTHER-REPORTED OUTBURSTS.** This variable was similar to the previous one, except that the reports were again provided by friend or relative who saw the participant on a regular basis.

Table 5.1

Scores on Four Measures of Irritability as a Function of Amount of Caffeine Consumed (High-Caffeine Versus No-Caffeine): Significance Tests, Confidence Intervals, and Point-Biserial Correlations

Dependent variable	High caffeine[a] M_1	(SD_1)	No caffeine[b] M_2	(SD_2)	Difference between means $M_1 - M_2$ [95% CI]	$t(148)$	p	r_{pb}
Self-rated irritability	575.00	(35.49)	544.73	(36.13)	30.27 [18.71, 41.82]	5.18	<.001	.39
Other-rated irritability	556.46	(35.86)	546.27	(36.10)	10.19 [-1.42, 21.80]	1.73	.085	.14
Self-reported outbursts	6.24	(2.67)	5.03	(2.69)	1.21 [0.35, 2.08]	2.77	.006	.22
Other-reported outbursts	5.24	(2.67)	4.99	(2.65)	0.25 [-0.61, 1.11]	0.58	.561	.05

Note. $N = 150$. CI = Confidence interval; r_{pb} = Point-biserial correlation between the independent variable and dependent variable.
[a]$n = 75$. [b]$n = 75$.

Results from the Caffeine Study

Dr. O'Day used an independent-samples t test to analyze data from the experiment. She performed the analysis four times: once for each criterion variable. She published her findings in a research journal, and part of the *Results* section from the article is reproduced below.

For each dependent variable, an independent-samples t test was performed to determine whether the mean score for the high-caffeine condition was significantly different from the mean score for the no-caffeine condition. The results from these analyses are summarized in Table 5.1.

The first row of Table 5.1 presents results obtained for the t test performed on self-rated irritability. The table shows that the mean score for the high-caffeine condition (M = 575.00, SD = 35.49) was higher than the mean for the no-caffeine condition (M = 544.73, SD = 36.13). The observed difference between the two means was 30.27, 95% CI [18.71, 41.82], and this difference was statistically significant, $t(148)$ = 5.18, p <.001. As a measure of effect size, the point-biserial correlation was computed between the independent variable (amount of caffeine) and the dependent variable (self-rated irritability). For this dependent variable, r_{pb} = .39, and this fell just short of a large effect according to Cohen's (1988) criteria.

The bar chart in Figure 5.1 presents mean scores on self-rated irritability for the two treatment conditions. Error bars represent 95% confidence intervals.

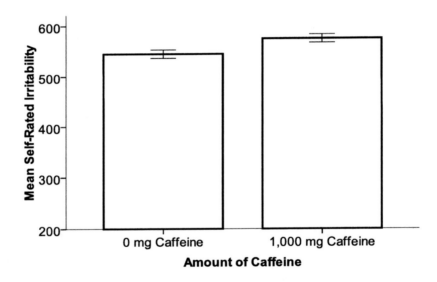

Figure 5.1. Mean self-rated irritability scores as a function of the amount of caffeine consumed (error bars represent 95% CI).

The remainder of this chapter will explore the results presented in this excerpt in some detail. We will see that the significance tests reported in the excerpt provide one perspective on the findings, the confidence intervals provide a different perspective, and the indices of effect size provide still another perspective.

This exploration begins where most researchers typically begin: with the null-hypothesis significance test.

Big-Three Results, Part 1: Statistical Significance

When you think *statistically significant* you should think *probability* or *likelihood*. When researchers say that results are statistically significant, they are saying the following:

> If the null hypothesis were true, it is unlikely that I would have obtained sample results such as these.

In general, a **null-hypothesis significance test** (abbreviation: NHST) is a procedure for determining whether the current results are statistically significant. It is a procedure for determining the probability that you would have obtained a sample statistic as large as the statistic you obtained (or an even larger statistic) if the null hypothesis were true.

Statistical Significance: Basic Concepts

To understand the meaning of statistical significance, you must first understand the meaning of the various questions and hypotheses that are investigated in a study. First, the **research question** is a concise summary of what the investigator hopes to learn from the study. For example, Dr. O'Day may have stated her initial research question as:

> Does caffeine cause people to be more irritable?

Next, the **research hypothesis** is an educated guess regarding the likely answer to the research question. It is a description of the relationships that are likely to exist between the variables of interest, typically stated in either the present tense or the future tense. Dr. O'Day might state her initial research hypothesis like this:

> Caffeine causes people to be more irritable.

The preceding paragraph described a research hypothesis as being an educated guess about the likely answer to the research question. To be a truly *educated* guess, the researcher should do her homework before stating it—she should thoroughly review theory and previous research that is relevant to the research question, and the research hypothesis that she states should be informed by this theory and research.

STATISTICAL NULL HYPOTHESIS. Eventually, the researcher will conduct an empirical investigation and will analyze data to determine whether the results are consistent with her research hypothesis. The statistical procedures that determine whether her results are "statistically significant" are actually tests of a specific type of hypothesis called the *statistical null hypothesis*. The **statistical null hypothesis** (symbol: H_0) may be defined in this way:

- It is a statement about some characteristic of the population(s) being studied.
- It is usually a statement of "no effect," "no difference," or "no relationship."

- It usually makes a prediction that is the opposite of the prediction made by the research hypothesis.

Here is the most important thing to remember about the statistical null hypothesis:

- The statistical null hypothesis is the hypothesis that the researcher typically hopes to *reject*.

The exact way that a researcher states a statistical null hypothesis is determined by the nature of the statistical test being performed, what outcome is predicted by the research hypothesis, and other factors. For the current caffeine study, Dr. O'Day might have stated the statistical null hypothesis in this manner:

$H_0: \mu_E - \mu_C = 0$. In the population, the difference between the mean irritability score for the experimental condition versus the mean irritability score for the control condition is equal to zero.

Notice that the preceding used the Greek letter μ as a symbol for "mean score in the population." More specifically, it used the symbol μ_E to represent the mean irritability score for the experimental condition (i.e., the high-caffeine condition) in the population, and it used the symbol μ_C to represent the mean irritability score for the control condition (i.e., the no-caffeine condition) in the population. Symbolically, the null hypothesis indicated that the difference between these two population means is equal to zero: $\mu_E - \mu_C = 0$.

Some researchers state exactly the same null hypothesis in a different way: by stating that the two population means are equal. Here is this equivalent version:

$H_0: \mu_E = \mu_C$. In the population, the mean irritability score for the experimental condition is equal to the mean irritability score for the control condition.

In almost all cases, the statistical null hypothesis is the hypothesis that the researcher hopes to *reject* at the conclusion of the statistical analysis. Think of it as a "straw man" that the researcher hopes to knock down. In the present case, the null hypothesis says that (with respect to irritability scores in the population), the mean score for the experimental condition is equal to the mean score for the control condition. Dr. O'Day very much hopes that this statement is not true—she hopes that the results that she obtains in her investigation will allow her to reject this idea.

STATISTICAL ALTERNATIVE HYPOTHESIS. The counterpart to the statistical null hypothesis is the statistical alternative hypothesis. The ***statistical alternative hypothesis*** (symbol: H_1 or H_A) may be defined in this way:

- It is a statement about some characteristic of the population(s) being studied.
- It is usually a statement that there *is* an effect, that there *is* a difference, or that there *is* a relationship.

- It usually makes a prediction that is consistent with the prediction made by the research hypothesis.

For the current study, Dr. O'Day might state the statistical alternative hypothesis in this manner:

$H_1: \mu_E - \mu_C \neq 0$. In the population, the difference between the mean irritability score for the experimental condition versus the mean irritability score for the control condition is not equal to zero.

Or, she might state an equivalent alternative hypothesis in this way:

$H_1: \mu_E \neq \mu_C$. In the population, the mean irritability score for the experimental condition is not equal to the mean irritability score for the control condition.

In most cases, the statistical alternative hypothesis is the hypothesis that researchers hope will be supported by their study's results. In this case, Dr. O'Day hopes to obtain evidence that the mean irritability score for the high-caffeine condition is not equal to the mean irritability score for the no-caffeine condition.

The Sampling Distribution of the Statistic

Dr. O'Day will determine whether the results from her study are statistically significant by:

- analyzing the sample data in order to compute an obtained t statistic (symbol: t_{obt}), and then
- determining where that t_{obt} statistic is located within a null-hypothesis sampling distribution.

This means that understanding the concept of a *sampling distribution* is essential if we are to understand what is meant by the words *statistically significant*. The **sampling distribution** for a given statistic (such as the t statistic) is:

- the distribution of all possible values of that statistic that would be obtained…
- if an infinite number of samples of the same size were drawn…
- from the populations described by the null hypothesis.

EXAMPLE. To make this concept a bit more concrete, imagine how we might theoretically create the sampling distribution for Dr. O'Day's study:

- First (and most important), assume that the null hypothesis is true (i.e., assume that caffeine has *no effect* on irritability in the population).
- Imagine that Dr. O'Day conducts the caffeine study described above. Specifically, she draws a sample of 150 participants, randomly assigns each subject to one of the two treatment conditions, and manipulates the independent variable. She computes the mean irritability score displayed by the 75 subjects in the high-caffeine condition (symbol: $\bar{X}_E$). She also

computes the mean irritability score displayed by the 75 subjects in the no-caffeine condition (symbol: $\bar{X}_C$). She subtracts $\bar{X}_C$ from $\bar{X}_E$, producing a *mean-difference* score. This mean difference is a single number which represents the size of the difference between two sample means. It is computed as follows:

$$\text{Mean difference} = \bar{X}_E - \bar{X}_C$$

- She adds this mean difference to the sampling distribution (at this point, the distribution contains only one mean-difference score).
- She repeats this entire process an infinite number of times. In other words, she conducts the experiment an infinite number of times, each time computing a new difference score and adding it to the sampling distribution).

The resulting distribution is called a *sampling distribution*. Figure 5.2 provides an illustration of the shape of this distribution. This sampling distribution contains the differences between means that we would expect to see if (a) caffeine has no effect on irritability, and (b) the experiment (with total $N = 150$) were conducted an infinite number of times.

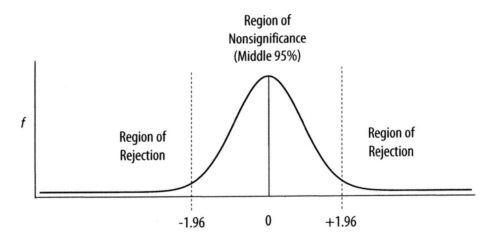

Figure 5.2. Sampling distribution of differences between means for the caffeine study.

The sampling distribution in Figure 5.2 looks very much like the standard normal distribution, or z distribution, which was covered in *Chapter 4*. Technically, it is actually a distinct distribution called the *t* distribution. With large samples, (such as the combined sample of $N = 150$ in the current investigation), however, the shape of the *t* distribution is almost identical to the shape of the z distribution, so the things that you learned about the z distribution in *Chapter 4* will also apply here.

The Critical Value and the Region of Rejection

Notice that the sampling distribution in the figure has been divided into two regions: (a) the middle 95% (labeled the "Region of Nonsignificance") and (b) most extreme 5% of the distribution, out in the two tails (labeled the "Region of Rejection"). The two vertical dashed lines that divide the distribution into these regions are labeled with the values "−1.96" and "+1.96." These values are called the *critical value of the statistic*.

THE CRITICAL VALUE OF THE STATISTIC. First, ***the critical value of the t statistic*** (symbol: t_{crit}) is the value that separates the region of rejection from the region of nonsignificance. When Dr. O'Day analyzes the data from her study, she will compute an obtained t statistic (t_{obt}), and will compare the size of this obtained statistic against the critical value of the statistic: t_{crit}., which in this case is equal to ±1.96. If t_{obt} is larger than ±1.96 in absolute value (such as t_{obt} = 5.18), this would indicate that the obtained statistic is in the region of rejection. In this case, Dr. O'Day will:

- Reject the null hypothesis.
- Tell the world that there is a statistically significant difference between the means of the two conditions.

However, Dr. O'Day may not be so fortunate. If the value of t_{obt} is smaller than ±1.96 in absolute value (such as t_{obt} = 0.20), this indicates that the obtained statistic is in the region of nonsignificance. In this case, Dr. O'Day will:

- Fail to reject the null hypothesis.
- Tell the world that there is a statistically nonsignificant difference between the means of the two conditions.

How did Dr. O'Day know that the critical value of the statistic was equal to ±1.96 for this analysis? She consulted a table of critical values of the t statistic. Such tables appear in the back of most statistics textbooks. One caveat: In this analysis, the t_{crit} is equal to ±1.96, but t_{crit} is likely to be a different value in a different investigation, especially if the number of subjects is different.

THE REGION OF NONSIGNIFICANCE. The middle 95% of the sampling distribution in Figure 5.2 is labeled the "Region of Nonsignificance." When the null hypothesis is true, there is a 95% probability that the t_{obt} statistic produced by a given investigation will fall in this section of the sampling distribution. Therefore, when researchers obtain a relatively small statistic that fall in the region of nonsignificance, they fail to reject the null hypothesis—they tell the world that their results are statistically nonsignificant.

THE REGION OF REJECTION. The most extreme 5% of the sampling distribution in Figure 5.2 is labeled the "Region of Rejection." When the null hypothesis is true, there is only a 5% probability that the t_{obt} statistic produced by a given investigation will fall in this section of the sampling distribution. Therefore, when the obtained statistic is large enough to be in the region of rejection, researchers reject the null hypothesis and tell the world that their results are statistically significant.

Sampling Error and the Standard Error

In general, **sampling error** is the difference between a population parameter versus a sample statistic attributable to the fact that samples are usually not perfectly representative of the populations from which they were drawn. Think about it this way: The null hypothesis said that there is no difference between the two population means (μ_E and μ_C). If this null hypothesis were true, and if all of the samples used in creating the sampling distribution had been perfectly representative of the populations from which they were drawn, then all of the differences between the pairs of sample means would have been exactly equal to zero: $\bar{X}_E - \bar{X}_C = 0$. If this had happened, then there would have been no variability in the sampling distribution—all of its scores would have been exactly equal to zero.

There was a problem, however. There were only 150 people in Dr. O'Day's study, and, even if they had been randomly sampled from their populations, we would not expect them to be perfectly representative of the populations from which they had been drawn. And many of those samples that were not perfectly representative of their populations would produce mean-difference scores just a bit different from zero—some would be a bit higher, and some a bit lower. That is why we see some variability in the sampling distribution depicted in Figure 5.2.

THE STANDARD ERROR. There is good news, however. We can compute a statistic called the *standard error* which tells us how much variability appears in the sampling distribution.

A **standard error** (symbol: *SE*) is a special type of standard deviation: It is a standard deviation of a sampling distribution. The specific name given to a standard error is determined by the nature of the investigation. In the present case, Dr. O'Day's sampling distribution consists of mean-difference scores, so she will call it the *standard error of the differences between means*. In a statistics textbook, this standard error is typically represented using the symbol $s_{\bar{X}_1 - \bar{X}_2}$, although this book will generally use SE_{diff} or just *SE*.

FORMULA. The formula for computing a standard error is determined by the nature of the specific investigation. If one is performing an independent-samples *t* test (and if certain statistical assumptions are met), the standard error may be computed using the following formula (from Howell, 2002):

$$SE_{\text{diff}} = \sqrt{\frac{s_1^2}{n_1} + \frac{s_2^2}{n_2}}$$

With the above formula: SE_{diff} = the standard error of the differences between means; s_1^2 = the estimated population variance of scores on the criterion variable in the first treatment condition; s_2^2 = the estimated population variance of scores on the criterion variable in the second treatment condition; n_1 = the number of participants in the first treatment condition; and n_2 = the number of participants in the second treatment condition.

Let's assume that Dr. O'Day computes the standard error of the differences between means for the current investigation as SE_{diff} = 5.848. Now that we have this value, we can compute the obtained *t* statistic for her investigation.

Computing the Obtained *t* Statistic

An **obtained statistic** is a value which indicates where the results from the current sample are located within the sampling distribution of the statistic. The way that researchers compute an obtained statistic depends on which statistic they are computing (e.g., a χ^2 test, an *F* test, etc.). Dr. O'Day is performing an independent-samples *t* test, and the definitional formula for this *t* statistic (in its simplest form) is as follows:

$$t_{obt} = \frac{\bar{X}_E - \bar{X}_C}{SE_{diff}}$$

With the preceding formula: t_{obt} = the obtained *t* statistic; $\bar{X}_E$ = the sample mean for the experimental condition; $\bar{X}_C$ = the sample mean for the control condition; SE_{diff} = the standard error of the differences between means.

Given the nature of the formula above, it is clear that the *t* statistic from an independent-samples *t* test is a measure of *difference*: It represents the size of the difference between the sample mean for the experimental group ($\bar{X}_E$) versus the sample mean for the control group ($\bar{X}_C$) as measured in standard errors (SE_{diff}).

Table 5.1 (presented earlier) provided mean scores for the dependent variable, *self-rated irritability*. The table indicated that the mean for the high-caffeine condition was $\bar{X}_E$ = 575.00 and the mean for the no-caffeine condition was $\bar{X}_C$ = 544.73. The standard error of the difference (not shown in the table) was SE_{diff} = 5.848. Inserting these values into the formula produces the following obtained *t* statistic:

$$t_{obt} = \frac{\bar{X}_E - \bar{X}_C}{SE_{diff}} = \frac{575.00 - 544.73}{5.848} = \frac{30.27}{5.848} = 5.176 = 5.18$$

So the obtained *t* statistic for the self-reported irritability dependent variable is t_{obt} = 5.18. The size of the statistic tells us that the mean score for the high-caffeine condition is 5.18 standard errors higher than the mean score for the no-caffeine condition.

Making a Decision Regarding the Null Hypothesis

Table 5.1 showed that the mean scores on self-rated irritability were in the direction that Dr. O'Day's research hypothesis had predicted, with the high-caffeine condition scoring higher than the no-caffeine condition (mean self-rated irritability scores were 575.00 and 544.73, respectively). To determine whether the difference between the means is large enough to be statistically significant, she must now review the results of the null-hypothesis significance test.

The preceding section showed that Dr. O'Day's obtained statistic is $t_{obt} = 5.18$. An earlier section of this chapter had indicated that the critical value of the statistic for this analysis was $t_{crit} = \pm 1.96$. Since this obtained statistic of $t_{obt} = 5.18$ is larger than the critical value of $t_{crit} = \pm 1.96$, Dr. O'Day may:

- Reject the null hypothesis
- Tell the world that there was a statistically significant difference between the two sample means
- Tell the world that the results of her investigation were consistent with her research hypothesis

What Statistical Significance Does and Does Not Mean

People often misinterpret just what is meant by *statistically significant* findings. To sharpen our understanding, this section provides some correct and incorrect interpretations of Dr. O'Day's current results.

Correct:

- *We may reject the null hypothesis.*
- *If the null hypothesis were true, there is only a 5% probability that we would have obtained a t statistic this large or larger* (i.e, as large as or larger than $t_{obt} = 5.18$, in this case).

Incorrect:

- *There is only a 5% chance that the mean score for the high-caffeine group is equal to the mean score for the no-caffeine group in the population* (This is incorrect because the statistical procedures described here allow us only to make probability statements about samples, not about populations).
- *There is only a 5% chance that the null hypothesis is true* (Come on! This communicates exactly the same idea as the previous incorrect statement!).
- *The amount of caffeine consumed had a strong effect on irritability scores* (This is incorrect because significance tests do not reveal the strength of the effect that an independent variable has had on a dependent variable; to learn this, we must instead compute an index of effect size).

Factors Affecting the Power of a Test

The **power** of a significance test is the probability that you will reject the null hypothesis and conclude that you have significant results when there really is an "effect" in the population. With a test of group differences (such as the current *t* test), the power of the test is the probability that you will conclude that you have significant results when there really is a difference between the means of the different conditions in the population.

Power is usually represented as a proportion or as a percentage, and an ideal level of power that many researchers strive for is .80. When power is equal to .80, the researcher has an 80% probability of obtaining significant results (if there really is an effect in the population, that is). The value of .80 is important because some funding agencies will award a research grant to investigators only if the investigators have demonstrated that the power of their significance tests will be .80 or higher.

The power of the test is influenced by a number of factors. Some of the more important factors are discussed here.

ACTUAL SIZE OF THE EFFECT IN THE POPULATION. In empirical investigations, one definition for *effect size* is *the strength of the relationship between a predictor variable and a criterion variable*. Other things held constant, the stronger the "effect" in the population, the greater the power of the significance test performed on a sample of data taken from that population.

ALPHA LEVEL. In this context, the ***alpha level*** (also called the ***significance level***; symbol: α) refers to the size of the region of rejection in a sampling distribution. You will recall that the region of rejection consists of the most extreme part of a sampling distribution, out in the tails. When the analysis is conducted correctly (i.e., when all assumptions are met), alpha is also the theoretical probability of making a Type I error (Type I errors will be discussed later).

The most popular alpha level is $\alpha = .05$ (meaning that the size of the region of rejection is the most extreme 5% of the distribution, out in the tails). Sometimes researchers set alpha at $\alpha = .01$ (meaning that the region of rejection is the most extreme 1% of the distribution), or $\alpha = .001$ (meaning that the region of rejection is the most extreme .1% of the distribution).

In general, researchers increase the power of the test (i.e., the likelihood that they will obtain significant results) if they set alpha at a large value, such as $\alpha = .05$. In general, they decrease the likelihood that they will obtain significant results if they set alpha at a small value, such as $\alpha = .001$.

ONE-TAILED TEST VERSUS TWO-TAILED TEST. The significance test for the caffeine study discussed earlier was a ***two-tailed significance test*** (also called a ***nondirectional significance test***). With a two-tailed test, the researcher makes a general prediction that one treatment condition will score higher than the other, but does not specifically predict which condition will score high and which will score low. With a nondirectional prediction, the region of rejection is divided into two tails. If the researcher has set alpha at $\alpha = .05$ and performs a two-tailed test, this means that the 5% that makes up the region of rejection is divided into two tails, with 2.5% in one tail, and 2.5% in the other.

The counterpart to a two-tailed test is a ***one-tailed significance test*** (also called a ***directional significance test***). With a one-tailed test, the researcher makes a specific prediction as to which condition will score higher than the other. This means that the region of rejection is concentrated in just one tail. For example, if Dr. O'Day had specifically predicted that the high-caffeine condition would score higher than the no-caffeine condition, she could have performed a directional test by concentrating the entire region of rejection in the right-side tale of the sampling distribution in Figure 5.2.

In general, one-tailed tests are more powerful than two-tailed tests, as long as the researcher correctly predicts the direction of the results. One-tail tests tend to be more powerful because the region of rejection is concentrated in just one tail, and the critical value of the statistic is therefore somewhat lower (and is therefore is easier to reach).

NUMBER OF SUBJECTS AND THE DEGREES OF FREEDOM. Other things held constant, the greater the number of participants in the study, the greater the power of the test. This means that (assuming that there really is an effect in the population) Dr. O'Day would be more likely to obtain significant results if she conducted her study with a sample of 300 participants, rather than a sample of just 30 participants.

One of the reasons that more subjects results in more power is because more subjects result in a larger value of the *degrees of freedom* for the analysis. In this context, ***degrees of freedom*** (symbol: *df*) may be defined as the number of observations used in the computation of a specific statistic (such as the *t* statistic), minus the number of restrictions that have been placed on the freedom of those observations to vary. For many statistical procedures, the degrees of freedom largely reflect the number of participants in the study. For example, the degrees of freedom for an independent-samples *t* test are computed as:

$$df = N - 2$$

With the preceding formula, N represents the total number of subjects in the study. In Dr. O'Day's caffeine study, the total sample size was 150, so the *df* were equal to 148.

Two Approaches for Determining Statistical Significance

Researchers typically determine whether their results are statistically significant by either (a) comparing their obtained statistic against a critical value of the statistic found in a table in the back of a statistics textbook, or (b) consulting the *p* value provided by a statistical application such as SAS or SPSS. This book refers to these approaches as the *old-school approach* and the *new-school approach*, respectively.

OLD SCHOOL: TABLES OF CRITICAL VALUES. Before computer applications such as SAS or SPSS were widely available, researchers routinely computed obtained statistics (such as the *t* statistic) by hand. To determine whether this obtained statistic was large enough to be in the region of rejection, they then looked up the critical value of the statistic that was appropriate for their data set. To make this possible, statistics textbooks routinely include tables of critical values for the *t* statistic, the *F* statistic, the chi-square statistic, and other procedures.

The size of the critical value of a statistic is determined by a variety of factors. A previous section indicated that the size of the critical value for an independent-samples *t* test is determined by (a) the degrees of freedom for the analysis, (b) the decision to perform either a one-tailed test or a two-tailed test, and (c) the alpha level selected by the researcher.

For example, assume that in Dr. O'Day's study, the df = 148, and she chose to perform a two-tailed test with alpha set at α = .05. A table of critical values in the back of a statistics textbook told her that, given these considerations, the critical value of the test (t_{crit}) was ±1.96.

One of the criterion variables in her investigation was *self-reported outbursts* (i.e., the number of temper tantrums that the subject displayed during the investigation, as reported by the subject). After analyzing the data, Dr. O'Day found that her obtained *t* statistic for this criterion variable was $t(148)$ = 2.77. Since this obtained statistic (2.77) was larger than the critical value (±1.96), she rejected the null hypothesis and told the world that her results for this variable were statistically significant.

Another criterion variable in her investigation was *other-reported irritability* (i.e., how irritable the subject was, according to friends or relatives). The obtained *t* statistic for this criterion variable computed as $t(148)$ = 1.73. Since this obtained statistic (1.73) was smaller than the critical value (±1.96), she failed to reject the null hypothesis. She told the world that her results for this variable were statistically nonsignificant.

NEW SCHOOL: CONSULTING THE *p* VALUE FROM COMPUTER OUTPUT. Today most researchers analyze their data using computer applications such as SAS or SPSS. Among the many advantages of this approach, is the fact that it is now seldom necessary to look up critical values of a statistic in the back of a statistics textbook. Instead, today's computer applications routinely report (a) the obtained statistic (such as the t_{obt} statistic), along with (b) the *p* value for that statistic. The *p* value tells the researcher whether the obtained statistic is significant.

A ***p value*** (i.e., ***probability value***) is the probability of obtaining a statistic the size of the current obtained statistic (or an even larger statistic) if the null hypothesis were true. In most studies, the null hypothesis is a hypothesis of no-differences between conditions, or no relationship between variables (in the population). This means that, in most cases, the *p* value provides the probability that the researcher would have obtained the current sample results if the null hypothesis of "no effect in the population" were true. In most cases, statistical applications provide precise *p* values, carried out to three or more decimal places, such as "p = .492," p = .022," or "p = .001." If the *p* value is small enough, the researcher rejects the null hypothesis and tells the world that the results are statistically significant.

Just how small must the *p* value be to reject the null hypothesis? It has to be smaller than the alpha level that the researcher has selected for the study. You will recall that the **alpha level** (symbol: α) is the size of the region of rejection in the sampling distribution. Although alpha may be set at .10, .01, 001, or just about any other value, the most popular value is α = .05. This means that, if an obtained *p* value is less than .05, most researchers will conclude that the results are statistically significant. In fact, this book will just about always use α = .05 as the criterion for making this decision.

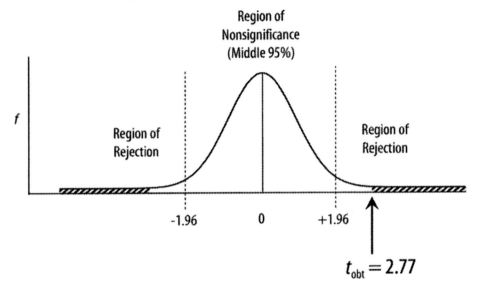

Figure 5.3. Location of an obtained *t* statistic of t_{obt} = 2.77 within the sampling distribution (these are statistically significant results; the area shaded with stripes represents the area beyond the obtained *t* statistic of t_{obt} = 2.77).

NEW SCHOOL EXAMPLE 1: SIGNIFICANT RESULTS. In addition to the definition provided above, here is an additional (and equivalent) way of defining a *p* value:

> The *p* value for an obtained statistic indicates the proportion of area in the sampling distribution which lies *beyond* that obtained statistic, out in the tail or tails.

For example, Table 5.1 (presented much earlier in the chapter) indicated that, for the criterion variable *self-reported outbursts*, the obtained *t* statistic was t_{obt} = 2.77 and the *p* value for this statistic was *p* = .006. This *p* value communicates the following message:

> The obtained *t* statistic (t_{obt} = 2.77) was so far away from the mean of the sampling distribution that only .006 of the sampling distribution was beyond this obtained statistic, out in the tails.

This is illustrated in Figure 5.3.

If the null hypothesis were true, we would expect Dr. O'Day's obtained t statistic to be $t_{obt} = 0$. This is represented by the solid vertical line in the center of the sampling distribution. But Dr. O'Day did not obtain a t statistic equal to zero. Instead, her obtained statistic was $t_{obt} = 2.77$, as represented by the single-headed arrow on the right side of the figure. The shaded (cross-hatched) area represents the proportion of the sampling distribution which lies beyond this obtained statistic, out in the two tails (the same amount of area was shaded on both the left tail and right tail, because Dr. O'Day had performed a two-tailed test).

If the null hypothesis were true, the probability is just .006 that Dr. O'Day would have obtained a t statistic as large (or larger) than 2.77. In the disciplines of research and statistics, this is considered a fairly low probability.

In summary, Dr. O'Day's obtained p value ($p = .006$) was smaller than the alpha level that she selected for the test ($\alpha = .05$). Therefore, she rejected the null hypothesis and told the world that there was a statistically significant difference between the two conditions with respect to their scores on self-reported outbursts.

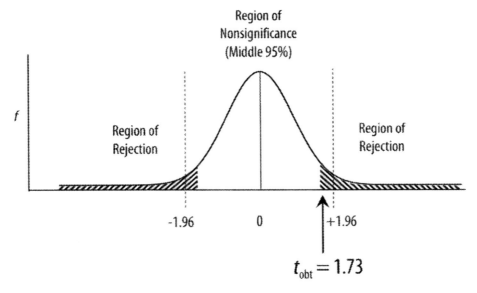

Figure 5.4. Location of an obtained t statistic of $t_{obt} = 1.73$ within the sampling distribution (these are statistically nonsignificant results; the area shaded with stripes represents the area beyond the obtained t statistic of $t_{obt} = 1.73$).

NEW SCHOOL EXAMPLE 2: NONSIGNIFICANT RESULTS. In Dr. O'Day's study, a different criterion variable was *other-rated irritability*. This variable indicated how irritable the subject was according to ratings made by the subject's friends or relatives. Table 5.1 (presented earlier in this chapter) indicated that, for this criterion variable, the obtained t statistic was $t_{obt} = 1.73$ and the corresponding p value was $p = .085$. These results are graphically illustrated in Figure 5.4.

Once again, if the null hypothesis were true, we would expect Dr. O'Day's obtained t statistic to be $t_{obt} = 0$. The single-headed arrow shows that for other-rated irritability, the obtained statistic was $t_{obt} = 1.73$. As before, the shaded area represents the proportion of area in the sampling distribution that lies beyond this obtained statistic, out in the two tails. The amount of shaded area totals to .085 of the sampling distribution.

If the null hypothesis were true, the probability is .085 that Dr. O'Day would have obtained a statistic as large (or larger) than 1.73. In the disciplines of research and statistics, this is considered a fairly high probability.

For this criterion variable, Dr. O'Day's obtained p value ($p = .085$) was larger than the alpha level of $\alpha = .05$. This meant that she failed to reject the null hypothesis and was forced to tell the world that there was a statistically nonsignificant difference between the two conditions with respect to their mean scores on other-reported irritability.

SUMMARY. There is no difference between the old-school approach and the new-school approach with respect to the final decision on the null hypothesis—if one approach indicates that the results are statistically significant, the other approach will arrive at the same conclusion. The main difference involves precision: The new-school approach easily produces a precise p value such as $p = .543$ or $p = .004$, whereas the old school approach does not. *The Publication Manual of the American Psychological Association 6th ed.* (2010) recommends reporting precise p values out to two or three decimal places whenever possible, and this means that the new-school approach is generally the preferred paradigm for performing null-hypothesis significance tests.

Type I Errors and Type II Errors

When testing null hypotheses, researchers sometimes make mistakes called *Type I errors* or *Type II errors*. This section describes the differences between the two types of mistakes, and summarizes the steps that can be taken to avoid them.

TYPE I ERRORS. A *Type I error* can be defined in the following ways:

- A Type I error involves rejecting a true null hypothesis.
- A Type I error occurs when a researcher tells the world that the results are *statistically significant* when (unbeknownst to the researcher) this is the wrong conclusion.

To expand upon these definitions a bit, think about how Dr. O'Day might make a Type I error in her caffeine study. Imagine that, in the population, caffeine has no effect on irritability at all. In other words, in the population there is no difference between the mean self-reported irritability scores for the high-caffeine condition versus the no-caffeine condition.

Imagine that Dr. O'Day conducts her caffeine study with 150 subjects. Before analyzing the data, she sets alpha at $\alpha = .05$.

Due to sampling error, the mean score for the high-caffeine condition happens to be much higher than the mean score for the no-caffeine condition—high enough so that the obtained t statistic is way out in the region of rejection (p = .002). Because the obtained p value (.002) was smaller than the selected level of alpha (.05), Dr. O'Day rejects the null hypothesis. She tells the world that there is a statistically significant difference between the two conditions.

Dr. O'Day rejected the null hypothesis when she should not have done so. She has committed a *Type I error*.

Making a Type I error is a bad thing, but researchers have a way of protecting themselves: They can select a conservative value for alpha (α) prior to analyzing the data. You will recall that **alpha** is the size of the region of rejection, out in the extreme tails of the distribution.

In addition to being the size of the region of rejection, alpha also represents the probability of making a Type I error. When a researcher sets alpha at .05 (and conducts the statistical analysis correctly), the probability of making a Type I error is .05. By setting alpha at a relatively small, conservative value (such as α = .001) rather than a large, liberal value (such as α = .05), the researcher decreases the likelihood of making a Type I error. If the researcher sets alpha at α = .001 and still obtains significant results, it is possible that she is making a Type I error, but the probability that she is making a Type I error is very small—it is less than .001.

COMPARISON-WISE VERSUS FAMILYWISE TYPE I ERROR RATE. The ***comparison-wise Type I error rate*** (also known as the ***per-comparison error rate***; symbol: α_{PC}) is the probability of making a Type I error when performing just one specific significance test (i.e, just one comparison). Imagine for a moment that Dr. O'Day has just one dependent variable in her study: scores on the *self-reported irritability index*. If she set alpha at α = .05 for the t test performed on this dependent variable, then the comparison-wise Type I error rate is equal to .05.

There is a problem, however. In the real world, researchers almost always have more than one dependent variable. Earlier in this chapter, you learned that Dr. O'Day actually measured four dependent variables in the caffeine study. She will perform a separate t test for each of these dependent variables. Unfortunately, this means that the probability of making at least one Type I error out of all of these tests will probably be greater than .05.

The ***familywise Type I error rate*** (symbol: α_{FW}; also known at the ***experiment-wise Type I error rate***) is the probability of making at least one Type I error out of all of the comparison tests performed on the same data set (Hays, 1988). When the individual significance tests are independent of one another and the same level of alpha is used for each comparison, the familywise Type I error rate can be estimated as:

$$\alpha_{FW} = 1 - (1 - \alpha_{PC})^K$$

With the above formula, α_{FW} = the familywise Type I error rate; α_{PC} = the level of alpha selected for each comparison; and K = the total number of comparisons to be made.

For example, imagine that Dr. O'Day plans to perform four t tests (one for each dependent variable), and that she will set alpha at $\alpha = .05$ for each comparison. The familywise error rate may be estimated as:

$$\alpha_{FW} = 1 - (1 - \alpha_{PC})^K = 1 - (1 - .05)^4 = 1 - (.95)^4 = 1 - (.815) = .185 = .19$$

So the familywise Type I error rate is .19. This means that Dr. O'Day has a 19% probability of making at least one Type I error out of all of the individual t tests that she plans to perform. *Ouch!*

Fortunately, there are a number of ways to deal with a familywise Type I error rate that exceeds the conventional level of .05. One popular (and fairly easy) approach involves using the *Bonferroni adjustment* (Howell, 2002; Warner, 2008). With the **Bonferroni adjustment** (also known as the **Bonferroni correction**), the researcher divides the desired familywise error rate (usually .05) by the number of comparisons to be made (four, in this case). The result is the *adjusted alpha level* (α_{ADJ}). In performing the individual comparisons, the researcher views a given comparison as being statistically significant only if the obtained p value is less than α_{ADJ}.

The general form for the Bonferroni adjustment is:

$$\alpha_{ADJ} = \alpha_{FW} / K$$

With the preceding formula, α_{ADJ} = the adjusted value of alpha which will be used for each individual significance test; α_{FW} = the desired value for the familywise Type I error rate; and K = the number of comparisons (i.e., the number of significance tests) to be performed.

Again, let's assume that Dr. O'Day wants the familywise Type I error rate to be $\alpha_{FW} = .05$ and she must make four comparisons. Inserting these values into the preceding formula results in the following adjusted alpha level:

$$\alpha_{ADJ} = \alpha_{FW} / K = .05 / 4 = .0125$$

This means that, when performing the individual t tests, Dr. O'Day would consider the results for a given dependent variable to be significant only if the p value for that test is less than .0125. By using this Bonferroni adjustment, the familywise Type I error rate will remain at $\alpha_{FW} = .05$.

TYPE II ERRORS. A *Type II error* can be defined in the following ways:

- A Type II error involves failing to reject a false null hypothesis.
- A Type II error occurs when a researcher tells the world that the results are *not* statistically significant when (unbeknownst to the researcher) this is the wrong conclusion.

To make this a bit more concrete, consider this twist of events in Dr. O'Day's study: Imagine that, in the population, caffeine *does* have a significant and substantial effect on irritability. In other words, in the population there *is* a difference between the mean self-reported irritability scores for the high-caffeine condition versus the no-caffeine condition.

Imagine that Dr. O'Day conducts her caffeine study with 150 subjects and, due to sampling error, the mean score for the high-caffeine condition happens to be very similar to the mean score for the no-caffeine condition—so similar that the obtained t statistic is close to zero, solidly in the region of nonsignificance. Let's imagine that $t_{obt} = 0.20$, $p = .841$.

Because the obtained p value (.841) is larger than the selected value of alpha (.05), Dr. O'Day fails to reject the null hypothesis. She tells the world that there is not a statistically significant difference between the two conditions.

Dr. O'Day failed to reject the null hypothesis when she should have rejected it. She has made a *Type II error*.

Making a Type II error is a bad thing, But researchers know that they can protect themselves against Type II errors by maximizing the *power* of their statistical procedures. In the previous section titled "Factors Affecting the Power of a Test" you learned about several strategies that researchers can use to increase the power of their tests. These strategies included using a larger sample, setting alpha at $\alpha = .05$ rather than $\alpha = .001$, and related tactics.

How Significance is Reported in Articles

Researchers follow a fairly standard set of conventions when presenting the results of significance tests in published reports. This section introduces some of those conventions.

USING PRECISE p VALUES TO REPORT SIGNIFICANCE. The *Publication Manual of the American Psychological Association 6th ed.* (2010) recommends that, when feasible, researchers should present the precise p values from significance tests, carried to two or three decimal places. This approach had been used in Table 5.1, a revised version of which is reproduced here as Table 5.2.

The information relevant to the significance tests appears under the general heading, "Difference between means." The column headed "$t(148)$" provides the obtained t statistic for each dependent variable. The entry "(148)" in this heading indicates that the degrees of freedom for each t test was $df = 148$. This column shows that the obtained t statistic for self-rated irritability was $t_{obt} = 5.18$, with corresponding p value of $p < .001$ (in the "p" column). The p value is less than .05, so Dr. O'Day rejected the null hypothesis for self-rated irritability. The results for the remaining criterion variables can be interpreted in the same way.

Table 5.2

Scores on Four Measures of Irritability as a Function of Amount of Caffeine Consumed (High-Caffeine versus No-Caffeine): Results from Independent-samples t Tests with p Values

Dependent variable	High caffeine[a]		No caffeine[b]		Difference between means			
	M_1	(SD_1)	M_2	(SD_2)	$M_1 - M_2$	SE	$t(148)$	p
Self-rated irritability	575.00	(35.49)	544.73	(36.13)	30.27	5.85	5.18	<.001
Other-rated irritability	556.46	(35.86)	546.27	(36.10)	10.19	5.88	1.73	.085
Self-reported outbursts	6.24	(2.67)	5.03	(2.69)	1.21	0.44	2.77	.006
Other-reported outbursts	5.24	(2.67)	4.99	(2.65)	0.25	0.43	0.58	.561

Note. $N = 150$.
[a]$n = 75$. [b]$n = 75$.

Within the text of the published article, authors are expected to highlight the investigation's most important findings. Dr. O'Day does this in the following excerpt:

> Because the study included four dependent variables, the Bonferroni adjustment was used to maintain a familywise Type I error rate of .05. The resulting Bonferroni-adjusted alpha level was $\alpha_{ADJ} = .0125$.
>
> Using this adjusted level of alpha, an independent-samples *t* test revealed a significant difference between the high-caffeine condition and the no-caffeine condition for self-rated irritability, $t(148) = 5.18$, $p < .001$, and self-reported outbursts, $t(148) = 2.77$, $p = .006$. However, the *t* test revealed a nonsignificant difference for other-rated irritability, $t(148) = 1.73$, $p = .085$, and other-reported outbursts, $t(148) = 0.58$, $p = .561$.

Some notes regarding the preceding excerpt:

- Early in the *Results* section, investigators should indicate the level of alpha that was selected prior to performing the significance tests. Dr. O'Day does this in the first paragraph of the preceding excerpt.

- When reporting the obtained *p* values, Dr. O'Day generally uses the "=" symbol and reports the *p* value to three decimal places, as recommended by the *Publication Manual of the APA* (2010). For example, with self-reported outbursts, she indicates "$p = .006$." The one exception is when she reports the *p* value for self-rated irritability. The output from the statistical application indicated that the actual *p* value was a very small probability—smaller than .0001. The publication manual of the APA recommends reporting *p* values to no more than three decimal places, so Dr. O'Day presented it as "$p < .001$." Be warned that statistical applications sometimes report *p* values as "$p = .0000$." In these cases, the statistical application is attempting to tell you that the *p* value is a very small number (such as .0000001). However, in these instances you should *not* report the *p* value as "$p = .0000$" in your article. If you reported it in this way, you would be telling the reader "if the null hypothesis were true, the probability that I would have obtained the current sample results is zero." Such a statement is not logical—in the field of statistics, certain events are unlikely, but no event is impossible. When statistical applications indicate that "$p = .0000$," it is best to report it as "$p < .001$," as Dr. O'Day did in her excerpt.

Table 5.3

Scores on Four Measures of Irritability as a Function of Amount of Caffeine Consumed (High-Caffeine versus No-Caffeine): Results from Independent-samples t Tests with Asterisks

Dependent variable	High caffeine[a]		No caffeine[b]		Difference between means		
	M_1	(SD_1)	M_2	(SD_2)	$M_1 - M_2$	SE	$t(148)$
Self-rated irritability	575.00	(35.49)	544.73	(36.13)	30.27	5.85	5.18**
Other-rated irritability	556.46	(35.86)	546.27	(36.10)	10.19	5.88	1.73
Self-reported outbursts	6.24	(2.67)	5.03	(2.69)	1.21	0.44	2.77*
Other-reported outbursts	5.24	(2.67)	4.99	(2.65)	0.25	0.43	0.58

Note. $N = 150$.
[a]$n = 75$. [b]$n = 75$.
*$p < .01$. **$p < .001$.

USING ASTERISKS TO REPORT SIGNIFICANCE. Prior to the availability of computer programs that provided *p* values, it was standard for researchers to use asterisks in tables to "flag" (i.e., to identify) statistics that were significant. Today, the use of asterisks is declining, in part because of the recommendation from the APA's *Publication Manual* that researchers should report precise *p* values whenever feasible. The *Publication Manual* does, however, allow researchers to use asterisks rather than *p* values in certain cases (e.g., when there is not enough room in the table for precise *p* values). Table 5.3 shows how Table 5.2 could be revised to make use of asterisks rather than *p* values.

In Table 5.3, asterisks are placed next to obtained *t* statistics that are significant. The note at the bottom of Table 5.3 shows that one asterisk (*) identifies statistics that are significant at the .01 level, whereas two asterisks (**) identify statistics that are significant at the .001 level. As the reader, you should always consult the notes at the bottom of the table to determine what level of significance is signified by the number of asterisks.

USING THE $p <$ AND $p >$ CONVENTIONS TO REPORT SIGNIFICANCE. Within the text of the published article, authors have traditionally used the "less than" symbol ("<") and "greater than" symbol (">") to identify statistics that are significant or nonsignificant. The *Publication Manual of the American Psychological Association 6th ed.* (2010) discourages this practice, advising that researchers should instead report precise *p* values using the "=" symbol whenever possible (e.g., "$p = .006$"). However, you should expect to encounter the use of the "$p <$" and "$p >$" conventions fairly often, especially when you read older articles. This section provides a short guide to their interpretation.

The "less than" symbol "<" is used to identify statistics that are significant at a given level of alpha. For example, if the researchers had set alpha at $\alpha = .05$ prior to analyzing the data and then found that the obtained statistic was significant, they might report it in this way:

>The results of the independent-samples *t* test were significant, $t(148) = 4.12$, $p < .05$

If the researchers had instead set alpha at $\alpha = .001$ prior to analyzing the data and then found that the obtained statistic was significant, they would report it in this way:

>The results of the independent-samples *t* test were significant, $t(148) = 4.12$, $p < .001$

In essence, the preceding sentence is communicating the following message:

>My obtained statistic was $t_{obt} = 4.12$. If the null hypothesis were true, the probability that I would have obtained a sample statistic this large (or larger) is less than .001. Therefore, I rejected the null hypothesis.

In contrast, the "greater than" symbol (">") is used to identify statistics that are not significant at a given level of alpha. For example, if the researchers had set alpha at $\alpha = .05$ prior to analyzing the data and then found that the obtained statistic was not significant, they might report it this way:

>The results of the independent-samples *t* test were nonsignificant, $t(148) = 1.10$, $p > .05$

In essence, the preceding sentence communicates the following:

>My obtained statistic was $t_{obt} = 1.10$. If the null hypothesis were true, the probability that I would have obtained a sample statistic this large (or larger) is greater than .05. Therefore, I failed to reject the null hypothesis.

I, for one, absolutely *hate* the use of "*p* <" and "*p* >" to identify significant versus nonsignificant findings. While reading an article, I am forced to stop in mid-sentence and try to recall whether the ">" symbol means *significant* or *nonsignificant*. In an act of great mercy, the *Publication Manual* of the APA also allows the use of the abbreviation *ns* to identify nonsignificant statistics. Authors using this abbreviation should be sure to indicate the level of alpha that had been chosen for the analysis. To use this *ns* abbreviation, a researcher might re-write one of the preceding sentences in this way:

> With alpha set at α = .05, the analysis revealed a nonsignificant difference between the means of the two conditions, $t(148) = 1.10$, *ns*.

The Controversy Regarding Significance Testing

Up until this point, most of this chapter has focused on just one type of statistical procedure: the *null-hypothesis significance test* (abbreviation: NHST). Most statistics textbooks spend a good deal of time on this procedure because it is so widely-reported in journal articles.

CRITICISMS OF SIGNIFICANCE TESTS. In recent decades, however, many researchers have criticized the role that the NHST currently plays in the social and behavioral sciences. These critics have argued that significance testing is illogical, does not tell researchers what they really want to know, produces results that are often misinterpreted, and has therefore slowed the pace of scientific advancement (e.g., Cohen, 1994; Fidler & Cumming, 2008; Kirk, 1996; Kline, 2004). Some well-respected researchers have suggested that the practice of significance testing should be abolished entirely (e.g., Hunter, 1997).

EXAMPLE: MISINTERPRETATION OF SMALL *p* VALUES. One of the most common criticisms of the NHST involves the fact that many people mistakenly believe that, if the obtained *p* value for an analysis is very small (such as $p < .001$), it indicates that there is a strong relationship between the predictor variable and the criterion variable. In reality, a small *p* value does not necessarily mean this at all.

For example, imagine that Dr. O'Day had performed her caffeine investigation using very large samples—say, 1,000 participants in each condition. Imagine that the mean number of self-reported outbursts per week was equal to 5.10 for the high-caffeine condition and 5.00 for the no-caffeine condition. The difference between the two means is computed as 5.10 − 5.00 = 0.10. This difference is fairly trivial in size—it shows that the difference between the two groups is equal to one-tenth of one temper tantrum per week. Big deal.

There is a problem, however. In an earlier section, you learned that researchers can increase the power of a significance test by using larger samples. By using 1,000 participants in each condition, Dr. O'Day might have made her test so powerful that even a trivial difference between the means could result in a large obtained *t* statistic (say $t_{obt} = 4.00$) with a really small corresponding *p* value (say, $p < .001$). Dr. O'Day might then summarize her findings in this way:

> Analyses revealed a highly significant difference between the two means, $t(1{,}998) = 4.00$, $p < .001$.

Many readers would be misled by this finding. When they see the tiny *p* value of "$p < .001$," many readers will think "*Yikes!* Caffeine must really have a big effect on self-reported outbursts." And if they never bothered to check the means, they would never know any better.

Okay—no one would really say "*Yikes!*" But they would be misled.

By the way, you should just about never use expressions such as "highly significant" in an article. The decision regarding statistical significance is a dichotomous decision: The results are either significant, or they are not. Telling a reader that your results are "highly significant" is like telling someone that a mutual friend's newborn baby not only is a boy, but "He is *highly* a boy!"

RESPONSE TO CRITICISM FROM THE APA. In response to these and other criticisms, the American Psychological Association formed a task force of well-known researchers and statisticians to review the role of significance testing in psychological research and to make recommendations for the future. In its report (Wilkinson & the Task Force on Statistical Inference, 1999), the task force did not recommend that researchers abandon significance testing. Instead, it recommended that researchers should supplement significance tests with additional statistical procedures in order to provide a clearer, more comprehensive understanding of the results of investigations. These additional procedures include confidence intervals as well as indices of effect size. Much of the remainder of this chapter is devoted to these latter two procedures.

An Alternative to the NHST: Replication Statistics and p_{rep}

In part due to the controversy described above, many researchers have turned away from null-hypothesis significance testing, and have replaced these procedures with **replication statistics:** procedures which estimate the probability that another investigation using the same research method and population of participants would replicate the current investigation's results (Killeen, 2005a, 2005b, 2008). This movement has been spurred, in part, by the fact that some prestigious journals (such as *Psychological Science*) encourage authors to report p_{rep} rather than the traditional *p* value.

The exact meaning of p_{rep} depends on how the researcher defines "replicate," and this definition may vary from study to study. In general, the **p_{rep} statistic** that is computed for a specific, individual study estimates the long-run probability that an exact replication of that study will support a specific research claim—that it will produce an effect in a specified direction (Killeen, 2008). Values of p_{rep} may range from 0.00 (meaning there is no probability that another study would replicate these findings) to 1.00 (meaning that there is a 100% probability that another study would replicate these findings).

For example, imagine that Dr. O'Day predicts that consuming caffeine will cause an increase in self-reported irritability. She conducts the study reported earlier, finds that the mean self-reported irritability score for the high-caffeine condition is in fact higher than the mean for the no-caffeine condition, and computes p_{rep} = .93. This indicates that, if another researcher conducted exactly the same study with a different sample from the same population, she estimates that there is a 93% probability that the second study would display the same trend in the results, with the high-caffeine condition displaying higher levels of irritability than the no-caffeine condition.

How large does p_{rep} have to be for the researcher to conclude that the study's results provide strong support for their research hypothesis? At present, there do not appear to be hard-and-fast criteria, although Killeen (2008) mentions p_{rep} = .90 as being in the "respectable range" (p. 114).

Many researchers are attracted to the p_{rep} concept, as it is readily interpretable and addresses the issue of replication, which is always important in empirical research. As more graduate programs incorporate replication statistics as part of their curriculum (and as more computer applications make it easy to compute p_{rep}), you should see this alternative to the null-hypothesis significance test appear more and more frequently in research articles.

Final Thoughts on Significance Testing

Remember: when you think *statistical significance*, think *probability* or *likelihood*. When results are statistically significant, it does not necessarily tell us that there is a strong relationship between the predictor variable and the criterion variable, and it does not necessarily tell us that the statistics computed from this sample are precise estimates of the corresponding population parameters. It merely tells us that, if the null hypothesis were true, it is *unlikely* that we would have obtained sample results such as these.

Big-Three Results, Part 2: Confidence Intervals

When you think *confidence interval*, you should think *precision*. A confidence interval communicates whether your estimate of a population parameter (based on the sample data) is relatively precise, or is relatively imprecise.

Confidence Intervals: Basic Concepts

A ***parameter*** is some characteristic of a population (such as the population mean). In most cases, the actual value of a population parameter is unknown, and this forces us to rely on *estimation*. ***Estimation*** is the process through which we analyze sample data in order to arrive at a "best guess" as to what this population parameter is likely to be.

When estimating population parameters, researchers may report point estimates, interval estimates, or both. A ***point estimate*** is a single number that is used as the best estimate of the corresponding population parameter (Hays, 1988). In contrast, a ***confidence interval*** (symbol: CI) is a *range* of plausible values for the population parameter being estimated (Cumming & Finch, 2005). Confidence intervals may be computed for a mean, for a difference between means, for correlation coefficients, or for other statistics.

A confidence interval extends from a lower confidence limit (symbol: *LL*) to an upper confidence limit (symbol: *UL*), and the values contained within the confidence interval represent plausible values for the population parameter being estimated. Confidence intervals are usually stated with a specified level of confidence, such as "the 95% confidence interval" or "the 99% confidence interval." The larger the confidence level (e.g., 95%, 99%), the more confident you can be that the interval contains the population parameter of interest. This means that it is more plausible that a 99% CI contains the true population parameter, and it is slightly less plausible that a 95% CI does so. The 95% CI appears to be more popular than the 99% CI.

The Confidence Interval for a Sample Mean

As an illustration, imagine that you have invented a tonic that you hope will raise the IQ scores of adults. Assume that you have administered this tonic to a sample of 20 participants, and afterward you measured the IQ of each subject, finding that the mean for the sample was $\bar{X} = 110$. You computed the 95% confidence interval around this sample mean, and found that it extended from a lower limit of 105 to an upper limit 115. Remember that one of the things that you hope to accomplish in your research is to estimate the effect that your tonic would likely have in the population (not just in your sample of 20 subjects). Given the fictitious results described here, you might interpret these results in the following manner:

> The sample mean was *M* = 110, 95% CI [105, 115], indicating that the best estimate for the population mean was also 110. Given the 95% confidence interval, it was plausible that the sample was drawn from a population whose mean was somewhere between 105 and 115.

Be warned that the interpretation of confidence intervals is a tricky business. A later section will discuss correct (as well as incorrect) interpretations of these interval estimates.

WHY CONFIDENCE INTERVALS ARE USEFUL. Interval estimates (i.e. confidence intervals) tend to be more useful than point estimates because only interval estimates communicate *precision*. A confidence interval communicates whether the parameter estimates you have computed from your sample data are relatively precise—or relatively imprecise—estimates.

The problem with point estimates (when presented alone, without confidence intervals) is that they provide no information about precision, and can therefore be misleading. For example, imagine that we rewrote the preceding excerpt so that it presented only the point estimate, and left out the confidence interval:

> The sample mean was *M* = 110, indicating that the best estimate for the population mean was also 110.

Many readers would interpret the preceding as saying that mean in the population is exactly equal to 110, period. They would not realize that some error was almost certainly involved in this estimate.

In contrast, confidence intervals are very up-front in acknowledging the amount of error that is involved when we estimate population parameters from sample data. For example, consider the IQ study described above: If your sample mean was 110 and the 95% CI extended from 108 to 112, this would be seen as a fairly narrow range of scores. You could, therefore, assume that your estimate of the mean in the population (110) was a relatively precise estimate.

For purposes of contrast, consider a different outcome in which the mean was the same number (110) but the 95% CI was much wider—extending from 90 to 130. Because your confidence interval covered such a wide range of values, you would be forced to assume that there was a good deal of error associated with your point estimate. In other words, your point estimate of 110 could no longer be viewed as a terribly precise estimate of the population mean.

CORRECT (AND INCORRECT) INTERPRETATIONS OF A CONFIDENCE INTERVAL. The concept of *confidence interval* can be defined in a number of different ways. This section provides some correct definitions and/or interpretations for *confidence interval* from well-regarded sources. These definitions refer to situations in which the researcher is using sample data to estimate μ, a population's mean.

Cumming and Finch (2005) provide the following: "Our CI is a range of plausible values for μ. Values outside the CI are relatively implausible" (p.174). Aron et al. (2006) offer this: "Roughly speaking, [the CI is] the region of scores...that is likely to include the true population mean; more precisely, the region of possible population means for which it is not highly unlikely that one could have obtained one's sample" (p. 711). Warner (2008) defines a confidence interval as "...a range of values above and below a sample statistic that is used as an interval estimate of a corresponding population parameter" (p.1001).

Based on these definitions, a *correct* interpretation of the 95% CI for the IQ study described above might go something like this:

Correct:

> The sample mean was *M* = 110, 95% CI [105, 115]. Given the confidence interval, it was plausible that the sample was drawn from a population whose mean was somewhere between 105 and 115.

Notice that the preceding included the word "plausible;" it did *not* include the word "probability." This is because it is generally a bad idea to use the word "probability" any time that you interpret a confidence interval. For example, below is an *incorrect* interpretation of the CI for the IQ study:

Incorrect:

> The sample mean was $M = 110$, 95% CI [105, 115]. Given the confidence interval, there was a 95% probability that the true mean in the population (μ) was somewhere between 105 and 115.

What was wrong with the second interpretation? The problem is that the "95%" in "95% confidence interval" does *not* tell you that there is a 95% probability that the true population parameter μ is included within the interval. A confidence interval does not make a probability statement about a *population*—it makes a probability statement about your *sample*. In this case, the confidence interval tells you that there is a 95% probability that this confidence interval (based on the current sample) is one of the confidence intervals that actually captured the true population parameter, out of the infinite number of confidence intervals that theoretically could be computed if the study were conducted an infinite number of times (Field, 2009a; Warner, 2008). Understanding this concept requires a little imagination:

- Imagine that, in the population of people taking this IQ tonic, the population mean really is 110 (that is, $\mu = 110$).

- Assume that you draw a sample of 20 people from this population and compute their mean IQ score. Assume that the mean for this sample is $\bar{X} = 108$, and the 95% confidence interval extends from 103 to 113. In Figure 5.5, the mean and confidence interval for this mini-study are represented by the small box-and-whisker plot that appears to the right of the label "Study 1." The small black box represents the mean of 108, and the whiskers extending out from the box represents the confidence interval.

- Assume that you draw a new sample of 20 different people from this same population and compute their mean IQ score. Assume that the mean for this sample is $\bar{X} = 106$, and the 95% confidence interval extends from 101 to 111. This mean and confidence interval appears in Figure 5.5, to the right of the label "Study 2."

- Finally, imagine that you repeat this process a total of 20 times, each time computing the sample mean and the 95% confidence interval. These means their confidence intervals are represented by the remaining box-and-whisker plots in Figure 5.5.

Remember that (unbeknownst to you), the actual mean in the population was $\mu = 110$. If you closely inspect the 20 confidence intervals in Figure 5.5, you will see that:

- 19 of the 20 confidence intervals (that is, 95% of the intervals) actually "capture" the true population mean of $\mu = 110$, and

- 1 of the 20 confidence intervals (that is, 5% of the intervals) does not "capture" the true population mean of µ = 110. The study that did not capture the true population mean was Study 10—notice that the whiskers for this boxplot do not overlap the vertical line that represents a score of 110.

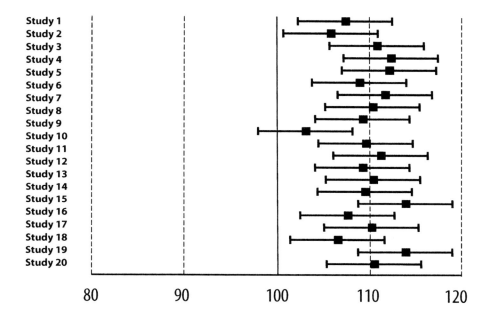

Figure 5.5. Mean IQ scores (represented by boxes) and their 95% confidence intervals (represented by whiskers) for 20 studies; *n* = 20 for each study.

When we use a 95% confidence interval and conduct the study an infinite number of times, we expect that 95% of the confidence intervals will capture the true population mean. That is why 95% (i.e., 19 out of 20) of the confidence intervals in Figure 5.5 capture the true population mean of 110.

Remember that a confidence interval allows you to make a probability statement about your *sample confidence interval*. It does *not* allow you to make a probability statement about the *population parameter* (such as µ) that you are trying to estimate. This is why this text recommends that you use words such as "plausible" or "confidence" rather than the word "probability" when discussing a CI.

COMPUTING THE 95% CI FOR A SAMPLE MEAN. The specific nature of the formula that will be used to compute a confidence interval is determined by a variety of factors (e.g., the desired level of precision, the nature of the investigation, as well as other considerations). Below is the formula for computing the 95% CI for a sample mean:

$$95\% \text{ CI} = \bar{X} \pm (SE_M)(t_{crit})$$

With the preceding formula, $\bar{X}$ = the observed sample mean; SE_M = the standard error of the mean (i.e., the standard deviation of the sampling distribution of means); and t_{crit} = the critical value of the t statistic, found in a table in the back of a statistics textbook.

Using Confidence Intervals to Test the Null Hypothesis

Seasoned consumers of research can review a confidence interval and immediately know whether the results are statistically significant. This section shows how it is done.

THE RULE. Using a confidence interval to determine significance is fairly simple, as long as you know the null hypothesis being tested. Here is the rule:

> If the population parameter described in the null hypothesis is outside of the obtained confidence interval, the results are statistically significant.

In other words, *outside = significant*. This following section explains why this is so.

AN EXAMPLE WITH SIGNIFICANT RESULTS. For example, consider the single-sample study in which you developed a tonic that you believed would increase IQ scores. In your investigation, you measured intelligence with a test that produces a mean IQ score of 100 in the population of "normal" adults (that is, adults who have not taken your tonic). This means that, in the population of normal adults, the mean IQ score is $\mu = 100$.

You are now ready to conduct your study in which you will administer the tonic to a random sample of 20 adults, wait two weeks (for the tonic to have its effect), and then measure the IQ of each subject in your sample. You hope that your tonic will cause your sample to display a mean IQ score that is higher than the mean of the normal population. Therefore, when you analyze your data, you will test the following statistical null hypothesis:

> $H_0: \mu = 100$. In the population, the mean IQ score is equal to 100.

Remember that the null hypothesis is the hypothesis that you (the researcher) hope to *reject* once you have analyzed your data. In this case, you hope that the mean IQ for the 20 subjects who took the tonic will be substantially higher than the mean IQ of 100 that is typically seen in the no-tonic population.

At the end of the investigation, you assess the IQ of each participant in the sample who took the tonic. To your delight, you find that the sample mean IQ for these 20 individuals is $\bar{X} = 120$, which is very high. You compute the 95% confidence interval around this sample mean, and find that the 95% CI extends from 110 to 130.

Knowing only these facts, most researchers and statisticians would understand that your results are statistically significant ($\alpha = .05$). They would understand that the sample of 20 individuals who took the tonic displayed a mean IQ score that was significantly higher than the population of regular people who have not taken the tonic.

How would they know this? Because the confidence interval did not contain the population parameter that had been described in the null hypothesis. This state of affairs is illustrated graphically in Figure 5.6.

The horizontal line in Figure 5.6 represents scores on a measure of IQ. In the population of "normal" adults (adults who have not taken your tonic), the mean IQ score is $\mu = 100$. In the figure, a dashed arrow points to this population mean of 100. You began with the null hypothesis that your 20 subjects came from a population in which the mean IQ score is $\mu = 100$. By the end of your investigation, you hope to reject this null hypothesis. By the end of your study, you hope to have evidence that the mean IQ score for the 20 people who took your tonic is significantly different from a mere 100 points.

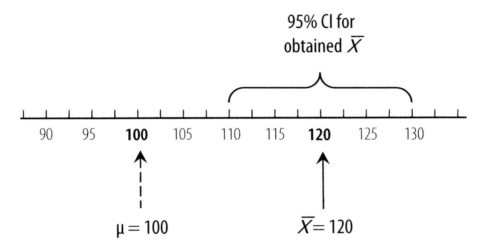

Figure 5.6. Statistically significant results: Mean IQ score for the null-hypothesis population (represented by the dashed arrow), mean IQ score for the sample of 20 participants who took the tonic (represented by the solid arrow), and 95% CI for this sample mean (represented by the bracket).

For the 20 participants who took the IQ tonic, the sample mean was a fairly high value: $\bar{X} = 120$. This sample mean is represented by the solid arrow. The 95% confidence interval around the sample mean is represented by the horizontal bracket centered above $\bar{X}$. Notice that this bracket extends from the lower limit of the confidence interval ($LL = 110$) to the upper limit ($UL = 130$). Notice also that the bracket representing the confidence interval does *not* contain the null-hypothesis population mean of 100. This tells you that it is unlikely that your sample of 20 tonic-drinkers were drawn from a population that displayed a mean IQ score of just 100. In other words, it tells you that there is a statistically significant difference between the sample mean of $\bar{X} = 120$ versus the null-hypothesis population mean of $\mu = 100$. Because of this, you may reject the null hypothesis and tell the world that you have statistically significant results.

AN EXAMPLE WITH NONSIGNIFICANT RESULTS. For the sake of contrast, imagine that you conducted the same study but instead obtained very different results. Imagine that the sample of 20 participants who took the IQ tonic now display a lower mean score—a mean score of just $\bar{X} = 105$. The 95% confidence interval for this sample mean extends from 95 to 115. In this second case, the confidence interval now contains the population mean described under the null hypothesis ($\mu = 100$).

This state of affairs is illustrated in Figure 5.7. In the figure, the obtained sample-mean for the 20 tonic-drinkers ($\bar{X} = 105$) is now pretty close to the null-hypothesis population mean ($\mu = 100$). The two means are so close to one another that the null-hypothesis population mean ($\mu = 100$) is now contained within the 95% confidence interval for the sample mean. This tells you that there is *not* a statistically significant difference between the sample mean of $\bar{X} = 105$ versus the null-hypothesis population mean of $\mu = 100$. Because of all this, you fail to reject the null hypothesis—you tell the world that you do not have statistically significant results.

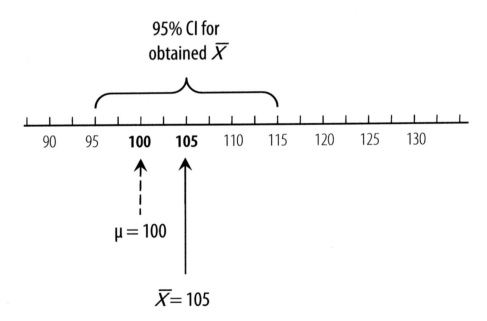

Figure 5.7. Statistically nonsignificant results: Mean IQ score for the null-hypothesis population (represented by the dashed arrow), mean IQ score for the sample of 20 participants who took the tonic (represented by the solid arrow), and 95% CI for this sample mean (represented by the bracket).

CONFIDENCE INTERVALS AND CORRESPONDING ALPHA LEVELS. The first of the two preceding examples illustrated statistically significant results. In a research article, you might have summarized these results in this way:

For the 20 participants who drank the IQ tonic, the sample mean relatively high, $M = 120$, 95% CI [110, 130]. The 95% CI did not capture the population mean described by the null hypothesis ($\mu = 100$), indicating that the results were statistically significant ($\alpha = .05$).

This sentence ends with the symbol, "$\alpha = .05$," indicating that the results were statistically significant with alpha set at .05. You will recall that the *alpha level* (or *significance level*) refers to the size of the region of rejection in a sampling distribution. When you use confidence intervals to test a null hypothesis, the size of the CI determines the alpha level for the significance test. In general, it works like this:

- A 90% confidence interval corresponds to a significance test with $\alpha = .10$
- A 95% confidence interval corresponds to a significance test with $\alpha = .05$
- A 99% confidence interval corresponds to a significance test with $\alpha = .01$

Confidence Interval for the Difference Between Two Means

Preceding sections of this chapter have discussed confidence intervals that have been computed around sample means. Researchers also sometimes compute confidence intervals around *differences* between sample means. This is typically done when subjects have been assigned to two or more treatment conditions, and the researcher has computed the mean score for each condition on some criterion variable. The researcher then computes a *mean-difference score:* a single number that represents the size of the difference between the two treatment means. Finally, the researcher computes the confidence interval around this mean-difference score. This section discusses how these confidence intervals for differences may be interpreted, and shows how you may use them to determine whether the observed difference between the two means is statistically significant.

POINT ESTIMATE FOR THE DIFFERENCE SCORE. To illustrate these concepts, this section will refer to the caffeine study that was described earlier in this chapter. You may recall that one of the criterion variables in the caffeine study was *self-reported outbursts*: the number of angry outbursts (i.e., temper tantrums) per week that were reported by each participant. In this fictitious study, participants in the high-caffeine condition reported 6.24 outbursts per week, whereas subjects in the no-caffeine condition reported 5.03 outbursts. The observed mean-difference score is computed as:

$$\bar{X}_{diff} = \bar{X}_E - \bar{X}_C$$
$$\bar{X}_{diff} = 6.24 - 5.03$$
$$\bar{X}_{diff} = 1.21$$

So the observed difference between sample means was equal to 1.21. It is useful to think of this "1.21" as a *point estimate*. Given this finding in the sample, you estimate that the mean-difference score is also equal to 1.21 in the population.

CONFIDENCE INTERVAL FOR THE DIFFERENCE SCORE. A confidence interval can be constructed around a mean-difference score, just as one can be constructed around a mean. Assume that Dr. O'Day does this and reports it in the following excerpt:

> On the criterion variable *self-reported outbursts*, the mean score for the high caffeine condition was 1.21 units higher than the mean for the no-caffeine condition, 95% CI [0.36, 2.08].

The above indicates that the observed difference between the means of the two samples was 1.21, and the 95% confidence interval for this difference extends from 0.36 to 2.08. We can now use this confidence interval to determine whether there is a significant difference between the two sample means.

USING THE CI FOR THE MEAN DIFFERENCE TO TEST THE NULL HYPOTHESIS. The same rule that had been used to perform the significance test for a single-sample mean is also used to test the significance of the difference between two means. The rule is once again reproduced below:

> If the population parameter described in the null hypothesis is outside of the obtained confidence interval, the results are statistically significant.

The null hypothesis for the current caffeine study could be presented in this way:

> $H_0 : \mu_E - \mu_C = 0$. In the population, the difference between the mean outburst score for the experimental condition versus the mean outburst score for the control condition is equal to zero.

This null hypothesis says that the actual difference between the two means is equal to zero. If we compute a confidence interval for the differences between means in the sample and this CI does not contain the value of zero, then we may reject this null hypothesis.

AN EXAMPLE WITH SIGNIFICANT RESULTS. Table 5.4 presents the some of the same results that had appeared in other tables earlier in this chapter. We will use the results in this table to perform some significance tests.

The third dependent variable in Table 5.4 is *self-reported outbursts*. Table 5.4 shows that subtracting the mean outburst score for the no-caffeine condition from the mean for the high-caffeine condition produces a difference score of 1.21, and the 95% confidence interval for this difference extends from 0.35 to 2.08. This confidence interval does not contain the value of zero. Using the rule provided above, this tells us that there was a statistically significant difference between the two conditions.

Table 5.4

Scores on Four Measures of Irritability as a Function of Amount of Caffeine Consumed (High-Caffeine versus No-Caffeine): Observed Differences Between Means with 95% Confidence Intervals

	Means			95% CI	
	High-caffeine	No-caffeine			
Dependent variable	(n = 75)	(n = 75)	Difference[a]	LL	UL
Self-reported irritability	575.00	544.73	30.27**	18.71	41.82
Other-reported irritability	556.46	546.27	10.19	-1.42	21.80
Self-reported outbursts	6.24	5.03	1.21*	0.35	2.08
Other-reported outbursts	5.24	4.99	0.25	-0.61	1.11

Note. N = 150. CI = confidence interval for difference between means; LL = lower confidence limit; UL = upper confidence limit.
[a]Difference was computed by subtracting mean for the no-caffeine sample from the mean for the high-caffeine sample. Differences flagged with an asterisk are statistically significant based on a t test for independent samples (df = 148).
*p < .01; **p < .001.

These concepts are easiest to understand if we can see them illustrated graphically. Figure 5.8 displays the mean-difference score and the confidence interval just described.

The horizontal line in Figure 5.8 represents *difference scores*—scores obtained when the mean for the control condition is subtracted from the mean for the experimental condition ($\bar{X}_E - \bar{X}_C$). In the figure, the dashed line points to the difference score that we would expect to see if there was really no difference between the two means in the population: 0.00. Think about why this makes sense: If the high-caffeine groups did not display any more outbursts than the no-caffeine group, then the difference between the means for these two groups should be equal to 0.00.

However, that is not what Dr. O'Day found. In her study, the difference between the two sample means was actually 1.21. This observed difference is represented by the solid line in the figure. A mean difference of 1.21 is pretty far away from the value of 0.00 which had been predicted by the null hypothesis. But is it far enough away for Dr. O'Day to conclude that her results are statistically significant? To find out, she reviewed the confidence interval for the difference.

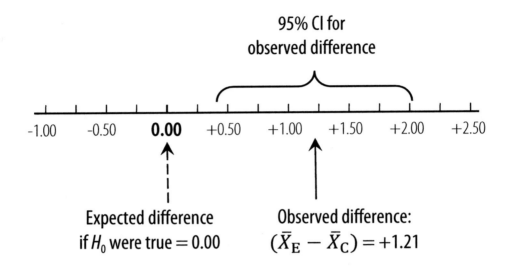

Figure 5.8. Statistically significant results from Dr. O'Day's analysis of self-reported outbursts: observed difference between means, 95% confidence interval for the observed difference, and difference that would be expected if the null hypothesis were true.

The confidence interval for the difference is represented by the bracket that appears above the horizontal line. You can see that the bracket extends from a low difference score of 0.35 (the lower limit for the CI) to a high difference score of 2.08 (the upper limit). This confidence interval tells us that it is plausible that the actual difference in the population might be as low as 0.35, or as high as 2.08. Notice that the bracket does not contain a difference score of 0.00. This tells us that Dr. O'Day's difference of 1.21 is significantly *different* from 0.00. In other words, there was a statistically significant difference between the two sample means.

AN EXAMPLE WITH NONSIGNIFICANT RESULTS. The fourth dependent variable listed in Table 5.4 is *other-reported outbursts*: the number of temper tantrums displayed by the participant, according to people who know the participant. The table shows that subtracting the mean outburst score for the no-caffeine condition from the mean for the high-caffeine condition produced a difference score of 0.25, and that the 95% confidence interval or this difference extended from –0.61 to 1.11. This confidence interval *does* contain the value of 0.00—notice that it extends from a negative number (–0.61) to a positive number (+1.11). Using the general rule provided above, the fact that this confidence interval contains the value of zero tells us that there was *not* a statistically significant difference between the two conditions. These results are illustrated graphically in Figure 5.9.

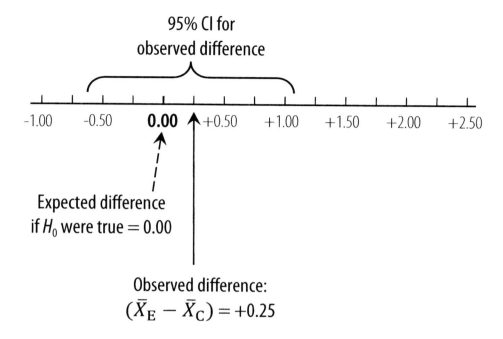

Figure 5.9. Statistically nonsignificant results from Dr. O'Day's analysis of other-reported outbursts: observed difference between means, 95% confidence interval for the observed difference, and difference that would be expected if the null hypothesis were true.

The solid line shows that, for this new criterion variable (other-reported outbursts), the difference between the mean for the two groups was only 0.25. This is not terribly far away from the difference of zero that had been predicted by the null hypothesis.

To determine whether it represents a statistically significant difference, Dr. O'Day reviewed the confidence interval represented by the bracket above the horizontal line. This bracket extends from –0.61 to 1.11. The bracket very clearly "captures" the null-hypothesis value of 0.00, and this tells Dr. O'Day that the difference between means is not statistically significant.

In summary: Whenever you are reviewing a confidence interval for the difference between two means, you can use the following simplified rule for determining whether the observed difference is statistically significant:

> If the confidence interval *contains* zero, it is plausible that the actual difference *is* zero.

How Confidence Intervals are Presented in Articles

The *Publication Manual of the American Psychological Association 6th ed.* (2010) recommends the inclusion of confidence interval for obtained statistics whenever feasible. It indicates that researchers should always report the confidence level for the interval (i.e., researchers should indicate whether it is a 90%, a 95%, or a 99% confidence interval). When feasible, researchers should use the same confidence level for each statistic in the *Results* section (i.e., if you report a 95% CI for one statistic, in most cases you should report a 95% CI for all statistics in that article).

PRESENTING CONFIDENCE INTERVALS IN TABLES. Many research articles present confidence intervals in tables under the heading "95% CI," as was done in Table 5.4, above. This table also illustrates the standard convention of using the headings *"LL"* and *"UL"* to identify lower limits and upper limits, respectively.

An alternative approach for presenting confidence intervals in a table is used with Table 5.1, which appeared near the beginning of this chapter. For convenience, it is reproduced below.

Table 5.5

Scores on Four Measures of Irritability as a Function of Amount of Caffeine Consumed (High-Caffeine versus No-Caffeine): Significance Tests, Confidence Intervals, and Point-Biserial Correlations

Dependent variable	High caffeine[a] M_1 (SD_1)	No caffeine[b] M_2 (SD_2)	Difference between means M_1-M_2 [95% CI]	$t(148)$	p	r_{pb}
Self-rated irritability	575.00 (35.49)	544.73 (36.13)	30.27 [18.71, 41.82]	5.18	<.001	.39
Other-rated irritability	556.46 (35.86)	546.27 (36.10)	10.19 [-1.42, 21.80]	1.73	.085	.14
Self-reported outbursts	6.24 (2.67)	5.03 (2.69)	1.21 [0.35, 2.08]	2.77	.006	.22
Other-reported outbursts	5.24 (2.67)	4.99 (2.65)	0.25 [-0.61, 1.11]	0.58	.561	.05

Note. N = 150. CI = Confidence interval; r_{pb} = Point-biserial correlation between the independent variable and dependent variable.
[a]n = 75. [b]n = 75.

In Table 5.5, the observed difference between means for each criterion variable appears below the heading "M_1-M_2." The confidence intervals for these differences appear below the heading "[95% CI]." Notice that the confidence intervals themselves are presented within brackets so that they are easily distinguished from other results in the table.

PRESENTING CONFIDENCE INTERVALS IN TEXT. The *Publication Manual of the American Psychological Association 6th ed.* (2010) recommends that brackets be used to enclose confidence intervals when they are presented within the text of an article. Below is an example of how Dr. O'Day might summarize one of the statistically significant results from her caffeine study (notice that the lower limit of the confidence interval is separated from the upper limit by a comma):

> One of the study's dependent variables was self-reported irritability. For this variable, the mean score displayed in the high-caffeine condition ($M = 575.00$, $SD = 35.49$) was higher than the mean displayed in the no-caffeine condition ($M = 544.73$, $SD = 36.13$). The observed difference between means was 30.27, 95% CI [18.71, 41.82], and this difference was statistically significant, $t(148) = 5.18$, $p < .001$.

In the same study, there was a nonsignificant difference between conditions on the criterion variable labeled *other-reported irritability*. Below is an example of how Dr. O'Day might have described this in the text of an article:

> One variable in the investigation was other-reported irritability. For this variable, the mean score displayed in the high-caffeine condition ($M = 556.46$, $SD = 35.86$) was only slightly higher than the mean displayed in the no-caffeine condition ($M = 546.27$, $SD = 36.10$). The observed difference between means was 10.19, 95% CI [−1.42, 21.80], and this difference was not statistically significant, $t(148) = 1.73$, $p = .085$.

Final Thoughts on Confidence Intervals

Remember: when you think *confidence interval*, think *precision*. A narrow confidence interval tells you that the estimate is fairly precise, while a wide confidence interval tells you that it is relatively imprecise.

Big-Three Results, Part 3: Effect Size

When you think *effect size*, you should think *strength*. An **index of effect size** is a number that represents the strength of the relationship between predictor variables and criterion variables. With most indices, higher values (in absolute terms) indicate a stronger association between the predictor variables and the criterion variables.

Statisticians and researchers have developed many different indices of effect size. Examples include Cohen's *d* statistic, eta squared (η^2), partial eta squared (η_p^2), omega squared (ω^2), and the r^2 and R^2 indices from regression. These indices differ with respect to (a) the types of data sets for which they are appropriate, and (b) the information that they provide.

Effect Size: Basic Concepts

Indices of effect size are important because they answer some of the questions that are of greatest interest to researchers:

- In this experiment, did the independent variable have a powerful effect on the dependent variable?

- In this correlational study, how strong was the relationship between these variables?
- If I use scores on this *X* variable to predict scores on this *Y* variable, to what extent will it decrease my errors of prediction?

In addition, indices of effect size may be less likely to be misinterpreted, compared to the null-hypothesis significance tests that have traditionally been emphasized in research. For example, an earlier section of this chapter pointed out that, when a study includes a large number of participants, the analysis might indicate that there is a statistically significant difference between two groups even when the actual size of the difference is quite small, and that this "statistically significant" outcome can mislead readers into believing that the difference is substantial when it is actually quite trivial. Because indices of effect size are not influenced by sample size in the way that significance tests are, many researchers feel that they are less likely to be misinterpreted. Because of this and other advantages, the *Publication Manual of the American Psychological Association 6th ed.* (2010) now strongly recommends that researchers should report effect size as part of their results.

A CAVEAT REGARDING THE WORD "EFFECT." Do not be misled by the word "effect," as it is used in this book. If a researcher has conducted a true experiment with strong internal validity, then an index of effect size may indeed be interpreted as the size of the causal effect that the independent variable has had on the dependent variable—the larger the value, the stronger the causal effect. However, if the study was a nonexperimental study, then the index of effect size should merely be interpreted as indicating the strength of the association (i.e., the correlation) between the predictor variable and the criterion variable. Although this book may use the term "effect size" with respect to correlational studies as well as experimental studies, remember that this term merely refers to the strength of the association between variables, not necessarily the size of a *causal* effect.

TWO APPROACHES TO COMPUTING EFFECT SIZE. An earlier section has already indicated that there are many different indices of effect size that are regularly reported in research articles: d, η^2, ω^2, r^2, R^2, and others. Most of these indices may be classified as falling into one of two categories. The first category includes the *standardized difference* indices, for which Cohen's *d* statistic is probably the best-known example. Rosenthal (1994) refers to this category as the *d family* of indices. The second category includes the *variance accounted for indices*, such as r^2, R^2, and eta-squared (η^2). Rosenthal (1994) refers to this as the *r family*. The next two sections discuss these two categories in more detail.

Standardized-Difference Indices of Effect Size

A ***standardized difference*** index of effect size indicates the size of the difference between two means, as measured in standard deviations. Examples include Cohen's *d* statistic, Hedges' *g* statistic, and Glass's Δ statistic (Rosenthal, 1991; 1994; 1995).

EXAMPLE: COHEN'S d STATISTIC. The precise way that a standardized difference index is computed depends on the nature and purpose of the investigation. However, most standardized difference indices are computed by (a) subtracting the mean obtained in one treatment condition from the mean obtained in a different treatment condition, and (b) dividing this difference by an estimate of the population standard deviation.

A good example of a standardized-difference index of effect size is Cohen's d statistic (Cohen, 1988). For a study with two independent samples (such as the caffeine study, described above), the d statistic is computed as:

$$d = \frac{\bar{X}_1 - \bar{X}_2}{s_p}$$

With this formula, d = Cohen's d statistic; $\bar{X}_1$ = the sample mean for one group; $\bar{X}_2$ = the sample mean for the second group; and s_p = pooled estimate of the population standard deviation (this is also called the within-group standard deviation, symbol: s_{within}).

LABELING AN EFFECT AS SMALL, MEDIUM, OR LARGE. The late Jacob Cohen did much to educate researchers on the importance of reporting effect size in research articles. He provided the following advice to researchers interested in interpreting the size of the effect obtained in a given study (e.g., Cohen, 1988, p. 12, 25):

1) In general, researchers should review theory and previous research findings that are relevant to the current study, and interpret the index of effect size obtained in the current study by comparing it to the indices observed in these previous investigations.

2) When previous research does not provide sufficient guidance along these lines, researchers can compare the size of the obtained effect against Cohen's own operational definitions for small, medium, and large effects.

As an illustration, Cohen's criteria for evaluating his d statistic are presented in Table 5.6.

To help researchers put these criteria in perspective, Cohen provided several real-world examples of differences that would be classified as being small, medium, or large according to this system. One example involves measurements of the height of girls, where the population standard deviation is about $\sigma = 2.1$ inches. Using this data set, Cohen (1988) offered the following examples:

- An example of a *small effect* would be the difference between the height of the average 16-year-old girl versus the average 15-year-old girl ($d = 0.24$). Cohen indicated that a small effect should be a difference that is "difficult to detect" (Cohen, 1988, p. 25).

- An example of a *medium effect* would be the difference between the height of the average 18-year-old girl versus the average 14-year-old girl ($d = 0.48$). He indicated that a medium effect should be "one large enough to be visible to the naked eye" (Cohen, 1988, p. 25).

- An example of a *large effect* would be the difference between the height of the average 18-year-old girl versus the average 13-year-old girl ($d = 0.81$). He felt that a large effect should be a difference that is "grossly perceptible" (Cohen, 1988, p. 27).

Table 5.6

Cohen's (1988) Criteria for Interpreting the Size of the d Statistic

d	Size of the effect
±.20	Small
±.50	Medium
±.80	Large

HEDGE'S g STATISTIC. In small samples, Cohen's d statistic tends to overestimate the actual effect size in the population. Multiplying Cohen's d statistic by a correction factor produces Hedge's g statistic: an *unbiased* estimate of the standardized difference in the population.

The correction causes Hedges' g to be slightly smaller than Cohen's d in absolute value. However, unless the sample is very small, the correction is slight and results in very little change in the size of Cohen's d (Borenstein, Hedges, Higgins, & Rothstein, 2009).

"Variance Accounted for" Indices of Effect Size

A *variance accounted for* index of effect size is a statistic that communicates the percent of variance in the criterion variable(s) that is accounted for by the predictor variable(s). Examples include the squared Pearson correlation coefficient (r^2), the squared point-biserial correlation coefficient (r_{pb}^2), the squared multiple correlation coefficient (R^2), eta-squared (η^2), and omega-squared (ω^2), among others. The following sections use the squared point-biserial correlation coefficient as an example of a variance-accounted-for index.

EXAMPLE: THE POINT-BISERIAL CORRELATION. The point-biserial correlation coefficient (symbol: r_{pb}) is used to investigate the relationship between a dichotomous variable and a continuous quantitative variable. Under ideal conditions, values of r_{pb} may range from –1.00 through 0.00 through +1.00, with values closer to zero indicating weaker relationships.

This is illustrated in Table 5.7, which presents results from Dr. O'Day's caffeine study. The column headed r_{pb} presents the point-biserial correlations between the dichotomous independent variable ("high-caffeine" group versus "no-caffeine" group) and each of the four quantitative dependent variables (self-reported irritability, other-reported irritability, etc.). A separate correlation coefficient is reported for each variable.

Table 5.7

Scores on Four Measures of Irritability as a Function of Amount of Caffeine Consumed (High-Caffeine versus No-Caffeine): Means, t Tests, and Point-Biserial Correlations

Dependent variable	Means		$t(148)$	p	r_{pb}	r_{pb}^2
	High-caffeine ($n = 75$)	No-caffeine ($n = 75$)				
Self-reported irritability	575.00	544.73	5.18	<.001	.39	.15
Other-reported irritability	556.46	546.27	1.73	.085	.14	.02
Self-reported outbursts	6.24	5.03	2.77	.006	.22	.05
Other-reported outbursts	5.24	4.99	0.58	.561	.05	.00

Note. $N = 150$; r_{pb} = point-biserial correlation between the dichotomous independent variable (amount of caffeine consumed) and quantitative dependent variable.

When the point-biserial correlation coefficient is squared, it indicates the percent of variance in the continuous criterion variable that is accounted for by the dichotomous predictor variable. When the squared point-biserial correlation is equal to zero, it means that there is no relationship between the two variables; larger absolute values indicate stronger relationships.

In Table 5.7, the column headed r_{pb}^2 contains the squared point-biserial correlation for each of the four criterion variables. The values in this column were computed by squaring the correlation coefficients in the r_{pb} column. Given the relative size of the r_{pb}^2 values, it appears that the amount of caffeine consumed had the strongest effect on self-reported irritability ($r_{pb}^2 = .15$), and the weakest effect on other-reported outbursts ($r_{pb}^2 = .00$).

PERCENT IMPROVEMENT IN PREDICTIVE ACCURACY. The numerical value of a variance accounted for index communicates the proportional improvement in accuracy that is attained by capitalizing on the observed relationship between the predictor variable and the criterion variable (Heiman, 2006). This value will range from 0% (when the predictor does a poor job of predicting scores on the criterion variable) to 100% (when the predictor does a perfect job of predicting scores on the criterion).

For example, consider the current study in which one criterion variable was self-reported irritability, and the predictor variable was the amount of caffeine consumed (high-caffeine versus no-caffeine). At the end of the investigation, the combined mean on self-reported irritability was $M_{combined} = 559.87$ (here, *combined mean* indicates the mean based on all 150 subjects, making no distinction between the high-caffeine sample versus the no-caffeine sample).

Imagine that you now wish to predict the score that was displayed by each of the 150 participants in the study. That is, for Participant 1, you must predict where he or she scored on self-rated irritability; for Participant 2, you must predict where he or she scored, and so forth. Imagine further that you must make these predictions under two different circumstances: *Situation 1* and *Situation 2*. These circumstances are described below.

SITUATION 1. In Situation 1, you have no idea whether a given subject was in the high-caffeine condition or the no-caffeine condition. Given this limitation, if you wish to minimize the average amount of error that is associated with the predictions that you are making, your "best guess" for each of 150 subjects would be the combined mean: The overall mean based on all 150 subjects. An earlier paragraph indicated that this combined mean was $M_{combined} = 559.87$. This indicates that, for Participant 1, you will guess that this subject's score on the criterion variable was 559.87. For Participant 2, will guess that this subject's score was also 559.87. In this way, you will guess each subject's score on the criterion variable, and your guess will be exactly the same value (559.87) for each person.

If you use the combined mean of 559.87 as your guess for each participant's score, your predictions will display a good deal of error. For many of your subjects, your guess of 559.87 will be too low and with other subjects, your guess of 559.87 will be too high.

SITUATION 2. Imagine that, in Situation 2, your circumstances have now changed: Imagine that—for each subject—you have now been told whether that participant was in the high-caffeine condition or the no-caffeine condition. Further, imagine that you now know that the mean irritability score for the high-caffeine (experimental) condition was $M_E = 575.00$, and the mean irritability score for the no-caffeine (control) condition was $M_C = 544.73$. Once again, for each participant you are required to guess that participant's score on the criterion variable (self-reported irritability). However, now that you know which condition each participant was in, it is no longer wise for you to use the combined mean score ($M_{combined} = 559.87$) as you are making your guesses. Instead, for each subject, you will now use the mean score of the treatment condition that the subject was in: That is, for each of the 75 subjects who were in the high-caffeine condition, your guess will be 575.00 (because the mean score for the high-caffeine condition was $M_E = 575.00$). Similarly, for each of the 75 subjects who were in the no-caffeine condition, your guess will be 544.73 (because the mean score for the no-caffeine condition was $M_C = 544.73$).

It makes sense that the accuracy of your predictions should be better in Situation 2 than they had been in Situation 1. This is because (a) in Situation 2, you are using the mean score of the subject's treatment condition as your best guess for that subject, and (b) we know that there was a significant difference between the two treatment conditions with respect to their scores on self-reported irritability. That is, we have already established that the high-caffeine condition scored significantly higher on self-reported irritability, compared to the no-caffeine condition.

So your predictions in Situation 2 will be more accurate than your predictions from Situation 1, but exactly *how much* more accurate will they be? In this case, we know that you have increased the accuracy of your predictions by 15%. Why 15%? Because the squared point-biserial correlation between this study's predictor variable (amount of caffeine consumed) and the criterion variable (self-reported irritability) was $r_{pb}^2 = .15$. This was shown in Table 5.7. The squared point-biserial correlation tells us the percent improvement in predictive accuracy that is achieved by capitalizing on the observed relationship between the predictor variable and the criterion variable. In this case, you "capitalized on the relationship" by using the mean of the condition that a subject was in as your best guess for that subject (rather than the combined mean based on all 150 participants).

If there had been a larger difference between the mean scores displayed by the two treatment conditions, the squared point-biserial correlation would have been a larger value, perhaps $r_{pb}^2 = .25$. If this had been the case, you would have increased the accuracy of your predictions by 25% by capitalizing on the relationship between the predictor and the criterion.

By the same reasoning, if there had been no difference between the mean scores for the two conditions, the squared point-biserial correlation would have been $r_{pb}^2 = .00$. In this case, capitalizing on the relationship between the two variables would not have increased the accuracy of your prediction at all.

OTHER VARIANCE-ACCOUNTED-FOR INDICES. This section has focused on just one "variance-accounted-for" index of effect size: the squared point-biserial correlation coefficient. However, it is important to remember that the basic concepts discussed here apply with the other variance-accounted-for indices listed at the beginning of this section: the squared Pearson correlation coefficient (r^2), the squared multiple correlation coefficient (R^2), eta-squared (η^2), and omega-squared (ω^2).

LABELING AN EFFECT AS SMALL, MEDIUM, OR LARGE. An earlier section reviewed Cohen's criteria for evaluating standardized-difference indices of effect size (such as the *d* statistic) to determine whether they should be labeled as representing a small, medium, or large effect. In the same book, Cohen (1988) also provides a different set of criteria for evaluating various types of correlation coefficients. For example, Table 5.8 presents criteria for evaluating the Pearson correlation coefficient and the point-biserial correlation coefficient (Cohen, 1988; Kline, 2004). Notice that these criteria apply to correlation coefficients that have *not* been squared.

Table 5.8

Cohen's (1988) Criteria for Interpreting the Size of r (the Pearson Correlation Coefficient) and r_{pb} (the Point-Biserial Correlation Coefficient)

r	r_{pb}	Size of the effect
±.10	±.10	Small
±.30	±.25	Medium
±.50	±.40	Large

Interpreting the Statistics Other than *d* or *r*

This section has provided guidelines for interpreting the relative size of the *d* statistic, the *r* statistic, and the r_{pb} statistic. But what if you are reading about a study that did not produce a simple statistic such as *d* or *r*? What if the study reports an *F* statistic (as is obtained with analysis of variance) or some other statistic? Is it still possible to interpret the size of the effect?

The answer is *yes*. First, Cohen's (1988) oft-cited book on power analysis is the most obvious reference for interpreting effect size based on a wide variety of statistics. In addition, some statistics textbooks provide guidelines (often adapted from Cohen, 1988) that can be used to interpret statistics from more advanced research designs. For example, Newton and Rudestam (1999) provide guidelines for interpreting the size of R^2 (from multiple regression) and eta-squared (from ANOVA), along with other statistics (see, for example, the table on page 76 of their book).

In addition, a number of publications provide formulas that can be used to convert certain statistics (such as *F*) into the *r* statistic, the *d* statistic, or some other common metric. See, for example, Hedges and Becker (1986) and Rosnow and Rosenthal (1996). Books on meta-analysis are particularly rich sources for these conversion formulas (e.g., Borenstein et al., 2009; Hedges & Olkin, 1985; Hunter & Schmidt, 2004).

But it may not be necessary to turn to other references at all. Most of the chapters in this book (the one that you are now reading) deal with specific statistical procedures, and many of these chapters discuss the indices of effect size that are typically used for that procedure. Wherever possible, these chapters also provide guidelines for labeling a given effect as small, medium, or large.

Final Thoughts on Effect Size

Remember: When you think *effect size*, think *strength*. An index of effect size communicates the strength of the relationship between predictor variables and criterion variables. When research articles report indices of effect size, it decreases the likelihood that readers will be misled by the results: Although a researcher might describe a result from a large-sample study as being "highly-significant" according to a significance test, you will be able to use the index of effect size to evaluate the actual strength of the relationship between the variables in question.

Chapter Conclusion

The big-three results covered in this chapter are important, both individually and collectively. Individually, they each provide a different *type* of information regarding the investigation's findings: the significance test provides information about probability, the confidence intervals provide information about the precision of the estimates, and the indices of effect size provide information about the strength of the relationships. Collectively, they allow us to see the big picture regarding the statistical and practical significance of the results. When conducting your own research, you should report all three whenever feasible.

When reading about the empirical research conducted by others, however, don't be shocked to see that they have omitted one or more of the big-three. In some cases, this is appropriate, as it is not possible to compute all three types of results with some statistical procedures. In other cases, however, it is likely that the researchers just didn't know how to compute all three results. This is particularly likely with older articles published prior to the development of statistical applications such as SPSS or SAS.

The vast majority of empirical research articles *will* present null-hypothesis significance tests. Despite the current controversy, most graduate programs continue to teach significance tests, and most journals continue to require them.

Depending on the type of journal you are reading, you may have only a 50-50 chance of encountering confidence intervals. Confidence intervals provide a lot of information: They not only tell you whether the results are statistically significant, but also convey the precision of the sample-based estimates. Because of this, the *Publication Manual of the American Psychological Association 6th ed.* (2010) strongly endorses their use. However, graduate programs have not emphasized confidence intervals to the extent that they have emphasized significance testing, and many data-analysis applications do not compute confidence intervals for many statistics.

Indices of effect size occupy a position somewhere between significance tests and confidence intervals. Popular statistical applications compute indices of effect size for many statistics, and some journals have required them for decades. As a result, you are fairly likely to encounter indices such as R^2 and η_p^2 in your reading, and the popularity of these indices is only likely to increase.

All of this is in transition. With the passage of time, graduate programs may increasingly emphasize confidence intervals and effect size, and statistical applications may make it easier to compute these indices for a wide variety of statistical procedures. If this occurs, tomorrow's researchers are likely to be as comfortable with the big-three results as today's researchers are with simple significance tests.

Chapter 6

Bivariate Correlation

This Chapter's *Terra Incognita*

You are reading about a study in which the researcher investigated the relationship between *health motivation* (the participant's desire to be healthy) and *exercise sessions* (the number of times per week that they engage in aerobic exercise). The published article reports the following:

> The correlation between exercise sessions and health motivation was statistically significant, $r = .29$, 95% CI [.17, .40], $p < .01$. According to the criteria provided by Cohen (1988), the size of this correlation fell just short of a medium effect. The sign of the correlation was positive, which was consistent with direction that was hypothesized by the four-factor model.

Have no fear. All shall be revealed.

The Basics of this Procedure

This chapter introduces ***bivariate correlation coefficients:*** statistics that convey the nature of the relationship between two variables. These statistics are called ***bivariate correlations*** because each coefficient is based on just two variables: a single predictor variable and a single criterion variable. The term *bivariate* distinguishes these procedures from *multiple correlation,* a procedure in which a single criterion variable is predicted from several predictor variables, all in a single analysis (you will learn about multiple correlation in *Chapter 9*).

The majority of this chapter focuses on the ***Pearson correlation coefficient*** (also known as the ***Pearson product-moment correlation,*** or the ***zero-order correlation coefficient;*** symbol: *r).* In the social and behavioral sciences, the Pearson *r* is the most widely-used correlation coefficient. Figure 6.1 illustrates the nature of the variables typically included in the computation of *r.*

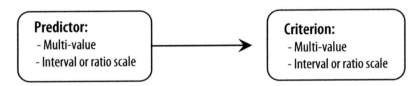

FIGURE *6.1.* Generic analysis model: Nature of the variables included in the typical computation of a Pearson correlation coefficient.

Figure 6.1 shows that researchers typically compute the Pearson correlation coefficient when the analysis involves (a) a single continuous criterion variable measured on an interval scale or a ratio scale, and (b) a single continuous predictor variable, also assessed on an interval or ratio scale. Both variables are usually multi-value variables. Additional assumptions are listed at the end of this chapter.

When one or both of the variables included in the analysis does not satisfy the assumptions underlying the Pearson *r,* researchers may report some alternative statistic, such as the Spearman correlation coefficient. A section near the end of this chapter briefly covers some of these alternative procedures.

Illustrative Study

Assume that a fictitious health psychologist named Dr. O'Day wants to identify variables that predict the frequency with which people engage in aerobic exercise. In her current study, the criterion variable is *exercise sessions*: the number of times that a participant engages in aerobic exercise per week. This is a quantitative multi-value variable assessed on a ratio scale of measurement.

THE MODEL. After reviewing previous research on aerobic exercise, Dr. O'Day has developed a sort of mini-theory that we will call the *four-factor model of exercise behavior.* We describe this as a "mini-theory" because it attempts to identify the most important determinants of exercise behavior. Her model is illustrated in

Figure 6.2 (by the way, don't bother trying to find this *four-factor model of exercise behavior* in PsycINFO® or any other database—it is a fictitious model created just for this book).

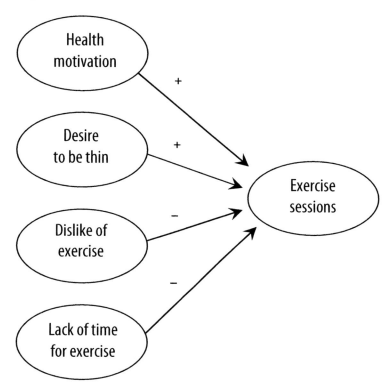

Figure 6.2. The four-factor model of exercise behavior.

The causal model illustrated in Figure 6.2 is called the *four-factor model* because it hypothesizes that the criterion variable of interest (number of exercise sessions per week) is determined by four predictor variables. The positive symbols (+) and negative symbols (–) in the figure illustrate whether a given predictor variable is hypothesized to have a positive or negative effect on criterion variable. For example, the arrow from health motivation to exercise sessions is identified with a positive symbol, meaning that an increase in health motivation is expected to cause an increase in the number of exercise sessions. In contrast, the arrow from dislike of exercise to exercise sessions is identified with a negative symbol, meaning that an increase in dislike of exercise is expected to cause a decrease in the number of exercise sessions.

Please remember that, although Dr. O'Day's model predicts causal relationships between the variables, this does not mean that the correlational study described here is capable of producing strong evidence that the relationships observed (if any) are actually *causal* in nature. In *Chapter 2*, you learned that nonexperimental studies (such as the one to be described here) typically provide only weak evidence of cause-and-effect.

Nevertheless, this scenario is still fairly realistic in the sense that researchers often use correlational studies at the early stages of a program of research dealing with a causal model. If these early investigations reveal a pattern of correlations that is consistent with the predicted causal relationships, the researchers typically then move on to the experimental studies which are capable of providing stronger evidence of cause-and-effect.

VARIABLES INVESTIGATED. Each of the four predictor variables in Dr. O'Day's model is assessed by a multiple-item summated rating scale that appears in the *Questionnaire on Eating and Exercising* in Appendix A of this book. With each of these summated rating scales, possible scores may range from 1.00 to 7.00, with higher scores representing higher levels of the construct being assessed.

The four variables that constitute the four-factor model of exercise are as follows:

- *Health motivation.* When individuals score high on *health motivation*, it means that they are willing to sacrifice and work hard in order to stay healthy. This variable is assessed by the summated rating scale that consists of items 27-29 in the *Questionnaire on Eating and Exercising*. The four-factor model hypothesizes that health motivation should display a positive relationship with the number of exercise sessions.

- *Desire to be thin.* When individuals score high on *desire to be thin*, it means that it is important to them to have a fairly low body weight and a lean physical appearance. This variable is assessed by questionnaire items 31-33, and the four-factor model hypothesizes that the desire to be thin should also display a positive relationship with the number of exercise sessions.

- *Dislike of exercise.* When individuals score high on *dislike of exercise*, it means that they hate aerobic exercise and find it to be unpleasant. This variable is assessed by questionnaire items 47-49, and the model hypothesizes that it should display a negative relationship with the number of exercise sessions.

- *Lack of time for exercise.* When individuals score high *lack of time for exercise,* it means that they believe that they are too busy to engage in frequent aerobic exercise. This variable is assessed by questionnaire items 51-53, and the model hypothesizes that it should also display a negative relationship with the number of exercise sessions.

METHOD. To investigate her model, Dr. O'Day administers the *Questionnaire on Eating and Exercising* to 260 students enrolled in various sections of General Psychology at a regional state university in the Midwest, and receives usable responses from 238 participants. Her main criterion variable is *exercise sessions*, which is assessed by item 76 on the Questionnaire (reproduced below):

How many times each week do you engage in a session of aerobic exercise that lasts at least 15 minutes? (Write zero if you never engage in aerobic exercise).

In analyzing her data, Dr. O'Day computes all possible bivariate correlations between this criterion variable and the four predictor variables described above. She determines whether the correlations are significantly different from zero, computes confidence intervals, and evaluates them in terms of relative effect size.

X Variables and Y Variables

A specific Pearson correlation coefficient is computed for just one pair of variables. When investigating the four-factor model presented above, Dr. O'Day would compute one correlation coefficient to understand the relationship between health motivation and exercise sessions, a separate correlation coefficient to understand the relationship between the desire to be thin and exercise sessions, and so forth.

When computing a specific correlation coefficient, the predictor variable is often referred to as the **X variable** (health motivation is an example of an X variable in the current study). The criterion variable is often referred to as the **Y variable** (exercise sessions is the Y variable here).

Research Questions Addressed

Below are examples of the types of questions that can be answered by computing bivariate correlation coefficients.

- What is the direction of the relationship between the predictor variable and the criterion variable?
- Is this correlation coefficient significantly different from zero?
- According to the confidence interval, how precise is this sample correlation when viewed as an estimate of the population correlation?
- How large is the effect size represented by this correlation coefficient?
- If the correlation has been computed between several predictor variables and the same criterion variable, which predictor variables appear to display the strongest relationship with the criterion?

Details, Details, Details

A Pearson correlation coefficient conveys a lot of information. Briefly inspect this two-digit number, and you immediately know a lot about the trend in the data, and the strength of the relationship between the two variables.

The Sign of the Correlation.

Under typical conditions, the Pearson correlation coefficient ranges in size from −1.00 through 0.00 through +1.00. The sign of the correlation (+ versus −) tells you whether there is a positive relationship or a negative relationship between the two variables.

The relationship between two quantitative multi-value variables is often illustrated using a graph called a ***scatterplot*** (also called *a scatter diagram* or *scatterplot*). In a scatterplot, values on the predictor variable are typically represented as points on the horizontal axis (i.e., the X axis), and values on the criterion variable are typically represented as points on the vertical axis (i.e., the Y axis).Each dot in a scatterplot represents an individual participant—by seeing where the dot falls in the plot, you can determine where that subject fell (simultaneously) on both the predictor and criterion.

Figure 6.3 presents an example of a scatterplot. In this plot, the predictor variable is health motivation and the criterion variable is exercise sessions per week. The data presented in the figure are fictitious.

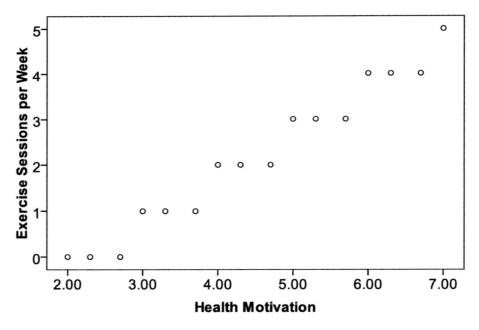

Figure 6.3. (Fictitious) positive correlation between health motivation and exercise sessions per week.

Figure 6.3 illustrates a positive correlation. With a ***positive correlation*** (e.g., r = +.20, r = +.73), low values on the predictor variable tend to be associated with low values on the criterion variable, and high values on the predictor tend to be associated with high values on the criterion. You can see how the scatterplot included in Figure 6.3 illustrates this trend: Subjects with low scores on health motivation tend to have low scores on exercise sessions per week, and subjects with high scores on health motivation tend to have high scores on exercise sessions.

The counterpart to a positive correlation is a ***negative correlation*** (e.g., $r = -.22$, $r = -.68$; also called an ***inverse correlation***). With a negative correlation, low values on the predictor variable tend to be associated with high values on the criterion variable, and high values on the predictor tend to be associated with low values on the criterion. Figure 6.4 uses fictitious data to present a negative correlation between two variables.

In Figure 6.4, the predictor variable is self-rated *dislike of exercise*. Higher scores on this variable indicate that the subject does not like to engage in aerobic exercise—it means that the subject finds exercise to be unpleasant or punishing. The criterion variable is (once again) exercise sessions per week.

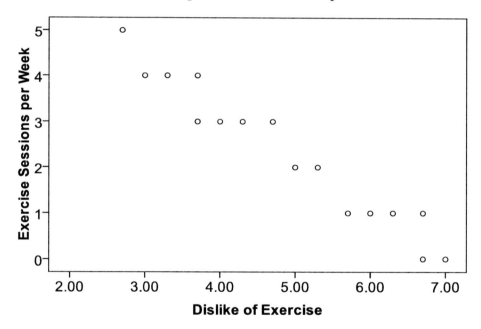

Figure 6.4. (Fictitious) negative correlation between dislike of exercise and exercise sessions per week.

You can see how Figure 6.4 illustrates a negative correlation: Subjects with low scores on dislike of exercise tend to have high scores on exercise sessions per week, and subjects with high scores on dislike of exercise tend to have low scores on exercise sessions.

Finally, a ***zero correlation*** (e.g., $r = 0.00$) means that there is no relationship between the predictor variable and the criterion variable. With zero correlation, as predictor variable score increase, the criterion variable scores do not change in any systematic way. Figure 6.5 uses fictitious data to illustrate a zero correlation between two variables.

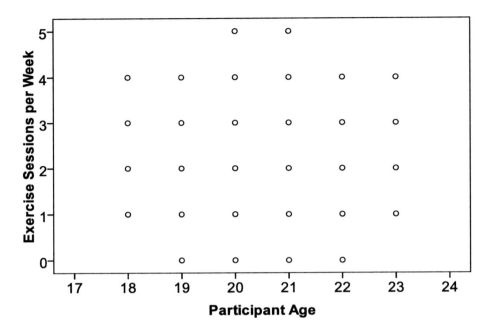

Figure 6.5. (Fictitious) zero correlation between participant age and exercise sessions per week.

This scatterplot in Figure 6.5 illustrates the relationship between participant age (in years) and exercise sessions. You can see how this scatterplot illustrates a zero correlation, as defined above: There is no systematic trend showing that younger people (say, at age 19) engage in more exercise sessions per week, compared to older people (say, at age 24).

The Size of the Correlation

As was stated above, under typical research conditions, a Pearson correlation coefficient usually ranges from −1.00 through 0.00 through +1.00. The absolute value of the correlation indicates the strength of the relationship between the two variables (the ***absolute value*** of a number is the magnitude of the number itself, disregarding whether it is a positive value or a negative value). This means that correlation coefficients that are close to zero (such as −.04, .03, or .06) indicate a weaker relationship between the predictor variable and the criterion variable, whereas correlations closer to ±1.00 (such as −.88, .92, or .98) indicate a stronger relationship. The section titled "How Strong is the Relationship?" (later in this chapter), discusses different approaches that may be used to determine whether a specific correlation coefficient should be labeled as large, small, or somewhere in-between.

The sign of the correlation is not relevant to the strength of the association. This means that a correlation of −.75 is just as strong as a correlation of +.75.

Definitional Formula for the Pearson *r*

The definitional formula for the Pearson correlation coefficient is as follows (from Gravetter & Wallnau, 2000):

$$r = \frac{\Sigma z_X z_Y}{N}$$

With the preceding formula, r = the Pearson correlation coefficient; z_X = the predictor variable in z-score form (more on this in a moment); z_Y = the criterion variable in z-score form; and N = the number of participants in the sample.

In *Chapter 4: z Scores and Area Under the Normal Curve*, you learned that a distribution of z scores is a standardized distribution—in this case, a distribution that has been transformed so that it has a mean of zero and a standard deviation of one. This formula shows that the Pearson *r* is merely the average cross-product of scores on the predictor and criterion, when these variables are in z-score form.

Results from the Investigation

The bulk of this chapter illustrates different ways that researchers report bivariate correlations in articles. You will see how they report significance tests, confidence intervals, and effect size, and you will see how this is done within tables as well as within the text of the research article.

Reporting All Possible Correlations in a Table

First up is the *correlation matrix*. When researchers compute every possible correlation between a set of variables, it is standard to organize these correlations in the form of a matrix. Table 6.1 shows how Dr. O'Day might present the bivariate (Pearson) correlations for the five variables assessed in her study.

PARTS OF THE TABLE. The column headed "Variable" in Table 6.1 provides the name for each variable, and identifies each variable with a number (such as "1," "2," and so forth). These numbers will be important when later locating specific correlation coefficients within the table.

In Table 6.1, the columns headed "*M*" and "*SD*" present means and standard deviations (respectively) for each variable. You can see that the mean score for exercise sessions was 2.93, indicating that the typical participant engaged in about three exercise sessions each week.

LOCATING SPECIFIC CORRELATIONS IN THE MATRIX. Bivariate correlations for all possible pairs of variables are presented under the heading "Pearson correlations." There are five columns (running up and down) in this section, and each column represents one of the five variables in the analysis. To save space, it is common that numbers (rather than the variable name) are used to identify each variable. The column headed "1" represents *exercise sessions per week*—you know this because exercise sessions is identified with the number "1" on the left side of the table. The headings for the remaining variables may be interpreted in the same way.

Table 6.1

Pearson Correlations (and Coefficient Alpha Reliability Estimates) for the Study's Variables, Decimals Omitted

			Pearson correlations				
Variable	M	SD	1	2	3	4	5
1. Exercise sessions per week	2.93	2.31	(--)				
2. Health motivation	5.76	0.98	29	(83)			
3. Desire to be thin	5.30	1.35	15	30	(85)		
4. Dislike of exercise	2.36	1.31	-37	-31	-15	(92)	
5. Lack of time for exercise	3.67	1.71	-46	-20	-08	39	(95)

Note. N = 238. For variables 2 through 5, possible scores could range from 1.00 to 7.00 with higher scores representing higher levels of the construct being measured. Correlations are significant at $p < .05$ for $r > \pm.127$; correlations are significant at $p < .01$ for $r > \pm.167$. Values on diagonal are coefficient alpha reliability estimates for multiple-item summated rating scales.

In a table such as Table 6.1, the location where the column for one variable intersects with the row for a second variable is called a **cell**. To locate the correlation between two specific variables, you must find the cell where the column for the first variable intersects with the row for the second variable. For example, to locate the correlation between exercise sessions and health motivation, find the location where the column headed "1" (for "Exercise sessions") intersects with the row headed "2" for "Health motivation." At this location, you will find the number "29," which means that the Pearson correlation coefficient for these two variables is $r = +.29$ (the title for the table tells us that decimal points have been omitted from all correlations; whenever the sign of a number is omitted, you can assume that it is positive).

The correlation between exercise sessions and health motivation is positive, meaning that higher scores on health motivation are associated with higher scores on exercise session. This is consistent with the positive relationship that was hypothesized by the "+" symbol in Figure 6.2. Correlation coefficients for the remaining variables may be interpreted in the same way.

ADVANTAGES OF PRESENTING A MATRIX OF CORRELATIONS. Presenting a table of correlations such as Table 6.1 serves two purposes. First, it provides readers with an understanding of the basic relationships between the variables: which predictor variables were strongly correlated with one another, which predictor variables were relatively independent of one another, and so forth. Second, it allows readers to see whether the bivariate correlations are consistent with the researcher's theoretical model which, in this case, had been presented in Figure 6.2.

VALUES ON THE DIAGONAL. On the diagonal of the correlation matrix, the table presents the coefficient alpha reliability estimate for each variable that had been assessed using a multiple-item summated rating scale. You will recall from *Chapter 3* that values of coefficient alpha typically range from .00 to 1.00, with higher values indicating better reliability. Values above .70 are usually interpreted as representing minimally acceptable levels of reliability. Coefficient alpha is not reported for exercise sessions per week, because this variable had not been assessed using a multiple-item scale.

Are the Relationships Statistically Significant?

When researchers test correlation coefficients for ***statistical significance***, they are determining the probability that they would have obtained the current sample results if the null hypothesis were true. This section describes the specific null hypothesis that is typically being tested, and shows how researchers usually indicate whether their correlations are statistically significant.

THE NULL HYPOTHESIS BEING TESTED. When researchers test a correlation coefficient for statistical significance, they are almost always testing the following null hypothesis:

> H_0: ρ = 0. In the population, the correlation between the predictor variable and the criterion variable is equal to zero.

Just to keep things as confusing as possible, the symbol "ρ" in the preceding null hypothesis is *not* a lower case "P." It is actually the small letter "rho" from the Greek alphabet. It is the symbol for the correlation coefficient in the population. This null hypothesis says that, in the population, the correlation between the predictor variable and the criterion variable is equal to zero. If the actual correlation (r) based on the sample data is large enough, the researcher will reject the null hypothesis and tell the world that the correlation is significantly different from zero.

REPORTING STATISTICAL SIGNIFICANCE IN THE TEXT OF THE ARTICLE. Although it is possible to use a *z* test to determine whether a correlation coefficient is significantly different from zero, it is more common to instead use a *t* test with $N - 2$ degrees of freedom

(N represents the number of subjects in the sample). If the researchers set alpha at $\alpha = .05$ and the obtained p value is less than .05, the correlation is viewed as being significantly different from zero. Of course, researchers may set alpha at a more conservative level such as $\alpha = .01$ or $\alpha = .001$, if they are so inclined.

In published articles, it is relatively uncommon for researchers to present the actual t statistic that was computed for this significance test. Instead, it is more common to simply present the obtained correlation coefficient, the degrees of freedom (in parentheses), and the obtained p value for the test. For example, Dr. O'Day analyzed data from 238 subjects, found the correlation between health motivation and exercise session to be $r = .29$, and found that the p value for the t statistic was $p < .01$. Therefore, she might report the finding this way within the body of the article:

$r(236) = .29, p < .01.$

The above tells the reader:

> The observed correlation between the two variables in the sample was equal to .29. If the null hypothesis of zero correlation in the population were true, the probability is less than .01 that I would have observed a correlation this large (or larger) in the sample.

REPORTING STATISTICAL SIGNIFICANCE IN TABLES. Table 6.1 presents every possible correlation between the five variables in Dr. O'Day's study. Researchers sometimes use asterisks (*) to flag statistically significant correlation coefficients in tables such as this. For example, they may use one asterisk to flag correlations that are significant at the .05 level, two asterisks to flag correlations that are significant at the .01 level, and so forth. If a table contains a large number of correlations, however, this takes up too much space. In such situations, the researchers may instead rely on notes such as the one at the bottom of the current table:

> Correlations are significant at $p < .05$ for $r > \pm.127$; correlations are significant at $p < .01$ for $r > \pm.167$.

The observed correlation between exercise sessions and health motivation is .29. Since this value is larger than the criterion of "±.167" indicated in the note, we know that this correlation is significantly different from zero at the $\alpha = .01$ level.

In contrast, the observed correlation between desire to be thin (Variable 3) and lack of time for exercise (Variable 5) was −.08. Because this value is less than ±.167, we know that it is not significantly different from zero with alpha set at $\alpha = .01$. Further, because it is less than ±.127, we know that it is also not significantly different from zero with alpha set at $\alpha = .05$. In other words, that correlation coefficient was not statistically significant at all.

What is the Nature of the Relationships?

When we ask about the *nature of the relationship* between two continuous variables, we are primarily interested in the direction of the relationship. This is revealed by the sign of the Pearson correlation coefficient—whether it is positive (such as $r = +.30$) or negative (such as $r = -.30$).

You learned about the sign of correlation coefficients in the *Details, Details, Details* section presented earlier. If we apply those concepts to the obtained correlations appearing in Table 6.1, we can draw the following conclusions:

1) There is a positive correlation between health motivation and exercise sessions, $r = +.29$. This indicates that subjects who score high on health motivation tend to engage in more exercise sessions.

2) There is a positive correlation between desire to be thin and exercise sessions, $r = +.15$. This indicates that subjects who score high on desire to be thin tend to engage in more exercise sessions.

3) There is a negative correlation between dislike of exercise and exercise sessions, $r = -.37$. This indicates that subjects who score high on dislike of exercise tend to engage in fewer exercise sessions.

4) There is a negative correlation between lack of time for exercise and exercise sessions, $r = -.46$. This indicates that subjects who score high on lack of time for exercise tend to engage in fewer exercise sessions.

Are the Confidence Intervals Precise?

Table 6.1 presented all possible correlations between all five variables in Dr. O'Day's study. Presenting all possible correlations is generally the preferred approach, as it provides a good deal of information to the reader (such as the intercorrelations between all of the predictor variables, for example). However, if the researcher is really interested in just the correlations between the predictor variables and the criterion variable, it may make more sense to instead prepare a table that includes only those correlations. By focusing on a subset of relationships, the researcher is able to present more information relevant to each correlation (such as confidence intervals). This approach is used with Table 6.2.

In Table 6.2, the names of the Dr. O'Day's four predictor variables appear below the heading "Predictor variable." The columns headed "*M*," "*SD*," and "*r*" provide means, standard deviations, and Pearson correlation coefficient, for each variable. The columns headed "95% CI" and "*p*" provide 95% confidence intervals and probability values for the Pearson correlations. The following section discusses the confidence intervals in greater detail.

CONFIDENCE INTERVALS FOR *r*. A ***confidence interval*** is an interval that contains plausible values for the population parameter being estimated. In this case, the population parameter is the correlation between variables in the population. In this context, confidence intervals provide the reader with some idea as to just how precise these sample correlation coefficients are, when we view them as estimates of the population correlations.

Table 6.2

Pearson Correlations (with Confidence Intervals) for the Relationships Between Exercise Sessions per Week and the Four Predictor Variables

Predictor variable	M	SD	r	95% CI	p
1. Health motivation	5.76	0.98	.29	[.17, .40]	<.001
2. Desire to be thin	5.30	1.35	.15	[.02, .27]	.011
3. Dislike of exercise	2.36	1.31	−.37	[−.26,−.47]	<.001
4. Lack of time for exercise	3.67	1.71	−.46	[−.35,−.55]	<.001

Note. N = 238. For variables 1 through 4, possible scores could range from 1.00 to 7.00 with higher scores representing higher levels of the construct being measured. r = Pearson product-moment correlation coefficient. 95% CI = 95% confidence interval for the correlation coefficient.

For example, Table 6.2 shows that the correlation between health motivation and exercise sessions (in the sample) is r = .29. This means that our best estimate for the actual correlation between these two variables in the population is also .29. The 95% confidence interval for this correlation ranges from .17 to 40, meaning that it is plausible that the population correlation might be as low as .17, and it might be as high as .40. The correlation coefficients and confidence intervals for the remaining predictors may be interpreted in the same way.

When interpreting confidence intervals, it is important to avoid the common misconception that a confidence interval provides a probability estimate about the population correlation. If a researcher computes a 95% confidence interval as extending from .17 to .40, it is a misconception to interpret this as meaning that "there is a 95% probability that the actual population correlation is somewhere between .17 and .40." It is more accurate to say "there is a 95% probability that this sample confidence interval is one of the confidence intervals that actually captured the population correlation." Remember that a confidence interval makes a probability statement about your *sample results*, not about the larger population. The correct and incorrect way to interpret a confidence interval was discussed in greater detail in *Chapter 5*.

USING THE CI TO EVALUATE THE PRECISION OF THE ESTIMATES. An earlier section indicated that one of the most important uses of confidence intervals involves the concept of precision: When confidence intervals are relatively narrow, you can be more confident that the point estimate (the correlation coefficient, in this case), is a fairly precise estimate of the population parameter being estimated. Many researchers would view the confidence intervals in Table 6.2 as not being very narrow. Notice, for example, the information provided for the predictor variable, *desire to be thin*. The researcher's "best estimate" for the correlation between desire to be thin and exercise sessions is .15 (because the sample correlation was r = .15). The 95% CI extended from .02 to .27, meaning that it is plausible that the population correlation coefficient could be as low as .02, or as high as .27. This is a fairly wide range. Given the relatively wide range seen for each of the confidence intervals in the table, many researchers would not be very confident that the obtained correlation coefficients were terribly precise estimates of the actual population correlations.

The really bad news about the preceding conclusion is that the present correlations were based on a moderately-large sample: N = 238. In general, as samples get larger, the confidence intervals for correlation coefficient tend to get smaller (more precise). Despite having a fairly large sample, the CI for each of these correlations was not terribly narrow. The broader lesson is this: Confidence intervals for correlation coefficients tend to be pretty wide unless the sample size is really, really large. This means that, with sample sizes of 200 or so participants (which is common in the social and behavioral sciences), the obtained correlation coefficients should be taken with a grain of salt, and not necessarily be accepted as highly-precise estimates of the correlation in the population.

USING THE CI TO DETERMINE STATISTICAL SIGNIFICANCE. None of the 95% confidence intervals for the correlations in Table 6.2 contain the value of zero, and this tells you that each of the obtained correlation coefficients are significantly different from zero with alpha set at α = .05. Of course, you already knew that each correlation was statistically significant, because the p values for each coefficient (reported in the column headed "p") were each smaller than .05.

How Strong are the Relationships?

An ***index of effect size*** is a statistic that conveys the strength of the association between a predictor variable and a criterion variable. When interpreting the size of the effect represented by a bivariate correlation, researchers typically focus on either the correlation coefficient itself (r), or the coefficient of determination, which is the square of the correlation coefficient (r^2). We shall take each approach in turn.

EFFECT SIZE: THE CORRELATION COEFFICIENT, r. Researchers often use the correlation coefficient itself (r) as the "common-metric" index of effect size when performing meta-analyses (to be covered in *Chapter 19*). The larger the absolute value of the correlation coefficient (i.e., disregarding the sign), the larger the "effect."

How large does a correlation coefficient have to be in order to be considered a large effect? Cohen (1988) recommends that the first resort should be to review previous investigations dealing with the same variables, and compare the current correlation coefficient to those found in this previous research.

When this is not possible (e.g., when there are no previous studies on this topic), Cohen offers a last resort: Compare the current correlation coefficient against his criteria small, medium, and large effects. Cohen's (1988) criteria are as follows:

$r = \pm.10$ represents a small effect

$r = \pm.30$ represents a medium effect

$r = \pm.50$ represents a large effect

Using these criteria, we can see that the observed correlation between exercise sessions and health motivation ($r = .29$) falls just short of representing a medium effect. You can use the same guidelines to interpret each of the remaining coefficients in Table 6.2.

EFFECT SIZE: THE COEFFICIENT OF DETERMINATION, r^2. To produce an effect-size statistic that provides a somewhat different perspective on the results, researchers often square correlation coefficients to produce r^2: the coefficient of determination. The *coefficient of determination* (symbol: r^2) represents the percent of variance in the criterion variable (Y) that is accounted for by the predictor variable (X). Because it is the result of squaring a number, the coefficient of determination is always a positive value. It may range from 0.00 (indicating that X does not account for any of the variance in Y) to 1.00 (indicating that X accounts for all of the variance in Y; in other words, values of X may be used to predict values of Y with perfect accuracy).

It is acceptable to use either r or r^2 as an index of effect size. When they wish to refer to Cohen's (1988) criteria (or when performing a meta-analysis), researchers typically discuss r. On the other hand, when they wish to discuss the percent of variance accounted for, they instead refer to r^2.

Reporting Results in the Text of an Article

Dr. O'Day's four-factor model made predictions about just four relationships: the relationships between (a) exercise sessions and (b) each of the four predictor variables. Therefore, it is likely that she will emphasize just these four correlations when she prepares the *Results* section of her article.

In a published article, it is unlikely that an author would present (a) Table 6.1 (which presents every possible correlation between all five variables), plus (b) Table 6.2 (which provides correlations, confidence intervals, and additional information for just the four relationships of greatest interest), plus (c) a detailed description of these correlations and confidence intervals within the text of the article itself. Presenting all three would be redundant, and would not represent an economical use of the very precious real estate in a published article.

If Dr. O'Day were my colleague, I would recommend that she (a) include Table 6.1, which provides every possible correlation between all five variables, and (b) provide the confidence intervals for the most important correlations within the body of the article itself. Below is an example of how this might be done.

Table 6.1 provides descriptive statistics and intercorrelations for the study's five variables. All statistics are based on $N = 238$.

Analyses revealed a statistically significant correlation between exercise sessions and health motivation, $r = .29$, 95% CI [.17, .40], $p < .01$. According to the criteria provided by Cohen (1988), the size of this correlation fell just short of a medium effect. The sign of the correlation was positive, which was consistent with direction that was hypothesized by the four-factor model. The correlation between exercise sessions and desire to be thin was also statistically significant, $r = .15$, 95% CI [.02, .27], $p < .05$, and fell somewhere between a small effect and a medium effect. The sign of the correlation was positive, which was consistent with the sign that was hypothesized by the four-factor model. The correlation between exercise sessions and dislike of exercise was statistically significant, $r = -.37$, 95% CI [-.26, -.47], $p < .01$, and fell somewhere between a medium effect and a large effect. The sign of the correlation was negative, consistent with the sign that was hypothesized by the model. Finally, the correlation between exercise sessions and lack of time for exercise was also statistically significant, $r = -.46$, 95% CI [-.35, -.55], $p < .01$, and fell somewhere between a medium effect and a large effect. The sign of the correlation was negative, consistent with the sign that was hypothesized by the model.

Additional Issues Related to this Procedure

There is much that you, as a reader of research, need to know in order to evaluate research that investigates the relationship between a predictor variable and a criterion variable. This section covers two important topics: (a) factors that can cause correlations based on samples to be misleading estimates of the corresponding population correlations, and (b) correlation coefficients that serve as alternatives to the Pearson r.

Factors that Cause Obtained Coefficients To Be Misleading

Researchers typically hope that the correlation coefficients that they compute based on relatively small samples will be good estimates of the corresponding correlations in the populations. Unfortunately, a number of factors can cause sample coefficients to be misleadingly large or misleadingly small. This section briefly describes a few of these factors.

As an illustration: Imagine that Dr. O'Day wishes to understand the nature of the relationship between *participant age* (as measured in years), and healthy-diet behavior. In this case, *healthy-diet behavior* is the extent to which a person eats a nutritious diet—a diet that includes a lot of fruits and vegetables and limits quantities of saturated fats. She measures healthy-diet behavior using a multiple-item summated rating scale consisting of items 21-25 on the *Questionnaire on Eating and Exercising.*

Imagine that, in the population, the actual correlation between these two variables is equal to +.50 (indicating that older people tend to score higher on healthy-diet behavior). However, if Dr. O'Day measures these two variables in a sample of, say 200 participants, the correlation coefficient that she computes based on this sample might be misleadingly larger than the actual population correlation, or it might be misleadingly smaller. The following sections describe some of the circumstances that can lead to such undesirable outcomes.

RESTRICTION OF RANGE ON THE PREDICTOR AND/OR CRITERION. The *restriction of range problem* occurs when the full range of scores that are possible in the population is not displayed in the sample. For example, imagine that, in the population, scores on the predictor variable (participant age) could theoretically range from 0 years through 90 years. If Dr. O'Day had obtained a large sample in which the subject's ages had ranged from 0 through 90, she would have obtained a sample correlation coefficient pretty close to the actual population correlation of +.50.

However, if she instead conducted her study using only college students as subjects, the ages of the participants in her sample would have likely been restricted to the range of 18 through 23. This means that she now has a restriction of range on the predictor variable (age).

When there is a restriction of range on either the predictor variable or the criterion variable, it typically causes the sample correlation to be a misleadingly small estimate of the actual correlation in the population. In the current example, Dr. O'Day might have obtained a sample correlation of +.30, whereas the correlation in the population was actually +.50

POOR RELIABILITY OF MEASURES. *Reliability* refers to the stability or consistency in a set of scores. Reliability is usually estimated by computing some type of reliability coefficient. These coefficients usually range from .00 through 1.00, with higher values indicating better reliability. One popular reliability estimate is the coefficient alpha reliability estimate (symbol: α). Values of coefficient alpha that exceed α = .70 are generally viewed as being acceptable (as was explained in *Chapter 3*).

When the variables used in an investigation are not perfectly reliable, it typically causes the resulting correlation coefficients to be underestimates of the actual correlations between two constructs in the population. For example, imagine that if participant age and healthy-diet behavior were both measured with perfect reliability, the correlation between them would be +.50. However, in her investigation, Dr. O'Day measured healthy-diet behavior using a scale with a poor reliability estimate of just α = .60. As a result, when she computed the correlation between age and healthy-diet behavior in a sample of 200 participants, she obtained a sample correlation of just r = +.30. Unbeknownst to her, the sample correlation was misleadingly small because one of her variables was measured with poor reliability.

CURVILINEAR RELATIONSHIPS. When there is a *linear relationship* between two variables, it means that it is possible to fit a best-fitting *straight* line through the scatterplot. The counterpart to a linear relationship is a curvilinear relationship. With a *curvilinear relationship* (also called a *nonlinear relationship*), it is possible to draw a best-fitting *curved* line through the scatterplot. Figure 6.6

illustrates a fictitious curvilinear relationship between participant age and healthy-diet behavior.

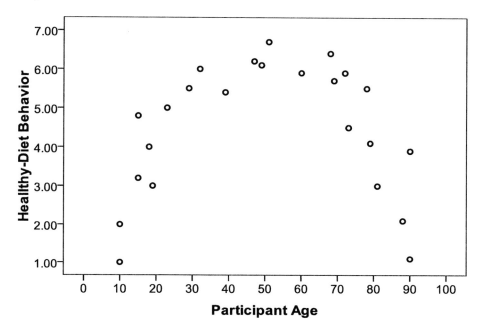

Figure 6.6. Scatterplot with fictitious data showing a curvilinear relationship between healthy-diet behavior and participant age.

Figure 6.6 reveals that there is actually a fairly strong relationship between participant age and healthy-diet behavior: If we know where an individual stands on the age variable, we can predict where that participant probably stands on the healthy-diet behavior variable with fairly good accuracy.

Nevertheless, if Dr. O'Day used the Pearson correlation coefficient to assess the strength of the relationship between the two variables depicted in Figure 6.6, it would result in a very small obtained statistic: $r = .07$. This is because one of the assumptions underlying the Pearson r is that there must be a linear relationship between the predictor and criterion. When we attempt to use the Pearson r to investigate curvilinear relationships, it typically underestimates the actual strength of the relationship.

When there is a curvilinear relationship between two variables, researchers may quantify the relationship by computing an alternative correlation coefficient: ***eta,*** the ***correlation ratio*** (symbol: η). As an illustration, if Dr. O'Day had computed eta to quantify the nature of the relationship between the two variables in Figure 6.6, she would have found $\eta = .96$ (pretty close to the upper limit of $\eta = 1.00$). In other words, eta correctly showed that there was an almost perfect relationship between the two variables, whereas the Pearson r incorrectly suggested that there was essentially no relationship. The moral of this story: Do not use the Pearson r when there is a curvilinear relationship between X and Y.

Other Types of Correlation Coefficients

This chapter has focused on just one type of correlation coefficient: The Pearson product-moment correlation. It is appropriate to compute this correlation when both the predictor variable and the criterion variable are continuous quantitative variables measured on an interval scale or ratio scale (and certain other assumptions are met). With the Pearson correlation, both variables are typically multi-value variables.

Other types of correlation coefficient are available for other types of data. The selection of the correct statistic for a given situation depends upon the nature of the variables being assessed, the purpose of the analysis, and other factors. A few of the more popular correlation coefficients are briefly described here.

POINT-BISERIAL CORRELATION COEFFICIENT. The ***point-biserial correlation coefficient*** (symbol: r_{pb}) is appropriate when one variable is a roughly continuous, multi-value variable assessed on an interval or ratio scale, and the other variable is a dichotomous variable (i.e., a variable that displays just two values). For example, imagine that Dr. O'Day created a new criterion variable: *sport-team status*. With this new variable, participants who are members of a university-related athletic team (such as the university basketball team) could be coded with a value of "1," and those who are not a member of any university-related athletic team could be coded with a "0" (coding a categorical variable with 1s and 0s in this way is called *dummy coding*). If Dr. O'Day then computed the Pearson correlation between this new sport-team status variable and a continuous variable such as health motivation, the resulting correlation coefficient would be called a *point-biserial correlation* (yes, she could use the same formula or application that would normally be used for the Pearson *r*, and the result would be called a point-biserial correlation; Howell, 2002).

BISERIAL CORRELATION COEFFICIENT. On the surface, the ***biserial correlation coefficient*** (symbol: r_b or r_{bis}) appears to be appropriate in exactly the same situation where the point-biserial correlation would be used: to investigate the relationship between one continuous variable and one dichotomous variable. There are, however, important conceptual differences between the two coefficients. The point-biserial correlation coefficient is appropriate when the dichotomous variable is a ***true dichotomy:*** a nominal-scale variable that consists of exactly two categories with no continuous underlying distribution. Examples of variables that are true dichotomies include subject sex (male versus female) and pregnancy status (pregnant versus not pregnant).

In contrast, the biserial correlation coefficient is appropriate when the dichotomous variable is not a true dichotomy; it is appropriate when the dichotomous variable represents an underlying normally-distributed continuous variable.

For example, imagine that the Marine Corps administers a physical fitness test to recruits prior to admitting them to basic training. Each recruit earns points on a number of sub-tests, adding up to 100 possible points. Recruits who earn 75 or more points "pass," and those who earn less than 75 points "fail." It is true that the "pass/fail" variable is a dichotomous variable, but it is not a *true* dichotomy—it is a dichotomy which represents an underlying continuous variable.

When researchers compute the biserial correlation, they estimate what the correlation would have been if the dichotomous variable had been continuous instead of dichotomous. This computation of a biserial correlation involves a number of semi-tedious steps, and is not for the faint of heart—see Field (2009a) for an example.

PHI COEFFICIENT. The ***phi coefficient*** (symbol: ϕ) is appropriate when both variables are true dichotomous variables. For example, if Dr. O'Day wanted to investigate the relationship between *subject sex* (male versus female) with *sport-team status* (team-member versus not a team-member), she might use the phi coefficient. As such, it is similar to the point-biserial correlation, except that both variables are dichotomous, instead of just one.

TETRACHORIC CORRELATION COEFFICIENT. For the sake of completeness (and not because you are actually likely to encounter one of these): the ***tetrachoric correlation coefficient*** (symbol: r_t) is appropriate when investigating the relationship between two dichotomous variables, both of which have continuous underlying distributions. This makes it similar to the phi coefficient, except that the phi coefficient is used with truly dichotomous variables (Howell, 2002).

SPEARMAN CORRELATION COEFFICIENT. The ***Spearman correlation coefficient for ranked data*** (also known as Spearman's rho, symbol: r_s) displays the correlation between two variables whose values have been ranked. The process of ranking a variable involves (a) ordering the values from the lowest to the highest; (b) assigning a rank of "1" to the lowest value, a rank of "2" to the next lowest value, and so forth; and (c) assigning average rank values to tied scores (scores that appear more than once). The formula for the Pearson *r* is then applied to the two ranked variables, and the result is the Spearman correlation coefficient.

For example, imagine that Dr. O'Day rank-orders each subject in a sample of 10 subjects according to how frequently they exercise, with "1" representing the subject who exercises the most, and "10" representing the subject who exercises the least. She then rank-orders the same 10 subjects with respect to their level of health motivation. Finally, she investigates the nature of the relationship between the two variables by computing the Spearman correlation.

Although the preceding was the "textbook" example of a data set appropriate for the Spearman correlation coefficient, there are actually three situations in which it is often computed in the real world: (a) when both variables are assessed on an ordinal-scale of measurement (as in the previous example), (b) when one variable is assessed on an interval scale or ratio scale, and the other variable is assessed on an ordinal scale, and (c) when both variables are assessed on an interval or ratio scale, and one or both variables display a marked departure from normality (such as a badly skewed distribution, thus making the Pearson correlation inappropriate).

ETA, THE CORRELATION RATIO. As was mentioned before, *eta* (symbol: η; also called the *correlation ratio*) can be used to quantify curvilinear relationship between two variables. Eta is computed by dividing the between-groups sum of squares (SS_{betw}) by the total sum of squares (SS_{tot}), and taking the square root of the result. Values of eta may range from 0.00 to 1.00, with higher values indicating a stronger relationship.

When eta is squared (i.e., $η^2$), it is interpreted in the same way that we interpret the coefficient of determination (r^2). That is, *eta squared* represents the percent of variance in the criterion variable that is accounted for by the predictor variable.

To compute eta, the criterion variable should be a quantitative variable assessed on an interval scale or ratio scale. But now we get to the part about why eta is so flexible:

- To compute eta, the predictor variable may be either a categorical variable or a continuous quantitative variable (Grissom & Kim, 2005), and
- If the predictor is quantitative, the relationship between the predictor and the criterion may be either linear or nonlinear (Warner, 2008).

The "nonlinear" part is what makes eta so useful. When there is a curvilinear relationship between two variables, the size of the Pearson *r* will typically underestimate the actual strength of the relationship. But eta will continue to give an accurate index of the relationship, whether the relationship is linear or nonlinear. In fact, eta cannot be any smaller than the Pearson *r* when both statistics are computed for the same sample of data (Yaremko et al., 1982).

Assumptions Underlying this Procedure

The assumptions underlying a correlation analysis is determined by (a) which specific coefficient being computed, and (b) what specific research questions are being addressed. To keep things simple, this section will summarize the assumptions that should be met when testing a Pearson *r* for statistical significance. Most of the assumptions mentioned below are described in greater detail in *Chapter 2: Basic Concepts in Statistics and Research*.

INTERVAL-SCALE OR RATIO-SCALE MEASUREMENT. Both variables should be quantitative variables. Both should be assessed on an interval scale or ratio scale.

NORMALLY-DISTRIBUTED SAMPLING DISTRIBUTION. The sampling distribution of the statistic should be normally distributed. This assumption is more likely to be met if the sample data are approximately normal, or if the sample is large.

RANDOM SAMPLE FROM BIVARIATE NORMAL DISTRIBUTION. The data points used to compute the Pearson *r* are the pairs of scores on the predictor and criterion variables. These data points should be a random sample drawn from a bivariate normal distribution.

LINEAR RELATIONSHIP. The relationship between the predictor variable and the criterion variable should be linear. The relationship between two variables is linear if the best-fitting line for their scatterplot is a straight line, as opposed to a curved line.

Chapter 7

Bivariate Regression

This Chapter's *Terra Incognita*

You are reading an article about an empirical investigation in which the researcher has obtained correlational data from 238 participants. She is using bivariate regression to predict scores on a criterion variable called *healthy-diet behavior* from a predictor variable called *healthy-diet intention*. The article reports the following:

> When healthy-diet behavior was regressed on healthy-diet intention, the resulting unstandardized regression coefficient was b = 0.691, 95% CI [0 .602, 0.780]. The value of this coefficient showed that, for every one-unit increase in healthy-diet intention, there was an increase of 0.691 units in healthy-diet behavior. The Pearson correlation for the two variables was r(236) = .707, p < .001. Using Cohen's (1988) criteria, this correlation represented a large effect size.
> The R^2 value for the regression model was R^2 = .499, indicating that healthy-diet intention accounted for about 50% of the variance in healthy-diet behavior. This value was significantly different from zero, F(1, 236) = 235.51, p < .001.

Be a good soldier, now. All shall be revealed.

The Basics of this Procedure

This chapter introduces basic concepts in regression and prediction. **Regression** is the process of estimating a best-fitting best line that summarizes the relationship between a predictor variable and a criterion variable. In a **regression analysis**, researchers fit a regression line to a sample of data, estimate the parameters of the regression equation (i.e., the constant and regression coefficient), and use the resulting equation to predict scores on a criterion variable. A major goal of most regression analysis involves assessing the fit between model and data. Regression analyses are typically part of correlational investigations (as opposed to experimental investigations).

The majority of this chapter deals with *linear bivariate regression*. The term **bivariate** means that the analyses discussed here include just two variables: a predictor variable (the *X* variable), and a criterion variable (the *Y* variable). The term **linear** refers to the fact that, when the *Y* scores are plotted against the *X* scores, it should be possible to fit a best-fitting *straight* line through the center of the scatterplot, as opposed to a best-fitting *curved* line. Figure 7.1 illustrates the nature of the variables included in a typical bivariate regression analysis.

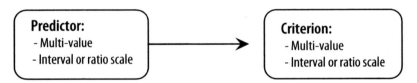

Figure 7.1. Generic analysis model: Nature of the variables included in the typical bivariate regression analysis.

Figure 7.1 shows that a bivariate regression analysis typically involves (a) a single continuous criterion variable measured on an interval scale or a ratio scale, and (b) a single continuous predictor variable, also assessed on an interval or ratio scale. Both variables are usually multi-value variables. Additional assumptions are listed at the end of this chapter.

Illustrative Investigation

Imagine that Dr. O'Day is building a program of research based on a conceptual model called the *Theory of Planned Behavior* (Azjen & Madden, 1986). This theory discusses the circumstances under which behavior can be predicted from beliefs, attitudes, and related constructs.

RESEARCH HYPOTHESIS. Among other things, the Theory of Planned Behavior says that *behavior* (the things that an individual actually does) can typically be predicted from *behavioral intentions* (the planned behavior that the individual intends to pursue). In her current program of research, the behavior that Dr. O'Day is interested in is *healthy-diet behavior*: the extent to which the individual actually engages in the behaviors of eating vegetables and whole grains. If the Theory of

Planned Behavior is valid, it should be possible to predict this variable from a separate variable: *healthy-diet intention*: the extent to which the subject *plans* to eat mostly vegetables and whole grains. The predicted relationship between these variables is illustrated in Figure 7.2.

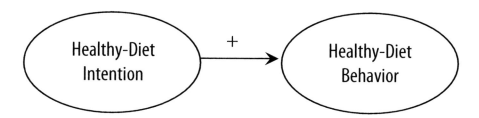

Figure 7.2. Predicted relationship between healthy-diet intention and healthy-diet behavior.

VARIABLES INVESTIGATED. Dr. O'Day measures the two variables by administering the *Questionnaire on Eating and Exercising* (which appears in *Appendix A* of this book) to a sample of over 200 college students. In her regression analyses, the predictor variable is *healthy-diet intention*, which consists of scores on a multiple-item summated rating scale created from items 16-19 on the *Questionnaire*. Scores on this variable can range from 1.00 to 7.00, with higher scores indicating a greater intention to eat vegetables, grains, and low-fat foods.

In these analyses, the criterion variable is *healthy-diet behavior*, which consists of scores on a multiple-item summated rating scale created from items 21-25 on the *Questionnaire*. Scores on this variable may also range from 1.00 to 7.00, with higher scores indicating that the subject actually consumed a healthier diet.

METHOD. Dr. O'Day administers the *Questionnaire on Eating and Exercising* to 260 students at a regional state university in the Midwest. She receives usable responses from 238 participants. She performs a bivariate regression analysis on the two variables described above.

Research Questions Addressed

Below are examples of the types of questions that can be answered in a bivariate regression analysis:

- Is there a relationship between the predictor variable (healthy-diet intention, in this case) and the criterion variable (healthy-diet behavior)? Is it a positive relationship, as is predicted by the Theory of Planned Behavior?

- How strong is the relationship between the two variables? Can scores on the predictor variable be used to predict scores on the criterion variable with much accuracy?

- When healthy-diet behavior is regressed on healthy-diet intention, what is the numeric value for *b* (the regression coefficient)? Is this regression coefficient significantly different from zero? Does the confidence interval indicate that this is a very precise estimate?

- If a subject has a low score (say 2.00) on healthy-diet intention, what is our best estimate for that subject's likely score on healthy-diet behavior? How about if a subject has a relatively high score (such as 6.00) on healthy-diet intention?

- Does this model display a good fit to the data?

Details, Details, Details

When you read research journals, you will encounter a bivariate regression analysis from time to time, but not often. Nevertheless, this chapter is important.

The concepts that you will learn here are the foundation concepts that are central to the advanced statistics that you *will* often encounter in research journals—advanced statistics to be covered later in this book. The prime example is *multiple regression*. Multiple regression is probably the most widely used advanced statistic for analyzing data from nonexperimental studies. And you will not understand multiple regression if you do not first understand the basics of bivariate regression: unstandardized regression coefficients, standardized regression coefficients, residuals of prediction, synthetic variables, and the principle of least squares.

Many of these concepts are easier to understand if they are illustrated visually in a scatterplot. The following section refreshes your memory regarding the basic features of a scatterplot.

The Scatterplot

In *Chapter 6*, you learned that a **scatterplot** (i.e., *scatter diagram*, or *scattergram*) is a graph that illustrates the nature of the relationship between two quantitative variables. In a scatterplot, the horizontal axis typically represents the predictor variable (i.e, the *X* variable), and the vertical axis typically represents the criterion variable (the *Y* variable). The dots appearing within the body of the figure show where each individual participant scored on the two variables simultaneously. Figure 7.3 presents a scatterplot that illustrates the results obtained by Dr. O'Day.

UNDERSTANDING THE SCATTERPLOT. Consider the dot that appears in the lower left corner of the figure. This point is labeled "210," because it represents Subject 210 (out of the 238 subjects). From the position of the dot, we can see that this subject:

- displayed a score of 1.00 on healthy-diet intention (we know this because the subject's dot is directly above the "1.00" on the horizontal axis), and also

- displayed a score of 1.20 on healthy-diet behavior (we know this because the subject's dot is directly to the right of "1.20" on the vertical axis).

The data points for the remaining participants in the figure may be interpreted in the same way.

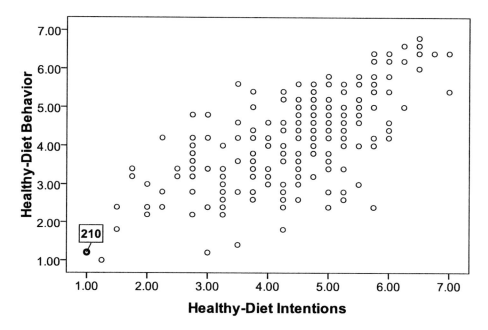

Figure 7.3. Scatterplot showing healthy-diet behavior as a function of healthy-diet intention ($N = 238$).

An earlier paragraph indicated that each dot represents an individual subject, but this is not necessarily true—some dots actually represent multiple subjects. This happens when more than one participant displays exactly the same scores on the predictor and criterion variables.

THE TREND DISPLAYED BY THIS SCATTERPLOT. Now step back and take in the "big picture" presented by the scatterplot in Figure 7.3. Notice that there appears to be a *positive relationship* between healthy-diet intention and healthy-diet behavior: Participants who have low scores on intention tend to have low scores on behavior, and participants who have high scores on intention tend to have high scores on behavior.

Dr. O'Day computes the Pearson correlation between healthy-diet intention and healthy-diet behavior, and finds $r = .71$. This value of r means that there is a fairly strong positive correlation between the two variables.

The Regression Line

The scatterplot is sort a big blob. Wouldn't it be better if we could draw a line through the center of it to give us a better sense of the general trend in the data? Yes it would, and we would call such a line a *regression line*.

A *linear regression line* is a straight line that summarizes the relationship between a predictor variable and a criterion variable A regression line is a *line of best fit* in the sense that, on the average, it passes through the center of the Y scores at each value of X.

Figure 7.4 presents the same graph that had been presented earlier. A best-fitting linear regression line has now been drawn through the center of the scatterplot.

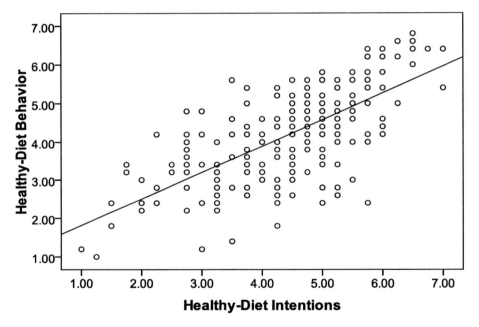

Figure 7.4. Scatterplot with best-fitting regression line.

The regression line in Figure 7.4 was drawn so as to satisfy the *principle of least squares*—a mathematical principle that ensures that the line comes as close as possible to passing through the center of the Y scores at each value of X. We will revisit this concept later (in the section titled *The Principle of Least Squares*), but for the moment, let's actually use the regression line for one of its most important purposes: prediction.

Predicting Y Scores from X Scores

Prediction is the process of using a specific subject's score on the X variable to estimate where that subject is likely to stand on the Y variable. This may be done using the *visual approach* (which is not terribly precise, but does a good job of showing what the concept of prediction is all about), or the *equation-based approach* (which results in more precise estimates).

THE VISUAL APPROACH TO PREDICTION. Figure 7.5 illustrates the scatterplot and regression line from the sample of data that Dr. O'Day collected. For the sake of making a point, let's pretend that this data set had been collected last year. This year, Dr. O'Day administers her healthy-diet intention scale to a new subject, and this subject displays a score of 3.00 on the variable. Given this score on the *X* variable, what is our best estimate for the score that the participant is likely to display on the *Y* variable (healthy-diet behavior?).

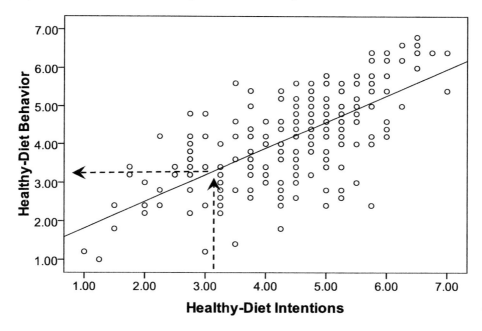

Figure 7.5. Using the visual approach to predict a participant's likely score on healthy-diet behavior, given a score of 3.00 on healthy-diet intention.

Remember that the new subject has a score of 3.00 on healthy-diet intention. To predict where the new subject will probably score on healthy-diet behavior, Dr. O'Day uses the data set and regression line created last year. She does the following:

- She locates this score of 3.00 on the *X* axis (the horizontal axis, which represents the *intention* variable).
- From the score of 3.00, she draws a vertical straight line up, and stops at the point where this vertical line touches the regression line (see the vertical dotted-line arrow in Figure 7.5).
- From that point on the regression line, she draws a horizontal line to the left, stopping at the *Y* axis (the *behavior* variable). This is illustrated by the horizontal dotted-line arrow.

- She reads the score on the *Y* axis where the horizontal dotted-line arrow touches it. This score represents her prediction of where the new subject will probably score on the *Y* variable. In this case, the predicted score is about 3.20.

In other words, if a new subject has a score of 3.00 on healthy-diet intention, Dr. O'Day's best estimate is that the subject will probably have a score of about 3.20 on healthy-diet behavior.

Using this approach, we can make the following predictions based on the following scores on X:

- If a subject had a score of 4.00 on *X* (intention), we would predict that the subject will probably have a score of about 3.90 on *Y* (behavior).

- If a subject had a score of 6.00 on *X* (intention), we would predict that the subject will probably have a score of about 5.30 on *Y* (behavior).

But all of this arrow-drawing seems kind of sloppy and unscientific. Wouldn't it be better if we could just plug the *X* scores into a mathematical formula and compute more exact values for the predicted *Y* scores?

THE EQUATION-BASED APPROACH TO PREDICTION. Much earlier in your education, you learned that a straight line appearing in two-dimensional space can be described in the form of a simple equation. And this holds true for a regression line in a scatterplot as well. Below is the general form for the bivariate regression equation:

$$Y' = a + b(X)$$

This equation consists of the following components:

Y′, the predicted score on the criterion variable. The symbol Y' (pronounced "*Y*-prime") represents the **predicted Y score:** our prediction of where a given subject is likely to stand on the criterion variable (healthy-diet behavior, in this case), given that subject's actual score on the predictor variable (healthy-diet intention, in this case).

a, the constant, or intercept. The symbol *a* represents the **constant,** or **intercept,** of the regression equation. It represents the predicted score on *Y* when the *X* variable displays a score of zero (it is called the "intercept" because it is the location on the *Y* axis where the regression line intersects it when *X* is equal to zero). This term is really not of much interest to researchers—it is simply a number that must be included in the equation so that the predicted *Y* scores will come out on the same metric that was used with the original *Y* scores.

b, the unstandardized regression coefficient. The symbol *b* represents the **unstandardized regression coefficient** for the predictor variable (it is also called the **nonstandardized regression coefficient,** the **raw-score regression coefficient**, the **regression weight**, and the **slope**). It represents the amount of change in *Y* that is associated with a one-unit change in *X* when both variables are in raw-score form (the concept of "raw-score form" will be explained later). Some textbooks call it the **regression weight**, because it indicates the amount of weight that should be given to the *X* variable in the prediction of the *Y* variable. Other textbooks refer to it as the ***slope***, because it represents the amount of slope (i.e., the amount of angle) displayed by the regression line. When *b* is a positive value (such as $b = +.80$), it means that the regression line displays a ***positive slope*** (i.e., there is a positive relationship between *X* and *Y,* as is illustrated in Figure 7.5). When *b* is a negative value (such as $b = -.80$), it means that the regression line displays a ***negative slope*** (this would be a regression line that slopes from the upper left corner down to the lower right corner of a scatterplot). Finally, when *b* is equal to zero, it means that the regression line displays ***zero slope***. This means that the line does not display any slope at all—it is a flat, horizontal line.

X, the observed score on the predictor variable. The ***observed X score*** is the score that a given participant actually displays on the predictor variable. We insert this actual *X* score into the regression equation, and the equation estimates the *Y* score this participant is most likely to display.

When a researcher performs a regression analysis, the statistical application (such as SAS or SPSS) computes values for *a* and *b*. These are sometimes called *parameter estimates*, because they are the researcher's best estimates for the likely values of the intercept and regression coefficient in the population.

For example, when Dr. O'Day analyzed the data illustrated in the scatterplot of Figure 7.5, the computer application arrived at the following regression equation:

$$Y' = a + b(X)$$

$$Y' = 1.124 + .691(X)$$

In other words, the application estimated that the constant of her equation should be $a = 1.124$, and the unstandardized regression coefficient should be $b = 0.691$.

WHAT DOES THE SIZE OF *b* REPRESENT? Earlier, the regression coefficient, *b*, was defined as the amount of change in *Y* that is associated with a one-unit change in X. What, exactly, is meant by this? Remember that in this study, *a healthy-diet intention (the X variable)* was measured using a scale that ranged from 1.00 to 7.00, and *healthy-diet behavior* (the *Y* variable) was also measured using a scale that ranged from 1.00 to 7.00. Since the regression coefficient was $b = 0.691$, we know that, if a subject increases his score in the intention variable by one unit (say, from 4.00 to 5.00), his score on the behavior variable would be expected to increase by 0.691 units (for example, it might be expected to increase from 4.00 to 4.691). The size of the increase corresponds to the amount of slope (angle) that we see in the regression line.

USING THE EQUATION TO PREDICT Y. It is time to see how the regression equation can actually be used to predict participants' scores on *Y*. Earlier, we used the visual approach to predict the most likely score on *Y* for a subject who had a score of

3.00 on X. By inserting this X score of 3.00 into her regression equation, Dr. O'Day can arrive at a more precise estimate of where the subject probably stands on Y:

$$Y' = a + b(X) = 1.124 + .691(X) = 1.124 + .691(3.00) = 1.124 + 2.073 = 3.197$$

So, if an individual has a score of 3.00 on healthy-diet intention, our best guess is that the subject will probably have a score of 3.197 on healthy-diet behavior. That is a bit more precise than the score of 3.20 that we arrived at earlier by using the visual approach.

What if a subject displays a relatively high score on X such as 6.00? Below we compute Y' for this individual:

$$Y' = a + b(X) = 1.124 + .691(X) = 1.124 + .691(6.00) = 1.124 + 4.146 = 5.270$$

So if an individual displays an actual score of 6.00 on healthy-diet intention, we predict that the individual will probably display a score of 5.270 on healthy-diet behavior. Again, that is a bit more precise than the predicted Y score of 5.30 that we estimated using the visual approach earlier.

The Y' Variable as a Synthetic Variable

Understanding bivariate regression (and therefore the more sophisticated procedures to follow such as multiple regression) requires understanding that Y' is a *variable*, and because it is a variable we can do things with it (like correlate it with other variables). This section introduces the concept of *synthetic variables* (which you will encounter often in this book) and uses Y' as a simple example.

COMPUTING Y' FOR EACH PARTICIPANT. You will recall that the regression equation from Dr. O'Day's healthy-diet study was:

$$Y' = a + b(X)$$

$$Y' = 1.124 + .691(X)$$

Imagine that, for each subject in the healthy-diet study, you inserted that subject's score on X (healthy-diet intention) into this formula, and computed the subject's predicted score on Y (i.e., predicted score on healthy-diet behavior, Y'). Table 7.1 shows the results obtained for a few of the 238 subjects in the study.

In Table 7.1, participants' observed scores on X (intention) appear under the heading X, and their predicted scores on Y (behavior) appear under the heading Y'. In the table, you can see that Subject 1 had a score of 3.00 on X. Inserting this score in the equation resulted in a predicted score on Y of Y' = 3.197 (the steps followed in this computation were presented earlier). This means that, since Subject 1 had a score of 3.00 on intention, our best guess is that he would have a score of 3.197 on behavior. The X scores and Y' scores for the remaining participants can be interpreted in the same way.

Table 7.1

Illustration of X Scores, Y Scores, Y' Scores, and Residual Scores for Participants in the Healthy-Diet Study

Participant	X	Y'	Y	Y – Y'
Subject 1	3.00	3.197	4.00	0.803
Subject 2	4.00	3.888	5.00	1.112
Subject 3	5.00	4.579	4.20	-0.379
.	.	.	.	.
.	.	.	.	.
.	.	.	.	.
Subject 238	6.00	5.270	5.00	-0.270

Note. $N = 238$. X = each participant's observed score on X (healthy-diet intention). Y' = each participant's predicted score on Y (healthy-diet behavior). Y = each subject's actual score on Y (healthy-diet behavior). $Y - Y'$ = the residuals of prediction.

THINKING OF Y' AS A SYNTHETIC VARIABLE. Y' is an example of a *synthetic variable*. Understanding this concept requires that we make a distinction between real-world variables versus synthetic variables. A ***real-world variable*** is a raw variable obtained directly from participants, prior to being transformed in any way. In contrast, a ***synthetic variable*** is an artificial variable, typically created as an optimally-weighted linear combination of one or more real-world variables. In Dr. O'Day's study, Y' was created by applying an optimal weight (b) to just one real-world variable: healthy-diet intention. It was accomplished with this relatively simple formula:

$$Y' = a + b(X)$$
$$Y' = 1.124 + .691(X)$$

The important point is this: Y' is a *variable*, and each subject in the study has a score on it. Do not be put off by the fact that it is a synthetic variable and therefore seems sort of artificial. Artificial or not, we can treat Y' as we would treat any other variable in a data set: correlate it with other variables, compute its mean, or perform any operation we like. And one of the most important operations that we will perform involves using Y' to compute the residuals of prediction.

Residuals of Prediction

A ***residual of prediction*** goes by a number of different names in different textbooks: *residual, error or prediction, deviation,* and *deviance*. A residual of prediction is a difference score; it represents the difference between Y (a subject's actual score on the criterion variable, Y) minus Y' (the subject's predicted score on Y). It is computed in this way:

$$Residual = Y - Y'$$

COMPUTING THE RESIDUAL FOR AN INDIVIDUAL PARTICIPANT. For example, consider Subject 1 in Table 7.1. For this subject, the predicted score on Y (in the column headed Y') was 3.197, whereas her actual score on Y was 4.00. The residual of prediction for this subject was therefore computed as:

$$Residual = Y - Y' = 4.00 - 3.197 = 0.803$$

This means that, for Subject 1, the Y' score was an underprediction. The regression equation produced a predicted score on healthy-diet behavior that was .803 points lower than her actual score. This residual score appears in Table 7.1 in the column headed $Y - Y'$.

COMPUTING RESIDUALS FOR EACH PARTICIPANT IN THE STUDY. In the same way, it is possible to create a residual-of-prediction score for each of the 238 participants in the study. For each participant, we subtract the participant's predicted score on Y from the participant's actual score on Y. This produces a single number that estimates the extent to which the regression equation made an error in predicting Y for that participant.

Table 7.1 presents residuals of prediction for four of the 238 subjects in the healthy-diet study. These scores appear in the column headed $Y - Y'$.

THE STANDARD ERROR OF ESTIMATE. If each subject has a score in the $Y - Y'$ column, it means that this "residuals of prediction" variable is yet another synthetic variable. And because it is a variable, this means that we can treat it as we would treat any other variable.

For example, if we computed the standard deviation of the residual scores that appear in the $Y - Y'$ column, we would have an estimate of how much variability is displayed by these errors of prediction. This statistic is called the ***standard error of the estimate*** (symbol: SE_{est} or $S_{Y-Y'}$ or $s_{Y \cdot X}$ or $\sigma_{Y \cdot X}$). The standard error of the estimate is the standard deviation of the actual Y scores around the predicted Y scores.

The standard error of the estimate tells us the accuracy of our predictions. In general, when the standard error of the estimate is a relatively small value, it means that the predictions are fairly accurate, and when it is a relatively large value, it means that the predictions are less accurate. It works like this:

- Predictions tend to be more accurate when the correlation between the two variables is a larger value (that is, when *r* is closer to ±1.00). In this situation, the standard error of the estimate tends to be a smaller value, because the actual *Y* values are closer to the regression line (the predicted *Y* values).

- In contrast, predictions tend to be less accurate when the correlation between the two variables is a smaller value (that is, when *r* is closer to 0.00). In this situation, the standard error of the estimate tends to be a larger value, because the actual *Y* values are farther away from the regression line (the predicted *Y* values).

The Principle of Least Squares

Earlier, we saw that, when Dr. O'Day performed the regression analysis, the computer application indicated that her regression equation should include the following parameter estimates:

$$Y' = a + b(X)$$

$$Y' = 1.124 + .691(X)$$

In other words, the application indicated that the constant should be $a = 1.124$ and the regression coefficient should be $b = .691$. But why, exactly, must they be these values? Why can't the regression coefficient instead be $b = .670$? Or perhaps $b = .712$?

THE LEAST-SQUARES CRITERION. The answer is that the computer application used a relatively simple algebraic formula to arrive at *optimal* estimates for *a* and *b*: Estimates which best satisfy the principle of least squares. The **principle of least squares** says that the "optimal" values for *a* and *b* are the values which minimize the sum of the squared residuals of prediction. This approach called the *least-squares criterion* or the *ordinary least-squares solution*. It is important because it is at the heart of many advanced statistical procedures such as multiple regression.

EXAMPLE. To see how it works, let's revisit the table of residuals presented earlier. Table 7.2 presents scores on *X*, *Y'*, and *Y* for four of the 238 participants in Dr. O'Day's study.

Most of the columns in the table are identical to the columns that appeared in Table 7.1, earlier. The only new column is the one headed $(Y - Y')^2$. This column contains the squared residual of prediction for each participant. We computed the participants' scores in this column by subtracting their predicted *Y* score from their actual *Y* score, and then squaring the resulting difference.

Table 7.2

Illustration of X Scores, Y Scores, Y' Scores, and Residual Scores for Participants in the Healthy-Diet Study

Participant	X	Y'	Y	Y – Y'	(Y – Y')²
Subject 1	3.00	3.197	4.00	0.803	0.645
Subject 2	4.00	3.888	5.00	1.112	1.237
Subject 3	5.00	4.579	4.20	-0.379	0.144
.	.	.	.	.	.
.	.	.	.	.	.
.	.	.	.	.	.
Subject 238	6.00	5.270	5.00	-0.270	0.073

Note. $N = 238$. X = each participant's observed score on X (healthy-diet intention). Y' = each participant's predicted score on Y (healthy-diet behavior). Y = each subject's actual score on Y (healthy-diet behavior). $Y - Y'$ = the residuals of prediction.

When a statistical application is based on a least-squares criterion, it minimizes the values that appear in the column headed $(Y - Y')^2$. Or, more specifically, it minimizes the *sum* of those values:

$$\Sigma(Y - Y')^2$$

When the statistical application estimates the regression coefficient as $b = .691$ and the constant as $a = 1.124$, it is telling us that these are the optimal values for these terms if we want to ensure that the quantity "$\Sigma(Y - Y')^2$" is as small as possible. In other words, it is telling us that these are the parameter estimates which best satisfy the principle of least squares.

Evaluating the Fit between Model and Data

With the majority of the empirical studies that you will read about in journals, a central question that the researcher is trying to answer is this:

> Is there a good fit between my model and the data?

The exact meaning of "fit between model and data" will vary from study to study, but concept typically involves questions such as:

- Does my theoretical model do a good job of reproducing the data that I actually obtained?
- Is there a good relationship between how the data *should* appear (according to my theoretical model), and how they *actually* appear?
- If I know where someone stands on the predictor variable(s) in my model, can I predict where he or she stands on the criterion variable(s) with much accuracy?

The specific statistic that researchers consult in order to assess the fit between model and data depends on which specific data-analysis procedure was used in that analysis. With linear regression, the fit between model and data is often assessed by reviewing the *coefficient of determination*, r^2.

The **coefficient of determination** indicates the percent of variance in the criterion variable that is accounted for by the predictor variable. In a study that includes just one criterion variable and one predictor variable, it is computed by squaring the Pearson correlation between the two variables: r^2. In the current study, the Pearson correlation between healthy-diet intention and healthy diet behavior was $r = .707$. The coefficient of determination was therefore computed by squaring this correlation coefficient:

$$r^2 = (.707)^2 = .499$$

This value of .499 rounds to .50. This means that about 50% of the variance in healthy-diet behavior is accounted for by the scores on healthy-diet intention.

When there is no fit at all between model and data, the coefficient of determination is $r^2 = .00$, which means that 0% of the variance in Y is accounted for by X. This tells us that knowing where someone stands on X is of no help at all when attempting to predict where they stand on Y. On the other hand, when there is a perfect fit between model and data, the coefficient of determination is $r^2 = 1.00$, which means that 100% of the variance in Y is accounted for by X. This is just another way of saying that, if we know where someone stands on X, we can predict where they stand on Y with no errors of prediction at all.

Based on the preceding discussion, it should come as no surprise to learn that the coefficient of determination belongs to a family of statistics called *variance accounted for* indices. Other members of this family include R^2 (from multiple regression) and eta-squared, η^2 (from ANOVA).

When you see r^2 (or any other member of this family), you should think *percent improvement in predictive accuracy*. To understand what this means, imagine that Dr. O'Day must predict where subjects stand on the criterion variable, and she must make these predictions under two different sets of circumstances: Situation 1 and Situation 2.

SITUATION 1. In *Situation 1*, Dr. O'Day must predict—for each individual participant—the participant's score on the criterion variable: healthy-diet behavior (*Y*). Prior to making these predictions, Dr. O'Day is told that the mean score on this criterion variable is $\bar{Y} = 4.22$.

If she knows nothing else about the data set, Dr. O'Day's best strategy is to use the mean score on the criterion variable as her prediction for each individual participant. That means that for Subject 1 she should predict 4.22, for Subject 2 she should predict 4.22, and so forth, for every participant.

Situation 1 is illustrated in Figure 7.6.

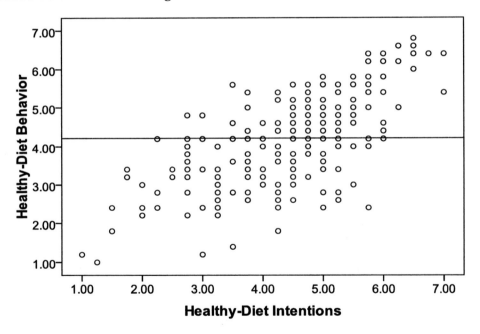

Figure 7.6. Scatterplot with horizontal line representing the mean score on Y (healthy-diet behavior).

Figure 7.6 contains a scatterplot for healthy-diet intention and healthy-diet behavior. The horizontal line running through the middle of the scatterplot represents the mean score on the criterion variable, healthy-diet behavior ($\bar{Y} = 4.22$). Therefore, the horizontal line also represents the prediction that Dr. O'Day made for each of her 238 subjects in *Situation 1*.

If she used the mean as her prediction for each subject, she would make a lot of errors of prediction, right? The size of the errors of prediction can be seen by inspecting Figure 7.6. If the dot for a given subject lies fairly close to the horizontal line in the figure (which represents the mean on Y), then her prediction would be fairly accurate for that subject. However, if the dot for a given subject lies far away from that horizontal line, then her prediction would be way off. The scatterplot shows that a lot of the dots are far away from the horizontal line. This tells you that, if Dr. O'Day used the mean of Y as her prediction, the typical error of prediction will be fairly large.

SITUATION 2. In *Situation 2*, Dr. O'Day must once again predict scores on Y for each participant. But in this second situation, she is given more useful information:

- She is told that there is a relatively strong relationship between healthy-diet intention scores (X) and healthy-diet behavior scores (Y),
- She is given the regression equation that describes that relationship (i.e., she is given the regression coefficient, b, and the constant, a), and
- She is given the X score for each individual participant.

For each participant, Dr. O'Day inserts that subject's X score into the formula and computes the corresponding Y' score. In Situation 2, Dr. O'Day will no longer use the mean Y score as her prediction for where each participant will stand on the criterion variable. Instead, she will use the Y' score that she computed for each individual subject (the Y' scores that she computed with the regression equation). These Y' scores fall directly on the regression line in Figure 7.7.

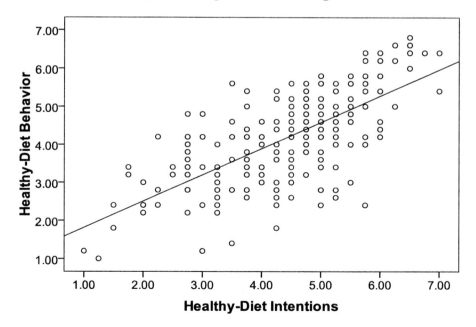

Figure 7.7. Scatterplot with sloping regression line.

You can see that the predictions that she makes in Situation 2 will be more accurate, compared to the predictions that she had made in Situation 1. In Situation 2, her predictions are represented by the sloping regression line that runs through the middle of the scatterplot. Notice that this sloping regression line comes a bit closer to the actual *Y* scores (i.e., the "dots") in the scatterplot. This tells us:

- Dr. O'Day's errors of prediction will be smaller in Situation 2 than in Situation 1.
- Therefore, the accuracy of her predictions will be better in Situation 2 than in Situation 1.

How much more accurate will her predictions be in Situation 2, compared to Situation 1? They will be 50% more accurate. We know this because the coefficient of determination for this investigation was computed as:

$$r^2 = (.707)^2 = .499 = .50$$

In summary, the coefficient of determination indicates the percent improvement in predictive accuracy that is achieved by capitalizing on the relationship between *X* and *Y* (as was done in Situation 2) as opposed to just ignoring the relationship between *X* and *Y* (as was done in Situation 1).

SUMMARY. The fit between model and data is going to be a recurring issue for most of the statistical procedures covered in this book. Researchers inspect different statistics to assess fit, depending on which data analysis procedure was used. Bivariate regression is one of the simplest procedures, so it provides a good point of departure for understanding basic concepts related to "fit."

Summarizing Results in a Published Article

We have covered the basic concepts, so we are now ready to see how the results from Dr. O'Day's analysis might be presented in tables and text within a research article. Here, you will learn about the various pieces of information that are presented as part of a typical bivariate regression analysis.

Summarizing Results in a Table

Remember that in her study, the criterion variable was healthy-diet behavior and the predictor variable was healthy-diet intention. Table 7.3 shows how Dr. O'Day might summarize her results in a table.

The truth is that a researcher would be more likely to report the results of a bivariate regression analysis within the text of the article, rather than in a table such as Table 7.3. However, researchers *are* likely to present results in a table when they use a more sophisticated procedure such as multiple regression, so learning the basics at this point will make it easier to understand such tables when we cover multiple regression in *Chapters 9* and *10*.

Table 7.3

Results from the Bivariate Regression of Healthy-Diet Behavior on Healthy-Diet Intention

Term in the equation	Unstandardized parameter estimates				Standardized parameter estimates (β)
	b	SE	95% CI	p	
(Constant: *a*)	1.124	0.209	[0.713, 1.535]	<.001	
Healthy-diet intention	0.691	0.045	[0.602, 0.780]	<.001	.707

Note. N = 238. Model R = .707, Model R^2 = .499, $F(1, 236)$ = 235.513, $p < .001$. b = unstandardized regression coefficient (*b* weight). SE = standard error. 95% CI = 95% confidence interval. p = probability value. β = standardized regression coefficient (beta weight).

SECTIONS OF THE TABLE. Table 7.3 includes the following sections and columns:

- *Terms in the equation.* Below this heading, you will find the names of the terms that appear in the regression equation. In the current analysis, there was only one predictor variable, and this means that the only two terms that will appear in the table will be the constant (or intercept, symbol: a) and the single predictor variable (which in this case is healthy-diet intention).

- *Unstandardized parameter estimates.* This section reports the various statistics that would be computed if the variables were in raw-score form. We make a distinction between raw variables versus standardized variables. **Raw variables** are the untransformed variables as they had been originally measured. In contrast, **standardized variables** are the same variables after they have been converted to z scores (that is, after they have been standardized to have a mean of zero and a standard deviation of one). Later (in the chapters dealing with multiple regression), we shall see that the results based on the raw variables are useful for some purposes, and the results based on the standardized variables are useful for other purposes.

- *The column headed "b."* Under the heading "*b*," you will find the unstandardized parameter estimates for the terms in the equation. In this context, a **parameter estimate** is the same thing as an **inferential statistic:** our best estimate of the likely value in the population, based on this sample of data.

Where the column headed "*b*" intersects with the row headed "Healthy-diet intention," you can see that the unstandardized regression coefficient for this predictor variable is $b = 0.691$ (this is the same regression coefficient presented earlier in the chapter). In other words, our best estimate for the regression coefficient in the population is $b = 0.691$. You will recall that an ***unstandardized regression coefficient*** indicates the amount of change in *Y* that is associated with a one-unit change in *X* while the variables are in raw-score form.

Where the column headed "*b*" intersects with the row headed "Constant: *a*," you can see that the unstandardized constant for the regression equation is $a = 1.124$ (this is the same constant presented earlier in the chapter).

Right now some of you are saying "Wait a minute! The symbol for the constant is *a*! You can't put the obtained constant in a column headed *b*!" But yes I can.

Some textbooks and research articles represent the constant using the symbol "b_0" (rather than the symbol "*a*"). They then use symbols such as b_1" to represent the regression coefficient for the predictor variable. So do not be surprised when you see a constant reported in a column headed "*b*."

And while we are talking about the constant, one more caveat: The constant is usually not of much interest to the typical reader of a research article anyway. Even though the line for the constant contains information such as a confidence interval and a *p* value, no one pays much attention to this information when it is for the constant. Readers *do*, however, pay attention to this information when it pertains to the predictor variable (healthy-diet intention, in this case), so most of this section will discuss results pertaining to the predictor variable, not the constant.

- *The column headed "SE."* We have already established that the unstandardized regression coefficient for healthy-diet intention was $b = 0.691$ (this was reported under the heading "*b*"). To move on: Under the heading "*SE*," you will find the standard error for this unstandardized regression coefficient. You will recall from *Chapter 5* that a ***standard error*** is the standard deviation for the sampling distribution of a statistic. In this case, the statistic being estimated is the regression coefficient *(b)* for healthy-diet intention. Therefore, the value reported in the *SE* column ($SE = 0.045$) must be the standard deviation of the sampling distribution for this *b* coefficient. Researchers use these standard errors to test the *b* coefficients for statistical significance, to compute confidence intervals around the *b* coefficient, and for other purposes. These standard errors are interpreted in a way similar to the way that we interpret confidence intervals: In general, if the standard error is a relatively small value, we can have more confidence that the sample regression coefficient is pretty close to the actual regression coefficient in the population.

- *The column headed "95% CI."* Under the heading "95% CI," you will find the 95% confidence intervals for the unstandardized regression coefficient. You will recall that a **confidence interval** is an interval that contains plausible values for some population parameter being estimated. In this case, the population parameter being estimated is the regression coefficient. The column headed *"b"* provides our single "best estimate" for this parameter (i.e,. the point estimate), and the column headed "95% CI" provides a range of plausible values for this parameter (i.e., the interval estimate). When the CI is relatively narrow, we can have more confidence that the sample estimate should be fairly close to the actual population parameter. Table 7.3 shows that the 95% confidence interval for the current regression coefficient extends from 0.602 to 0.780, which most researchers would view as being fairly narrow. When the CI does not contain zero, this means that the regression coefficient is significantly different from zero. The current confidence interval [0.602, 0.780] does not contain zero, so we know that the regression coefficient for healthy-diet intention is statistically significant.

- *The column headed "p."* Under the heading "p," you will find the probability value for the unstandardized regression coefficient. As you recall from *Chapter 5*, a **probability value** is the probability that we would have obtained the current sample results if the null hypothesis were true. In a regression analysis, the null hypothesis typically states that the regression coefficient is equal to zero in the population. In a specific study, if the p value for a given predictor is less than .05, it means that we can reject this null hypothesis, and conclude that the regression coefficient for that predictor is significantly different from zero. Table 7.3 shows that the probability value for the current regression coefficient is "$p < .001$," which is less than the standard criterion of $\alpha = .05$. This means that Dr. O'Day can tell the world that the regression coefficient for healthy-diet intention is statistically significant (but we already knew this when we reviewed the confidence interval earlier, didn't we?).

- *The column headed "Standardized parameter estimates (β)."* Under the heading "β," you will find the **standardized regression coefficient** (also called the **beta weight**) for the predictor variable. This coefficient indicates the amount of change in Y that is associated with a one-unit change in X while both variables are in standard-score (z-score) form. When a regression analysis includes just one predictor variable (as it does in this case), the beta weight is identical to the Pearson correlation coefficient (r) for the two variables. Table 7.3 shows that the standardized regression coefficient is $\beta = .707$. This should come as no surprise, as we learned earlier that the Pearson r for these variables was $r = .707$.

You may have noticed that there is no separate p value for the standardized β coefficient. This is because it is not necessary: A p value has already been reported for the unstandardized b coefficient. The p value for the standardized β coefficient would be identical to the p value for the unstandardized b coefficient.

INTERPRETING RESULTS RELATED TO MODEL FIT. Notes at the bottom of Table 7.3 provide information regarding the overall fit between model and data. This is a common convention, particularly when the analysis involves bivariate or multiple regression. The most important information from the note at the bottom of Table 7.3 is reproduced below:

Note. $N = 238$. Model $R = .707$, Model $R^2 = .499$, $F(1, 236) = 235.513$, $p < .001$.

The following sections discuss how this information should be interpreted:

- *The model R statistic.* The note includes the entry: "Model $R = .707$." When performing bivariate regression, the model R statistic is essentially equal to the Pearson correlation (r) between the predictor and criterion variable (except that the model R may only be a positive number, whereas the Pearson correlation may be either positive or negative). In the present case, the model R is .707, which is consistent with the earlier statement that the Pearson correlation between intention and behavior was $r = .707$.

- *The model R^2 statistic.* The note also includes the entry: "Model $R^2 = .499$." The model R^2 indicates the percent of variance in the criterion variable that is accounted for by the predictor variable. When performing bivariate regression, this statistic is equivalent to the coefficient of determination (r^2) that is obtained by squaring the Pearson correlation coefficient. Values for R^2 may range from 0.00 (indicating that the X variable was a poor predictor of the Y variable) to 1.00 (indicating that the X variable predicted the Y variable with perfect accuracy).

- *The F test for the R statistic.* The note also includes the following F statistic:

    $F(1, 236) = 235.513$, $p < .001$.

    The F statistic is a null-hypothesis significance test that is widely used with the family of statistical procedures called *analysis of variance* (abbreviation: ANOVA*).* As we shall see, the F statistic can be used with regression as well as ANOVA. In this case, the F statistic tests the null hypothesis that $R = 0.00$ in the population (it therefore also serves as a test of the null hypothesis that $R^2 = 0.00$ in the population). The note at the bottom of the table indicates that the p value for this F test is $p < .001$. This means that Dr. O'Day may reject the null hypothesis, and conclude that the obtained model R of .707 is significantly different from zero. When a regression equation contains just one predictor variable (as it does here), the F test for the model R statistic is equivalent to the significance test for the regression coefficient, b, which was discussed earlier.

The numbers that appear within parentheses next to the F symbol (1, 236) are the degrees of freedom for the test. With this F test, the first value represents the degrees of freedom for the numerator. This value is equal to k, the number of predictor variables in the equation ($k = 1$ for the present analysis). The second number within parentheses represents the degrees of freedom for the denominator (the error term). This value is computed as $N - k - 1$, where N = the total number of subjects in the study. For the current analysis, the degrees of freedom for the denominator are computed as $238 - 1 - 1 = 236$.

Summarizing Results in the Text of an Article

An earlier section indicated that researchers are more likely to report bivariate regression results in the text of an article, rather than in a table. This section shows how this might be done when reporting a significant relationship, as well as when reporting a nonsignificant relationship.

DESCRIBING A SIGNIFICANT RELATIONSHIP. You probably recall that the relationship between healthy-diet intention and healthy-diet behavior was statistically significant and substantial in size. Dr. O'Day might have used the following approach to describe those results in a journal article:

> When healthy-diet behavior was regressed on healthy-diet intention, the resulting unstandardized regression coefficient was $b = 0.691$, 95% CI [0.602, 0.780]. The value of this coefficient showed that, for every one-unit increase in healthy-diet intention, there was an increase of 0.691 units in healthy-diet behavior. The Pearson correlation for the two variables was $r(236) = .707$, $p < .001$. Using Cohen's (1988) criteria, this correlation represented a large effect size.
>
> The R^2 value for the regression model was .499, indicating that healthy-diet intention accounted for about 50% of the variance in healthy-diet behavior. This value was significantly different from zero, $F(1, 236) = 235.51$, $p < .001$.

The preceding summary reported the Pearson correlation as "$r(236) = .707$." In *Chapter 6*, you learned that the value within parentheses (the "236") represents the degrees of freedom for the correlation coefficient, which are computed as $N - 2$ (i.e., $238 - 2 = 236$, in this case).

You may have noticed that the preceding summary did not provide a p value that indicated specifically whether the b coefficient was significantly different from zero. Such a p value was not really necessary for two reasons:

- First, the 95% confidence interval reported in the first paragraph did not include a value of zero, and this tells us that the b coefficient was significantly different from zero with alpha set at $\alpha = .05$.

- Second, the F statistic in the second paragraph indicated that the R^2 statistic for the model was significantly different from zero ($p < .001$). When a regression equation includes just one predictor variable (as is the case here), this F test is equivalent to a significance test for the regression coefficient.

DESCRIBING A NONSIGNIFICANT RELATIONSHIP. When there is a near-zero relationship between a predictor and a criterion, it means that there is essentially no association between the two variables. Neither high values nor low values on the predictor are consistently associated with either high or low values on the criterion.

It is important to emphasize that, when you wish to determine whether there is a "near-zero relationship" between the two variables, you should *not* review the unstandardized *b* coefficient, but should instead review either the Pearson *r* correlation coefficient or the standardized regression coefficient, β. You should not review the unstandardized *b* coefficient (the "raw-score" *b* coefficient) because it is possible that this coefficient may seem to be close to zero (such as *b* = .04) and still be a statistically significant and meaningful predictor. This is because the size of the unstandardized *b* coefficient is sort of arbitrary and is influenced by factors other than the strength of the relationship. In particular, the size of *b* is heavily influenced by the metric used with the raw variables: If the raw variables include very large numbers (such as 200, 500, and so forth), the unstandardized *b* coefficients may be near-zero and still be significant and meaningful. In short, when deciding whether a coefficient is near-zero, you should review the Pearson *r* or the standardized β coefficient, but not the unstandardized *b* coefficient.

To illustrate a near-zero regression coefficient, imagine that one of the variables that Dr. O'Day has measured in her study was *lack of time for food preparation*. This variable consisted of scores from a multiple-item summated rating scale assessed by items 43-45 on the *Questionnaire on Eating and Exercising* (in *Appendix A*). Scores on this variable may range from 1.00 to 7.00. When subjects score high on this variable, it means that they believe that they are too busy to cook and prepare meals that require a lot of time.

Logically, it makes sense that this variable should display a pretty strong *negative* relationship with healthy-diet behavior. After all, how can someone eat healthy food if they do not have time for all the chopping and dicing that healthy-foods require? This reasoning led Dr. O'Day to state the following research hypothesis:

> There is a negative linear relationship between lack of time for food preparation and healthy-diet behavior.

Dr. O'Day performs a regression analysis in which the criterion variable is healthy-diet behavior, and the predictor variable is lack of time for food preparation. Figure 7.8 presents the scatterplot for the observed relationship between these two variables.

Dr. O'Day predicted that there would be a negative relationship between lack of time for food preparation and healthy-diet behavior. The scatterplot in Figure 7.8 shows that there is a slight negative trend to the data, although the data-analysis results (to be presented below) will show that the relationship is not statistically significant, and does not represent a large effect size.

Bivariate Regression | 219

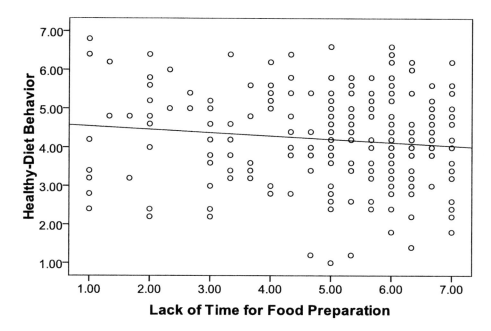

Figure 7.8. Scatterplot illustrating scores on healthy-diet behavior as a function of lack of time for food preparation (a near-zero relationship).

A best-fitting regression line has been drawn through the scatterplot in Figure 7.8. The slope and intercept of this regression line is described by the following equation:

$Y' = a + b(X)$

$Y' = 4.630 - .084(X)$

In the text of a journal article, Dr. O'Day might summarize the results in this way:

> When healthy-diet behavior was regressed on lack of time for food preparation, the resulting unstandardized regression coefficient was $b = -0.084$, 95% CI [−0.175, 0.006]. The value of this coefficient showed that, for every one-unit increase in lack of time for food preparation, there was a decrease of 0.084 units in healthy-diet behavior. The Pearson correlation for the two variables was $r(236) = -.119$, $p = .067$. Using Cohen's (1988) criteria, this correlation represented a small effect size.
>
> The R^2 value for the regression model was .014, indicating that lack of time for food preparation accounted for about 1% of the variance in healthy-diet behavior. This value was not significantly different from zero, $F(1, 236) = 3.38$, $p = .067$.

Let's consider the fit between model and data observed in this most recent analysis. In her summary, Dr. O'Day indicated that the R^2 value for the regression model was .014, which we can round to .01. Remember that the value of R^2 represents the percent improvement in predictive accuracy. Dr. O'Day found that

R^2 is equal to .01 (or about 1%), in this analysis. What exactly, does it mean when we say that R^2 is equal to .01? It means the following:

- If Dr. O'Day had used the sample mean on healthy-diet behavior ($\bar{Y}$) as her prediction as to where each subject probably falls on this variable, her predictions would not be terribly accurate, in general.
- Alternatively, she could instead compute the regression line that summarizes the relationship between healthy-diet behavior and lack of time for food preparation, and use points on this regression line to predict where subjects probably fall on the criterion variable.
- However, if she used this second "regression line" approach, her predictions would be only 1% more accurate than they would have been if she had instead used the mean on the criterion variable for her predictions.

The fact that her predictions would be only 1% more accurate by using the regression-line approach tells us that there is a relatively poor fit between model and data. In other words, the variable *lack of time for food preparation* is a poor predictor of healthy-diet behavior.

Assumptions Underlying Linear Bivariate Regression

The assumptions listed below (taken largely from Keith, 2006) are the assumptions that should be met if bivariate regression is used for purposes of prediction. Some of the following assumptions are discussed in greater detail in *Chapter 2* of this book.

Linearity
The criterion variable should be a linear function of the predictor variable. In other words, it should be possible to fit a best-fitting *straight* line through the scatterplot (as opposed to a best-fitting *curved* line).

Independence
Each observation included in the sample should be drawn independently from the population of interest. Among other things, this means that the researcher should not have taken repeated measures on the same variable from the same participant.

Homogeneity of Variance (Homoscedasticity)
The variance of the Y scores around the regression line should remain fairly constant at all values of X. For example, choose a low value of X and notice how spread-out the Y scores are directly above this X value. Now choose a high value of X and notice how spread-out the Y scores are directly above this X value. Is the dispersion of the Y scores pretty much the same at both of these two values of X? Is the dispersion of the Y scores pretty much the same at *each possible value* of X?

If yes, it means that the dataset displays ***homoscedasticity*** (which is a good thing). If not, it has violated this assumption, and instead displays ***heteroscedasticity*** (which is a bad thing).

Normality

The residuals of prediction should be normally distributed. Violation of this assumption appears to be serious only with relatively small samples, as regression is relatively robust against violations of this assumption (Kline, 1998).

Bivariate Normality

If the researcher wishes to test the statistical significance of the Pearson r correlation coefficient, the X-Y pairs that constitute the sample must be randomly sampled from a bivariate normal distribution (Howell, 2002). Bivariate normality means that, for any specific score on one of the variables, scores on the other variable should follow a normal distribution.

Chapter 8

Partial Correlation and Statistical Control

This Chapter's *Terra Incognita*

You are reading a sports-psychology journal. The article describes a study in which the central criterion variable was *exercise sessions:* the number of times that a participant engages in aerobic exercise each week. The researcher had investigated the relationship between this criterion variable and the following predictor variables: *health motivation, desire to be thin, dislike of exercise,* and *lack of time for exercise.*

The results section reports the following:

> The zero-order Pearson correlation between exercise sessions and health motivation was $r_{Y1} = .29$, $p < .001$. However, after statistically controlling for the other three variables (desire to be thin, dislike of exercise, and lack of time for exercise), the higher-order semipartial correlation was substantially smaller at $r_{Y(1\cdot234)} = .16$, $p = .013$.

Calm down. All shall be revealed.

The Basics of this Procedure

This chapter explains how researchers use regression procedures to statistically control for nuisance variables in correlational research. It introduces two statistics that capitalize on these methods of statistical control: the *partial correlation coefficient* and the *semipartial correlation coefficient*.

A **partial correlation coefficient** (symbol: *pr* or $r_{Y1 \cdot 2}$ or r_p) is a statistic that estimates the correlation between two variables after variance shared with one or more "control" variables has been partialed out of both the predictor variable and the criterion variable. It provides what an estimate of the Pearson correlation between the two variables would have been if all subjects had the same score on the control variables.

In contrast, **semipartial correlation** (symbol: *sr* or r_{sp} or $r_{Y(1 \cdot 2)}$) estimates the correlation between two variables after variance shared with one or more control variables has been partialed out of just the predictor variable (but not the criterion variable). A semipartial correlation is sometimes called a **part correlation**.

Partial correlation is generally performed using nonexperimental (correlational) data. The typical analysis involves (a) one criterion variable, (b) one predictor variable, and (c) one or more control variables. Figure 8.1 illustrates the nature of the variables included in the typical analysis.

FIGURE 8.1. Generic analysis model: Nature of the variables included in a typical partial or semipartial correlation.

Figure 8.1 shows that researchers generally compute a partial or semi-partial correlation coefficient when the analysis involves (a) a single continuous criterion variable measured on an interval scale or a ratio scale, and (b) a single continuous predictor variable, also assessed on an interval or ratio scale. Both variables are usually multi-value variables. Figure 8.1 does not display the control variables that play an important role in the analysis. These control variables are typically similar to the predictor and criterion variables: continuous, multi-value variables assessed on an interval or ratio scale. These coefficients may also be computed if either the predictor variable or the control variable is a dichotomous variable (Warner, 2008). Additional assumptions underlying partial and semipartial correlation are listed at the end of this chapter.

Illustrative Investigation

Imagine that a fictitious health psychologist named Dr. O'Day is studying the criterion variable *exercise sessions:* the number of times that a participant engages in aerobic exercise each week. Dr. O'Day believes that one of the variables that has a causal effect on exercise sessions is *health motivation*: the extent to which the individual is willing to sacrifice and work hard in order to stay healthy. Health motivation is measured by items 27-29 on the *Questionnaire on Eating and Exercising*, which appears in *Appendix A*. The nature of her hypothesis is illustrated in Figure 8.2.

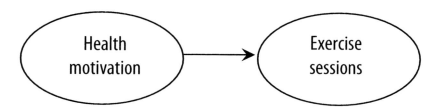

Figure 8.2. Dr. O'Day's hypothesized causal relationship between health motivation and exercise sessions.

THE ZERO-ORDER CORRELATION. Dr. O'Day administers the *Questionnaire* to 238 research participants. She computes the Pearson correlation between health motivation and exercise sessions and finds that the Pearson correlation between these two variables is, in fact, statistically significant, $r = .29$, $p < .001$.

This statistic is called a **zero-order correlation** because it represents a simple correlation between a predictor variable and a criterion variable, without statistically controlling for any potential confounding variables (here's a memory device: "zero control variables" means that it is a "zero-order correlation").

DR. O'DAY'S QUESTIONABLE INITIAL EXPLANATION. Pleased with the results, Dr. O'Day presented her findings at a research symposium. She showed the audience her proposed causal model (presented earlier as Figure 8.2), and described the study's results in this way:

> The Pearson correlation between health motivation and exercise sessions was $r = .29$. This correlation was statistically significant ($p < .001$), and fell just short of representing a medium effect according to Cohen's (1988) guidelines. Because the study used a correlational research design, it is not possible to draw a strong inference of cause-and-effect. Nevertheless, the observed correlation was consistent with the hypothesis that health motivation has a positive effect on exercise sessions.

DR. GREY'S ALTERNATIVE EXPLANATION. "*Poppycock!*" shouted Dr. Grey, another presenter at the symposium. He continued:

> I do not doubt that there is a positive correlation between health motivation and exercise sessions, but it is not due to the causal relationship that Dr. O'Day proposes. Instead, the correlation between these two variables is *spurious*—they are correlated only because they share the same underlying third variable: the desire to be thin.

On the screen at the front of the room, Dr. Grey replaced Dr. O'Day's figure with his own proposed model, presented here as Figure 8.3:

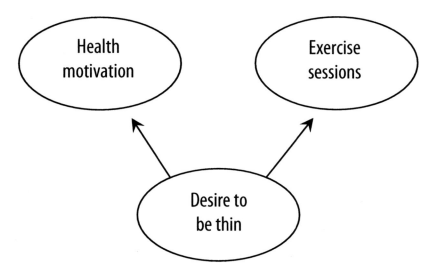

Figure 8.3. Dr. Grey's hypothesis that desire to be thin may be operating as an underlying third variable.

Dr. Grey continued:

> As you can see, the only reason that health motivation is correlated with exercise sessions is because they are both influenced by the same underlying third variable: desire to be thin. *Desire to be thin* is the extent to which an individual wishes to have a low body weight and a lean physical appearance.
>
> The spurious correlation works like this:
> - When people desire to be thin, this causes them to develop higher levels of health motivation. This is because they believe that being healthier will help them lose weight.
> - Quite separately, when people desire to be thin, this also causes them to engage in more exercise sessions. Again, this is because they believe that exercising more frequently will help them lose weight.
>
> In summary, health motivation never causes people to engage in more exercise sessions. The wish for good health is just not a very powerful motivator—not nearly powerful enough to make people go out and exercise more frequently. In our appearance-conscious society, only the *desire to be thin* has enough incentive power to cause people actually go out and exercise and sweat. Therefore, the correlation that Dr. O'Day reports between health motivation and exercise sessions is spurious. Her variables are correlated only because they share the same underlying third variable: the desire to be thin. If Dr. O'Day had statistically controlled for the desire to be thin, the correlation between health motivation and exercise sessions would disappear.

The audience gasped, shocked by by Dr. Grey's attack on Dr. O'Day's work. *Spurious correlations? Underlying third variables?* Had Dr. O'Day anticipated these criticisms?

And what was she going to do?

Research Questions Addressed

We will return to Dr. Grey's concerns a bit later. First, however, we will learn about the basic methods of statistical control and will see how researchers use these methods to compute partial and semipartial correlations. Ultimately, we will see how researchers use these methods to answer questions such as:

- Is there a substantial zero-order correlation between this predictor variable and this criterion variable? Is the correlation still substantial in size after I have statistically controlled for a possible confounding variable?

- Is the correlation still substantial in size after I have statistically controlled for *several* possible confounding variables?

Details, Details, Details

This section introduces the concept of *statistical control* and explains why it is important to researchers. It explains some basic terms and concepts: spurious correlation, confounding variables, the concept of experimental control, and the subscript notation used with partial correlations.

Procedures that Make Use of Statistical Control

Scientific articles often describe results which were analyzed using methods of statistical control. These articles include expressions such as *controlling for, holding constant, partialing,* and *residualizing* (Licht, 1995). A very short and very incomplete list of procedures that employ methods of statistical control includes the following:

- Partial and semipartial correlation
- Multiple regression
- Logistic regression
- Analysis of covariance (ANCOVA)
- Structural equation modeling (SEM)

Basic Concepts in Statistical Control

Methods of statistical control use mathematical formulas to estimate what the relationship between a predictor variable and a criterion variable would have been if one or more confounding variables had been experimentally controlled. Methods of statistical control typically make use of regression techniques to remove from the predictor variable, the criterion variable (or both), the variance that they share with one or more potential confounding variables (more on these regression techniques a bit later).

CONFOUNDING VARIABLES. Researchers typically use methods of statistical control when their investigation suffers from one or more confounding variables. In this context, a *confounding variable* (also called a *nuisance variable*) can be defined as a third variable that is correlated with the predictor and/or criterion variable, and whose presence makes it difficult to determine the actual nature of the causal relationship (if any) between the predictor and criterion. Some textbooks use the term *confounding variable* only within the context of a true experiment, but this book will define the term more broadly to refer to true experiments as well as nonexperimental studies.

You will recall that Dr. O'Day initially interpreted the correlation between health motivation and exercise sessions as evidence that motivation had a causal effect on exercise. Dr. Grey (correctly) reminded her that a simple bivariate correlation typically provides only weak evidence of cause-and-effect. He pointed out the possibility that the observed correlation between the two variables may merely be due to the presence of a potential confounding variable: the desire to be thin. He referred to this construct as an *underlying third variable*, but it may also be viewed as a confounding variable.

SPURIOUS CORRELATIONS. You will recall that Dr. Grey referred to the correlation that Dr. O'Day observed between health motivation and exercise sessions as being a spurious correlation. When the relationship between a predictor and a criterion variable is a *spurious correlation*, it means that neither variable has a causal effect on the other variable; instead, it means that the relationship between them is due to their shared relationship with one or more other variables (often referred to as *third variables*).

Experimental Versus Statistical Control

Researchers typically attempt to deal with confounding variables and possible spurious correlations through either experimental control or statistical control. Experimental control typically results in the strongest evidence of cause-and-effect, but statistical control can still be useful when experimental control is not feasible.

EXPERIMENTALLY CONTROLLING FOR CONFOUNDING VARIABLES. *Experimental control* refers to the researcher's attempts to manipulate or hold constant environmental conditions and/or subject characteristics that might affect scores on the criterion variable. There are many methods of experimental control, but this section will focus on just one: *physically holding constant the potential confounding variable.* This is the tactic of experimental control that is most analogous to the concept of statistical control.

Imagine that Dr. O'Day conducts the simple correlational study described above: she measures health motivation and exercise behavior in a single sample of participants and computes the Pearson correlation between these two variables.

She knows that, even if she obtains a strong positive correlation, Dr. Grey will argue that the only reason that there is a correlation between health motivation and exercise sessions is because they are both influenced by the same underlying third variable: desire to be thin.

But what if the desire to be thin was not a variable at all? What if the desire to be thin instead was a *constant?* (When a construct is a **constant**, it means that there is zero variability in its values—all of the participants in the sample have exactly the same value on the construct.) If desire to be thin were a constant, no one could argue that it is having a causal effect on anything in the study.

For example, imagine that the measure of desire to be thin consisted of just seven different scores on a scale. Subjects with a score of "1" displayed the lowest level of desire to be thin, and subjects with a score of "7" displayed the highest level. What if Dr. O'Day administered this scale to a really large pool of potential subjects (say, $N = 2,000$) and then selected for her study only those individuals who had a score of exactly "4" on this variable? (We will assume that "4" represented the mean score on the scale). For example, maybe just 238 people had a score of exactly "4" on this measure of the desire to be thin, and therefore only these 238 people were included in the statistical analyses.

In this sample of 238 participants, Dr. O'Day computed the Pearson correlation between health motivation and exercise sessions and found it to be $r = .26$, $p < .001$. This is how she shared her findings at the symposium:

> I have controlled for the possible effects of desire to be thin by physically holding it constant. After doing this, I *still* obtained a correlation of $r = .26$ between health motivation and exercise sessions. This shows that the correlation between these two variables could not possibly be due to the desire to be thin operating as an underlying third variable. So I don't want to hear one *peep* out of Dr. Grey.

Researchers often hold variables constant in investigations, although seldom in the way described here. A more typical example would involve holding constant a categorical variable, such as *subject sex*. For example, if a pharmaceutical firm investigated a new blood-pressure medication by using only men as subjects, they would be holding subject sex constant.

In summary: There are a number of ways in which researchers use methods of experimental control to deal with confounding variables. One very basic approach is to physically hold the offending variable constant, thus preventing it from having any effect on the criterion variable.

STATISTICALLY CONTROLLING FOR CONFOUNDING VARIABLES. The preceding example was pretty far-fetched, as it is seldom possible to physically hold constant a quantitative variable such as the desire to be thin. When the confounding variable is quantitative, the researcher is more likely to use some method of statistical control. These methods typically use regression techniques to remove from the predictor variable, the criterion variable (or both), the variance that they share with one or more potential confounding variables. A later section will show how this is done in some detail.

Subscript Notation

The subscript notation that is sometimes used for partial correlation coefficients requires some explanation. An example of a subscript would be the "*Y1·2*" subscript that appears with the symbol: $r_{Y1 \cdot 2}$. The system of subscripts used here assumes that there is just one criterion variable (represented by the symbol "*Y*"), but there are several predictor variables, represented by the symbols $X_1, X_2, X_3, X_4,$ and X_5.

First, remember that a ***zero-order correlation*** represents the correlation between two variables without controlling for any other variables at all. The Pearson correlations that you learned about in *Chapter 7* were zero-order correlations. *Chapter 7* had used the symbol *r* to represent a zero-order correlation, but the current chapter might instead represent a zero-order correlation with the following symbol: r_{Y1}. The subscript "*Y1*" tells the reader: "This is the zero-order correlation between the criterion variable, *Y*, and the predictor variable, X_1. With this convention, the subscript symbol on the left ("*Y*," in this case) identifies the criterion variable, and the subscript symbol on the right ("1," in this case) identifies the predictor variable: X_1.

Next, you have already learned that a ***partial correlation*** is a statistic that estimates the correlation between two variables after variance shared with one or more control variables has been partialed out of both the predictor variable and the criterion variable. This chapter might represent a partial correlation with the symbol: $r_{Y1 \cdot 2}$. With this symbol, the subscript is "*Y1·2*." This subscript tells the reader: "This is the partial correlation between the criterion variable (*Y*), and predictor variable (X_1), with the variance shared with control variable (X_2) partialed out of both *Y* and X_1.

The general convention is that (a) the numbers that appear to the left of the middle dot (the "·") identify the two variables being correlated, and (b) the numbers that appear to the right of the middle dot identify the variables that are being statistically controlled. And, yes, it is possible to statistically control for more than one variable (as we shall see in a later section dealing with higher-order partial correlations). Some examples of these conventions:

> $r_{Y1 \cdot 2}$ means "This is the correlation between *Y* and X_1, with X_2 partialed out of both *Y* and X_1."
>
> $r_{Y2 \cdot 1}$ means "This is the correlation between *Y* and X_2, with X_1 partialed out of both *Y* and X_2."
>
> $r_{Y1 \cdot 2345}$ means "This is the correlation between *Y* and X_1, with $X_2, X_3, X_4,$ and X_5 partialed out of both *Y* and X_1."

The preceding were all examples of symbols that might be used to represent partial correlations. A later section of this chapter shows how this system is slightly modified to represent semipartial correlations.

Size and Sign of the Coefficient

The size and sign of a partial or semipartial correlation coefficient is interpreted in the same way that the size and sign of a Pearson correlation coefficient (r) is interpreted (this was discussed in some detail in *Chapter 6*). Under typical conditions, these coefficients range in size from -1.00 through 0.00 through $+.100$, with values closer to 0.00 indicating weaker relationships.

Regarding the sign: (a) a *positive correlation* (such as $r_{Y1\cdot 2} = +.40$) means that, as values of X_1 increase, values of Y also tend to increase (after controlling for the confounding variable, X_2); (b) a *negative correlation* (such as $r_{Y1\cdot 2} = -.40$) means that, as values of X_1 increase, values of Y tend to decrease (after controlling for the confounding variable, X_2); and (c) *zero correlation* (such as $r_{Y1\cdot 2} = .00$) means that there is no relationship between X_1 and Y (after controlling for the confounding variable, X_2). The preceding were all examples of partial correlations, but the sign and size of semipartial correlations are interpreted in a similar manner.

Computing Partial Correlations from Residuals

Imagine that you wish to compute the partial correlation between just one predictor variable and just one criterion variable while statistically controlling for just one confounding variable (we shall refer to this confounding variable as the *control variable*). In such a situation, you could use simple bivariate regression procedures (as covered in *Chapter 7*) to compute the partial correlation. Theoretically, this would involve the following steps:

- **Step 1.** Regress the predictor variable on the control variable and compute a residual of prediction (e_p) for each participant.

- **Step 2.** Regress the criterion variable on the control variable and compute a residual of prediction (e_c) for each participant.

- **Step 3.** Correlate the residuals of prediction for the predictor variable (e_p) with the residuals of prediction for the criterion variable (e_c). And *viola!* The result is the partial correlation between the predictor and the criterion, statistically controlling for the variance that they shared with the control variable.

The following section illustrates these steps by showing how they could be used to compute a partial correlation relevant to Dr. O'Day's investigation. Most of what follows is based on Pedhazur (1982, pp. 101-107).

Partialing Desire To Be Thin Out of Health Motivation

A previous section had described *health motivation* as being a predictor variable (i.e., an *X* variable). However, for the current analysis, we must instead think of health motivation as being a criterion variable—a *Y* variable. In the analysis, the predictor variable will be the confounding variable that Dr. O'Day hopes to statistically control: *the desire to be thin*.

PERFORMING THE REGRESSION ANALYSES. Imagine that Dr. O'Day performs a bivariate regression in which the criterion variable is health motivation (symbol: Y_m) and the predictor variable is desire to be thin (symbol: X). This analysis would produce the following bivariate regression equation:

$$Y'_m = a + b(X)$$

$$Y'_m = 4.607 + .218(X)$$

Now that she has the values for a and b in the regression equation, Dr. O'Day can compute the predicted score for each participant on health motivation (symbol: Y'_m), based on their actual score on desire to be thin. These scores are reported for a few fictitious subjects in Table 8.1.

Table 8.1

Residualized Variables Used in Computing Partial Correlations

Subject	X	Predicting health motivation (Y_m) from desire to be thin (X)			Predicting exercise sessions (Y_s) from desire to be thin (X)		
		Y_m	Y'_m	e_m	Y_s	Y'_s	e_s
1.	1.5	5.1	4.9	0.2	2.4	2.0	0.4
2.	2.3	5.2	5.1	0.1	2.0	2.2	-0.2
3.	3.0	5.0	5.3	-0.3	2.3	2.3	0.0
.	.	.	.	.	.	.	.
.	.	.	.	.	.	.	.
.	.	.	.	.	.	.	.
238.	6.5	7.0	6.0	1.0	4.7	3.2	1.5

Note. X = Participant's actual score on desire to be thin. Y_m = Actual score on health motivation. Y'_m = Predicted score on health motivation. e_m = residuals of prediction for health motivation ($Y_m - Y'_m$). Y_s = Actual score on exercise sessions. Y'_s = Predicted score on exercise sessions. e_s = residuals of prediction for exercise sessions ($Y_s - Y'_s$).

In Table 8.1, the column headed X contains subject actual scores on desire to be thin. You can see that Subject 1 had a score of 1.5, on this variable; Subject 2 had a score of 2.3, and so forth.

Inserting these X scores in the regression equation presented earlier allows us to compute each participant's predicted score on health motivation. These predicted scores appear in the column headed Y'_m in Table 8.1. You can see that

Subject 1 had a predicted health motivation score of 4.9, Subject 2 had a predicted score of 5.1, and so forth.

The participants' actual scores on health motivation appear in the column headed Y_m. You can see that Subject 1 had an actual score of 5.1 on health motivation, and Subject 2 had an actual score of 5.2.

COMPUTING RESIDUAL OF PREDICTION SCORES. It is now possible to compute a residual of prediction score for each subject on the health motivation variable. You will recall from *Chapter 7* that a **residual of prediction** is an *error score*—it represents the difference between what we predicted a subject would display on some variable versus what they actually displayed. These residual scores appear in the column headed e_m (the "e" stands for "error of prediction"). Residuals of prediction are computed as:

$$e_m = Y_m - Y'_m$$

For example, Subject 1 had an actual score of 5.1 on health motivation, and a predicted score of 4.9 on health motivation. The residual of prediction score is therefore computed as:

$$e_m = Y_m - Y'_m$$
$$e_m = 5.1 - 4.9$$
$$e_m = 0.2$$

So Subject 1 has a residual of prediction score of 0.2 for the health motivation variable. Residual scores for the remaining participants may be computed in the same way.

We have just created a new variable: the *health-motivation residuals* variable (symbol: e_m) . This new construct is a **synthetic variable**: an artificial variable created by performing mathematical operations on one or more real-world variables. In *Chapter 7*, you learned that you can work with synthetic variables in the same way that you work with real-world variables—you can compute the standard deviation of their scores, and you can even correlate them with other variables (as we shall see in a moment).

WHAT IS MEASURED BY THE ORIGINAL HEALTH MOTIVATION VARIABLE? For the sake of brevity, we will call this new e_m variable *residualized health motivation*. To understand what e_m represents, let's start by comparing it to the variable that contributed to its birth: The original health motivation variable.

The original health motivation variable was measured by items 27-29 on the *Questionnaire on Eating and Exercising*. Like many variables measured by multiple-item summated rating scales, it was probably a somewhat complex variable (more complex than the researcher wanted it to be, at any rate). For the most part, it measured the construct *health motivation*, as it was intended to do. But many variables created from multiple-item summated rating scales may also (unintentionally) measure additional constructs beyond the constructs that they were intended to measure. For example, it is possible that this scale also measured the construct *desire to be thin* to some small extent. In their minds, many subjects probably have a difficult time separating the construct *health motivation* from the construct *desire to be thin*, and, as a result, their scores on the health motivation scale might also reflect where they stand on desire to be thin, to some small extent.

Wouldn't it be great if we could rid the health motivation scale of any last trace of the desire to be thin? That's what we attempted to do when we created the residualized health motivation scale, which is discussed next.

WHAT IS MEASURED BY THE RESIDUALIZED HEALTH MOTIVATION VARIABLE? Within the context of partial correlation, a **residualized variable** is a variable that has been transformed so that it no longer shares any variance with the control variable(s). When Dr. O'Day created the *residualized health motivation* variable (e_m), she stripped from health motivation any variance that it had shared with the desire to be thin. As a result, the new residualized health motivation variable is now totally uncorrelated with the desire to be thin (yes—if you computed the Pearson correlation between e_m and the desire to be thin scale, the correlation would be $r = .00$). Think of e_m as a variable that still measures health motivation, but in a cleaner way, in a way completely unbesmirched by the desire to be thin.

CAVEAT EMPTOR REGARDING "STATISTICAL CONTROL." Statisticians no doubt cringe at the oversimplified description presented in the preceding paragraph ("Measuring variables in a *cleaner* way. *Indeed!*"). The purpose of the preceding section was to convey an understanding of what researchers *hope* to accomplish by partialing one variable out of a second variable. Unfortunately, the actual practice of statistical partialing is not quite as straightforward as vacuuming the dog fur out of an oriental rug.

You should always take partial correlations and other procedures that rely heavily on methods of statistical control with a grain of salt. Remember that a partial correlation coefficient provides an *estimate* of what the correlation between *X* and *Y* would have been if we had used experimental methods to hold constant some confounding variable—this estimate may be fairly accurate, or it may be fairly misleading. A later section of this chapter titled "Potential Problems with Statistical Control" explains some of the reasons for this. Readers interested in a more nuanced discussion of regression, residualized variables, and similar issues should also read Licht (1995), especially pages 35-38 and 42-44.

The short version goes like this: If you wish to obtain research results that are clear-cut, unambiguous, and (relatively) immune from criticism, then methods of experimental control are almost always superior to methods of statistical control. However, when methods of experimental control are not possible, then methods of statistical control (such as the methods described here) are useful tools for refining our data analyses.

Partialing the Desire To Be Thin Out of Exercise Sessions

Having created a residualized version of the health motivation variable, we will now create a residualized version of the exercise sessions variable. When finished, the new *residualized exercise sessions* variable will also be stripped of any variance that it shared with the *desire to be thin* variable.

Assume that Dr. O'Day performs another bivariate regression analysis. In this analysis, the criterion variable is exercise sessions, and the predictor variable is once again the control variable: the desire to be thin. This analysis produces the following bivariate regression equation (notice that the obtained values for *a* and *b* are different from those displayed in the previous equation):

$$Y'_S = a + b(X)$$
$$Y'_S = 1.584 + .254(X)$$

By inserting a subject's score on desire to be thin (X) into the equation, we can compute the subject's predicted score on exercise sessions (Y'_S). These values have already been computed for a few subjects, and appeared in Table 8.1 (presented earlier). That table showed that:

- Subject 1 had a score of 1.5 on desire to be thin, and therefore had a predicted score of 2.0 on exercise sessions.

- Subject 2 had a score of 2.3 on desire to be thin, and therefore had a predicted score of 2.2 on exercise sessions.

- Subject 3 had a score of 3.0 on desire to be thin, and therefore had a predicted score of 2.3 on exercise sessions.

Because we know where each subject actually stood on the exercise sessions variable (Y_S), it is possible to compute a residual of prediction score (e_S) for each of them. For example, Table 8.1 shows that Subject 1 had an actual score on exercise sessions of 2.4, and a predicted score of 2.0. Therefore, the residual of prediction score is computed as:

$$e_S = Y_S - Y'_S$$
$$e_S = 2.4 - 2.0$$
$$e_S = 0.4$$

Residual scores for the remaining subjects may be computed in the same way.

We have just created another new variable: *residualized exercise sessions* (symbol: e_S). This new variable is another synthetic variable, and may be interpreted in the same way that we interpreted the residualized health motivation variable. The residualized exercise sessions variable is a transformed version of the original variable: It represents what is left over when all of the variance associated with desire to be thin has been partialed out of exercise sessions. As a result, the new residualized exercise sessions variable (e_S) is no longer correlated with the desire to be thin (this is what we mean when we say that the desire to be thin has been *partialed out* of it). If we computed the Pearson correlation between the residualized exercise sessions variable and the desire to be thin, it would be $r = .00$.

Because e_s is a variable just like any other variable, we can correlate it with other variables if we like. And that is just what we will do when we compute the partial correlation coefficient in the next section.

Computing the Partial Correlation Coefficient

If we compute the correlation between e_m and e_s, the result is the partial correlation between health motivation and exercise sessions, while statistically controlling for the desire to be thin. This partial correlation coefficient is our best estimate of what the correlation between health motivation and exercise sessions would be if the desire to be thin were held constant (i.e., if all subjects had exactly the same score on the desire to be thin). Think about why this makes sense:

- The variable e_m is *residualized health motivation*. It is a new version of health motivation, from which the variance shared with desire to be thin has been statistically removed.

- The variable e_s is *residualized exercise sessions*. It is a new version of exercise sessions, from which the variance shared with desire to be thin has been statistically removed.

- If we correlate e_m with e_s, we are correlating a residualized version of health motivation with a residualized version of exercise sessions. The resulting **partial correlation coefficient** (symbol: pr or $r_{Y1 \cdot 2}$ or r_p) is an estimate of what the correlation between health motivation and exercise sessions might have been if all of the subjects had the same score on the desire to be thin. In the current case, this partial correlation coefficient was found to be $r_{Y1 \cdot 2} = .26$.

So (*finally!*) we have computed the partial correlation coefficient and have found it to be $r_{Y1 \cdot 2} = .26$. Is this good news for Dr. O'Day or for good news for Dr. Grey? As the following section shows, the answer is "it depends."

Comparing the Two Correlation Coefficients

In most cases, the size of a partial correlation coefficient is meaningful only if it is compared to the size of the corresponding zero-order correlation coefficient. You will recall that the zero-order correlation is the original Pearson correlation between the predictor and criterion—the correlation that did not involve any partialing or any attempts at statistical control.

This section describes two fictitious possible outcomes from Dr. O'Day's computation of a partial correlation. We will see how one outcome would support Dr. Grey's hypotheses, whereas a different outcome would support Dr. O'Day's.

FICTITIOUS OUTCOME 1. An earlier section in this chapter indicated that the zero-order correlation between health motivation and exercise sessions was $r_{Y1} = .29$. Imagine that the partial correlation between health motivation and exercise sessions (while statistically controlling for desire to be thin) was $r_{Y1 \cdot 2} = .00$. Many researchers would see this outcome as supporting Dr. Grey's position. Dr. Grey had predicted that the observed correlation of $r_{Y1} = .29$ between health motivation and exercise sessions was spurious due to the shared influence of an underlying third variable: the desire to be thin. Sure enough, once Dr. O'Day statistically controlled for this latter variable, the correlation between health motivation and exercise sessions disappeared. Many researchers would see this outcome as evidence that Dr. Grey was correct.

FICTITIOUS OUTCOME 2. Once again, assume that the zero-order correlation between health motivation and exercise sessions was $r_{Y1} = .29$. In this second example, however, imagine that the partial correlation between health motivation and exercise sessions (while statistically controlling for desire to be thin) was $r_{Y1 \cdot 2} = .29$. In this second example, statistically controlling for the desire to be thin did not have any effect at all on the observed correlation. Dr. O'Day had predicted that the observed correlation between health motivation and exercise sessions was real—it was not merely a spurious relationship brought about by an underlying third variable. If the partial correlation ($r_{Y1 \cdot 2} = .29$) were the same size as the zero-order correlation ($r_{Y1} = .29$), most researchers would conclude that Dr. O'Day was correct.

THE ACTUAL OUTCOME. As was indicated earlier, the partial correlation between health motivation and exercise sessions (after controlling for the desire to be thin) was actually $r_{Y1 \cdot 2} = .26$. This indicates that controlling for desire to be thin did not have much of an effect on the size of the correlation: The zero-order correlation had been $r_{Y1} = .29$ whereas the partial correlation had been only a teeny bit smaller at $r_{Y1 \cdot 2} = .26$. Most researchers would see these results as being more consistent with Dr. O'Day's hypothesis than with Dr. Grey's hypothesis.

First-Order and Higher-Order Partial Correlations

The motto for the field of statistics is:

> Anything worth doing is worth *overdoing*.

And nowhere is this truer than in the area of partial correlation. Until this point, this chapter has discussed the situation in which the researcher wishes to control for just one confounding variable. But what if she wants to control for two confounding variables? Or six? We need new terms to describe the resulting statistics, and this section introduces the lexicon (from Pedhazur, 1982; Warner, 2008).

A ***first-order partial correlation*** is a correlation between two variables with just one additional variable partialed out of both. The correlation between health motivation and exercise sessions, with the desire to be thin partialed out of both of them is an example of a first-order partial correlation. The symbol $r_{Y1 \cdot 2}$ represents this partial correlation. In the subscript, the fact that there was just *one variable* to the right of the dot told us that we were looking at a *first-order* partial correlation.

In contrast, a ***higher-order partial correlation*** is a correlation between two variables with more than one control variable partialed out of both. For example, if we computed the correlation between Y and X_1 after partialing X_2, X_3, X_4, and X_5 out of both variables, the result would be a higher-order partial correlation, and the symbol would be $r_{Y1 \cdot 2345}$. In subscript of $r_{Y1 \cdot 2345}$, the fact that there were *several* *variables* to the right of the dot told us that we were looking at a *higher-order* partial correlation.

There are different ways of computing higher-order partial correlations. If you have a pocket calculator and nothing else to do for the next few hours, they can be computed by hand using formulas such as those provided by Pedhazur (1982, pp. 103-110). In practice, however, researchers almost always rely on statistical applications such as SAS or SPSS.

Semipartial (Part) Correlation

You have already learned that **partial correlation** is a statistical procedure that estimates the correlation between a predictor variable and a criterion variable after variance shared with one or more "control" variables has been partialed out of both the predictor and criterion. In contrast, **semipartial correlation** estimates correlation between a predictor variable and a criterion variable after variance shared with one or more control variables has been partialed out of just the predictor (but not the criterion). Semipartial correlation is sometimes called **part correlation**. Typical symbols include *sr* or r_{sp} or $r_{Y(1 \cdot 2)}$.

The good news is that most of the things that you have just learned about partial correlation coefficients will also apply to semipartial correlation coefficients. This means that this chapter can devote a bit less space to semipartial correlations.

Subscript Notation

A typical symbol for a semipartial correlation coefficient is $r_{Y(1 \cdot 2)}$. The *Y(1·2)* subscript that appears as part of this symbol tells the reader: "This is the semipartial correlation between the criterion variable (*Y*), and predictor variable (X_1) with the variance shared with a control variable (X_2) partialed out of just the X_1 variable."

As was the case with the partial correlation coefficient, the numbers that appear to the left of the middle dot (the "·") identify the two variables being correlated, and the numbers that appear to the right of the middle dot identify the variables that are being statistically controlled. However, the fact that the parentheses enclose the "1" but not the "Y" tells the reader that the control variable is being partialed out of just the X_1 variable, not the *Y* variable. Here is a hint: Anytime the subscript for an *r* statistic includes parentheses, you are probably looking at a semipartial correlation rather than a partial correlation.

As was the case with partial correlation, it is possible to statistically control for more than one variable, and the subscript tells you how many. Some examples of these subscript conventions:

$r_{Y(1\cdot 2)}$ means "This is the correlation between Y and X_1, with X_2 partialed out of just X_1."

$r_{Y(1\cdot 2345)}$ means "This is the correlation between Y and X_1, with X_2, X_3, X_4 and X_5 partialed out of just X_1."

Typical Research Questions

Below are some examples of the types of research questions that may be addressed using semipartial correlation:

> Is there a substantial correlation between this predictor variable and this criterion variable? Is the correlation still substantial in size after I have statistically removed from the predictor variable any variance that it shared with this potential confounding variable?
>
> Is the correlation still substantial in size after I have statistically removed from the predictor variable any variance that it shared with *several* potential confounding variables?

When Appropriate

As with partial correlations, semipartial correlations are almost always computed for correlational studies, and are almost always computed using quantitative, multi-value variables assessed on an interval scale or a ratio scale. However, semipartial correlations are used to investigate somewhat different research questions.

In general, semipartial correlation allows researchers to explore the nature of the relationship between a predictor and a criterion when (a) the criterion is left unchanged, but (b) all variance shared with a control variable has been partialed out of the predictor variable. Essentially, the researcher is asking, "If I transform this predictor variable so that it is no longer correlated with this control variable, will the predictor still be correlated with the criterion?"

Illustrative Investigation

For example, imagine that a new researcher named Dr. Silver makes this criticism of Dr. O'Day's study on exercise behavior:

> I disagree with Dr. Grey's claim that is it necessary to partial the desire to be thin out of both health motivation and exercise behavior. But I do believe that it is appropriate to partial it out of the health motivation variable.
>
> The problem is this: People are completely unable to psychologically separate the concept of *health motivation* from the concept of *desire to be thin*. They think that the question, "How motivated are you to be healthy?" means exactly the same thing as "How much do you desire to be thin?"

In other words, measuring health motivation in a sample of individuals does not tell us anything beyond what we already know by measuring the desire to be thin. The apparent correlation of $r = .29$ between health motivation and exercise sessions is an illusion. If you partial out of the health motivation variable the variance that it shares with the desire to be thin variable, the correlation between health motivation and exercise sessions will drop to zero.

Given Dr. Silver's criticisms, it may make sense for Dr. O'Day to compute the semipartial correlation between health motivation and exercise sessions with the desire to be thin partialed out of health motivation. If the resulting semipartial correlation is small (relative to the zero-order correlation of .29 between health motivation and exercise sessions, reported above), this result would lend support to Dr. Silver's criticism that the variable of health motivation does not tell us anything beyond what we already know by measuring the desire to be thin.

Dr. O'Day will, in fact, compute this semipartial correlation. Her findings will be reported later in the chapter.

Why Semipartial Correlation Is Important

Now that this chapter has spent so much time on partial correlation, it is time to face the ugly truth: Semipartial correlation is probably a much more important statistic, compared to partial correlation.

One of the most popular statistical procedures in research is multiple regression (the topic of *Chapter 9* and *Chapter 10*). One of the reasons that researchers use multiple regression is to determine whether a specific predictor variable accounts for a significant amount of variance in a criterion variable, beyond the variance already accounted for by a number of control variables. When they perform this analysis, they may refer to the result as the "increment in R^2 due to the predictor variable," but what they are really doing is computing a squared semipartial correlation coefficient. Multiple regression is important because researchers often wish to know whether a predictor variable is still related to a criterion variable, after statistically controlling for one or more control variables. And that is exactly what is achieved with a semipartial correlation coefficient.

Computing Semipartial Correlation from Residuals

Following up on Dr. Silver's criticisms, imagine that Dr. O'Day now wishes to compute the semipartial correlation between health motivation and exercise sessions, with the desire to be thin partialed out of health motivation (but not out of exercise sessions). In the real world she would do this with a few clicks in a statistical application. But since this is a textbook, we will instead show how it is done conceptually, by computing residuals. The steps are as follows:

- **Step 1**. Regress the predictor variable (health motivation) on the control variable (desire to be thin) and compute a residual of prediction (e_m) for each participant.

- **Step 2**. Leave the criterion variable (exercise sessions) as it is—do *not* compute a residualized version of this variable.
- **Step 3**. Correlate the residuals of prediction for health motivation (e_m) with the unchanged exercise sessions variable (Y_s). The result is the semipartial correlation between health motivation and exercise sessions, after statistically controlling for the variance that health motivation shares with the desire to be thin.

Dr. O'Day computed the semipartial correlation coefficient in this way and found it to be $r_{Y(1\cdot2)} = .26$. In the next section, we will evaluate this coefficient by comparing it to the corresponding zero-order correlation.

Comparing the Two Coefficients

Let's put Dr. O'Day's semipartial correlation of $r_{Y(1\cdot2)} = .26$ in context. The original zero-order correlation between health motivation and exercise sessions had been $r_{Y1} = .29$. The current semipartial correlation shows that, when variance shared with the desire to be thin is partialed out of health motivation, it does not cause much of a reduction: the correlation decreases just a bit, from $r_{Y1} = .29$ to $r_{Y(1\cdot2)} = .26$.

These results are not consistent with the argument put forward by Dr. Silver. He claimed that the observed correlation of $r_{Y1} = .29$ between health motivation and exercise sessions was misleading, due to the underlying influence of the desire to be thin. He predicted that if you partial from health motivation the variance that it shares with the desire to be thin, the correlation between health motivation and exercise sessions would drop to zero. The correlation dropped a bit (from .29 to .26), but not to zero.

These results suggest that health motivation may, in fact, be a useful predictor of exercise sessions. They suggest that, when we assess where people stand on health motivation, we learn something more (something different) than what we learn when we merely ask them how much they desire to be thin. These results are consistent with Dr. O'Day's hypothesis that the observed relationship between health motivation and exercise sessions is a real relationship, not a spurious correlation brought about by the potential confounding variable identified by Dr. Silver.

Relative Size of Partial Versus Semipartial Correlations

Those of you with really good memories may have notices that the size of the semipartial correlation just computed ($r_{Y(1\cdot2)} = .26$) was similar to the size of the partial correlation reported earlier ($r_{Y1\cdot2} = .26$). This was a fluke—sometimes a semipartial correlation will be similar in size to a partial correlation based on the same data, and sometimes it will not. Cohen and Cohen (1983) show that, according to their formulas, partial correlations should almost always be larger, and can never be smaller, than semipartial correlations based on the same variables.

Higher-Order Semipartial Correlations

When researchers compute **higher-order semipartial correlations**, they compute the correlation between a predictor variable and a criterion variable after variance shared with more than one control variable has been partialed out of just the predictor variable. To get specific: When one control variable is partialed out, it is called a ***first-order semipartial correlation***; when two variables are partialed out, it is called a ***second-order semipartial correlation***, and so forth.

The subscript convention introduced earlier is particularly useful in representing higher-order semipartial correlations. For example, assume that a researcher computed the semipartial correlation between the criterion variable (Y) and predictor variable X_1 after the variance shared with control variables X_2, X_3, X_4, and X_5 have been partialed out of X_1. This statistic might be represented as:

$$r_{Y(1 \cdot 2345)}$$

With the above subscript, the numbers 2, 3, 4, and 5 appear to the right of the dot ("·"), telling us that these are the control variables that are being partialed out. Notice that a set of parentheses encloses the "1" along with the "2345." This tells us that the variables X_2, X_3, X_4, and X_5 are being partialed out of X_1 variable, but not out of the Y variable (because the Y is not enclosed by the parentheses).

As an illustration, Dr. O'Day computed the semipartial correlation between health motivation and exercise sessions with three other variables (desire to be thin, dislike of exercise, and lack of time for exercise) partialed out of health motivation. The resulting semipartial correlation coefficient was $r_{Y(1 \cdot 2345)} = .14$, $p = .013$. This shows us that there is still a non-zero correlation between health motivation and exercise sessions, even after controlling for the other three variables.

Squared Semipartial Correlations

A ***squared semipartial correlation coefficient*** (symbol: sr^2) indicates the percent of variance in a criterion variable that is accounted for by a predictor variable, above and beyond the variance already accounted for by the control variables. This means that the interpretation of sr^2 is analogous to the interpretation of the coefficient of determination (r^2) from bivariate correlation. You will recall that r^2 indicates the percent of variance in the criterion variable that is accounted for by the predictor variable. In contrast, sr^2 is an index of ***incremental variance accounted for***—it indicates the percent of variance in the criterion variable that is accounted for by this predictor variable, beyond the variance already accounted for by the control variables.

The preceding section indicated that Dr. O'Day computed the semipartial correlation between health motivation and exercise sessions, controlling for desire to be thin, dislike of exercise, and lack of time for exercise. She found the resulting semipartial correlation coefficient to be $r_{Y(1 \cdot 2345)} = .14$. Squaring this value results in $sr^2 = (.14)^2 = .0196 = .02$.

This tells us that health motivation accounts for about 2% of the variance in exercise sessions, beyond the variance already accounted for by the three control variables. This percentage may be large enough to be statistically significant (if the sample is large enough), but most researchers would not view it as being a terribly large value in terms of meaningfulness.

Presenting the Results in Journal Articles

The following section includes a table which presents the three types of correlation coefficients discussed in this chapter. It also shows how Dr. O'Day might discuss these within the text of a journal article.

Table 8.2

Zero-Order Correlations (r), Partial Correlations (r_p), and Semipartial Correlations (r_{sp}) Between Four Predictor Variables and Exercise Sessions per Week

Predictor	r	r_p	r_{sp}
Health motivation	.293***	.163*	.140*
Desire to be thin	.148*	.052	.044
Dislike of exercise	−.373***	−.187**	−.162**
Lack of time for exercise	−.458***	−.358***	−.325***

Note. $N = 238$. For each predictor variable, higher scores represent higher levels of the construct being measured. r represents the zero-order Pearson correlation between the predictor variable and exercise sessions. r_p represents the partial correlation between the predictor variable and exercise sessions, with variance shared with the other three predictors partialed out of both variables. r_{sp} represents the semipartial correlation between the predictor variable and exercise sessions with variance shared with the other three variables partialed out of just the predictor variable.
  * $p < .05$. ** $p < .01$. *** $p < .001$

    Table 8.2 presents correlations between each the four predictor variables and the study's criterion variable, exercise sessions. The column headed r reports zero-order Pearson correlation coefficients (not controlling for any other variables). The column headed r_p reports the partial correlation between a given predictor variable and exercise sessions, with the remaining three predictor variables partialed out of both the predictor variable and the exercise sessions variable. Finally, the column headed r_{sp} reports the semipartial (part) correlation between a given predictor variable and the exercise sessions variable, with the remaining three variables partialed out of just the predictor variable.

The general trend illustrated in Table 8.2 shows that the variable *lack of time for exercise* tended to display the strongest correlation with exercise sessions per week. The zero-order correlation was $r = -.458$, $p < .001$. When controlling for the other three predictor variables, partial and semipartial correlations decreased to $-.358$ and $-.325$, respectively (for each, $p < .01$).

The variables *health motivation* and *dislike of exercise* each displayed somewhat weaker correlations with exercise sessions ($r = .293$, $p < .001$, and $r = -.373$, $p < .001$, respectively). After controlling for the other predictors, these variables each displayed partial and semipartial correlations that were still significantly different from zero, $p < .05$.

Finally, *desire to be thin* displayed the weakest zero-order correlation with exercise sessions, $r = .148$, $p < .05$. After controlling for the other three predictor variables, partial and semipartial correlations decreased to .052 and .044, respectively. With alpha set at .05, neither of the latter two statistics was significantly different from zero.

Partial correlation coefficients and semipartial correlation coefficients can be tested for statistical significance, just like zero-order correlations. In most cases, the null hypothesis is:

> In the population, the partial [or semipartial] correlation coefficient is equal to zero.

If the obtained p value for the coefficient is less than the selected value of alpha, the null hypothesis is rejected and the coefficient is said to be statistically significant. In the preceding table and text, Dr. O'Day followed the popular convention of indicating whether a given coefficient was statistically significant at the $p < .05$ level, the $p < .01$ level, or the $p < .001$ level.

Additional Issues Related to These Procedures

Methods of statistical control provide powerful tools for investigating correlational relationships, but researchers and readers of research should be aware of their limitations. This section discusses (a) the relative effectiveness of statistical control versus experimental control, and (b) common circumstances that can cause methods of statistical control to give misleading results.

Is Statistical Control as Good as Experimental Control?

No. When researchers wish to obtain strong evidence that one variable has a causal effect on a second variable, true experiments that make effective use of experimental control will almost always provide stronger evidence.

When used correctly, methods of statistical control can provide our best estimate of what the relationship between the predictor variable and criterion variable would have been if the confounding variable had been experimentally held constant, but the key term here is the word "estimate." In most cases, using sophisticated statistics to achieved high levels of statistical control will not produce results as convincing as might have been produced by a relatively simple true experiment.

Despite this, methods of statistical control continue to be useful tools and have led to important breakthroughs in a variety of disciplines. Their value lies in one simple truth: *Not all important topics can be studied using true experiments.* Due to ethical concerns and physical limitations, some phenomena can only be studied by investigating the naturally occurring relationships between naturally occurring variables. And methods of statistical control allow researchers to do science in those situations.

Potential Problems with Statistical Control

Computer applications such as SPSS and SAS have made it easy for just about anyone to use sophisticated data-analysis procedures that employ methods of statistical control. This is a good thing. It is also a bad thing.

Part of the problem is attributable to authors such as myself who say things like "Partial correlation allows us to investigate the relationship between X and Y while statistically controlling for the confounding variable Z." A reasonable person might read such a statement and respond "Oh. So it's okay if I run an experiment with an obvious confounding variable—I'll just statistically *control* for it later."

Although it has become easy to use methods of statistical control, it has become even easier to misuse them and misinterpret the results that they produce. There are many circumstances that can cause researchers to draw inappropriate conclusions from results produced by partial correlation, semipartial correlation, multiple regression, and similar procedures. This section discusses just three.

LACK OF A CLEAR THEORETICAL MODEL MAY LEAD TO MISLEADING RESULTS. In most cases, methods of statistical control should be applied only when researchers are very familiar with theory and previous research on the topic—sufficiently familiar so that they understand the likely causal relationships between the variables of interest. Gordon (1968) points out that when researchers present all possible higher-order partial correlations for a set of variables, it is a good sign that they probably have no such theory.

There are causal models for which partial correlations are appropriate and causal models for which they are not, and too few researchers know the difference. For more on this, see the *Causal Assumptions* section in *Chapter 5* of Pedhazur (1982). For a good introductory-level discussion of how theory and research should inform the choice of statistical control variables, see Cohen and Cohen (1983), especially pages 120-123.

FAILING TO CONTROL FOR ALL RELEVANT CONFOUNDING VARIABLES MAY LEAD TO MISLEADING RESULTS. In the *Results* section of her article, Dr. O'Day seemed to imply that *lack of time for exercise* was the most important predictor of *exercise sessions*. After all, this predictor not only displayed the strongest zero-order correlation with exercise sessions ($r = -.458$), but also displayed the strongest partial correlation ($r_p = -.358$) and the strongest semipartial correlation ($r_{sp} = -.325$). But is lack of time for exercise really the most important predictor of exercise sessions?

Maybe not. The size of a partial or semipartial correlation coefficient is very much influenced by which other control variables happen to be included in the analysis (Licht, 1995). It is possible that there is an additional confounding variable that Dr. O'Day failed to include in her investigation. If that variable had been included, it may have caused the coefficients for lack of time for exercise to become much smaller—in absolute size, as well as in relation to the coefficients for the other predictors. Again, this possibility points to the importance of having a solid, well-informed theoretical model prior to conducting the study.

UNRELIABLE MEASURES MAY LEAD TO MISLEADING RESULTS. Oh yes, one more minor point: Regression procedures (such as the ones related to partial correlation) typically assume that predictor variables are measured without error (Pedhazur, 1982, pp. 33-34). The extent to which variables are free from error is reflected in the reliability estimates for those variables—the lower the reliability, the greater the error. Reliability is usually represented as a correlation coefficient with a symbol such as r_{xx}, or as a statistic similar to a correlation coefficient, such as the coefficient alpha reliability estimate that was discussed in *Chapter 3* (symbol: α). Reliability estimates usually range from .00 through 1.00, with higher values indicating better reliability.

With zero-order correlations, poor reliability tends to result in sample correlation coefficients that underestimate the actual relationship between the variables in the population. Unfortunately, the situation is more complex with partial correlations, for which poor reliability may result in sample coefficients that either underestimate or overestimate the actual relationships. Pedhazur (1982, pp. 112-114) provides formulas that allow researchers to correct partial correlation coefficients for unreliability of measures.

The moral of the story is this: You should always take partial and semipartial correlations with a grain of salt, but you should take them with an extra-large grain of salt when the variables included in the analysis do not display really high reliability coefficients. Many would define "really high" reliability coefficients as those that exceed .80, or possibly .90.

In recent decades, many researchers have addressed this issue of reliability by performing a relatively sophisticated data-analysis procedure called *structural equation modeling* (abbreviation: SEM). SEM allows researchers to take into account the reliability of their measures, so that the parameter estimates that they compute are less biased. Structural equation modeling is covered in this book in *Chapter 16*, *Chapter 17*, and *Chapter 18*.

Assumptions Underlying these Procedures

Warner (2008) provides the following assumptions underlying partial correlation. Some of these assumptions are described in greater detail in *Chapter 2* of the current book.

QUANTITATIVE VARIABLES. The criterion variable should be a quantitative variable. Although the predictor variable and the control variable are typically quantitative variables, either of these may alternatively be a dichotomous variable.

NORMAL DISTRIBUTIONS. Scores on the criterion, predictor, and control variables should display an approximate normal distribution.

LINEARITY. There should be a linear relationship between each pair of variables.

BIVARIATE NORMALITY. The joint distribution of scores for each pair of variables should be bivariate normal.

HOMOGENEITY OF VARIANCE. For each pair of variables, scores on the criterion variable should be constant across all values of the predictor variable.

HOMOGENEITY OF THE REGRESSION LINES. The regression coefficient that describes the relationship between the criterion variable and the predictor variable should be the same at each level of the control variable.

Chapter 9

Multiple Regression I: Basic Concepts

This Chapter's *Terra Incognita*

You are reading an article in a sport-psychology journal. It describes a correlational study in which the criterion variable was the *number of aerobic exercise sessions* that participants engaged in each week. The predictor variables were *health motivation, desire to be thin, dislike of exercise,* and *lack of time for exercise* (all measured using Likert-type summated rating scales). The *Results* section says the following:

> Data were analyzed using multiple regression with simultaneous entry of predictor variables. In this analysis, the criterion variable was exercise sessions per week, and the predictor variables were the four constructs that constitute the four-factor model of exercise.
>
> Results showed that the linear combination of predictor variables accounted for 28% of the variance in exercise sessions, $R^2 = .282$, $F(4, 233) = 22.84$, $p < .001$. According to Cohen's (1988) guidelines, this qualifies as a large effect.
>
> Standardized multiple regression coefficients for the four predictor variables were as follows: (a) health motivation, $\beta = .153$, $p = .013$; (b) desire to be thin, $\beta = .046$, $p = .430$; (c) dislike of exercise, $\beta = -.182$, $p = .004$; and (d) lack of time for exercise, $\beta = -.354$, $p < .001$.

Have courage. All shall be revealed.

The Basics of this Procedure

Multiple regression is arguably the single most important statistical procedure used to analyze data from correlational studies. This claim may surprise you, as many students complete their elementary statistics course without even hearing the term *multiple regression*. But this procedure is important, and the reason is simple: It is an extremely flexible procedure, capable of analyzing a variety of different types of variables from a variety of research designs. This flexibility is reflected in the generic analysis model for multiple regression, presented as Figure 9.1.

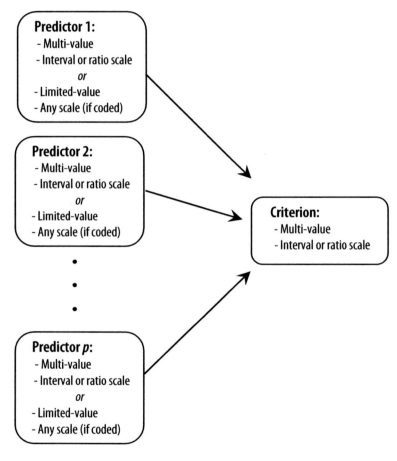

FIGURE 9.1. Generic analysis model: Nature of the variables included in a typical multiple regression analysis.

Figure 9.1 shows that, in the prototypical analysis, multiple regression involves a single criterion variable and two or more predictor variables. In the figure, the last predictor variable is named "Predictor p," where "p" represents the total number of predictor variables in the analysis. Theoretically there might be any number of predictors, although it is uncommon to see more than a dozen or so in practice.

The criterion variable must be a quantitative variable assessed on an interval scale or ratio scale, and it is usually a ***multi-value variable*** (a variable that displays seven or more values). If the criterion variable is ***dichotomous*** (i.e., if it displays just two values), it may be more appropriate to analyze the data using logistic regression (*Chapter 12*) or possibly discriminant analysis (*Chapter 11*).

Figure 9.1 shows that two types of variables may serve as predictor variables in multiple regression. In the prototypical multiple regression analysis, each predictor variable is a multi-value quantitative variable assessed on an interval or ratio scale of measurement. The current chapter shows how multiple regression is performed and interpreted when all predictors are of this type.

However, it is also possible that one or more of the predictors may be a ***limited-value variable***—a variable that contains just two to six categories. The boxes for the predictor variables in Figure 9.1 include the note "*if coded.*" This refers to the fact that these limited-value variables may be assessed on any scale of measurement as long as they have been properly transformed using dummy-coding, effect coding, or some related procedure. Most of the widely used statistical applications such as SAS or SPSS can easily transform a limited-value variable for inclusion in a multiple regression analysis. *Chapter 10* shows the results of a multiple regression analysis that includes dummy-coded variables.

The current chapter covers the most basic application of multiple regression. This procedure goes by a number of names, including:

- direct multiple regression,
- standard multiple regression,
- multiple regression with forced-entry of predictor variables, and
- multiple regression with simultaneous entry of predictor variables.

In this context, *simultaneous entry* simply means that the predictor variables are entered into the multiple regression equation all at once, in a single step. We make a distinction between this approach versus the approach used with *multiple regression with hierarchical entry*. With hierarchical entry, the predictor variables are entered into the equation in a sequence of multiple steps, with the order of entry pre-determined by the researcher. *Chapter 10* covers multiple regression with hierarchical entry.

Illustrative Investigation

Assume that a fictitious health psychologist named Dr. O'Day wants to identify variables that predict the frequency with which people engage in aerobic exercise. In her current study, the criterion variable is *exercise sessions*: the number of times that a participant engages in aerobic exercise each week. This is a quantitative variable assessed on a ratio scale of measurement.

THE MODEL. After reviewing relevant theory and research, Dr. O'Day developed the *four-factor model of exercise behavior*. This model is a type of "mini-theory" that attempts to identify the most important determinants of exercise behavior. It is illustrated in Figure 9.2 (please remember that this four-factor model of exercise behavior is fictitious—you will not find it in the actual research literature).

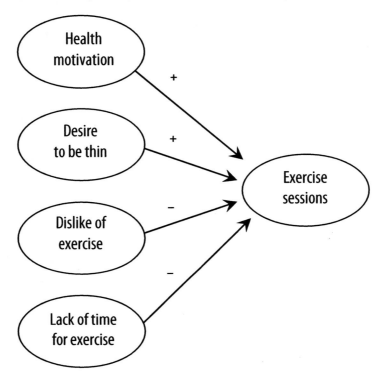

Figure 9.2. The four-factor model of exercise behavior.

It is called the *four-factor model* because it hypothesizes that the criterion variable of interest (number of exercise sessions per week) is determined by four predictor variables. The positive symbols (+) and negative symbols (–) in the figure indicate whether a given predictor variable is hypothesized to have a positive effect or a negative effect on the criterion variable.

Remember that, although Dr. O'Day's model predicts causal relationships between the variables, this does not mean that the multiple regression analyses described here are capable of producing strong evidence that these relationships are *actually* causal in nature. In *Chapter 2*, you learned that nonexperimental studies (such as the one to be described here) typically provide only weak evidence of cause-and-effect.

MEASURING THE VARIABLES. Each of the four predictor variables in Dr. O'Day's model is assessed by a multiple-item summated rating scale constructed from items on the *Questionnaire on Eating and Exercising,* which appears in *Appendix A* of this book. With each of these summated rating scales, possible scores may range from

1.00 to 7.00, with higher scores representing higher levels of the construct being assessed.

The four variables that constitute the four-factor model of exercise were described in some detail in *Chapter 6,* and are once again briefly described below. The item numbers for the questionnaire items measuring a given variable are provided within parentheses:

- *Health motivation.* When individuals score high on *health motivation*, it means that they are willing to sacrifice and work hard in order to stay healthy (items 27-29 on the *Questionnaire on Eating and Exercising*).

- *Desire to be thin.* When individuals score high on *desire to be thin*, it means that it is important to them to have a fairly low body weight and a lean physical appearance (items 31-33).

- *Dislike of exercise.* When individuals score high on *dislike of exercise*, it means that they hate aerobic exercise and find it to be unpleasant (items 47-49).

- *Lack of time for exercise.* When individuals score high *lack of time for exercise,* it means that they believe that they are too busy to engage in frequent aerobic exercise (items 51-53).

This chapter will continue the standard convention of using the symbol Y to represent the criterion variable (exercise sessions, in this case), and using the symbol X_i to represent predictor variables (e.g., X_1 will represent health motivation, X_2 will represent desire to be thin, and so forth).

Research Questions Addressed

Multiple regression is a very flexible procedure, capable of answering a wide array of research questions. Below are the most basic questions addressed in the typical analysis:

- Is there a statistically significant relationship between the criterion variable and the predictor variables, taken as a set?

- How strong is the relationship between the criterion variable and the predictor variables, taken as a set?

- What is the nature of the relationship between the criterion variable and the individual predictor variables in the multiple regression equation? Specifically:

    o What is the sign of the multiple regression coefficient for each predictor?

    o How large is the multiple regression coefficient for each predictor?

    o How precise is the confidence interval for each multiple regression coefficient?

Results from the Investigation

Imagine that Dr. O'Day analyzed the data from the study described above, and published her findings in a scholarly journal. Excerpts from the resulting fictitious article are presented and explained in this section.

Bivariate Correlations

When researchers analyze data using a more advanced procedure such as multiple regression, a standard convention is to compute simple bivariate Pearson correlations for all possible pairings of the quantitative variables, and to present these correlations early in the *Results* section. Dr. O'Day does this in the following excerpt:

Table 9.1

Pearson Correlations (and Coefficient Alpha Reliability Estimates) for the Study's Variables, Decimals Omitted

Variable	M	SD	Pearson correlations				
			1	2	3	4	5
1. Exercise sessions per week	2.93	2.31	(--)				
2. Health motivation	5.76	0.98	29	(83)			
3. Desire to be thin	5.30	1.35	15	30	(85)		
4. Dislike of exercise	2.36	1.31	-37	-31	-15	(92)	
5. Lack of time for exercise	3.67	1.71	-46	-20	-08	39	(95)

Note. N = 238. For variables 2 through 5, possible scores could range from 1.00 to 7.00, with higher scores representing higher levels of the construct being measured. Correlations are significant at $p < .05$ for $r > \pm.127$; correlations are significant at $p < .01$ for $r > \pm.167$. Values on diagonal are coefficient alpha reliability estimates for multiple-item summated rating scales.

> **Bivariate correlations** Pearson correlations were computed for all possible pairings of the study's five variables. These correlations, along with means, standard deviations, and coefficient alpha reliability estimates are presented in Table 9.1.

The size of the "effect" represented by each correlation in Table 9.1 was determined using Cohen's (1988) criteria in which $r = \pm.10$ represents a small effect, $r = \pm.30$ represents a medium effect and $r = \pm.50$ represents a large effect. Table 9.1 shows that exercise sessions demonstrated the strongest bivariate correlation with lack of time for exercise ($r = -.46$, $p < .01$, a medium effect), followed by dislike of exercise ($r = -.37$, $p < .01$, a medium effect), health motivation ($r = .29$, $p < .01$, just short of a medium effect), and desire to be thin ($r = .15$, $p < .05$, a small effect). The sign for each of these Pearson correlations was in the predicted direction, and each correlation was statistically significant with alpha set at $\alpha = .05$.

ADVANTAGES OF PRESENTING A CORRELATION MATRIX. Presenting a table of correlations such as Table 9.1 serves two purposes. First, it provides readers with a understanding of the basic relationships between the variables: which predictor variables were strongly correlated with one another, which predictor variables were relatively independent of one another, and so forth. Second, it allows readers to see whether the bivariate correlations are consistent with the researcher's theoretical model which, in this case, had been presented in Figure 9.2. In her excerpt, Dr. O'Day indicates that the signs of the correlations were consistent with the signs that had been predicted in Figure 9.2, although some of the correlations were fairly weak.

ADDITIONAL STATISTICS REPORTED IN THE TABLE. In addition to presenting the Pearson correlations for all pairings of variables, Table 9.1 also presents each variable's mean under the heading "*M*" and each variable's standard deviation under the heading "*SD*." On the diagonal of the correlation matrix, it presents the coefficient alpha reliability estimate for each variable that had been assessed using a multiple-item summated rating scale. In *Chapter 3*, you learned that **coefficient alpha** is an estimate of internal-consistency reliability, and that its values typically range from .00 to 1.00, with higher values indicating better reliability. Values above .70 are generally viewed as being minimally acceptable. Coefficient alpha is not reported for exercise sessions per week, because this variable had not been assessed using a multiple-item scale.

Results from the Multiple Regression Analysis

Having covered the simple relationships between the variables, Dr. O'Day now turns to the analysis of central importance: the multiple regression procedure. She reports this in the following excerpt:

> **Multiple regression analysis.** Results were analyzed using multiple regression with simultaneous entry of predictor variables. In this analysis, the criterion variable was exercise sessions per week, and the predictor variables were the four predictors constituting the four-factor model of exercise behavior. Results of this regression are presented in Table 9.2.
>
> Table 9.2 shows that the linear combination of predictor variables accounted for 28% of the variance in exercise sessions, $R^2 = .282$, $F(4, 233) = 22.84$, $p < .001$. This qualifies as a large effect according to Cohen's (1988) guidelines.

The sign of the multiple regression coefficient for each predictor variable was in the predicted direction, although one of the predictor variables (desire to be thin) displayed a standardized multiple regression coefficient that was not statistically significant ($\beta = .046$, $p = .430$). The squared semipartial correlation coefficient for desire to be thin showed that it accounted for essentially no variance in the criterion variable, beyond the variance already accounted for by the other three predictors, ($r_{sp}^2 = .002$, ns).

The remaining three predictor variables each displayed a multiple regression coefficient that was statistically significant ($p < .05$). Squared semipartial correlation coefficients showed that each of the remaining predictors accounted for 2% to 11% of the variance in exercise sessions beyond the variance accounted for by the other predictors.

Table 9.2

Multiple Regression Results: Predicting Exercise Sessions per Week from Four Predictor Variables

Predictor variable	b	SE	95% CI	p	β	r_{sp}^2
(Constant)	2.938	1.004	[.959, 4.917]	.004		
1. Health motivation	0.361	0.144	[.078, .644]	.013	.153	.020
2. Desire to be thin	0.079	0.100	[-.118, .276]	.430	.046	.002
3. Dislike of exercise	-0.320	0.110	[-.537, -.103]	.004	-.182	.026
4. Lack of time for exercise	-0.478	0.082	[-.638, -.317]	<.001	-.354	.106

Note. $N = 238$. Model $R = .531$, Model $R^2 = .282$, $F(4, 233) = 22.835$, $p < .001$. Adjusted $R^2 = .269$. b = unstandardized multiple regression coefficient. SE = standard error. 95% CI = 95% confidence interval. p = probability value. β = standardized multiple regression coefficient. r_{sp}^2 = squared semipartial correlation coefficient (partialing out remaining three predictor variables).

There is a lot to digest in the preceding excerpt, so we will take things one step at a time. First, the reader needs to know whether there is a significant and substantial relationship between the criterion variable and the four predictor variables taken as a set. This relationship is discussed in the following section.

Interpreting the R^2 Statistic

As a reader of this article, your first priority is to understand the big picture: Is there much of a relationship between (a) exercise sessions, and (b) the four predictor variables taken as a set? To find out, you would review the R^2 statistic for the full multiple regression equation. This R^2 statistic is sometimes call the ***coefficient of multiple determination***. It indicates the percent of variance in the criterion variable that is accounted for by the linear combination of predictor variables (we will see what is meant by this term *linear combination* a bit later). Values of R^2 may range from 0.00 to 1.00. A value of 0.00 means that there is no relationship between the criterion and the set of predictors, and higher values indicate stronger relationships.

Dr. O'Day provided the value of R^2 for the current analysis in the note at the bottom of Table 9.2, as well as in the text of her article. The latter is reproduced once again below:

> Table 9.2 shows that the linear combination of predictor variables accounted for 28% of the variance in exercise sessions, $R^2 = .282$, $F(4, 233) = 22.84$, $p < .001$.

INTERPRETING THE STATISTICAL SIGNIFICANCE OF THE R^2 VALUE. First we will determine whether the obtained coefficient of multiple determination is statistically significant. When researchers report R^2, they typically also report an F statistic that tests the following null hypothesis:

> H_0: $R^2 = .00$ .In the population, the linear combination of predictor variables accounts for 0% of the variance in the criterion variable.

The preceding excerpt indicated that $R^2 = .282$, which means that the four predictor variables account for about 28% of the variance in exercise sessions. An F statistic is used to determine whether the obtained value of R^2 ($R^2 = .282$, in this case) is significantly larger than 0.00. Dr. O'Day reported her F statistic as

$F(4, 233) = 22.835$, $p < .001$.

In this excerpt, the numbers within parentheses are the degrees of freedom for the F test. The first number (the "4" within the parentheses) represents the degrees of freedom for the numerator (symbol: df_{num}). For this test, the df_{num} is equal to k, the number of predictor variables included in the multiple regression equation. Dr. O'Day's multiple regression equation included four predictor variables, so $df_{num} = 4$.

The second number (the "233" within the parentheses), represents the degrees of freedom for the denominator (symbol: df_{den}). With this F test, the degrees of freedom for the denominator are equal to $N - k - 1$, where $N =$ the total number of subjects in the study. For the current study, $N = 238$. This means that the degrees of freedom for the denominator are therefore computed as $df_{den} = 238 - 4 - 1 = 233$.

The excerpt indicated "$F(4, 233) = 22.835$," and this tells us that the obtained F statistic was 22.835. When the null hypothesis is true (e.g., when $R^2 = 0.00$ in the population), we expect the F statistic to be approximately equal to 1.00; the larger the obtained value of the F statistic, the lower the probability that this sample came from a population in which R^2 is equal to 0.00. This obtained F statistic of 22.835 is pretty far away from the F of 1.00 that would have been expected if the null hypothesis were true. To determine whether it is far enough away from an F of 1.00 to reject the null hypothesis, we consult the obtained p value.

A ***p value***, or ***probability value***, provides the probability that the researcher would have obtained the current sample results if the null hypothesis were true. In Dr. O'Day's study, the p value for the F statistic is "$p < .001$," which is less than the standard criterion of $\alpha = .05$. Therefore, Dr. O'Day rejects the null hypothesis and concludes that her obtained R^2 value of .282 is significantly different from zero.

INTERPRETING THE ADJUSTED R^2 VALUE. The last part of the preceding excerpt indicated "adjusted $R^2 = .269$." This reminds us that the obtained R^2 value from a multiple regression analysis is usually ***positively biased***. This means that the obtained R^2 value (based on the sample) usually overestimates the actual value of R^2 in the population. This overestimation is typically worse when the sample is relatively small, and when the number of predictor variables is relatively large. Fortunately, formulas are available to compute ***adjusted R^2***: an unbiased estimate of what R^2 is likely to be in the population. Dr. O'Day reports that adjusted $R^2 = .269$, which is just a bit smaller than the unadjusted value of $R^2 = .282$ that was reported earlier.

Table 9.3

Criteria for Evaluating Effect Size Represented by the R^2 obtained in a Multiple Regression Analyses

Label for effect size	R^2
Small	.02
Medium	.13
Large	.26

INTERPRETING R^2 AS AN INDEX OF EFFECT SIZE. To label the R^2 obtained in a study as representing a small versus a medium versus a large effect, Cohen (1988) recommends that researchers identify previous studies that investigated the same phenomena, and compare the R^2 value for the current study against the R^2 values obtained in previous research. When this is not possible, he offers the criteria presented in Table 9.3 as a last resort.

Dr. O'Day obtained a value or $R^2 = .282$ for the multiple regression equation. This constitutes a large effect, according to the criteria in Table 9.3.

You should not use the cut scores in Table 9.3 if you are evaluating the incremental R^2 (symbol: ΔR^2) that you have computed when using hierarchical multiple regression. Hierarchical (also called *block-wise*) multiple regression involves adding predictor variables to an equation in a series of predetermined steps. This approach to multiple regression will be covered in *Chapter 10*.

Interpreting the Multiple Regression Coefficients

Now that we have determined that the overall relationship between the criterion variable and the set of predictor variables is both significant and substantial, we can turn our attention to the individual predictor variables. We will be especially interested in the multiple regression coefficient (symbol b or β) for each predictor. The **multiple regression coefficient** for a given predictor variable represents the amount of change in the criterion variable that is associated with a one-unit change in the predictor, while statistically holding constant the other predictor variables.

THE MULTIPLE REGRESSION EQUATION (GENERAL FORM). When discussing multiple regression equations, a standard convention is to use the symbol "*Y*" to represent the criterion variable, and the symbol "*X*" to represent a given predictor variable. *Chapter 7* of this book covered bivariate regression, and in that chapter you learned the form of a basic regression equation that contained just one predictor variable:

$$Y' = a + b(X)$$

Multiple regression is conceptually similar to bivariate regression, except that multiple regression allows the equation to have more than one predictor variable (and this makes a *big* difference in the usefulness of multiple regression, as opposed to bivariate regression).

Below is the general form for the multiple regression equation:

$$Y' = a + b_1(X_1) + b_2(X_2) + b_3(X_3) \ldots + b_p(X_p)$$

Where:

> Y' represents a given subject's predicted score on the criterion variable. Y' is a *synthetic variable*—it is a linear composite of the optimally weighted predictor variables. Each subject has a score on this synthetic variable, and this score represents the equation's "best guess" as to where that subject probably stands on the criterion.
>
> a represents the **constant** (i.e., the intercept) of the regression equation. As was the case with bivariate regression, this constant is typically not of much interest to researchers—it is simply a constant number that must be included in the equation so that the Y' scores will be on the same metric

as was used with the actual criterion. Some textbooks and articles use the symbol b_0 to represent the intercept.

b_1 represents the unstandardized **multiple regression coefficient** for the first predictor variable (e.g., the first X variable: X_1). These coefficients are sometimes called *b weights*. The *b* coefficient for a given X variable represents the amount of change in Y that is associated with a one-unit change in X, while statistically holding constant the other X variables. Because the other X variables have been held constant, multiple regression coefficients are also called *partial regression coefficients* (Licht, 1995).

X_1 represents a given subject's actual score on the first predictor variable.

b_p represents the unstandardized multiple regression coefficient for the pth (final) predictor variable in the equation.

X_p represents the pth (final) predictor variable in the equation.

THE MULTIPLE REGRESSION EQUATION (DR. O'DAY'S RESULTS). When Dr. O'Day analyzed data from the 238 participants in her study, the analysis produced the following multiple regression equation (this is the unstandardized version of the equation):

$$Y' = 2.938 + 0.361(X_1) + 0.079(X_2) - 0.320(X_3) - 0.478(X_4)$$

With this equation:

Y' represents the participants' *predicted scores on Y* (number of exercise sessions engaged in each week).

The "2.938" is the constant, or intercept, for the equation.

The "0.361" is the unstandardized multiple regression coefficient for X_1 (the first predictor variable).

The "0.079" is the unstandardized multiple regression coefficient for X_2 (the second predictor variable).

And the remaining regression coefficients may be interpreted in the same way.

Of the various terms presented here, those of greatest interest are usually the unstandardized multiple regression coefficients (i.e., the *b* weights) for the predictor variables. Readers of research articles are particularly interested in (a) whether a given coefficient has a positive sign or a negative sign, and (b) whether a given coefficient is statistically significant.

INTERPRETING THE SIGN OF A MULTIPLE REGRESSION COEFFICIENT. First, the sign of the regression coefficient (i.e., + versus −) reveals the direction of the relationship between a given predictor variable and the criterion variable. We interpret a sign of a multiple regression coefficient in a way analogous to the way that we interpret the sign of a bivariate regression coefficient:

- If the sign of the regression coefficient for a specific X variable is *positive* (e.g., $b_1 = +0.361$), it means that there is a positive relationship between that X variable and the Y variable after statistically controlling for the other X variables. In other words, as values on this X variable increase, values on the Y variable also tend to increase (after controlling for the other X variables).

- If the sign of the regression coefficient is *negative* (e.g., $b_3 = -0.320$), it means that there is a negative relationship between that X variable and the Y variable, after statistically controlling for the other X variables: As values on this X variable increase, values on the Y variable tend to decrease.

- If the regression coefficient for a specific X variable is equal to *zero* (e.g., $b_p = 0.000$), it means that there is *no relationship* between that X variable and the Y variable, after statistically controlling for the other X variables. In other words, as values on the X variable increase, values on the Y variable do not change in any systematic way (after controlling for the other X variables).

The nonstandardized b weights from Dr. O'Day's investigation had been presented earlier. They are again reproduced below so that we can interpret their sign:

$$Y' = 2.938 + 0.361(X_1) + 0.079(X_2) - 0.320(X_3) - 0.478(X_4)$$

In the preceding equation, X_1 represents the predictor variable *health motivation*, and the b weight for this predictor variable is +0.361. The sign and size of this coefficient indicates that:

- for every one-unit increase in health motivation,
- there is an increase of 0.361 units in exercise sessions,
- while statistically controlling for the remaining predictor variables.

Remember that health motivation was measured using a scale that ranged from 1.00 to 7.00. Given the size of the b weight for this predictor variable, the equation says that (for example), if an individual were to increase his score on health motivation from a score of 4.00 to a score of 5.00 (on the 7-point scale), we would expect to see him engage in 0.361 more exercise sessions per week. Further, we would expect to see this change even if we held constant the other three predictor variables.

In the same multiple regression equation, X_3 represents the predictor variable *dislike of exercise*, and the regression coefficient for that predictor is negative: −0.320. The sign and size of this coefficient indicates that:

- for every one-unit increase in dislike of exercise,
- there is a *decrease* of 0.320 units in exercise sessions,
- while statistically controlling for the remaining predictor variables.

DETERMINING WHICH COEFFICIENTS ARE STATISTICALLY SIGNIFICANT. Computer applications typically use a *t* test to determine whether an obtained multiple regression coefficient is significantly significant. This is almost always a test of the null hypothesis that the regression coefficient is equal to zero in the population. The degrees of freedom for the *t* test are computed as $df = N - k - 1$ where N represents the total sample size and k represents the number of predictor variables included in the multiple regression equation. For the current study, $N = 238$ and $k = 4$, so the *df* are computed as $df = N - k - 1 = 238 - 4 - 1 = 233$.

Table 9.2, which appeared earlier in this chapter, had reported the significance test for each of the multiple regression coefficients in Dr. O'Day's study. That table is reproduced here as Table 9.4.

Table 9.4

Predicting Exercise Sessions per Week: Results from Multiple Regression Analysis (Multiple Regression Coefficients and Related Statistics)

Predictor variable	b	SE	95% CI	p	β	r_{sp}^2
(Constant)	2.938	1.004	[.959, 4.917]	.004		
1. Health motivation	0.361	0.144	[.078, .644]	.013	.153	.020
2. Desire to be thin	0.079	0.100	[-.118, .276]	.430	.046	.002
3. Dislike of exercise	-0.320	0.110	[-.537, -.103]	.004	-.182	.026
4. Lack of time for exercise	-0.478	0.082	[-.638, -.317]	<.001	-.354	.106

Note. $N = 238$. Model $R = .531$, Model $R^2 = .282$, $F(4, 233) = 22.835$, $p < .001$. Adjusted $R^2 = .269$. b = unstandardized multiple regression coefficient. SE = standard error. 95% CI = 95% confidence interval. p = probability value. β = standardized multiple regression coefficient. r_{sp}^2 = squared semipartial correlation coefficient (partialing out remaining three predictor variables).

The unstandardized multiple regression coefficients appear under the heading "*b*." You can see that the *b* weight for health motivation is 0.361, the *b* weight for desire to be thin is 0.079, and so forth. The standard error used in the significance tests for the *b* weights appears under "*SE*." The *t* statistics themselves are not reported, but the *p* values for significance tests appear under the heading "*p*." With alpha set at α = .05, a multiple regression coefficient is viewed as being significantly different from zero if its *p* value is less than .05. Using this criterion, Table 9.4 shows that the *b* weight for desire to be thin is nonsignificant ($p = .430$), but that the *b* weight for each of the remaining three predictors is significant.

CONFIDENCE INTERVALS FOR MULTIPLE REGRESSION COEFFICIENTS. The 95% confidence intervals for the nonstandardized multiple regression coefficients appear under the heading 95% CI. These intervals provide plausible values for the actual values of the multiple regression coefficients in the population. They are useful because they give some sense regarding the precision of the sample *b* weights as estimates of the population *b* weights. Table 9.4 shows that the 95% CI for the *b* weight for health motivation extends from 0.078 to 0.644. Many researchers would interpret this as being a fairly wide interval. Confidence intervals for the remaining predictors may be interpreted in the same way.

DISADVANTAGES OF USING UNSTANDARDIZED *b* WEIGHTS AS AN INDEX OF "IMPORTANCE." Now that you can interpret the basic aspects of a *b* weight, it would be tempting to use the relative size of the coefficients as an index of their relative importance, assuming that the predictors with larger *b* weights (in absolute value) must be the more important predictors of the criterion variable. But this would be a mistake, as the *b* weights can be very misleading when interpreted this way.

In general, the unstandardized multiple regression coefficient for a given predictor variable should *not* be used as an index of that predictor variable's importance, because the size of the unstandardized *b* weights is very much influenced by the size of the numbers that are used as values for the predictor variable in question. Other things held constant, if a predictor variable consists of relatively large numbers (such as 250, 500, 700, and so forth), the unstandardized multiple regression coefficient will tend to be a relatively *small* number (such as *b* = .0034), regardless of how important it is as a predictor. On the other hand, if a predictor variable consists of small numbers (such as 0.2, 1.5, 3.7, and so forth), the unstandardized multiple regression coefficient will tend to be a relatively *large* number (such as *b* = 23.5565), again, regardless of how important it is as a predictor.

In other words, the size of an unstandardized coefficient tends to reflect the *metric used with the predictor variable*, not necessarily the importance variable as a predictor. Therefore, it is generally not appropriate to use unstandardized multiple regression coefficients (i.e., *b* weights) when evaluating the relative importance of predictors within a multiple regression equation (Newton & Rudestam, 1999). As an alternative, many researchers instead review the standardized multiple regression coefficients, which are discussed next.

STANDARDIZED MULTIPLE REGRESSION COEFFICIENTS. Because of the obvious problems associated with the unstandardized regression coefficients, many readers have turned to standardized multiple regression coefficients (i.e., β weights) to assess the relative importance of predictor variables. But don't get too excited just yet: Although the standardized coefficients have advantages over the nonstandardized coefficients, they are also problematic when used as measures of importance. Before discussing these problems, however, let's first define our terms.

The ***standardized multiple regression coefficient*** for a specific predictor variable (*X*) represents

- the amount of change in *Y* (as measured in standard deviations) that is associated with a one-standard deviation change in that *X* variable,
- while statistically controlling for the remaining *X* variables,
- while all of the variables are standardized to have a mean of zero and a standard deviation of one (in other words, when all variables are in *z*-score form).

Many books (including this book) use the Greek letter *beta* (β) to represent standardized multiple regression coefficients. Authors who use this symbol often refer to these statistics as ***beta weights***. The *Publication Manual of the American Psychological Association* (2010) recommends that the symbol b^* should be used to represent standardized regression coefficients, although it is not yet clear how long it will take for that recommendation to catch on among researchers.

Under typical research conditions, a beta weight may range in size from -1.00 through 0.00 through $+1.00$ (although, under certain conditions, an obtained beta weight may fall outside of these bounds). The sign of the coefficient is interpreted in the same way that we interpret the sign of the nonstandardized multiple regression coefficients (i.e., the *b* weights) described earlier. When the beta weight for a given *X* variable is close to zero, it means that the multiple regression equation gives little weight to that *X* variable in the prediction of *Y*. When the beta weight for a given *X* variable is relatively large in absolute value (i.e., closer to either -1.00 or $+1.00$), it means that the multiple regression equation gives a great deal of weight to that *X* variable.

Once a statistical application has performed significance tests for the unstandardized multiple regression coefficients (i.e., the *b* weights), it is not necessary to perform a separate significance test for the standardized coefficients (i.e., the beta weights). In other words, if the *b* weight for a given predictor is statistically significant, the beta weight for that predictor is also significant.

Many researchers interpret the size of the standardized multiple regression coefficient for a given predictor variable as an index of that predictor variable's importance, relative to the other predictor variables. With this convention, beta weights that are relatively close to zero are viewed as being less important predictors, whereas beta weights that are closer to 1.00 (in absolute value) are viewed as being more important predictors.

On the superficial level, beta weights would appear to be well-suited as indices of the relative importance of predictor variables precisely because they are *standardized* coefficients. You will recall that the preceding section indicated that the *b* weights (the unstandardized regression coefficients) for different predictor variables cannot be directly compared because their size is very much influenced by the metric used to measure the different variables. In contrast, beta weights are based on variables that have been standardized to have a mean of zero and a standard deviation of one. With beta weights, all of the variables are now based on the same metric, and it would therefore seem appropriate to compare the multiple regression coefficients for different predictor variables. Unfortunately, there is a total different set of problems that can cause beta weights to be misleading when they are interpreted as an index of a predictor variable's importance.

THE CORRELATED PREDICTOR VARIABLE PROBLEM. One reason that problems arise when beta weights are used as indices of importance is the so-called *correlated predictor variable problem.* This refers to the fact that, when predictor variables are correlated with one another in a nonexperimental study, it is typically impossible to draw unambiguous conclusions about the relative importance of the various predictors, even when working with standardized regression coefficients. Azen and Budescu (2009) provide a comprehensive discussion of the relevant issues, but the current chapter will attempt to convey the basic problem through the use of a quick fictitious example.

Imagine that you are conducting research on two different intelligence tests, *IQ Test A* and *IQ Test B*. When you administer these tests to 100 middle school students, the bivariate correlation between the two IQ tests is very strong: Pearson $r = .95$.

When you compute the bivariate correlation between IQ Test A and the grade point average (GPA) of the 100 students, the correlation is decent: Pearson $r = .40$. When you compute the bivariate correlation between IQ Test B and GPA, it is also decent, and just a smidgeon smaller: Pearson $r = .38$.

You now perform a multiple regression analysis in which the criterion variable is GPA and the two predictor variables are IQ Test A and IQ Test B. You find that the standardized multiple regression coefficient for IQ Test A is of a decent size: $\beta = .30$, but the standardized multiple regression coefficient for IQ Test B is essentially zero: $\beta = .02$. How could this be? The Pearson correlations showed that IQ Test B had a decent correlation with GPA (Pearson $r = .38$).

The problem was the fact that the two predictor variables (the two IQ tests) were so strongly correlated with one another. Because they were so strongly correlated with each other, the computer application determined that it was necessary to allocate a relatively large regression weight to just one of the predictors (IQ Test A, in this case) in order to do an adequate job of predicting the criterion, GPA. With this done, the computer application did not need to allocate any weight to the second predictor (IQ Test B). As a result, the beta weight for IQ Test A was large, whereas the beta weight for IQ Test B was near-zero.

The point is this: A near-zero multiple regression coefficient for a predictor variable does not necessarily mean that the variable is an unimportant predictor. This is true even when the coefficient is a standardized regression coefficient such as a beta weight (Thompson, 2000).

When researchers wish to evaluate the relative importance of predictor variables with correlational data, experts often recommend alternatives to the beta weights that are computed using multiple regression. These alternatives typically involve relatively new data-analysis procedures such as *dominance analysis* or *structural equation modeling* (Azen & Budescu, 2009; Pedhazur, 1982). These procedures are described in greater detail in the section titled "Newer Approaches for Studying the Importance of Predictors" near the end of this chapter.

EVALUATING THE SIZE OF A BETA WEIGHT. There do not appear to be any widely accepted criteria for labeling a standardized multiple regression coefficient as representing a small effect, a medium effect, and so forth. In part, this is because researchers understand that the size of *beta weights* is determined, in part, by the phenomena being studied—just how large a beta weight must be to be considered large will vary, depending on topic being studied.

Having said that, I could not resist the temptation to share one expert's criteria for evaluating the relative size of beta weights. In his well-regarded textbook on multiple regression and related procedures, Keith (2006) describes research that he has conducted in which the criterion variables were different measures of learning outcomes (variables such as *grades* and *academic achievement*), and the predictor variables were measures that were expected to be related to these outcomes (variables such as *time spent on homework, academic coursework,* and *parental involvement*). He used the following rules of thumb when evaluating the size of the standardized β coefficients obtained in this research:

- β < ±.05 were "too small to be considered meaningful influences on school learning, even when they are statistically significant."
- β > ±.05 were considered "small but meaningful."
- β > ±.10 were considered "moderate."
- β > ±.25 were considered large

Keith (2006) is very clear in warning that these criteria may not be appropriate for other areas of research. He encourages researchers to develop their own criteria for evaluating regression coefficients obtained from research involving different variables and different populations.

SQUARED SEMIPARTIAL CORRELATION COEFFICIENTS. In *Chapter 8*, you learned that a *semipartial correlation coefficient* is an estimate of the correlation between a predictor variable and a criterion variable after variance shared with one or more control variables has been partialed out of just the predictor variable (but not the criterion variable). Semipartial correlations are sometimes called *part correlations*, and are often represented using symbols such as *sr* or r_{sp} or $r_{Y(1\cdot2)}$.

When we square a semipartial correlation coefficient, the resulting statistic (r_{sp}^2) is sometimes referred to as the variable's *uniqueness* or *usefulness* (Darlington, 1968; Pedhazur, 1982), and may be represented using the abbreviation *UI* (for *usefulness index* or *uniqueness index*). The squared semipartial correlation coefficient for a given predictor variable represents the percent of variance in the criterion variable that is accounted for by that predictor, above and beyond the variance already accounted for by all of the other predictors.

Squared semipartial correlations are helpful because they provide information that is not provided by zero-order correlations (i.e., the Pearson correlations) or the multiple regression coefficients. Specifically, they can identify the predictors that account for *unique* variance—variance not accounted for by the other predictors.

The squared semipartial correlation coefficient for each of the four variables in Dr. O'Day's study had been presented earlier in Table 9.4, in the column headed r_{sp}^2. For convenience, those coefficients are presented again at this point, along with their p values:

- For health motivation, $r_{sp}^2 = .020$, $p = .013$.
- For desire to be thin, $r_{sp}^2 = .002$, $p = .430$.
- For dislike of exercise, $r_{sp}^2 = .026$, $p = .004$.
- For lack of time for exercise, $r_{sp}^2 = .106$, $p < .001$.

The preceding shows that the variable *lack of time for exercise* accounts for about 11% of variance in exercise sessions, beyond the variance accounted for by the other three predictor variables. The remaining three predictors each account for less than 3% of the unique variance. For the variable *desire to be thin*, $r_{sp}^2 = .002$, indicating that it accounts for essentially none of the variance in the criterion variable, beyond the variance accounted for by the other predictors.

Summary of Findings

In summary, the multiple regression analysis showed that there was a statistically significant relationship between (a) exercise sessions per week and (b) the four predictor variables, taken as a set, $R^2 = .282$, $F(4, 233) = 22.84$, $p < .001$. The strength of this relationship qualifies as a large effect according to Cohen's (1988) guidelines.

The sign of the multiple regression coefficient for each predictor variable was in the predicted direction, although one of the predictor variables (desire to be thin) displayed a standardized multiple regression coefficient that was not statistically significant ($\beta = .046$, $p = .430$). The squared semipartial correlation coefficient for desire to be thin showed that it accounted for essentially no variance in the criterion variable, beyond the variance already accounted for by the other three predictors, ($r_{sp}^2 = .002$, ns).

The remaining three predictor variables each displayed a multiple regression coefficient that was statistically significant ($p < .05$ or smaller). Squared semipartial correlation coefficients showed that each of the remaining predictors accounted for 2% to 11% of the variance in exercise sessions beyond the variance accounted for by the other predictors.

Additional Issues Related to this Procedure

But wait. *There's more!*

This section describes the differences between three different versions of multiple regression that you are likely to encounter in the research literature: simultaneous, hierarchical, and stepwise. It also introduces some newer approaches for determining the relative importance of predictor variables in a correlational study.

Simultaneous, Hierarchical, and Stepwise Entry of Predictors

Articles describing a multiple regression analysis may use the terms *simultaneous regression, hierarchical regression,* or *stepwise regression*. These names refer to different approaches for adding predictor variables to the multiple regression equation. As we shall see, the first two are scientifically respectable; the last one is something that you should use only when no one is looking.

SIMULTANEOUS MULTIPLE REGRESSION. With ***simultaneous multiple regression***, the predictor variables are added to the multiple regression equation all at once, in a single step. In some ways, this is the most elementary version of multiple regression: The researcher estimates just one regression model, and interprets the results. Because it is the most basic approach, this chapter focused on simultaneous multiple regression.

HIERARCHICAL MULTIPLE REGRESSION. In contrast, with ***hierarchical multiple regression***, predictor variables are entered into the regression equation in discrete steps, in a sequence *pre-determined* by the researcher (the *pre-determined* part is important). For example, the researcher might begin at Step 1 with a multiple regression equation that contains predictors X_1 and X_2. At Step 2, the researcher adds the predictor X_3 to the equation that already contains X_1 and X_2. At this point, the researcher determines whether adding X_3 resulted in a substantial improvement in the equation's ability to predict the criterion variable. Finally, at Step 3, the researcher might add predictors X_4 and X_5 to the equation that already contains X_1, X_2, and X_3. Again, the researcher determines whether adding these new variables resulted in a substantial improvement in the equation's ability to predict the criterion.

Hierarchical multiple regression goes by a number of names. Some sources refer to it as *sequential multiple regression, user-determined regression, multiple regression with hierarchical entry,* and *multiple regression with block-wise entry*. Hierarchical regression will be covered in some detail in *Chapter 10*.

STEPWISE MULTIPLE REGRESSION. With ***stepwise multiple regression***, the statistical application uses a mathematical algorithm to develop a type of "optimal" multiple regression model. It is typically used in situations in which the researchers have measured a relatively large number of predictor variables, and wish to reduce this to a relatively small number of predictors that account for a maximal amount of variance in the criterion. The statistical algorithms that do this have names such as the *forward selection method*, the *backward selection method*, the *stepwise selection method*, *all-possible regression*, and others.

Many researchers believe that this entire stepwise concept is a bad idea for people who want to do good science. There are least two reasons: (a) The entire process is driven by the data (and the mathematical algorithm), not by theory or the researcher's understanding of previous research, and (b) under typical research conditions, stepwise regression often capitalizes on chance characteristics of a sample and produces results that will not generalize to other samples.

Newer Approaches for Studying the Importance of Predictors

Much of this chapter has been devoted to making the point that it is difficult or impossible to determine the relative importance of a given predictor variable based on the results that are provided in a typical multiple regression analysis. But given the interest in this topic, new statistical procedures are constantly being developed for the purpose of better evaluating the relative importance of predictor variables. Two of the more highly recommended approaches—structural equation modeling and dominance analysis—are discussed here.

DOMINANCE ANALYSIS. A ***dominance analysis*** (Azen & Budescu, 2009) is a procedure that rank-orders the importance of predictors in a multiple regression equation by comparing uniqueness indices (*UI*) rather than standardized multiple regression coefficients (you will recall that a ***uniqueness index*** is essentially just a squared semipartial multiple correlation coefficient; it represents the percent of variance in the criterion that is accounted for by a specific predictor, beyond the variance accounted for by the other predictors).

A dominance analysis for a multiple regression equation begins with an *all-subsets regression*. This is a procedure in which a large number of different multiple regression equations are estimated. This set of multiple regression equations represents every possible combination of predictor variables, and every possible number of predictor variables (given the predictors that were included in the original equation). For each equation, the application computes the R^2 statistic.

A dominance analysis involves (a) using these R^2 values to determine the *UI* for each predictor variable when it is added last to each of the multiple regression equations, and (b) using these *UI*s to compare the predictors (one pair at a time), determining which predictors tend to dominate which other predictors. In this context, one predictor variable "dominates" another predictor variable if the *UI* for the first variable is larger than the *UI* for the second. With this approach, the "more important" predictors are the ones that dominate the "less important" predictors in these comparisons.

A dominance analysis results in the computation of a separate index for each predictor variable. The size of the index for a given variable reflects the relative importance of that predictor in dominating the other predictors.

STRUCTURAL EQUATION MODELING. Among others, Pedhazur (1982) has argued that the relatively simple application of multiple regression (e.g., inspecting regression coefficients from a single multiple regression equation) is not really the best procedure for identifying the relative size of the "effects" that antecedent variables have on a consequent variable. Better procedures for this purpose include *path analysis* (to be covered in *Chapter 16* of this book) and *structural equation modeling* (i.e., SEM, to be covered in *Chapter 18*).

Path analysis and SEM are based on regression procedures, but tend to be superior to standard multiple regression for investigating the importance of predictor variables because they require the researcher to become very familiar with theory and research relevant to the variables of interest. They require that the researcher formulate a causal model that clearly specifies the hypothesized causal relationships between antecedent variables, consequent variables, and the intervening variables that mediate the relations between them. Among other things, path analysis and SEM allow researchers to estimate **total effect coefficients**: statistics that estimate the overall size of the effect that an antecedent variable has on a consequent variable. As their name would suggest, these total effect coefficients provide a more comprehensive picture regarding the relative importance of predictor variables, compared to the somewhat limited picture provided by simple multiple regression coefficients.

Assumptions Underlying this Procedure

Yes, there is a *looooong* list of assumptions underlying multiple regression. Some of the following assumptions were discussed in greater detail in *Chapter 2* and *Chapter 7* of this book.

Assumptions Part I: Prediction

When multiple regression is being used for the simple purpose of *prediction* (as opposed to the investigation of possible causal effects), the following assumptions should be met:

INTERVAL-LEVEL MEASUREMENT. The criterion variable should be a continuous quantitative variable assessed on an interval-scale or a ratio scale. The predictor variables may be either (a) continuous quantitative variables assessed on an interval-scale or ratio-scale, or (b) categorical variables that have been appropriately transformed using dummy-coding or some similar procedure (Field, 2009a). Dummy coding will be discussed in *Chapter 10*.

LINEARITY. For every pairing of any variable with any other variable, it should be possible to fit a best-fitting *straight* line through the scatterplot, as opposed to a best-fitting *curved* line (Keith, 2006; Warner, 2008).

INDEPENDENT OBSERVATIONS. Each observation included in the sample should be drawn independently from the population of interest. Among other things, this means that the researcher should not take repeated measures on the same variable from the same participant (Field, 2009a; Keith, 2006).

INDEPENDENT ERRORS. For any pair of observations, the residual terms should be uncorrelated (Field, 2009a).

HOMOGENEITY OF VARIANCE (HOMOSCEDASTICITY). The variance of the Y scores around the regression line should remain fairly constant at all values of X. When this assumption is met, it means that the dataset displays *homoscedasticity*. If not, it instead displays *heteroscedasticity* (Field, 2009a; Keith, 2006).

NORMALITY. The residuals of prediction should be normally distributed. Violation of this assumption appears to be serious only with relatively small samples, as regression is relatively robust against violations of this assumption (Keith, 2006; Kline, 1998; Tabachnick & Fidell, 2007).

ABSENCE OF MULTICOLLINEARITY. *Multicollinearity* occurs when two or more predictor variables display very strong correlations with one another. An old rule of thumb is that two predictors may have a multicollinearity problem if the Pearson correlation between them exceeds .80 or .90 (Field, 2009a). Along with this rule of thumb, a number of texts provide more sophisticated approaches for investigating potential multicollinearity problems (e.g., Field, 2009a; Tabachnick & Fidell, 2007).

ABSENCE OF INTERACTION. The regression coefficient for the relationship between Y and any X variable should be constant across all values of the other X variables (Warner, 2008).

ADEQUATE RATIO OF OBSERVATIONS TO PREDICTOR VARIABLES. Multiple regression requires a relatively large number of observations in order to have reasonable levels of power. Green (1991) and Tabachnick and Fidell (2007) cite a number of rules of thumb for determining a minimally adequate sample size.

ABSENCE OF OUTLIERS ON THE PREDICTOR AND CRITERION VARIABLES. Nothing can bias an obtained regression coefficient quite as dramatically as an ***outlier*** (an unusual data point that clearly stands apart from the other data points in the sample). The good news is that computer applications provide a number of approaches for identifying and dealing with outliers (e.g., Tabachnick & Fidell, 2007).

Assumptions Part II: Explanation

The remaining assumptions are relevant to situations in which the researcher is using multiple regression for purposes of *explanation*, rather than simple prediction. In this context, *explanation* means that the researcher hypothesizes that the predictor variables have a causal effect on the criterion variable. For research of this sort, the following statistical assumptions should also be met (in addition to the previous assumptions):

CORRECT DIRECTION OF CAUSATION. The predictor variables included in the regression equation must, in fact, be causes of the criterion variable; it must not be the case that the criterion variable is actually the cause of any of the predictor variables (Keith, 2006).

ABSENCE OF MEASUREMENT ERROR. Technically, the predictor variables must be perfectly reliable—they must be measured without error (Keith, 2006). In practice, this assumption is met about as often as I bench press 300 pounds. This is one of the reasons that researchers are increasingly turning to structural equation modeling (SEM) when they wish to test causal models. SEM allows them to model the amount of measurement error displayed by their predictors, so that the resulting parameter estimates are less biased. Structural equation modeling is covered in *Chapter 16, Chapter 17,* and *Chapter 18* of this book.

ABSENCE OF SPECIFICATION ERRORS. There are actually different types of specification errors, but researchers worry the most about the type of specification error that involves *leaving out an important predictor variable*. Technically, the multiple regression equation should include all important common causes of the predictor variables and criterion variable in the equation. (Field, 2009a; Keith, 2006).

Chapter 10

Multiple Regression II: Advanced Concepts

This Chapter's *Terra Incognita*

You are reading an article in a sports psychology journal. In this study, the criterion variable is *exercise sessions:* the number of sessions of aerobic exercise that the subjects engage in each week. The researcher wants to determine whether this variable can be predicted from the so-called *four-factor model of exercise behavior* after statistically controlling for a number of demographic variables such as *subject sex*. The following excerpt describes the results:

> At Step 2, subject sex (dummy-coded) and athletic status (also dummy-coded) were added to the equation that already contained participant age and hours of employment. This resulted in a significant improvement in the equation's ability to predict exercise sessions, $\Delta R^2 = .121$, $F(2,233) = 16.15$, $p < .001$. For the model containing all four control variables, $R^2 = .130$, $F(4,233) = 8.68$, $p = .001$.
>
> Finally, at Step 3, the four variables that constitute the four-factor model of exercise behavior were added to the equation which already contained the four control variables. This resulted in a significant increase in incremental variance accounted for, $\Delta R^2 = .214$, $F(4, 229) = 18.71$, $p < .001$. The full model (including all four control variables plus the four variables of the four-factor model) accounted for 34% of the variance in exercise sessions, $R^2 = .344$, $F(8, 229) = 15.01$, $p < .001$.

Be not afraid. All shall be revealed.

The Basics of this Procedure

This chapter focuses on *hierarchical multiple regression:* a statistical procedure that allows researchers to add predictor variables to a multiple regression equation in a set of pre-determined steps. The predictor variables may be added in groups of variables (sometimes called *blocks of variables*), or they may be added one at a time.

By controlling the sequence in which the variables are added to the equation, the researcher is able to determine whether the variables added at a given step account for meaningful amounts of incremental variance in the criterion variable. *Incremental variance* refers to the amount of variance accounted for by a group of variables, beyond the variance already accounted for by the predictors that were already in the equation. Because it allows researchers to investigate the incremental variance associated with groups of predictor variables, hierarchical multiple regression is one of the most flexible and useful statistical procedures in the social and behavioral sciences. Among other things, it allows researchers to determine whether a specific group of predictor variables accounts for a meaningful amount of variance in the criterion variable after statistically controlling for a set of potential confounding variables. Most of the current chapter is devoted to this type of analysis.

Hierarchical multiple regression goes by a variety of names. It is sometimes called *multiple regression with hierarchical entry, sequential multiple regression, user-determined regression,* and *multiple regression with block-wise entry*. Once in a blue moon, an article will incorrectly refer to hierarchical multiple regression as *stepwise multiple regression,* but stepwise multiple regression is actually an entirely different procedure (as was explained in *Chapter 9*).

For the most part, the variables analyzed using hierarchical multiple regression are identical to those analyzed using simultaneous multiple regression, as covered in *Chapter 9*. This is illustrated in the generic analysis model in Figure 10.1.

Figure 10.1 shows that the criterion variable is typically a quantitative multi-value variable assessed on an interval scale or ratio scale or measurement. The figure shows that the predictor variables are also typically multi-value variables assessed on an interval scale or ratio scale.

However, Figure 10.1 shows that the predictor variables may also be *limited-value variables* (variables that display just two to six values in the sample) as long as they have been properly transformed using dummy coding or some similar procedure. If they have been dummy-coded or effect coded, these limited-value variables may be assessed on any scale of measurement—even the *nominal-scale*. This is what is meant by the note "*if coded,*" which appears in the boxes for the predictor-variables in Figure 10.1.

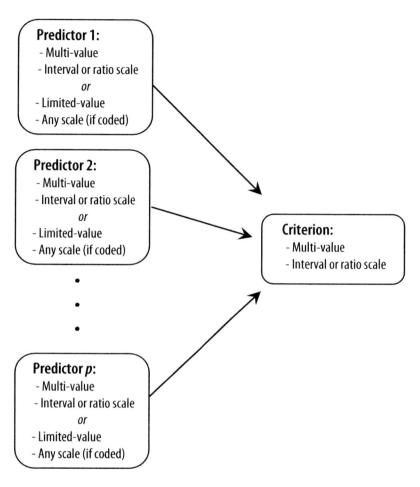

FIGURE 10.1. Generic analysis model: Nature of the variables included in a typical multiple regression analysis.

As a result, the set of predictors in a multiple regression equation may consist of (a) multi-value variables assessed on an interval or ratio scale, (b) limited-value variables assessed on any scale of measurement, or (c) a combination of both. These are the features that make hierarchical multiple regression such a flexible and useful statistical procedure.

Illustrative Study

The current chapter illustrates hierarchical regression by expanding the fictitious investigation introduced in *Chapter 9*. It presents two examples of hierarchical multiple regression. In the first example, all predictor variables are quantitative variables assessed on an interval or ratio scale. In the second example, some of the variables are quantitative variables, and some are categorical variables.

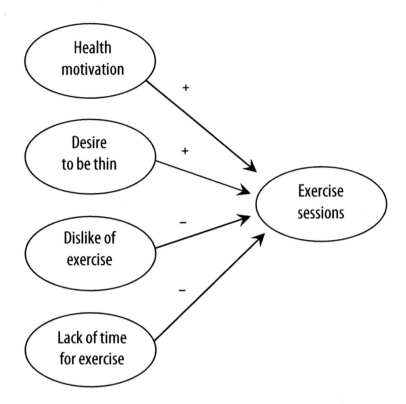

Figure 10.2. Predicted relationships according to the four-factor model of exercise behavior.

VARIABLES INVESTIGATED. To refresh your memory regarding our research scenario: A fictitious researcher named Dr. O'Day wishes to identify variables that can be used to predict *exercise sessions*: the number of times that people engage in aerobic exercise each week. Toward that end, she has developed a *four-factor model of exercise behavior*: a mini-theory describing the four variables that she believes may have a causal effect on exercise behavior. Each of these variables is measured by a multiple-item summated rating scale appearing on the *Questionnaire on Eating and Exercising* (located in *Appendix A*). The four variables that constitute her four-factor model are described below.

- *Health motivation:* the extent to which an individual is willing to sacrifice and work hard in order to stay healthy (assessed by items 27-29 on the *Questionnaire*).

- *Desire to be thin*: the extent to which it is important to the individual to have a fairly low body weight and a lean physical appearance (assessed by items 31-33).

- *Dislike of exercise*: the extent to which the participant hates aerobic exercise and finds it to be unpleasant (assessed by items 47-49).

- *Lack of time for exercise:* the extent to which the participant believes that he or she is too busy to engage in frequent aerobic exercise (assessed by items 51-53).

With each of the preceding variables, scores can range from 1.00 to 7.00. With each scale, higher scores indicate higher levels of the construct being assessed.

Dr. O'Day predicts that some of these variables will have positive relationships with exercise sessions, and others will have negative relationships. The nature of her predictions is illustrated in Figure 10.2.

The signs on the arrows in the figure reveal the nature of Dr. O'Day's research hypotheses:

- There will be a significant and substantial *positive* relationship between exercise sessions and (a) health motivation and (b) desire to be thin.

- There will be a significant and substantial *negative* relationship between exercise sessions and (c) dislike of exercise, and (d) lack of time for exercise.

Table 10.1

Standardized Multiple Regression Coefficients (Beta Weights) Obtained When Exercise Sessions Was Regressed on Four Predictor Variables Using Multiple Regression with Simultaneous Entry

Predictor variable	β	p
1. Health motivation	.153	.013
2. Desire to be thin	.046	.430
3. Dislike of exercise	-.182	.004
4. Lack of time for exercise	-.354	<.001

Note. N = 238. Model R = .531, Model R^2 = .282, $F(4, 233)$ = 22.835, p < .001. Adjusted R^2 = .269. β = standardized multiple regression coefficient. p = probability value for test of null hypothesis that β = 0 in population.

INITIAL RESULTS. *Chapter 9* showed how Dr. O'Day tested these hypotheses by measuring the relevant variables in a sample of 238 college students. In that investigation, she performed a simultaneous multiple regression analysis in which the criterion variable was exercise sessions and the predictor variables were the four variables of the four-factor model. Among the most important of her findings were the standardized multiple regression coefficients that are reproduced in Table 10.1.

POSSIBLE CONFOUNDING VARIABLES. Pleased with her results, Dr. O'Day shared them with her faculty colleague, Dr. Grey.

"Look," said Dr. O'Day as she showed Dr. Grey her computer output:

> The beta weight for lack of time for exercise was fairly large at −.354, and it was statistically significant. This finding is supportive of my hypothesis that *lack of time for exercise* has a causal effect on exercise behavior.

"Nonsense," scoffed Dr. Grey. He continued:

> The relationship between lack of time for exercise and exercise sessions is a spurious relationship, brought about by the fact that both variables are influenced by the same underlying third variable: *participant age*. If you statistically controlled for participant age, the beta weight for lack of time for exercise would drop to zero. In fact, if you controlled for participant age, the beta weight for each of the four variables in your four-factor model would drop to zero.

Ouch!

Dr. Grey will never win *Miss Congeniality*, but he has a point. Many readers will be concerned that each of the relationships that Dr. O'Day found may be *spurious relationships*. When there is a **spurious relationship** between Variable A and Variable B, it means that neither variable has a causal effect on the other. Instead, the relationship between the two variables is due to the influence of other variables in the situation. The concept of a spurious relationship (as well as some of the statistical techniques that are used to deal with them) was covered in *Chapter 8*.

The most persuasive way of dealing with a spurious relationship is to conduct a true experiment. When that is not feasible, researchers often use methods of statistical control as a second-best alternative. In this chapter, Dr. O'Day will use the latter approach.

TWO CONTROL VARIABLES. Dr. O'Day needs to determine whether there is still a relationship between (a) exercise sessions and (b) each of the four variables in her four-factor model after she has statistically controlled for participant age. If the relationships still exist after she has statistically controlled for this variable, it will strengthen her argument that the relationships are real and not spurious.

In addition to controlling for *participant age*, Dr. O'Day decides that she will also statistically control for a second potential confounding variable: *hours employed* (i.e., the number of hours that the subjects are employed at a job each week). Hierarchical multiple regression makes it easy to statistically control for multiple variables, and by including hours employed in the analysis, Dr. O'Day might preemptively squash any speculation that her observed relationships are due to this construct operating as an underlying third variable.

PLANNED HIERARCHICAL ANALYSES. Dr. O'Day plans to use hierarchical multiple regression in an attempt to statistically control for the two potential confounding variables described above. As before, the criterion variable in this regression analysis will be exercise sessions. She will achieve the statistical control that she needs by adding predictor variables to a multiple-regression equation in a sequence of pre-determined steps:

- At Step 1, she will add the two control variables: (a) hours employed and (b) participant age. She will note the percent of variance in exercise sessions that is accounted for by just these two control variables.

- At Step 2, she will add the four variables of theoretical interest to the multiple regression equation that already contains the two control variables. These "four variables of theoretical interest" will be the variables that constitute her four-factor model of exercise behavior. She will determine whether adding the four new variables results in a significant and substantial improvement in the equation's ability to predict the criterion variable. If it does, this outcome can be interpreted as evidence that the four-factor model *is* useful for predicting exercise sessions, even after statistically controlling for the two nuisance variables.

Research Questions Addressed

Using hierarchical multiple regression in the way described above will allow Dr. O'Day to answer a number of important research questions. They include the following:

- Do the two control variables account for a significant and substantial amount of variance in exercise sessions?

- When the variables that constitute the four-factor model are added to a multiple regression equation that already contains the control variables, does it result in a significant and substantial increase in the percent of variance accounted for?

- In the full equation (the multiple regression equation that contains the control variables as well as the variables of the four-factor model), are the regression coefficients for the variables of the four-factor model statistically significant? Are their signs in the predicted direction? Are they non-trivial in size?

Preview of Things To Come

This chapter is divided into two major sections. Part 1 presents results from the hierarchical multiple regression analysis described above. With this analysis, two continuous control variables are added to the equation at Step 1, and the variables of theoretical interest are added at Step 2.

Part 2 of this chapter presents results from a modified analysis in which a new set of variables are added to the equation in three steps instead of two. In the Part 2 analysis, some of the control variables are limited-value variables assessed on a nominal scale of measurement. This analysis highlights the fact that limited-value variables may be used in a multiple regression analysis, provided that they have been appropriately transformed using dummy-coding or some similar technique.

Part 1: Continuous Control Variables

This section illustrates a relatively elementary application of hierarchical multiple regression. In the following excerpt from the published article, Dr. O'Day summarizes her analysis and findings.

> Data were analyzed with hierarchical multiple regression. In all analyses, the criterion variable was exercise sessions per week. Predictor variables were added to the regression equation in two steps. At Step 1, the two control variables (hours employed and participant age) were added, creating Model 1. The results are presented in Table 10.2.
>
> In Table 10.2, The R^2 statistic for the model that contains only the two control variables appears at the location where the column headed "Model R^2" intersects with the row headed "Step 1." This R^2 value is .009, which means that the regression equation containing just the two control variables (hours employed and participant age) accounted for less than 1% of the variance in exercise sessions per week, $F(2, 235) = 1.07$, $p = .345$. Standardized multiple regression coefficients (beta weights) for these two predictors appear in the column headed "Model 1," below the heading "β." With alpha set at $\alpha = .05$, neither beta weight was significantly different from zero.
>
> At Step 2, the four variables that constitute the four-factor model of exercise behavior were added to the equation that contained the two control variables. The R^2 value for the resulting model was $R^2 = .285$, $F(6, 231) = 15.36$, $p < .001$. Adding these four variables at Step 2 resulted in an increase in R^2 of $\Delta R^2 = .276$, $F(4, 231) = 22.31$, $p < .001$. In other words, adding the four predictor variables at Step 2 resulted in a model that accounted for about 28% of the variance in exercise sessions, beyond the variance already accounted for by the two control variables. This increase corresponds to an index of effect size of $f^2 = .39$. According to criteria provided by Cohen (1988), this constitutes a large effect.
>
> Standardized multiple regression coefficients for the six predictors included in the final model appear in the column section headed "Model 2," below the heading "β." With respect to the four predictor variables that constitute the four-factor model of exercise behavior, these results showed that standardized multiple regression coefficients for three of the four variables were significantly different from zero ($p < .05$) and in the direction predicted by the four-factor model. Just one of these four predictor variables—the desire to be thin—displayed a beta weight that was not significantly different from zero, $\beta = .053$, $p = .365$.

Table 10.2

Results from Hierarchical Multiple Regression: Predicting Exercise Sessions per Week from Two Control Variables and the Four-Factor Model of Exercise Behavior

				Model 1		Model 2	
Step	Predictors added	Model R^2	ΔR^2	β	p	β	p
Step 1 (Creating Model 1)		.009	.009				
	Hours employed			−.099	.145	.033	.578
	Participant age			.028	.678	.043	.461
Step 2 (Creating Model 2)		.285***	.276***				
	Health motivation					.154	.013
	Desire to be thin					.053	.365
	Dislike of exercise					−.181	.004
	Lack of time for exercise					−.366	<.001

Note. $N = 238$. Model R^2 = Percent of variance in the criterion variable accounted for by all variables in the model. ΔR^2 = Increase in the percent of variance accounted for by the variables added at a specific step. β = Standardized multiple regression coefficient (beta weight).
*$p < .05$. ** $p < .01$. *** $p < .001$.

The preceding table and text presented a good deal of information. The next few sections will review their contents in more detail.

Model R^2 Statistics for Model 1 and Model 2

In this section, the ***model R^2 statistic*** represents the total percent of variance in the criterion variable accounted for by a specific multiple regression equation. In Dr. O'Day's table, these statistics appear in the column headed "Model R^2." Her table provides the model R^2 statistic for two separate models: one that contains just the control variables, and one that contains the control variables as well as the theoretical variables.

MODEL R^2 FOR MODEL 1 (JUST THE CONTROL VARIABLES). In Table 10.2, at the location where the column headed "Model R^2" intersects the row headed "Step 1," you can see the value ".009." This indicates that the equation that contains just the two control variables (hours employed and participant age) displays a model R^2 of just .009. This tells us that the equation containing only the two control variables accounts for 0.9% of the variance in exercise sessions (that's less than 1%).

There are no asterisks next to this model R^2 of .009, which this tells us that the R^2 value is not statistically significant. The null hypothesis being tested states that $R^2 = 0$ in the population, so when this statistic is nonsignificant, it means that the model R^2 is not significantly different from zero. As you learned in *Chapter 9*, the null hypothesis for the model R^2 is tested using an F statistic. Due to a lack of space, neither the F statistic itself nor its degrees of freedom are included in the current table, but are instead provided in the text of Dr. O'Day's article:

> This R^2 value is .009, which means that the regression equation containing just the two control variables (hours employed and participant age) accounted for less than 1% of the variance in exercise sessions per week, $F(2, 235) = 1.07$, $p = .345$.

To refresh your memory regarding the degrees of freedom: In this F test, the degrees of freedom for the numerator (df_{num}) are equal to k, where k = the number of predictor variables in the model. The degrees of freedom for the denominator (df_{den}) are equal to $N - k - 1$. You will recall that $N = 238$ for this study. At Step 1, just two predictor variables are included in the model, so $df_{num} = 2$ and $df_{den} = 235$.

MODEL R^2 FOR MODEL 2 (CONTROL VARIABLES PLUS THEORETICAL VARIABLES). At Step 2, the four theoretical predictors that constitute the four-factor model were added to the equation that already contained the two control variables. We shall refer to the resulting model containing all six predictor variables as the *full model*.

Under the heading "Model R^2," and to the right of the heading "Step 2," we see that the model $R^2 = .285$. This means that the full model (containing the two control variables as well as the four theoretical variables) accounted for nearly 29% of the variance in exercise sessions. There are three asterisks next to this statistic, indicating that the R^2 value for the full model containing all six predictor variables was significantly different from zero ($p < .001$). In the body of her article, Dr. O'Day reported the details of this significance test:

> The R^2 value for the resulting model was $R^2 = .285$, $F(6, 231) = 15.36$, $p < .001$.

Statistical Significance of ΔR^2 at Step 2

With hierarchical multiple regression, readers are typically most interested in the question, "Did the variables added at the current step result in a significant *improvement* in the equation's ability to predict the criterion variable?" For the answer to that question, we must consult the *incremental R^2 statistic*, which is represented by the symbol "ΔR^2." In the field of statistics, the Greek letter delta (Δ)

is often used to represent *change* or *increase*. In this case, it is used to represent the increase in variance accounted for. More specifically, the ***incremental R^2 statistic*** represents:

- the percent of variance in the criterion variable accounted for by the predictors added at the current step...
- *beyond* the variance accounted for by the predictors that were already in the equation.

In Table 10.2, incremental R^2 statistics appear under the heading "ΔR^2." The following sections discuss the values of ΔR^2 which are obtained as predictor variables and are added to the multiple regression equation in separate steps.

SIZE OF ΔR^2 AT STEP 1. At Step 1, there were no predictors "already in the equation." This means that, at Step 1, the value in the column headed "ΔR^2" will be always equal to the value in the column headed "Model R^2." You can see that this is the case in Table 10.2. At Step 1, Dr. O'Day added two control variables (hours employed and participant age) to a multiple regression equation that previously did not contain any predictor variables. Adding these two predictors to a previously empty equation caused R^2 to increase from .000 to .009—that is why the value in the "ΔR^2" column is .009. The total variance accounted by these two control variables is .009, and that is why the value in the "Model R^2" column is also .009.

SIZE OF ΔR^2 AT STEP 2. Things become more interesting at Step 2. In Table 10.2, find the location where the column headed "ΔR^2" intersects with the row headed "Step 2." At that location, you can see that $\Delta R^2 = .276$. This tells us that when Dr. O'Day added the four variables that constitute the four-factor model to the equation that already contained the two control variables, the increase in R^2 was equal to .276. In other words:

- the four variables that constitute the four-factor model accounted for 27.6% of the variance in exercise sessions...
- *beyond* the variance accounted for by the two control variables that were already in the equation.

STATISTICAL SIGNIFICANCE OF ΔR^2 AT STEP 2. A test is routinely performed to determine whether the increase in variance accounted for (ΔR^2) is statistically significant. The statistic is an F statistic, and it tests the null hypothesis that $\Delta R^2 = 0$ in the population. The degrees of freedom for the numerator are equal to the number of predictor variables *added to the equation* at *just that step* (notice that it is *not* the total number of predictor variables in the equation). The degrees of freedom for the denominator are once again equal to $N - k - 1$, where k again represents the total number of predictors in the equation (this includes the current step as well as the previous steps). This means that the F test for the ΔR^2 statistic at Step 2 had 4 and 231 degrees of freedom.

In Table 10.2, there are three asterisks next to the ΔR^2 of .276 at Step 2. This tells us that adding the four theoretical variables at Step 2 resulted in a significant increase in the equation's ability to predict exercise sessions ($p < .001$). There was not enough room in the table to present the F statistic and df relevant to this test, so Dr. O'Day presented them within the text of her article:

> Adding these four variables at Step 2 resulted in an increase in R^2 of $\Delta R^2 = .276$, $F(4, 231) = 22.31$, $p < .001$. In other words, adding the four predictor variables at Step 2 resulted in a model that accounted for about 28% of the variance in exercise sessions, beyond the variance already accounted for by the two control variables.

Index of Effect Size for ΔR^2 at Step 2

The preceding section indicated that adding the four theoretical variables at Step 2 resulted in a statistically significant increase in ΔR^2. This means that we can reject the null hypothesis of "zero increase," but it does not necessarily mean that the increase was large in practical terms. To evaluate the relative size of the increment, we must turn to some index of effect size.

In practice, many researchers simply review the ΔR^2 value itself as an index of effect size. When $\Delta R^2 = 0$, it is clear that the size the "effect" of adding the new predictor variables to the equation was zero. However, there are no widely accepted criteria telling us how large ΔR^2 must be to be considered a medium effect, a large effect, and so forth. Researchers who wish to do this have the option of converting the obtained value of ΔR^2 into Cohen's (1988) f^2 statistic and then comparing the resulting value against criteria that he provided.

EFFECT SIZE FOR TOTAL R^2. At Step 1, the ΔR^2 statistic is equivalent to the model R^2 statistic (remember that the *model R^2 statistic* is the R^2 for all of the predictor variables in the equation, as of that point in the analysis). Cohen (1988) provides the following formula for computing f^2 from the model R^2:

$$f^2 = \frac{R^2}{1 - R^2}$$

Table 10.2 shows that the model R^2 at Step 1 was $R^2 = .009$. We can therefore compute f^2 as:

$$f^2 = \frac{R^2}{1 - R^2} = \frac{.009}{1 - .009} = .00908 = .009$$

As with all indices of effect size, Cohen (1988) recommends that researchers identify previous studies that investigated the same phenomena and compare the f^2 value for the current study against the typical f^2 value obtained in previous research. When this is not possible, he offers the following criteria to use as a last resort:

> $f^2 = .02$ represents a *small* effect
>
> $f^2 = .15$ represents a *medium* effect
>
> $f^2 = .35$ represents a *large* effect (pp. 413-414).

So the effect size for the control variables added at Step 1 is less than a small effect, according to these guidelines.

EFFECT SIZE FOR ΔR^2 AT STEP 2. Researchers typically do not care much about the variance accounted for by the control variables added at Step 1. What they care about is the incremental variance accounted for by the theoretical variables added at Step 2 (or whatever step is the final step). Table 10.2 showed that at Step 2, the incremental variance accounted for by the four theoretical variables was $\Delta R^2 = .276$. How would Cohen classify the size of this effect?

This is where things get tricky. When we are evaluating the *incremental* variance accounted for, we do not use the same formula for f^2 that was presented above. Instead, Cohen (1988) provides an alternative formula, which I will paraphrase below. For those of you who are reading this with a copy of Cohen's book in your lap, the following is my translation of Cohen's *Case 1*, which appears on page 410 of his book.

To convert incremental R^2 into f^2, Cohen provides the following formula:

$$f^2 = \frac{\Delta R^2}{1 - R^2_{model}}$$

Where:

ΔR^2 = the incremental variance accounted for by just the predictor variables added at this step (beyond the variance already accounted for by predictors that were already in the equation).

R^2_{model} = the model R^2 statistic at this step (the percent of variance accounted for by all of the variables in the model at this step—the control variables added previously, as well as the theoretical variables added at the current step).

For example, what is the f^2 statistic for the incremental variance contributed by Dr. O'Day's four theoretical variables when they were added at Step 2? Table 10.2 shows that, at Step 2, the model $R^2 = .285$, and the $\Delta R^2 = .276$. Inserting these values in the formula produces the following:

$$f^2 = \frac{\Delta R^2}{1 - R^2_{model}} = \frac{.276}{1 - .285} = \frac{.276}{.715} = .386 = .39$$

So, the index of effect size for the four theoretical predictors added at Step 2 was = $f^2 = .39$. Cohen's (1988) criteria for evaluating f^2 (presented earlier) indicate that anything larger than $f^2 = .35$ may be considered a large effect. The incremental variance accounted for by Dr. O'Day's four-factor model may, therefore, be regarded as a large effect.

Interpreting the Multiple Regression Coefficients

To determine the nature of the relationship between the criterion variable and the predictor variables, researchers typically focus on the same three statistics that were covered in *Chapter 9*: (a) the zero-order bivariate correlations (such as the Pearson *r* coefficients), (b) the standardized multiple regression coefficients (i.e., the beta weights) for the final regression model, and (c) the squared semipartial correlation coefficients. Some researchers will go a step further and review the results of a dominance analysis (Azen & Budescu, 2009), as these results probably provide the most accurate index of the relative importance of the predictor variables.

Since *Chapter 9* has already covered these three statistics in some detail, the current chapter will save space by focusing on just one of them: the beta weights for the final model (the *final model* is the model created after the last variables have been added to the equation in the final step). In Table 10.2, these are the statistics that appear under the heading "Model 2."

THE STANDARDIZED MULTIPLE REGRESSION COEFFICIENTS. A ***beta weight*** for a given predictor variable estimates the size of the change in the criterion variable (as measured in standard deviations) that is associated with a one-standard deviation increase in that predictor variable while all of the other predictor variables are held constant. For example, the "Model 2" section of Table 10.2 shows that the beta weight for health motivation is $\beta = .154$. The size and sign of this beta weight indicates the following:

- for every increase of 1 standard deviation in health motivation,
- there is an increase of about .15 standard deviations in exercise sessions,
- while statistically holding constant the other predictors.

The *p* value ($p = .013$) for this beta weight is less than the standard criterion of $\alpha = .05$, indicating that it is significantly different from zero. The beta weights for the remaining predictor variables may be interpreted in the same way.

Table 10.2 shows that the beta weight for the desire to be thin is near-zero in size: $\beta = .053$. In addition, it is not significantly different from zero: $p = .365$. This finding is not supportive of Dr. O'Day's four-factor model.

However, the beta weights for the remaining three theoretical variables were all statistically significant and displayed the sign that had been predicted in Figure 10.2. Specifically, Table 10.2 shows that for health motivation, $\beta = .154, p = 013$; for dislike of exercise, $\beta = -.181, p = 004$; and for lack of time for exercise, $\beta = -.366, p = <.001$. These findings are supportive of Dr. O'Day's four-factor model.

IMPLICATIONS. Remember that Dr. Grey had said that the relationships between exercise sessions and the variables of the four-factor model were all spurious—he said that the only reason Dr. O'Day had found significant beta weights in her earlier analyses was because of the influence of underlying third variables (such as participant age). But the results for Model 2 in Table 10.2 do not support Dr. Grey's arguments. The results in the section headed "Model 2" show that three of Dr. O'Day's four predictors continue to display significant beta weights, even after statistically controlling for hours employed and participant age. In short, these results display greater support for Dr. O'Day's hypotheses than for Dr. Grey's hypotheses.

Part 2: Continuous and Categorical Control Variables

The analysis to be presented in this section will be similar to the previous analysis, except that two new control variables will be added to the multiple regression equation: *participant sex* (males versus females), and *athletic status* (members of sports teams versus non-members). Both of these variables are *limited-value variables:* variables that display just two to six values. They are also *categorical variables*: nominal-scale variables that do not reflect quantity, but instead simply indicate the group to which a participant belongs.

Categorical predictor variables can be included in a multiple regression analysis, but first they must be transformed using dummy-coding, effect-coding, or some similar procedure. This chapter illustrates dummy coding.

Categorical predictor variables may be used as control variables or as theoretical variables in a hierarchical multiple regression analysis. When used as control variables, they are entered into the equation in an early step, and when used as theoretical variables, they are entered at a later step. In the current example, they are used as control variables.

Developing a List of Control Variables

In the current investigation, Dr. O'Day will once again investigate the relationship between exercise sessions and the four variables that constitute her four-factor model of exercise behavior. In the course of designing her research, Dr. O'Day develops a list of all of the variables that she should use as control variables. This is a list of *nuisance variables:* constructs that could operate as underlying third variables, causing spurious relationships between exercise sessions and her four theoretical predictor variables.

Part 1 of this chapter focused on two potential nuisance variables: hours employed and participant age. The analyses reported in Part 1 showed that these two variables displayed very weak relationships with exercise sessions. Nevertheless, Dr. O'Day will use the same variables as control variables in the current section in order to address the potential concerns of critics like Dr. Grey.

Imagine that a review of the literature identifies two additional variables that could produce spurious correlations between the other variables in the investigation. These potential confounding variables are:

- *participant sex* (whether a given subject is male versus female), and
- *athletic status* (whether a given participant is a member of some sports-related team, or is not a member of such a team).

Dr. O'Day will include these two variables in an early step of her regression analysis in an effort to statistically control for their effects. Since both variables are limited-value variables assessed on a nominal scale, she realizes that they must be properly transformed if they are to be included in the analysis.

Dummy-Coding the Categorical Predictor Variables

When predictor variables are categorical (nominal-scale) variables, they must be transformed using the process of ***dummy coding***, effect coding, or some similar procedure before they can be included in a regression analysis. This section illustrates the simplest case: situations in which the predictor variables are dichotomous variables (variables with just two categories). This section illustrates how dichotomous variables can be dummy-coded.

First the good news: When you analyze your data using one of the popular data-analysis applications such as SPSS or SAS, you typically do not have to worry about the dummy-coding procedures described here. Most statistical applications allow you to identify the predictor variables that are classification (categorical) variables, and the application then does the dummy coding or effect coding for you. This section shows how you would perform this coding manually, but it does this merely to provide a conceptual understanding of what the statistical application is doing behind the scenes.

DUMMY CODING THE TWO DICHOTOMOUS VARIABLES. When dummy coding is used to represent a categorical variable that includes just two categories, the researcher creates a new variable that contains only the values of "0" and "1." The 0s are used to represent one of the two categories, and the 1s are used to represent the other category (the selection of which value is used to represent which category is usually arbitrary).

For example, in Table 10.3, a new dummy-coded version of the *athletic status* variable was created by using the value "1" to represent students who are members of a sports-related team, and the value "0" to represent students who are not members of any such team. The transformed version of the variable is called a ***vector***, and appears under the heading "Athletic status" "(dummy-coded)" in the table. This vector may now be used as a predictor variable in a multiple regression analysis.

Similarly, Table 10.3 also shows that a new dummy-coded version of the *participant sex* variable was created by using the value "1" to represent males and the value "0" to represent females. The vector that represents this variable appears under the heading "Participant sex" "(dummy-coded)".

DUMMY CODING CATEGORICAL VARIABLES WITH MORE THAN TWO CATEGORIES. The two categorical variables discussed so far were both *dichotomous variables*, meaning that they each consisted of just two categories. However, coding systems can also handle categorical variables with more than two categories. These systems typically utilize a set of multiple vectors to code a single categorical variable. The number of vectors required is equal to $k - 1$, where k = the number of categories included in the categorical variable. For details regarding the use of dummy coding and effect coding with variables consisting of more than two categories, see Pedhazur (1982).

Table 10.3

Dummy-Coded Versions of the Categorical Predictor Variables

	Athletic status		Participant sex	
Subject	(non-coded)	(dummy-coded)	(non-coded)	(dummy-coded)
1	Athlete	1	Female	0
2	Athlete	1	Male	1
3	Non-athlete	0	Male	1
4	Athlete	1	Female	0
5	Non-athlete	0	Male	1
6	Non-athlete	0	Female	0
7	Non-athlete	0	Female	0
8	Athlete	1	Female	0
9	Non-athlete	0	Male	1
10	Non-athlete	0	Female	0

Determining the Order for Entering the Variables

It occurs to Dr. O'Day that she actually has two sets of control variables. Two of the control variables (hours employed and participant age) were suggested to her by Dr. Grey, and the other two (participant sex and athletic status) were based on her own review of the literature. Why not add these sets of control variables to the equation in two separate steps? At Step 1, she could add Dr. Grey's control variables, at Step 2, she could add her own control variables, and at Step 3, she could add the four theoretical variables. This approach would allow her to see if the variables that she identified in her literature review displayed a stronger relationship with the criterion variable, compared to the two that had been suggested by Dr. Grey.

It will also result in a fairly crowded table, as you shall soon see.

Table 10.4

Results from Hierarchical Multiple Regression: Predicting Exercise Sessions per Week from Four Control Variables and the Four-Factor Model of Exercise Behavior

				Model 1		Model 2		Model 3	
Step	Predictors added	Model R^2	ΔR^2	β	p	β	p	β	p
Step 1 (Creating Model 1)		.009	.009						
	Hours employed			−.099	.145	−.040	.540	.056	.337
	Participant age			.028	.678	.022	.738	.036	.531
Step 2 (Creating Model 2)		.130***	.121***						
	Participant sex					.099	.119	.127	.028
	Athletic status					.317	<.001	.200	.001
Step 3 (Creating Model 3)		.344***	.214***						
	Health motivation							.116	.053
	Desire to be thin							.107	.069
	Dislike of exercise							−.186	.002
	Lack of time for exercise							−.304	<.001

Note. N = 238. Model R^2 = Percent of variance in the criterion variable accounted for by all variables in the model. ΔR^2 = Increase in the percent of variance accounted for by the variables added at a specific step. β = Standardized multiple regression coefficient (beta weight).
*p < .05. **p < .01. ***p < .001.

Analysis and Results

Dr. O'Day performs the analysis described above. The results are presented in Table 10.4.

PARTS OF THE TABLE. By now you should be familiar with the basic layout of the table:

- Below the headings "Step" and "Predictors added," you can see which predictor variables were added at Step 1 versus Step 2 versus Step 3.

- Below the heading "Model R^2," you can see the total percent of variance in the criterion variable that is accounted for by all of the predictor variables in the model (as of that specific step).

- Below the heading "ΔR^2," you can see the incremental percent of variance accounted for by just the predictor variables added to the model at that specific step (i.e., the variance accounted for by just these new predictors, beyond the variance accounted for by the predictors already in the equation).

- Below the heading "Model 1," you will find the beta weights (and their corresponding p values) for the variables that are in the equation as of Step 1.

- Below the heading "Model 2," you will find the beta weights (and their p values) for the variables that are in the equation as of Step 2.

- Below the heading "Model 3," you will find the beta weights (and their p values) for the variables that are in the equation as of Step 3.

INTERPRETING THE RESULTS. We have already gone through the process of interpreting a table such as this, back when we reviewed Table 10.2. Therefore, we can breeze through the current table a bit more quickly, just hitting the more important points.

- At Step 1, control variables suggested by Dr. Grey (hours employed and participant age) were added to the empty multiple regression equation. This resulted in a model R^2 of just .009, which was not statistically significant. This tells us that these two control variables were essentially unrelated to exercise sessions.

- At Step 2, the two control variables that Dr. O'Day identified in her review of the literature were added to the equation that already contained hours employed and participant age. The addition of the two new control variables resulted in a significant improvement in the equation's ability to predict exercise sessions, $\Delta R^2 = .121$, $p < .001$. The "model R^2" column shows that the equation containing all four control variables accounted for 13% of the variance in exercise sessions, $p < .001$.

- At Step 3, the four variables that constitute Dr. O'Day's four-factor model were added to the equation that already contained the four previously-described control variables. The addition of these four new theoretical variables resulted in a significant improvement in the equation's ability to predict exercise sessions, $\Delta R^2 = .214$, $p < .001$. The "model R^2" column shows that the final equation containing all eight variables accounted for about 34% of the variance in exercise sessions, $p < .001$.

The section headed "Model 3" presents beta weights (and their *p* values) for the full multiple regression equation that contains all eight predictors. These standardized multiple regression coefficients show the following:

- With alpha set at $\alpha = .05$, neither of the two control variables suggested by Dr. Grey was significantly different from zero. (*Hah!*)

- Both of the control variables located by Dr. O'Day's literature review displayed a beta weight that was significantly different from zero. The beta weight for participant sex was a positive value: $\beta = +.127$, $p = .028$. Remember that when Dr. O'Day dummy-coded this variable, she used 1s to represent males and 0s to represent females. Given this coding scheme, the fact that the beta weight is a positive number ($\beta = +.127$) tells us that males tended to score higher on exercise sessions, compared to females (after statistically controlling for the other variables).

- The beta weight for athletic status was also a positive value: $\beta = +.200$, $p = .001$. When Dr. O'Day dummy-coded this variable, she used 1s to represent team members, and 0s to represent non-members. Given this coding scheme, the fact that the beta weight is a positive number ($\beta = +.200$) tells us that team members tended to score higher on exercise sessions compared to non-members (after statistically controlling for the other variables).

Dr. O'Day will be most interested in the beta weights for the four variables that constitute her four-factor model of exercise behavior. These are the last four beta weights in the column headed "Model 3," and they reveal the following:

- Under the heading "Model 3," we can see that the beta weights for health motivation and desire to be thin were not significantly different from zero. For health motivation, $\beta = .116$, $p = .053$, and for desire to be thin, $\beta = .107$, $p = .069$. For both variables, the *p* value was slightly larger than the standard criterion of $\alpha = .05$.

- For dislike of exercise, the beta weight was statistically significant and the direction of the relationship was consistent with Dr. O'Day's theoretical model, $\beta = -.186$, $p = .002$. The same was true for lack of time for exercise, $\beta = -.304$, $p < .001$.

CAUTIONS REGARDING THIS INTERPRETATION. The preceding represents the kind of interpretation that would be made by most researchers in the social and behavioral sciences. However, some caveats are in order.

First, the preceding interpretation focused on the beta weights (i.e., the standardized multiple regression coefficients), and many experts warn that beta weights can be misleading when they are used to evaluate the relative importance of predictor variables in a multiple regression equation. This problem was discussed in *Chapter 9*, in the section titled "The Correlated Predictor Variable Problem." Some experts advise that, when the goal of the investigation is to evaluate the relative importance of predictors, it is best to either (a) perform a dominance analysis (Azen & Budescu, 2009) or (b) analyze the data using structural equation modeling (to be covered in *Chapter 16*, *Chapter 17*, and *Chapter 18* of this book).

Second, the preceding section implied that the predictor variables with statistically significant beta weights were important predictors, and the predictors with statistically nonsignificant beta weights were not important. As you learned in *Chapter 5*, the entire enterprise of significance testing is being questioned by many researchers. For these skeptics, the results presented in Table 10.4 might serve as *Exhibit A* in their case that significance testing should be abolished. It seems so arbitrary to view one predictor as being "important" because its p value is $p = .028$ and at the same time view another predictor as being "unimportant" because its p value is $p = .053$. I mean, *come on!*

Summary of Findings

The results from these analyses provide partial support for the theoretical model presented in Figure 10.2. After statistically controlling for four subject variables (hours employed, age, sex, and athletic status), the variables that constitute the four-factor model accounted for a significant amount of incremental variance in exercise sessions per week. In the final model that contained all eight variables, two of the theoretical variables displayed standardized multiple regression coefficients that were statistically significant and in the predicted direction. The full model accounted for about 34% of the variance in exercise sessions, $p < .001$.

These results suggest that two of the variables that constitute the four-factor model (dislike of exercise and lack of time for exercise) may be useful predictors of exercise behavior. The results suggest that the other two variables (health motivation and desire to be thin) may be less useful.

Assumptions Underlying this Procedure

The assumptions underlying multiple regression with simultaneous entry of predictor variables were listed in excruciating detail at the end of *Chapter 9* of this book. All of those assumptions apply to hierarchical multiple regression as well, and will therefore not be repeated here.

Chapter 11

Discriminant Analysis

This Chapter's *Terra Incognita*

You are reading about an empirical investigation in which the criterion variable is something called *team & exercise status*. This is a classification variable consisting of three groups: nonexercisers, individual exercisers, and team exercisers. The researcher wants to determine whether membership in these groups can be accurately predicted from a set of four predictor variables: health motivation, desire to be thin, dislike of exercise, and lack of time for exercise. She analyzes the data using a procedure called *discriminant analysis*. This is what she reports:

> Wilks' lambda for the first discriminant function was statistically significant: $\Lambda = .81$, $\chi^2 (8, N = 238) = 49.44$, $p < .001$, $R_c^2 = .17$. In contrast, Wilks' lambda for the second discriminant function was nonsignificant: $\Lambda = .97$, $\chi^2 (3, N = 238) = 6.17$, $p = .104$, $R_c^2 = .03$. These findings indicated that there was a significant difference between the subgroups, but only for the first discriminant function. Therefore, only the first function was further examined.
>
> The nature of the first discriminant function was interpreted by reviewing the standardized discriminant function coefficients and structure coefficients in conjunction with the group centroids. The pattern of results showed that compared to nonexercisers, the team exercisers (a) were motivated to stay healthy, (b) enjoyed exercising, and (c) had adequate time to exercise.

Courage! All shall be explained.

The Basics of this Procedure

This chapter covers ***discriminant analysis***, which is also called ***discriminant function analysis***. This procedure is used to investigate the relationship between (a) a single categorical criterion variable and (b) multiple quantitative predictor variables. Figure 11.1 illustrates the nature of the variables included in a typical discriminant analysis.

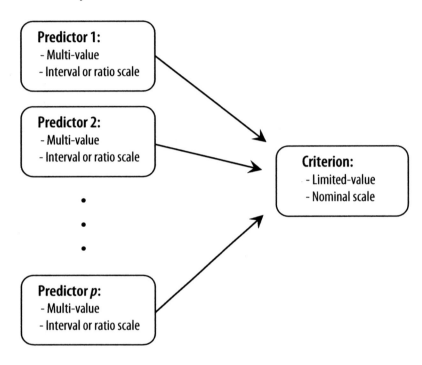

FIGURE 11.1. Generic analysis model: Nature of the variables included in a typical discriminant analysis.

In discriminant analysis, the criterion variable is also called the ***grouping variable***. Figure 11.1 shows that the grouping variable is typically a ***limited-value variable***: a variable that consists of just two to six categories. In discriminant analysis, the grouping variable is usually a ***nominal-scale variable***—a variable that does not measure quantity, but instead simply indicates the group to which an individual participant belongs. For example, if the grouping variable in a given investigation was *political party*, the groups might be Democrat versus Republican versus Independent.

The typical discriminant analysis includes from 2 to 10 predictor variables. These predictor variables are sometimes called ***attributes*** (Silva & Stam, 1995). Attributes are usually multi-value quantitative variables measured on an interval or ratio scale. Published studies sometimes also include nominal-scale predictor variables that have been dummy-coded. However, some experts have warned that categorical predictor variables are inappropriate when performing a discriminant analysis (e.g., Wright, 1995).

Criterion Variable or Predictor Variable?

In Figure 11.1, the group of quantitative variables is treated as a set of predictor variables (on the left side of the figure), and the single grouping variable is treated as the criterion variable (on the right side of the figure). However, this labeling of the variables could have been reversed—the group of quantitative variables could have been treated as criterion variables (on the right side of the figure), and the single grouping variable could have been treated as the predictor variable (on the left). The point is that discriminant analysis allows researchers to investigate the relationship between a set of quantitative variables and a single grouping variable regardless which is labeled "predictor" versus "criterion."

Illustrative Investigation

Assume that the fictitious researcher, Dr. O'Day, gathers data for the current investigation by administering the *Questionnaire on Eating and Exercising* (which appears in *Appendix A* of this book) to a sample of 260 college students. She obtains usable responses from 238 of these students.

CRITERION VARIABLE. The main focus of Dr. O'Day's investigation is a grouping variable called *team & exercise status*. Team & exercise status is a nominal-scale variable that consists of just three categories:

- *Nonexercisers* ($n = 35$): people who do not engage in aerobic exercise at all.

- *Individual exercisers* ($n = 143$): people who engage in aerobic exercise on their own, without joining any team.

- *Team exercisers* ($n = 60$): people who engage in aerobic exercise as members of organized university-related sports teams or intramural teams.

PREDICTOR VARIABLES. Team & exercise status serves as the criterion variable in Dr. O'Day's investigation. She wishes to determine whether it can be predicted from the four variables that constitute her four-factor model of exercise behavior. As was explained in earlier chapters, the *four-factor model of exercise behavior* is a sort of fictitious mini-theory. It consists of the four variables that, according to Dr. O'Day, are the most important determinants of an individual's decision to engage in aerobic exercise.

Each of the four variables that constitute the four-factor model was measured using a multiple-item summated rating scale, and each scale consisted of three Likert-type items. Participants indicated the extent to which they agreed with each item using a response format which ranged from 1 = *Extremely Disagree* to 7 = *Extremely Agree*. Scores on the resulting variables could range from 1.00 (meaning that the participant had less of the attribute) to 7.00 (meaning that the subject had more of the attribute). The name of each scale is provided below (along with a representative questionnaire item used in creating the scale):

- *Health motivation* ("It is important to me to be healthy.")
- *Desire to be thin* ("It is important to me to be thin.")
- *Dislike of exercise* ("I find aerobic exercise to be unpleasant.")
- *Lack of time for exercise* ("I am too busy to engage in frequent aerobic exercise.")

RESEARCH HYPOTHESIS. In performing the discriminant analysis, team & exercise status serves as the criterion variable, and the four variables which constitute the four-factor model serve as the predictor variables. This arrangement is illustrated graphically in Figure 11.2.

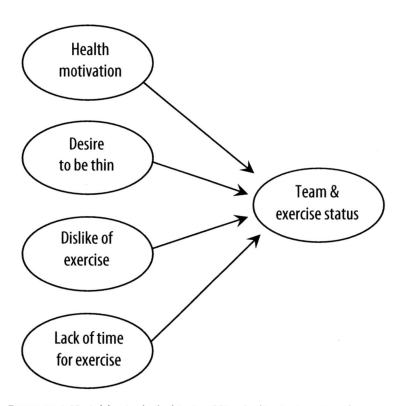

FIGURE 11.2. Variables included in Dr. O'Day's discriminant analysis.

The current investigation serves as yet another test of Dr. O'Day's four-factor model of exercise behavior. The study's main research hypothesis is stated below:

> There is a significant and substantial relationship between (a) team & exercise status and (b) the four variables of the four-factor model, taken as a set. In other words, the four variables of the four-factor model may be used to correctly predict whether a given participant is a nonexerciser versus an individual exerciser versus a team exerciser.

Research Questions Addressed

The research questions addressed by discriminant analysis are similar to those addressed by other statistical procedures that allow for a single criterion variable and multiple predictors. Below are some examples:

- Is there a statistically significant relationship between the variables? More specifically, is there a statistically significant relationship between the grouping variable and the attribute variables, taken as a set?

- What is the nature of the relationship? More specifically, which subgroups are being discriminated by the discriminant function? Which attribute variables appear to be the more important predictor variables?

- How strong is the relationship? More specifically, what is the strength of the association between the grouping variable and the attribute variables, taken as a set? When the results of the discriminant analysis are used to classify participants into subgroups, how accurate are the classifications?

Details, Details, Details

In a discriminant analysis, the computer software creates one or more ***discriminant functions*** (also called ***discriminant function variates***, ***canonical discriminant functions, canonical variates***, or ***canonical variables***). This is an *uber*-important concept—once you understand what a discriminant function is, the remaining concepts in the chapter will be much easier to grasp.

The Discriminant Function

A ***discriminant function*** is a *composite variable*—it is a linear composite of the optimally weighted attribute variables. You should think of a discriminant function as being a *variate*—a *synthetic variable* created by the software. Since a discriminant function is a variate, we will use the symbol V_i to represent it. Each subject in the study will have a score on this discriminant function, and these scores are called ***discriminant scores***. Discriminant scores assume values somewhat similar to *z* scores, taking on values such as +0.917, +0.030, and −0.692.

FORMULA FOR A DISCRIMINANT FUNCTION. A preceding section defined a discriminant function as being a linear composite of the optimally-weighted attribute variables. Sound familiar? This definition tells us that a discriminant function (symbol: V_i) in a discriminant analysis is similar to the predicted *Y* variable (symbol: Y') in a multiple regression analysis. In Chapter 9, you learned that, in linear multiple regression, each subject has a score on the Y' variable, and these predicted scores are created using a formula such as the following:

$$Y' = a + b_1(X_1) + b_2(X_2) + b_3(X_3) \ldots + b_p(X_p)$$

A similar process occurs in a discriminant analysis: The statistical application computes a different score for each subject on the discriminant function, V_i. The general form for an equation that computes discriminant scores is as follows:

$$V_i = a + b_1(X_1) + b_2(X_2) + b_3(X_3) \ldots + b_p(X_p)$$

Where:

V_i represents a given subject's discriminant function score (remember that the "V" stands for "*variate*," which is just another way of saying "*synthetic variable*").

a represents the constant of the equation. Some textbooks represent this constant by using the symbol b_0.

b_1 represents the unstandardized canonical discriminant function coefficient for the first predictor variable (e.g., the first attribute variable).

X_1 represents a given subject's actual score on the first attribute variable.

b_p represents the unstandardized discriminant function coefficient for the pth (final) attribute variable in the equation.

X_p represents the pth (final) attribute variable in the equation.

MAXIMIZING SEPARATION OF THE GROUPS. The computer software computes the terms of this equation (i.e., the constant and discriminant function coefficients) so as to *maximize separation* of the groups under the grouping variable. It creates the discriminant function (the synthetic variable) so that (a) the different groups under the grouping variable will have very different scores on this discriminant function and (b) people within the same group will have similar scores on the discriminant function.

Number and Nature of the Discriminant Functions

In the typical discriminant analysis, more than one discriminant function is created. The number of discriminant functions will be equal to either p (where p = the number of attribute variables) or $k - 1$ (where k = the number of groups), whichever is smaller. In Dr. O'Day's study, the grouping variable (team & exercise status) consists of three groups. This means that $k = 3$, and therefore $k - 1 = 2$. In her study there are four attribute variables, so $p = 4$. The smaller of the two values is 2, and this tells us that just two discriminant functions will be created.

In a study with several discriminant functions, there is sort of a law of diminishing returns: The first few functions tend to be more important in discriminating between groups, and the later functions tend to be less important. More specifically, the first discriminant function created by the software will maximize the separation between the groups. The next discriminant function will (a) be uncorrelated with the first discriminant function, and will (b) further maximize the separation between the groups (to the extent that this is possible, given that the first function has probably done most of the heavy lifting). In a discriminant analysis, it is common that the first one or two functions are statistically significant, but that the remaining functions are nonsignificant (because the first functions have already separated the groups as much as can be done). This means that it is common that the researcher will interpret just the first one or two functions, and will disregard the remaining functions.

Results from the Investigation

The preceding section made the point that in many investigations, more than one discriminant function is created in the course of the anaysis. Therefore, the researcher typically begins by determining which (if any) of these discriminant functions are statistically significant. For each significant function, the researcher then interprets its *meaning*. Among other things, this involves identifying the attribute variables that appear to be measuring the function. Finally, the researcher evaluates the strength of the relationship between the grouping variable and the attribute variables for each significant function. This section shows how each of these steps might be conducted in Dr. O'Day's investigation.

Is There a Significant Relationship Between the Variables?

Early in the *Results* section of her article, Dr. O'Day would probably address the issue of statistical significance. This is done in the following excerpt:

> Data were analyzed using discriminant function analysis. In this analysis, the grouping variable was team & exercise status, and the attribute variables were health motivation, desire to be thin, dislike of exercise, and lack of time for exercise. Two discriminant functions were created, and the significance tests for these functions are presented in Table 11.1.
>
> In a discriminant analysis in which more than one function is created, Wilks' lambda (Λ) for the first function serves as a significance test for that function and all remaining functions. More specifically, it tests the null hypothesis that, in the population, the means of all of the discriminant functions are equal in all of the subgroups (Norusis, 2005).
>
> Table 11.1 shows that Wilks' lambda for the first discriminant function was statistically significant: $\Lambda = .81$, $\chi^2 (8, N = 238) = 49.44$, $p < .001$. In contrast, Wilks' lambda for the second discriminant function was nonsignificant: $\Lambda = .97$, $\chi^2 (3, N = 238) = 6.17$, $p = .104$. These findings indicate that there was a significant difference between the subgroups, but only for the first discriminant function. Therefore, only the first function will be interpreted.

Table 11.1

Indices of Association and Significance Tests for the Discriminant Functions

	Discriminant function	
	1	2
R_c^2	.17	.03
Wilks' lambda, Λ	.81	.97
χ^2 for Wilks' lambda (df)	49.44(8)	6.17(3)
p for Wilks' lambda	<.001	.104

Note. N = 238. R_c^2 = Squared canonical correlation coefficient. df = Degrees of freedom.

In Table 11.1, the first column of statistics (under "Discriminant function 1") provides results relevant to the first discriminant function, and the second column (under "Discriminant function 2") provides statistics relevant to the second function. In the table, the symbol "R_c^2" represents the squared canonical correlation coefficient. This is an index of association that will be discussed later, in the section titled "*How Strong is the Relationship?*" The remaining statistics in the table are relevant to the statistical significance of the functions, and will be discussed in the current section.

WILKS' LAMBDA. The preceding excerpt focused on *Wilks' lambda* (symbol: Λ or λ), a statistic that is often reported as part of multivariate statistical procedures. Wilks' lambda is important, in part, because it is the basis for the null-hypothesis significance test in a discriminant analysis.

Wilks' lambda is a multivariate index of association. Its values may range from 0.00 to 1.00, with smaller values (closer to 0.00) representing a stronger relationship between the grouping variable and the attribute variables, taken as a set. Yes, you read that correctly—Wilks' lambda is one of the few statistics for which a *smaller* value indicates a *stronger* relationship. This is because lambda represents the ratio of error variance to total variance—the stronger the relationship, the smaller the error variance (and hence the smaller the value of Wilks' lambda).

THE NULL HYPOTHESIS. As the preceding excerpt indicates, the significance test for Wilks' lambda for the first function tests the following null hypothesis:

H_0: In the population, the means of all of the discriminant functions are equal in all of the subgroups.

In this case, "all of the discriminant functions" means "Discriminant Function 1 and Discriminant Function 2" (remember that there were only two functions in this analysis). The Wilks' lambda statistic for the first function essentially tests the significance of all of the functions simultaneously.

Remember that a null hypothesis is almost always a prediction of no difference or no relationship. The preceding null hypothesis stated "In the population, the means on all of the discriminant functions are equal in all of the subgroups." Think about what it would imply if all of the subgroups had exactly the same mean score on a given discriminant function—it would indicate that all of the subgroups probably had the same mean score on each of the quantitative predictor variables. If this were the case, it would not be possible to predict which subgroup a given subject was in, based on that subject's scores on the quantitative predictor variables. That would be bad news for Dr. O'Day, and that is why Dr. O'Day hopes to reject this null hypothesis.

THE SIGNIFICANCE TEST. To determine whether a given discriminant function is statistically significant, Wilks' lambda for that function is usually converted to either an F statistic or to a χ^2 statistic (in the current analysis, it was converted to χ^2). If the p value for this F or χ^2 statistic is less than the standard criterion of $\alpha = .05$, the null hypothesis is rejected and Wilks' lambda is said to be statistically significant. Sure enough, the preceding excerpt indicates that Wilks' lambda for the first function was significant: $\Lambda = .81$, $\chi^2 (8, N = 238) = 49.44$, $p < .001$. That is good news for Dr. O'Day—it means that she has significant results. Later in her article, she will interpret the nature of this first discriminant function.

Remember that a statistically significant result merely tells us that it is okay to reject the null hypothesis—it does *not* necessarily tell us that there is a *strong* relationship between the grouping variable and the attribute variables. To determine whether the relationship is strong, we must consult indices of effect size and the results of the classification analysis (to be presented later in this chapter).

Dr. O'Day's excerpt further states that Wilks' lambda for the second discriminant function is not statistically significant: $\Lambda = .97$, $\chi^2 (3, N = 238) = 6.17$, $p = .104$. This means that Dr. O'Day will not interpret the nature of the second function in her article. Remember, however, that the entire concept of statistical significance is controversial, and some researchers would go on to interpret a function even if it were not statistically significant, as long as the strength of the association were strong enough.

What is the Nature of the Relationship?

When we ask "what is the nature of the relationship?" we are asking questions such as "Which groups are being discriminated in the analysis?" and "Which attribute variables are the more important predictors of the grouping variable?" This is a tricky business, because we can never conclusively determine which predictors are the most "important" when they are correlated with one another (as they almost always are in nonexperimental research). Nevertheless, this chapter will describe the results that are frequently consulted for this purpose.

To aid in the interpretation of the results, Dr. O'Day prepares a table that presents group centroids, standardized discriminant function coefficients, and structure coefficients. The following excerpt presents this table, along with the text from her published article in which she discusses it.

> Table 11.2 shows that, in the column for "Discriminant function 1," the group centroid for the nonexercisers was relatively high at .946, whereas the centroid for the team exercisers was much lower at −.524. The centroid for the individual exercisers was near-zero at −.012. This indicates that the first discriminant function served to discriminate between the nonexercisers versus the team exercisers.
>
> The upper section of Table 11.2 provides standardized discriminant function coefficients and structure coefficients for the attribute variables. The pattern of these coefficients for the first discriminant function shows that (compared the nonexercisers), the team exercisers tended to:
> - Score high on health motivation.
> - Score low on dislike of exercise.
> - Score low on lack of time for exercise.
>
> The structure coefficient for the "desire to be thin" variable was relatively small in absolute value: −.189. For this reason, the desire to be thin was not included in the interpretation of the first discriminant function.
>
> An earlier section of this article indicated that Wilks' lambda for the second discriminant function was not statistically significant ($p = .104$). The second discriminant function was therefore not interpreted.

Table 11.2

Standardized Discriminant Function Coefficients, Structure Coefficients (in Parentheses), and Group Centroids for the Discriminant Functions

	Discriminant function	
	1	2
Standardized[a] (and structure[b]) coefficients		
Health motivation	-.340 (-.550)	.659 (.378)
Desire to be thin	.015 (-.189)	-.572 (-.436)
Dislike of exercise	.615 (.840)	.710 (.458)
Lack of time for exercise	.439 (.682)	-.513 (-.335)
Group centroids[c]		
Nonexercisers	.946	.189
Individual exercisers	-.012	-.132
Team exercisers	-.524	.206

Note. N = 238. df = Degrees of freedom.
[a]Standardized coefficients are the standardized canonical discriminant function coefficients for attribute variables. [b]Structure coefficients (in parentheses) are bivariate correlations between attribute variables and the discriminant function. [c]Group centroids are means of the discriminant function for a given subgroup.

Don't panic if the preceding table and text were confusing. Understanding the results from a discriminant analysis isn't that difficult, as long as you understand the steps to follow. Prior to the steps, let's define some of the statistics appearing in Table 11.2.

- First, a ***centroid*** is the mean for a subgroup on a discriminant function. In an earlier section of this article, you learned that a discriminant function is a variate—an artificial variable created by the software. Because it is a quantitative variable (on which the subjects may have scores such as +.232 or -.576), we should be able to use it to compute means, right? These means are called *centroids*.

- Next, Table 11.2 shows that there is a standardized discriminant function coefficient for each quantitative attribute variable. A ***standardized discriminant function coefficient*** indicates the relative amount of weight that was given to a specific attribute variable in the creation of a discriminant function, while statistically controlling for the other attribute variables. They are standardized so that they are all on the same scale—so that the relative size of the coefficients for different attribute variables may be directly compared. In this way, they are analogous to the beta weights (symbol: β) from linear multiple regression (from *Chapter 9*). The general idea is that attribute variables with larger standardized coefficients (in absolute value) are "measuring" that discriminant function. In practice, however, these coefficients have important weaknesses when used for this purpose—in some cases, the standardized discriminant function coefficient for a given attribute variable may be near-zero in size even when the attribute variable has a strong correlation with the discriminant function. Because of this weakness, many experts recommend that we should generally use *structure coefficients* (to be defined in the next section) rather than standardized discriminant function coefficients to identify the attribute variables that are measuring a given discriminant function.

- Finally, Table 11.2 also contains a structure coefficient for each attribute variable. A ***structure coefficient*** for a given attribute variable represents the bivariate correlation (similar to the Pearson's *r*) between that attribute variable and the discriminant function. The structure coefficient for a given attribute variable represents the relationship between that attribute variable and the discriminant function *without* statistically controlling for the other attribute variables. This is actually an advantage, when it comes to interpreting a discriminant function—we do not have to worry that a structure coefficient for a given predictor variable is misleadingly low because that attribute variable was correlated with other attribute variables. Attribute variables that have structure coefficients that are relatively large (in absolute value) are assumed to be "measuring" a discriminant function. A bit later, we will see how this is done.

STEP 1: FOCUS ON JUST THE FIRST DISCRIMINANT FUNCTION. At this point, we will focus on just the centroids and coefficients for Discriminant Function 1. We will disregard the other discriminant function for the time being.

STEP 2: REVIEW CENTROIDS IN CONJUNCTION WITH STRUCTURE COEFFICIENTS TO INTERPRET THE FIRST DISCRIMINANT FUNCTION. To interpret a specific discriminant function, it is necessary to (a) identify the subgroups that have relatively high centroids and relatively low centroids, and simultaneously (b) identify the attribute variables with structure coefficients that are relatively large in absolute value. Just to keep us on our toes, we also have to worry about the sign (+ versus -) for each structure coefficient.

We begin by identifying the specific subgroups that are being discriminated by Discriminant Function 1. To do this, we look for the subgroup that has the highest positive centroid, as well as the subgroup that has the lowest negative centroid. If we have more than one subgroup with really high or really low centroids, we will just lump those subgroups together.

Table 11.2 shows that the nonexercisers have the highest centroid (a centroid of +.946) and the team exercisers have the lowest centroid (a centroid of -.524). The individual exercisers have a near-zero centroid of -.012, so we will more or less ignore this subgroup for the moment. Combined, the centroids indicate that Discriminant Function 1 was created so as to discriminate between the nonexercisers versus the team exercisers. Given the signs of these centroids, we know that subjects who had high positive scores on Discriminant Function 1 were likely to be nonexercisers, and subjects who had low negative scores tended to be team exercisers.

To interpret the nature of this discriminant function, we must determine which attribute variables are measuring it. To do this, we will focus on the structure coefficients in Table 11.2. We will largely ignore the standardized discriminant function coefficients, due to the fact that they are sometimes misleading (as was explained above). As a somewhat arbitrary criterion, we will assume that a given attribute variable is "measuring" Discriminant Function 1 if the structure coefficient for that attribute exceeds ±.40 (in absolute value).

To be even more specific in interpreting the function, we can determine whether a specific subgroup tended to have high scores or low scores on a given attribute by reviewing the signs of the centroids in conjunction with the signs of the structure coefficients. It works like this:

- The subgroup that had a high positive centroid (such as +.946) tended to have high scores on attribute variables whose structure coefficient had a positive sign (+) and tended to have low scores on attribute variables whose structure coefficient had a negative sign (-).

- The opposite was true for the subgroup that had a low negative centroid (such as -.524): It tended to have low scores on attribute variables whose structure coefficient had a positive sign (+) and tended to have high scores on attribute variables whose structure coefficient had a negative sign (-).

With these rules in mind, the centroids and structure coefficients for Discriminant Function 1 in Table 11.2 reveal the following:

- The nonexercisers tended to have low scores on the health motivation scale. We know this because the centroid for nonexercisers was positive (+.946), and the structure coefficient for health motivation was negative (-.550).

- The nonexercisers tended to have high scores on the dislike of exercise scale. We know this because their centroid was positive (+.946), and the structure coefficient for dislike of exercise was also positive (+.840).

- The nonexercisers also tended to have high scores on the lack of time for exercise scale. We know this because their centroid was positive (+.946), and the structure coefficient for lack of time for exercise was also positive (+.682).

Once we have established the types of scores that the nonexercisers are likely to show on the attribute variables, it is relatively easy to predict the type of scores that the team exercisers will show. Since the centroid for the team exercisers showed the opposite sign of the centroid for the nonexercisers (-.524 versus +.946), we know that the team exercisers will show the opposite trend on each of the attribute variables. Specifically:

- The team exercisers will tend to have high scores on the health motivation scale (opposite of the low scores displayed by the nonexercisers).
- The team exercisers will tend to have low scores on the dislike of exercise scale (opposite of the high scores displayed by the nonexercisers).
- The team exercisers will tend to have low scores on the lack of time for exercise scale (opposite of the high scores displayed by the nonexercisers).

In summary, the team exercisers (a) are motivated to stay healthy, (b) like to exercise, and (c) say they have time to exercise. Isn't that pretty much what you would expect from college students who join sport teams?

STEP 3: NOW FOCUS ON THE NEXT DISCRIMINANT FUNCTION (IF APPROPRIATE). Once we have interpreted the first discriminant function, we would turn our attention to the second. If Discriminant Function 2 had been statistically significant, we would have interpreted it by following the same steps we had followed with Discriminant Function 1. However, an earlier section indicated that Wilks' lambda for Discriminant Function 2 was not statistically significant. For this reason, we will skip its interpretation.

How Strong is the Relationship?

Earlier, we established that there is a statistically significant relationship between the categorical grouping variable and the quantitative attribute variables in Dr. O'Day's study (for the first discriminant function, at any rate). However, the fact that the relationship is statistically significant does not necessarily mean that the relationship is strong in practical terms. To understand the strength of the relationship, researchers typically consult indices of effect size such as Wilks' lambda (Λ) or the squared canonical correlation coefficient (R_c^2), along with the results of a classification analysis.

MULTIVARIATE INDICES OF ASSOCIATION. An earlier section indicated that ***Wilks' lambda*** is a multivariate index of association that (in the present analysis) represents the strength of the relationship between a single grouping variable and multiple attribute variables. The most common symbol for Wilks' lambda is Λ (the Greek letter lambda), and its values may range from 0.00 to 1.00, with lower values (closer to 0.00) indicating a stronger relationship.

Table 11.1 (presented earlier) provided Wilks' lambda for each discriminant function created in Dr. O'Day's analysis. Below is the excerpt from the article in which she referred to this index:

> Table 11.1 shows that Wilks' lambda for the first discriminant function was statistically significant: $\Lambda = .81$, $\chi^2 (8, N = 238) = 49.44$, $p < .001$. In contrast, Wilks' lambda for the second discriminant function was nonsignificant: $\Lambda = .97$, $\chi^2 (3, N = 238) = 6.17$, $p = .104$.

The above indicates that, for the first discriminant function, $\Lambda = .81$. At first blush, you might assume that this means that there is only a weak relationship between the criterion variable and the predictor variables, since .81 is closer to 1.00 (which represents the weakest possible relationship) than it is to 0.00 (which represents the strongest possible relationship). But this is not necessarily the case. Remember that the oft-cited Cohen (1988) advises that researchers should compare their obtained indices of effect size against the indices obtained by other researchers who studied the same variables. Let's assume that Dr. O'Day has reviewed the research literature on her topic and found that a Wilks' lambda of $\Lambda = .81$ represents a fairly strong relationship compared to the relationships found by other researchers studying exercise behavior in college students. So far, so good.

A related index of effect size that is sometimes reported as part of a discriminant analysis is the ***squared canonical correlation coefficient*** (symbol: R_c^2). The R_c^2 statistic is the multivariate counterpart to the R^2 statistic used in linear multiple regression (from *Chapter 9* in this book). Values of R_c^2 may range from .00 to 1.00, with *higher* values indicating a stronger relationship between the grouping variable and the attribute variables (yes, the interpretation of R_c^2 is essentially the opposite of the interpretation of Wilks' lambda).

For Discriminant Function 1, Table 11.1 (presented much earlier in this chapter) showed that $R_c^2 = .17$. Again, let's assume that this is a fairly large value for R_c^2, compared to the values obtained by other researchers studying exercise behavior in college students.

CLASSIFICATION ANALYSIS. A different approach to understanding the strength of the relationship between the grouping variable and the attribute variables is to perform a classification analysis. Dr. O'Day presents the results of this analysis in the following excerpt:

A classification analysis was performed using relative subgroup sizes as prior probabilities. Results are presented in Table 11.3.

In Table 11.3, the rows represent the actual subgroups, and the columns represent the predicted subgroups. The frequencies for correct predictions appear in parentheses on the diagonal. The table shows that 8 of the 35 nonexercisers (23%) were correctly classified, 135 of the 143 individual exercisers (94%) were correctly classified, and 7 of the 60 team exercisers (12%) were correctly classified.

The overall percentage of correct classifications was 63%, and the leave-one-out cross-validated correct classification rate was 61%. However, these overall correct classification rates are misleadingly high because they are heavily influenced by the high correct classification rate of 94% for the largest subgroup: the individual exercisers ($n = 143$).

Table 11.3

Results of Classification Analysis Based on the Discriminant Function

	Predicted group			
Actual group	Nonexercisers	Individual exercisers	Team exercisers	Percent correctly classified
Nonexercisers ($n = 35$)	(8)	26	1	23%
Individual exercisers ($n = 143$)	2	(135)	6	94%
Team exerciser ($n = 60$)	2	51	(7)	12%

Note. $N = 238$. Values in table are frequencies unless otherwise indicated. Overall percentage of correct predictions = 63%. Leave-one-out cross-validated percentage of correct predictions = 61%.

The logic of a classification analysis goes something like this: If the discriminant function does a good job of discriminating between the subgroups of participants, we should be able to use that discriminant function to correctly predict the subgroup to which a given participant belongs. This means that, for a given participant, his or her scores on the attribute variable are inserted into an equation, and the resulting composite is used to predict the subgroup to which the participant is most likely to belong. Once these predictions are made, we can evaluate the discriminant function's **hit rate**: The percent of subjects who were correctly classified. If the hit rate is high, this means that the discriminant function did a good job of discriminating between groups. This, in turn, could be interpreted as evidence that there is a relatively strong relationship between the grouping variable and the attribute.

OVERALL CORRECT CLASSIFICATIONS. As Dr. O'Day indicated in the excerpt, the frequencies within parentheses on the diagonal of Table 11.3 represent "hits:" the number of people who were classified into the correct subgroup based on the discriminant function. The ***overall correct classification rate*** is computed as the total number of people correctly classified (i.e., the sum of the frequencies on the diagonal) divided by the total number of people in the study. Table 11.3 shows that 150 of the 238 participants in the investigation were correctly classified, resulting in an overall correct classification rate of 63%. Pretty good, huh?

BIAS DUE TO UNEQUAL GROUP SIZES. Well, maybe the overall correct classification rate is not as good as it appears. The problem, as Dr. O'Day points out in the excerpt, is that this 63% is misleadingly high. It is very much influenced by the fact that the hit rate for the largest subgroup (the individual exercisers) was quite high at 94%. Because this subgroup was much larger than the other two subgroups, it had a disproportionate effect on the overall hit rate, causing the overall hit rate to be misleadingly high. Notice that the hit rate was only 23% for the nonexercisers, and was only 12% for the team exercisers. *Ouch!*

BIAS DUE TO SAMPLE-SPECIFIC RELATIONSHIPS. But wait—it gets worse. The overall correct classification rate of 63% for the combined sample is also misleadingly high due to chance characteristics of the sample. Whenever a researcher uses a specific sample of participants to compute the discriminant function coefficients, and then uses the very same sample to perform the classification analysis, the overall hit rate is likely to be biased upwards due to chance relationships that are unique to that specific sample. The smaller the sample, the worse the bias is likely to be.

CROSS-VALIDATION. One solution to this problem of bias is to perform some type of *cross-validation*. For a discriminant analysis, a cross-validation typically involves (a) starting with a really large sample (say 500 or more subjects), (b) randomly dividing this sample into a calibration sample versus a validation sample, (c) developing the coefficients of the discriminant function based on the calibration sample, and then (d) performing the classification analysis by applying this discriminant function to the validation sample. The correct classification rate observed with the validation sample is likely to be less biased, compared to the correct classification rate observed when all analyses are based on just one sample.

LEAVE-ONE-OUT CROSS-VALIDATION. The problem, of course, is that researchers often do not have access to samples that are large enough to divide into calibration versus validation samples. This has led to an alternative procedure called the ***leave-one-out estimator*** (Lachelnbruch, 1967). The leave-one-out estimator allows researchers to (a) analyze data from just one sample, and (b) statistically estimate what the overall correct classification would have been if they had performed a cross-validation study using a calibration sample and a separate validation sample. The leave-one-out cross-validation approach is not perfect, but it is known to produce nearly unbiased estimates with just one sample. As a result, it is often included in researcher articles reporting discriminant analyses.

Here again is the excerpt in which Dr. O'Day reports her leave-one-out estimator:

> The overall percentage of correct classifications was 63%, and the leave-one-out cross-validated correct classification rate was 61%.

Summary of Findings

Wilks lambda revealed a statistically significant relationship between the grouping variable and the attribute variables, but only for the first discriminant function. Therefore, only the first discriminant function was interpreted.

For the first function, Wilks' lambda was $\Lambda = .81$ and the squared canonical correlation coefficient was $R_c^2 = .17$. These statistics revealed a moderately strong relationship between the grouping variable (team & exercise status) and the four attribute variables.

Group centroids showed that the first discriminant function discriminated between the nonexercisers versus the team exercisers. Compared to the nonexercisers, the team exercisers (a) scored higher on health motivation, (b) scored lower on dislike of exercise, and (c) scored lower on lack of time for exercise.

The classification analysis showed that the overall correct classification rate was 63%, and the leave-one-out cross-validated correct classification rate was 61%. However, these rates were misleadingly high because they were heavily influenced by the high correct classification rate obtained with the largest subgroup: the individual exercisers.

Additional Issues Related to this Procedure

If you conduct much research, sooner or later you will find yourself in a situation in which you have the choice of analyzing your data with either discriminant analysis (covered in the current chapter) or logistic regression (to be covered in *Chapter 12*). This section discusses the factors that should inform your decision and summarizes the major assumptions underlying discriminant analysis.

Discriminant Analysis or Logistic Regression?

In some situations, researchers have the choice of analyzing their data with either discriminant analysis or logistic regression. One example of such a situation is as follows:

- When there is a single criterion variable and this criterion is a dichotomous variable assessed on a nominal scale of measurement.
- When there is more than one predictor variable and each predictor is a multi-value variable assessed on an interval or ratio scale.

In some situations, the choice between discriminant analysis and logistic regression is fairly arbitrary, as they tend to give similar results when the two groups under the criterion variable are similar in size. On the other hand, when the prior probability of membership in one of the groups is close to 0 or close to 1, they may give different results (Press & Wilson, 1978; Wright, 1995).

There are some situations in which discriminant analysis is more appropriate than logistic regression. For example, discriminant analysis is appropriate when the criterion variable is assessed on a nominal scale and consists of more than two categories (Silva & Stam, 1995). The binary logistic regression procedures discussed in *Chapter 12* of this book are not appropriate in such a situation.

Despite this, some have reported a growing preference for logistic regression over discriminant analysis, especially in clinical prediction studies (Warner, 2008). Possible explanations for the growing popularity of logistic regression include the following:

- The statistical assumptions underlying logistic regression are easier to satisfy, compared to the assumptions underlying discriminant analysis (Wright, 1995).

- Logistic regression is appropriate for studies in which the criterion variable consists of three or more categories and is assessed on an ordinal scale of measurement; traditional discriminant analysis is less appropriate for criterion variables of this type (Silva & Stam, 1995).

- Some have argued that it is appropriate to include categorical predictor variables as part of a logistic regression analysis, but that categorical predictors are inappropriate with discriminant analysis (e.g., Wright, 1995, p. 241).

- When the two groups under the criterion variable are very different in size (e.g., 5% in one group and 95% in the other group), logistic regression is more likely to give correct results, compared to discriminant analysis (Tabachnick & Fidell, 2007; Warner, 2008).

Assumptions Underlying this Procedure

This section summarizes some of the more important assumptions underlying discriminant analysis. The list is taken from Tabachnick and Fidell (2007), who discuss each assumption in some detail.

MULTIVARIATE NORMALITY. Significance tests for the discriminant functions require that the attribute variables display multivariate normality. The tests are robust against some departure from multivariate normality as long as the subgroups are similar in size, the smallest subgroup has at least 20 observations, and the number of attribute variables is not greater than five or so.

ABSENCE OF OUTLIERS. Outliers are extreme, atypical observations (i.e., participants with scores that are very different from the scores displayed by most participants).

LINEAR RELATIONSHIP. The relationships between all pairs of attribute variables (within a subgroup) should be linear.

HOMOGENEITY OF THE VARIANCE-COVARIANCE MATRICES. When the subgroups are small or unequal, significance tests and classification analyses may be biased if the variance-covariance matrices are not homogeneous.

ABSENCE OF MULTICOLLINEARITY. *Multicollinearity* occurs when attribute variables are very highly correlated with one another. Approaches for detecting multicollinearity are described by Field (2009a) and Tabachnick and Fidell (2007).

Chapter 12

Logistic Regression

This Chapter's *Terra Incognita*

You are reading an article about a study in which the criterion variable is *exercise status*, a dichotomous variable that indicates whether a given participant is in the *exercisers* subgroup versus the *nonexercisers* subgroup. The researcher attempted to determine whether membership in these two groups can be predicted from four variables that constitute something called the *four-factor model of exercise behavior*. This is what she reports in the *Results* section:

> The logistic regression analysis showed that the model containing the constant plus the four predictor variables displayed a significantly better fit to the data, compared to the model containing just the constant: model χ^2(4, N = 238) = 35.128, $p < .001$. For the full model, the Cox & Snell R^2 = .137, and the Nagelkerke R^2 = .235.
>
> Wald χ^2 statistics showed that only two of the predictor variables (dislike of exercise and lack of time for exercise) displayed statistically significant logistic regression coefficients ($p < .05$). The adjusted odds ratio for dislike of exercise was AOR = 0.583, 95% CI [0.435, 0.781], and for lack of time for exercise was AOR = 0.743, 95% CI [0.576, 0.959]. For both predictors, the adjusted odds ratio was less than 1.0, indicating that participants with high scores on these two variables were less likely to be in the exercisers subgroup.

Be strong of heart. All shall be revealed.

The Basics of this Procedure

Binary logistic regression is a data-analysis procedure that allows researchers to investigate the nature of the relationship between a dichotomous criterion variable and a set of two or more predictor variables. A ***dichotomous variable*** is a variable that assumes just two values. With binary logistic regression, the criterion variable consists of two groups, and a given participant appears in one group or the other. This is illustrated in the generic analysis model in Figure 12.1.

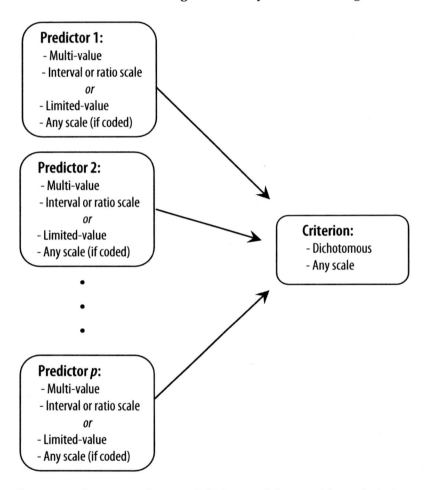

FIGURE 12.1. Generic analysis model: Nature of the variables included in a typical logistic regression analysis.

Figure 12.1 shows that a logistic regression analysis allows for multiple predictor variables. Theoretically, there is no limit to the number of predictors, although in practice it is unusual to see more than a dozen or so. The predictor variables are typically multi-value variables assessed on an interval or ratio scale. A ***multi-value variable*** is a variable that displays seven or more values.

Figure 12.1 shows that a predictor variable in logistic regression may also be a *limited-value variable*—a variable that displays just two to six values. In the figure, the boxes for the predictor variables include the note "*if coded.*" This indicates that limited-value variables may be assessed on any scale of measurement as long as they have been appropriately transformed using dummy-coding or some similar procedure (dummy-coding and related topics were covered in *Chapter 10*). Additional assumptions underlying logistic regression are presented at the end of this chapter.

In the analysis, the researcher estimates the *logistic regression equation*, which is also called the *logistic regression model*. The logistic regression equation consists of a constant (similar to the intercept in linear regression), and a separate logistic regression coefficient (symbol: *B*) for each predictor variable.

Logistic regression produces a *model χ^2 statistic*, and this statistic indicates whether there is a statistically significant relationship between the dichotomous criterion variable and the predictor variables, taken as a set. As a part of the analysis, researchers usually review the individual logistic regression coefficient for a given predictor variable to determine (a) whether the coefficient is statistically significant and (b) the nature of the relationship between that predictor variable and the dichotomous criterion variable. They typically review adjusted odds ratios (symbol: *OR* or *AOR)* for the same purpose.

Researchers may consult pseudo-R^2 statistics to evaluate the strength of the relationship between the dichotomous criterion variable and the predictor variables. In some cases, they may also perform a classification analysis to evaluate the equation's success in accurately predicting group membership (under the criterion variable) for the study's participants.

Illustrative Example: Predicting Exercise Status

In earlier chapters, you read about a fictitious health psychologist named Dr. O'Day who is interested in identifying variables that predict the frequency with which people engage in aerobic exercise. In *Chapter 9* and *Chapter 10*, the criterion variable was *exercise sessions*: the number of times that a participant engaged in aerobic exercise each week. In those chapters, *exercise sessions* was a multi-value quantitative variable assessed on a ratio scale of measurement.

The data analysis procedures covered in *Chapter 9* and *Chapter 10* are sometimes called *linear multiple regression*. In part, the word "linear" is used to distinguish those statistical procedures from the procedures to be covered in the current chapter: logistic multiple regression.

THE CRITERION VARIABLE. The analyses to be reported in this chapter will be similar to the previous analyses, except that the criterion variable will be *exercise status* rather than *exercise sessions*. In the current chapter, *exercise status* is a dichotomous variable that merely classifies participants regarding their membership in one of two subgroups. The *exercisers* subgroup includes those subjects who engage in at least one exercise session per week, and the *nonexercisers* subgroup includes those subjects who never engage in aerobic exercise at all.

THE PREDICTOR VARIABLES. In the logistic regression analyses to be reported here, the predictor variables will be the same predictor variables that had been used in the linear multiple regression analyses reported in *Chapter 9* and *Chapter 10*. The analyses will focus on Dr. O'Day's *four-factor model of exercise behavior*. This model hypothesizes that exercise behavior is determined by the four variables illustrated in Figure 12.2:

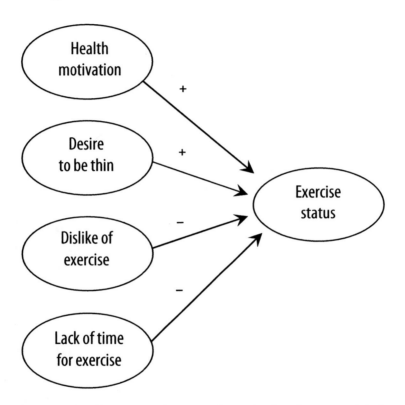

Figure 12.2. Predicting exercise status from the four-factor model of exercise behavior.

The four variables that constitute the four-factor model are:

- **HEALTH MOTIVATION**: the extent to which an individual is willing to sacrifice and work hard in order to stay healthy.

- **DESIRE TO BE THIN**: the extent to which it is important to an individual to have a fairly low body weight and a lean physical appearance.

- **DISLIKE OF EXERCISE**: the extent to which the individual hates aerobic exercise and finds it to be unpleasant.

- **LACK OF TIME FOR EXERCISE**: the extent to which the individual believes that he or she is too busy to engage in frequent aerobic exercise.

To investigate this model, Dr. O'Day administered the *Questionnaire on Eating and Exercising* (from *Appendix A*) to 260 students enrolled in various psychology courses at a regional state university in the Midwest. She obtained usable responses from 238 participants.

This chapter presents two detailed examples of logistic regression. Example 1 illustrates **direct logistic regression**: an analysis in which all predictor variables are added to a logistic regression equation in just one step. Example 2 illustrates **sequential logistic regression**: an analysis in which predictor variables are added to the equation in a sequence of pre-determined steps.

Typical Research Questions Addressed

Binary logistic regression allows researchers to investigate largely the same types of research questions that are addressed with linear multiple regression. In the current study, Dr. O'Day could pursue questions such as the following:

- Is there a statistically significant relationship between the criterion variable and the predictor variables, taken as a set? More specifically, is there a statistically significant relationship between (a) exercise status and (b) the four variables that constitute the four-factor model of exercise?

- What is the nature of the relationship between the criterion variable and the predictor variables? Of the four variables that constitute the four-factor model, which variables display logistic regression coefficients that are significantly different from zero? What is the sign for each of these coefficients?

- What is the strength of the relationship between the criterion variable and the predictor variables taken as a group? More specifically, what is the strength of the relationship between (a) exercise status and (b) the four variables of the four-factor model, taken as a group? When the logistic regression equation is used to classify participants into groups under the criterion variable, how accurate are these classifications?

Details, Details, Details

In a just world, logistic regression would be identical to the linear multiple regression that you learned about in *Chapter 9*, except that the criterion variable would be a dichotomous variable, rather than a multi-value variable. But we live on Planet Earth, so you can abandon that idea. For better or for worse, there are substantial differences between the two procedures, and this section will focus on just two: the nature of the predicted criterion variable and the mathematical procedure used to estimate the regression coefficients.

The Nature of the Criterion Variable in Logistic Regression

You have learned that in logistic regression, a criterion variable is a dichotomous variable. That, however, is the *actual* criterion variable, not the *predicted* criterion variable. In logistic regression, the predicted criterion variable is a somewhat more nebulous creature, not so easy to understand at first glance. With this procedure, we may think of predicted criterion variable as being two distinct (though related) animals: (a) as being a *predicted probability*, or, alternatively, (b) as being something called the *logit*.

Already lost? Let's take this one step at a time.

In logistic regression, the actual criterion variable is a dichotomous variable—a variable that can display just two values. For example, in the analyses to be reported in this chapter, the actual criterion variable is *exercise status*—a dichotomous variable that simply indicates whether a given participant is in the *nonexerciser* group (participants who never exercise) versus the *exerciser* group (participants who exercise at least once each week). The researcher will use the value "0" to represent nonexercisers, and "1" to represent exercisers. So the *actual* criterion variable simply consists of 0s and 1s.

In the eyes of statisticians, a dichotomous criterion variable consisting of 0s and 1s has a number of mathematically undesirable properties (trust me on this—you do not want to hear the details). Therefore, logistic regression converts it into a synthetic variable called the *logit*. The **logit** can be defined as the natural logarithm of the odds (Haddock, Rindskopf, & Shadish, 1998). Understanding this definition requires that we first understand the concepts of *probability* and *odds*.

PROBABILITY. The concept of ***probability*** is more or less synonymous with the concept of "likelihood." Probabilities may range from .00 to 1.00, with an infinite number of values in-between (such as .32 or .89).

With binary logistic regression, *probability* refers to the likelihood that a specific participant will display some specific "event." Remember that in the current study, the criterion variable (exercise status) consists of two categories: the "exerciser" group versus the "nonexerciser" group. In order to define our terms for Dr. O'Day's study, let's define the "event" as "being in the exerciser group."

As an illustration: Imagine some hypothetical subject who has extremely positive attitudes toward exercise and health. In Dr. O'Day's study, this would be a person who has the highest possible scores on the *health motivation* scale and the *desire to be thin* scale, and the lowest possible scores on the *dislike of exercise scale* and the *lack of time for exercise* scale. Imagine that a subject with this profile of scores turns out to be in the "exerciser" group 90% of the time. If this were the case, then the probability of the "event" for this kind of subject is .90. This probability can be represented in the following way:

$Prob(event) = .90$

THE ODDS. Now that we have the probability for this subject, we can convert it into an even more obtuse statistic called the *odds*. With binary logistic regression, the

odds of displaying the event of interest are equal to (a) the probability of the event occurring divided by (b) the probability of the event not occurring. The formula for computing the odds is as follows:

$$Odds = \frac{P(event)}{P(no\ event)}$$

For example, consider the subject described earlier—the one with very high scores on health motivation and desire to be thin and very low scores on dislike of exercise and lack of time for exercise. For this individual, the probability of the "event" (being an exerciser) was .90. From this, we can deduce that the probability of "no event" (not being an exerciser) must be .10. We know this because 1.00 − .90 = .10.

Given these probabilities, we can now compute the odds that this individual will be an exerciser:

$$Odds = \frac{P(event)}{P(no\ event)} = \frac{.90}{.10} = 9$$

So, the odds that this subject will be an exerciser are equal to 9. In other words, for this person, the probability of being an exerciser is 9 times greater than the probability of not being an exerciser.

Some facts regarding the concept of the odds, in general:

- Values for the odds may range from zero through positive infinity ($+\infty$). "Positive infinity" is just a fancy way of saying that, theoretically, there is no limit as to how large the odds may be. The odds for a given participant may display decimal values such as 0.14, 1.21, 6.59, and so forth.

- When the odds are equal to 1.0, it means that the probability of the event is equal to .50, and the probability of no-event is also equal to .50. For example, imagine a participant who has mediocre scores on all four of the predictor variables in Dr. O'Day's study. For this participant, there is a 50% chance that he is an exerciser, and a 50% chance that he is not an exerciser. For this individual, the odds that he is an exerciser will be equal to 1.0 (because .50 ÷ .50 = 1.0).

- When the odds are greater than 1.0 (e.g., 1.60, 4.53), it indicates that the probability of the event is greater than the probability of no-event, and the size of the odds indicates how much greater. For example, if the odds for a given subject in Dr. O'Day's study are equal to 2.73, it indicates that the probability of this person being an exerciser is 2.73 times greater than the probability of this person not being an exerciser.

- When the odds are less than 1.0 (e.g., 0.17, 0.62), it indicates that the probability of the event is less than the probability of no event. Unfortunately, when the odds are less than 1, the actual value of the odds does not have a relatively straightforward, common-sense interpretation.

THE LOGIT: THE NATURAL LOG OF THE ODDS. Statisticians do not like to use the odds as a criterion variable in some statistical analyses because the odds often do not follow the shape of a normal distribution, and the odds often display a nonlinear relationship with other variables. To correct these problems, statisticians perform a fairly standard transformation: They take the natural logarithm of the odds. This produces the criterion variable that is actually used in a logistic regression: the *logit* (now do you see why the logit was previously defined as "the natural log of the odds?").

Compared to the odds, the logit has no upper limit and no lower limit, is more likely to be normally distributed, and is more likely to display linear relationships with other variables. All of these characteristics make it desirable for use as the criterion variable in a logistic regression equation (Warner, 2008).

ESTIMATING SCORES ON THE LOGIT. Let's use the symbol L_i to represent the logit. The formula for estimating scores on the logit is as follows (from Warner, 2008):

$$L_i = B_0 + B_1 X_1 + B_2 X_2 + \ldots B_p X_p$$

Where:

L_i = the logit of Y, the criterion variable

B_0 = the constant

B_1 = the logistic regression coefficient for the first predictor variable

X_1 = the subject's observed score on the first predictor variable

B_p = the logistic regression coefficient for the last predictor variable

X_p = the subject's observed score on the last predictor variable

In the preceding formula, the symbol X represents the various predictor variables included in the analysis (such as health motivation, desire to be thin, and so forth). The symbol B represents the logistic regression coefficients for these predictor variables. These coefficients can be defined in this way:

> The ***logistic regression coefficient*** (B) represents the amount of change in the logit (i.e., the log odds) that is associated with a one-unit change in the predictor variable, while statistically holding constant the other predictor variables.

Logistic regression coefficients may display a positive sign or a negative sign. If a predictor variable is completely unrelated to the logit (after statistically controlling for the other predictors), its logistic regression coefficient will be equal to zero. The more the coefficient differs from zero, the more important it is typically assumed to be. A bit later, we will see how researchers determine whether a given coefficient is significantly different from zero.

SUMMARY. In summary, with logistic regression, the *actual* criterion variable is *dichotomous* (i.e., exerciser group versus nonexerciser group), whereas the *predicted* criterion variable is *continuous* (i.e., scores on the logit). Either directly or indirectly, these scores on the logit are used in the process of estimating logistic regression coefficients, computing residuals of prediction, classifying participants into groups, and for other purposes to be discussed below.

Maximum Likelihood Estimation of the Model

Although the terms of the logistic regression equation (B_0, B_1, B_2, and so forth) are similar to the constant and multiple regression coefficients that you learned about in the chapters on linear multiple regression, they are not estimated using the same ordinary least squares (OLS) method that is used with linear multiple regression. Instead, they are estimated using the maximum likelihood (ML) method. Maximum likelihood estimation is used with a number of advanced statistical procedures, including structural equation modeling and some types of factor analysis.

Maximum likelihood estimation is an *iterative* process, meaning that the software performs the estimation process multiple times in order to get closer and closer to the best values for the constant and logistic regression coefficients. The software begins with arbitrary values for the constant and regression coefficients and then determines the direction and magnitude that the coefficients must change in order to maximize the likelihood of obtaining the data that were actually obtained. It established new, better coefficients and then analyzes the residuals from this model to determine how the coefficients might be further improved. It repeats the process in order to inch closer and closer to the "best" values for coefficients. The process stops when convergence is achieved (***convergence*** means that the coefficients are changing very little from iteration to iteration—the size of the improvements in the coefficients has become negligible). The process is called *maximum likelihood estimation* because final coefficients are the coefficients that maximize the likelihood of obtaining the sample data that were actually obtained (Tabachnick & Fidell, 2007).

Using the $-2LL$ to Test Hypotheses Related to the Model

With most statistical procedures, *goodness-of-fit* refers the extent to which the predictor variables do a good job of predicting the criterion variable. In linear multiple regression, we used R^2 and the F statistic to test a variety of hypotheses related to goodness of fit. To test similar hypotheses in logistic regression, we will instead use an index abbreviated as $-2LL$, along with a chi-square test derived from the $-2LL$.

THE $-2LL$ STATISTIC. With logistic regression, the most important index of model fit is the *-2 log likelihood* statistic, which is also called the ***deviance***. It is typically represented by the symbol **$-2LL$**. The $-2LL$ is generally a positive number. Large values for the $-2LL$ indicate that the equation does a poor job of predicting the criterion variable, and smaller values indicate that the equation does a better job of predicting the criterion. When the equation predicts the criterion variable with perfect accuracy, the $-2LL$ is equal to zero (Wright, 1995).

THE LIKELIHOOD-RATIO TEST. The $-2LL$ by itself is not terribly useful. Its importance lies in the fact that it is a key ingredient in performing *likelihood ratio tests* (Norusis, 2005). The ***likelihood ratio test*** is a general-purpose procedure that determines whether adding predictor variables to a logistic regression equation results in a significant improvement in the equation's ability to predict the criterion variable. The likelihood ratio test produces a chi-square statistic called the *likelihood ratio statistic* (symbol: χ^2, G, or G^2).

The formula for computing the likelihood ratio χ^2 statistic is as follows:

$$\text{Likelihood ratio} = (-2LL \text{ for the reduced model}) - (-2LL \text{ for the full model})$$

Where:

- *Likelihood ratio* = the obtained value for the χ^2 statistic.
- The *reduced model* is the logistic regression equation containing the smaller number of terms (a *term* is a predictor variable or the constant).
- The *full model* is the logistic regression equation containing the larger number of terms.

The resulting *likelihood ratio statistic* is a chi-square statistic (symbol: χ^2). The degrees of freedom for this χ^2 test are equal to the difference between the number of terms in the full model versus the number of terms in the reduced model. If the chi-square statistic is significant, the researcher concludes that the full model provides a fit that is significantly better than the fit achieved by the reduced model.

This chapter shows how this likelihood ratio test can be used to perform two important analyses in logistic regression:

- The *model χ^2 test* is used to determine whether there is a significant relationship between (a) the dichotomous criterion variable and (b) the predictor variables included in the equation, taken as a set. This test is illustrated with Example 1, below.
- The *model-improvement χ^2 test* is used to determine whether adding a new set of predictor variables to an equation that already contains one or more predictor variables results in a significant improvement in the equation's ability to predict the criterion. This procedure is illustrated with Example 2, later in the chapter.

Example 1: Direct Logistic Regression

This section illustrates one of the most elementary applications of logistic regression: an analysis in which all predictor variables are entered into the equation in just one step. This approach is called *direct logistic regression, simultaneous logistic regression,* and *logistic regression with forced-entry of predictor variables*. It will be illustrated here by discussing Dr. O'Day's investigation in which she attempted to predict the dichotomous criterion variable *exercise status* from the four variables that constitute her four-factor model of exercise behavior (as illustrated in Figure 12.2).

Is There a Significant Relationship Between the Variables?

In the *Results* section of the article describing her study, Dr. O'Day would probably begin with a table that provides means, standard deviations, and intercorrelations for all quantitative variables, similar to the correlation-matrix tables that were covered in *Chapter 6*. This table could include point-biserial correlations between exercise status and the study's quantitative variables. To save space, that table will not be presented here.

THE MODEL χ^2 TEST. Next, Dr. O'Day would turn her attention to the logistic regression analysis itself. Articles reporting this procedure very often begin with a **model χ^2 test**. This procedure determines whether there is a statistically significant relationship between (a) the dichotomous criterion variable and (b) the equation's predictor variables, taken as a set. More specifically it tests the following null hypothesis:

$$H_0: \beta_1 = \beta_2 = \ldots \beta_p = 0.$$

The preceding null hypothesis states that, in the population, the logistic regression coefficients for all predictor variables in the logistic regression equation (other than the constant) are equal to zero. This null hypothesis is tested by computing a model χ^2 statistic. The model χ^2 statistic, in turn, is computed using the likelihood ratio test that was introduced in the previous *Details, Details, Details* section:

$$\text{Likelihood ratio} = (-2LL \text{ for the reduced model}) - (-2LL \text{ for the full model})$$

To compute a model χ^2 statistic that will test the null hypothesis stated above, the terms of the preceding formula should represent the following:

- The *likelihood ratio* is the model χ^2 statistic.
- The *reduced model* is the logistic regression equation that contains only the constant.
- The *full model* is the logistic regression equation that contains the constant plus all of the predictor variables.

The resulting model χ^2 statistic has degrees of freedom equal to the number of predictor variables in the full model (not including the constant). If this model χ^2 is statistically significant, it tells us that the full model (containing the constant plus all predictor variables) provides a fit to the data that is significantly better than the fit provided by the reduced model (the model that contained just the constant). If the model χ^2 is significant, the researcher rejects the null hypothesis that the *B* coefficients for all predictors are equal to zero, and concludes that at least one of the predictors must have a coefficient that is significantly different from zero. In other words, the researcher concludes that there is a statistically significant relationship between the criterion variable and the predictor variables, taken as a set.

In the following excerpt from her published article, Dr. O'Day discusses the results of the model χ^2 test for the current investigation:

> Data were analyzed using binary logistic regression. In this analysis, the criterion variable was *exercise status* (coded 0= nonexercisers, 1= exercisers), and the four predictor variables were *health motivation, desire to be thin, dislike of exercise,* and *lack of time for exercise.*
>
> The logistic regression analysis showed that the model containing the constant plus the four predictor variables displayed a significantly better fit to the data compared to the model containing just the constant, model $\chi^2(4, N = 238) = 35.128$, $p < .001$. This finding indicated that at least one of the predictors displayed a logistic regression coefficient (*B*) that was significantly different from zero.

The preceding excerpt presented the model χ^2 statistic in the following format, which is the standard format for presenting chi-square statistics in APA journals:

$$\chi^2(4, N = 238) = 35.128, p < .001$$

Two numbers are presented inside the parentheses in the preceding entry. The first number is 4, the degrees of freedom for the chi-square test. The second number is 238, the number of participants on which the analysis is based. The value 35.128 is the obtained chi-square statistic itself. Finally, the "$p < .001$" tells us that the model χ^2 statistic is significant.

These results indicate that there is a statistically significant relationship between the criterion variable and the group of predictors. Encouraged by this finding, Dr. O'Day now turns her attention to the regression coefficients for the individual predictor variables.

Nature of Relationship Part I: Logistic Regression Coefficients

The previous analysis established that there is a statistically significant relationship between exercise status and the four predictor variables, taken as a set. But this finding begs the question: Are all four of the predictor variables making significant contributions to the prediction of the criterion variable, or just a few? And, for each predictor, is the nature of the relationship in the predicted direction?

To answer these questions, the researcher may interpret one or both of the following:

- Results related to the *logistic regression coefficients* for each predictor.
- Results related to the *adjusted odds ratio* for each predictor.

The current section discusses the logistic regression coefficients. A subsequent section covers the adjusted odds ratios.

THE LOGISTIC REGRESSION COEFFICIENTS. An earlier section presented the following general form for the logistic regression equation:

$$L_i = B_0 + B_1X_1 + B_2X_2 + \ldots B_pX_p$$

In this equation, the logistic regression coefficients are represented with symbols such as B_1 and B_2 (although other sources may use lower-case letters such as b_1 and b_2). A ***logistic regression coefficient*** can be defined as the change in the logit (i.e., the change in the log odds) that is associated with a one-unit change in the predictor variable, while statistically holding constant the remaining predictor variables (Norusis, 2005).

As a reader of research, you will be most interested in two questions related to a given coefficient: (a) is it significantly different from zero?, and (b) does it display a positive sign or a negative sign? Here are some basic guidelines for interpreting logistic regression coefficients:

- The *zero-effect value* for a logistic regression coefficient is zero. In other words, if the logistic regression coefficient for a specific predictor is *equal to zero*, it means that the predictor is not related to the criterion variable (while statistically holding constant the other predictors).

- If the logistic regression coefficient is *positive*, it means that higher scores on the predictor are associated with being in the target group (i.e., being in the "event" group; the group coded 1 rather than 0).

- If the logistic regression coefficient is *negative*, it means that higher scores on the predictor are associated with being in the non-target group (i.e., being in the "no event" group; the group coded 0 rather than 1).

TESTING THE COEFFICIENTS USING THE z TEST FOR THE WALD STATISTIC. Significance tests for logistic regression coefficients almost always test the following null hypothesis:

H_0: $\beta = 0$. In the population, the logistic regression coefficient for this predictor variable is equal to zero.

Several different statistical procedures have been developed to test this null hypothesis. Some of these tests are based on the ***Wald statistic***, which is simply the logistic regression coefficient for a given predictor variable divided by the coefficient's estimated standard error (Field, 2009a):

$$Wald = \frac{B}{SE}$$

With large samples, the logistic regression coefficient divided by its standard error can also be interpreted as a z statistic:

$$z = \frac{B}{SE}$$

This z statistic is the *obtained z* statistic. This obtained z is compared against standard critical values (symbol: z_{crit}) for a two-tailed z test. If the obtained z statistic exceeds the critical value, the researcher rejects the null hypothesis that the logistic regression coefficient is equal to zero in the population (Wright, 1995). Standard criteria for determining the statistical significance of the obtained z statistics are as follows: if $z > \pm 1.96$; then $p < .05$; if $z > \pm 2.58$, then $p < .01$; and if $z > \pm 3.29$, then $p < .001$.

TESTING THE COEFFICIENTS USING THE WALD χ^2 STATISTIC. As an alternative, some articles instead report a Wald χ^2 as the significance test for a regression coefficient. The **Wald χ^2** is computed by dividing the squared regression coefficient by its squared standard error. The resulting statistic follows a chi-square distribution with 1 degree of freedom (Tabachnick & Fidell, 2007):

$$Wald\ \chi^2 = \frac{B^2}{SE^2}$$

TESTING THE COEFFICIENTS USING THE LIKELIHOOD RATIO TEST. When the logistic regression coefficient is large, however, the preceding significance tests should be interpreted with caution, as they tend to produce Type II errors in that situation (Menard, 1995). You will recall that a **Type II error** occurs when the researcher concludes that the statistic of interest is nonsignificant, but is wrong in doing so.

When the regression coefficients are large, it is generally best to test their significance with a specific application of the likelihood ratio test. You will recall that the likelihood ratio statistic is computed using this formula:

Likelihood ratio = $(-2LL$ for the reduced model$) - (-2LL$ for the full model$)$

When using this procedure to test the significance of the regression coefficient for a specific predictor variable, the *full model* is the logistic regression equation containing the constant and all predictor variables, while the *reduced model* is the equation containing all of the preceding terms except for the predictor variable of interest. The resulting likelihood-ratio statistic is distributed as chi-square with one degree of freedom. If this chi-square statistic displays an obtained p value that is less than the stated alpha criterion (say, if the obtained p value is less than $\alpha = .05$), then the logistic regression coefficient for the deleted predictor variable is viewed as being significantly different from zero.

In her investigation, Dr. O'Day would perform this likelihood ratio test four times. That is, she would perform one test for each of her four predictor variables.

DR. O'DAY'S RESULTS. Below, Dr. O'Day summarizes the results of her analyses in Table 12.1. She provides additional details in an excerpt from the text of her article.

> In Table 12.1, the logistic regression coefficients for the four predictor variables are presented in the column headed *B*. The Wald χ^2 statistic tests whether the logistic regression coefficient for a specific predictor variable is significantly different from zero. The p values for the Wald χ^2 statistic show that only two of the predictor variables (dislike of exercise and lack of time for exercise) displayed significant coefficients. The sign of the *B* coefficients indicated that participants were more

likely to be in the "exerciser" group if they indicated that they (a) liked to exercise and (b) had adequate free time for exercise.

The Wald statistic has been shown to sometimes give incorrect results, especially when the size of the B coefficient is large (Menard, 1995). Therefore, the statistical significance of each B coefficient was also assessed by performing a series of likelihood-ratio tests. These tests involved subtracting the −2 log likelihood for a full model (the model containing all four predictor variables) from the −2 log likelihood for a reduced model (the model containing all predictors except for the predictor of interest). These analyses showed that the B coefficient for dislike of exercise was still significant, $\chi^2(1, N = 238) = 13.859$, $p < .001$, as was the B coefficient for lack of time for exercise, $\chi^2(1, N = 238) = 5.427$, $p < .05$. With alpha set at $\alpha = .05$, the B coefficient for health motivation and desire to be thin were nonsignificant.

Table 12.1

Predicting Exercise Status: Results from Logistic Regression Analysis

Predictor variable	B	SE	Wald χ^2	p	OR	[95% CI]
1. Health motivation	0.254	0.203	1.554	.213	1.289	[0.865, 1.920]
2. Desire to be thin	0.071	0.147	0.236	.627	1.074	[0.805, 1.433]
3. Dislike of exercise	−0.540	0.149	13.103	<.001	0.583	[0.435, 0.781]
4. Lack of time for exercise	−0.297	0.130	5.216	.022	0.743	[0.576, 0.959]
(Constant)	2.579	1.386	3.462	.063	13.183	

Columns OR and [95% CI] are under heading "Adjusted odds ratios".

Note. N = 238. For the model containing only the constant, −2LL = 209.017; For the model containing the constant plus all predictors: −2LL = 173.889. Model χ^2 (4, N = 238) = 35.128, $p < .001$. Cox & Snell R^2 = .137. Nagelkerke R^2 = .235. B = Logistic coefficients. SE = Standard errors for logistic coefficients. Wald χ^2 = Test of null hypothesis that B = 0 (for each test, df = 1). p = Probability value for Wald χ^2. AOR = Adjusted odds ratio. 95% CI = 95% Confidence interval for the adjusted odds ratio.

In the preceding summary, Dr. O'Day does not refer to the z test for the logistic regression coefficients. Instead, she reports the results from Wald χ^2 test (in the table), as well as the results from the likelihood ratio tests (within the text). In this instance, all three procedures converged in the conclusion that that only the predictor variables *dislike of exercise* and *lack of time for exercise* were statistically significant.

You may have noticed that Table 12.1 provides a good deal of information in addition to the information about the logistic coefficients. Most of this additional information will be covered in later sections of this chapter.

Nature of Relationship Part II: The Odds Ratio

Because logistic regression coefficients can be conceptually difficult to understand, researchers sometimes supplement them by reporting an additional statistic— the *odds ratio*—for each predictor. To understand this new statistic, we shall begin by reviewing the distinction between the concept of *odds* versus the concept of *odds ratio*.

THE ODDS. Earlier you learned that, with binary logistic regression, the **odds** of displaying the event of interest are equal to (a) the probability of the event occurring divided by (b) the probability the event not occurring. Here is the formula:

$$Odds = \frac{P(event)}{P(no\ event)}$$

THE ADJUSTED ODDS RATIO. In contrast, the **odds ratio** for a given predictor variable (as used on logistic regression) estimates the multiplicative change in the odds of membership in the targeted group for every one-unit increase in that predictor variable (Tabachnick & Fidell, 2007, pp. 461-462; Wright, 1995, p. 243). Symbols for the odds ratio include *OR*, ψ, Exp(*B*), Exp(*b*), and e^b.

When the logistic regression equation contains more than one predictor, it is appropriate to refer to this statistic as an **adjusted odds ratio** (Huck, 2004). The adjusted odds ratio estimates the multiplicative change in the odds of membership in the targeted group for every one-unit increase in the predictor variable while statistically controlling for the other predictor variables.

Technically, when a logistic regression equation includes more than one predictor variable, researchers really should refer to this statistic as the *adjusted odds ratio*, and should represent it using the symbol *AOR* (Huck, 2004). Be warned, however, that many researchers simply refer to it as the *odds ratio*, and represent it using the symbol *OR* (despite the fact that the odds ratio has obviously been adjusted for the other predictors in the equation).

INTERPRETING THE SIZE OF THE ADJUSTED ODDS RATIO. If the preceding definitions for odds ratio and adjusted odds ratio are exceeding your brain's RAM, then take heart: Things are about to get easier. When you see these statistics reported in a research article, you may interpret them using these (relatively simple) criteria:

- The *no-effect value* for an odds ratio (or adjusted odds ratio) is 1.0. In other words, when the adjusted odds ratio for a given predictor variable is *equal to 1.0*, it means that this predictor variable was not related to the dichotomous criterion variable after statistically controlling for the other predictors.

- When the adjusted odds ratio for a given predictor is *greater than 1.0* (such as 1.5 or 3.8), it means that this predictor variable was positively related to the criterion variable after statistically controlling for the other predictors: As scores on the predictor variable increased, subjects were more likely to be assigned to the "event" category (i.e., the category coded as "1").

- When the adjusted odds ratio for a given predictor is *less than 1.0* (such as .15 or .23), it means that this predictor variable was negatively related to the criterion variable after statistically controlling for the other predictors: As scores on the predictor variable increased, subjects were less likely to be assigned to the "event" category (i.e., the category coded as "1" under the criterion variable).

Haddock, Rindskopf, and Shadish (1998) offer the following general rules of thumb for interpreting the strength of the relationships represented by odds ratios:

- Odds ratios close to 1.0 may be interpreted as representing weak relationships between predictor and criterion.

- Odds ratios greater than 3.0 may be interpreted as strong positive relationships.

- Odds ratios less than about 0.33 may be interpreted as strong negative relationships.

CONFIDENCE INTERVALS FOR THE ADJUSTED ODDS RATIO. Now you know how to interpret the direction of the adjusted odds ratio. But how do you know whether the adjusted odds ratio is statistically significant? You do this by consulting its *confidence interval*.

So far, this section has focused on the point estimate for the adjusted odds ratio. You will recall that a ***point estimate*** is a single number that serves as our best guess for the population value of interest. Along with this point estimate, research articles often report 95% confidence intervals (CIs) for the adjusted odds ratio as well. A ***confidence interval*** is a range of values that are plausible values for the population parameter being estimated. The confidence interval tells you how *precise* the estimate is: If the CI is relatively narrow, you can view the adjusted odds ratio as being a fairly precise estimate; if the CI is relatively wide, you should view the adjusted odds ratio as being a less precise estimate.

In previous chapters, you learned that one use of a confidence interval is to determine whether a given point estimate is statistically significant. In the case of an adjusted odds ratio, "statistically significant" means "significantly different from 1.0" (remember that an adjusted odds ratio of 1.0 means "no relationship"). When reviewing the 95% CI for the adjusted odds ratio for a given predictor variable, if this confidence interval does not contain the value of "1.0," it means that the adjusted odds ratio for this predictor variable is significantly different from 1.0 (with alpha set at $\alpha = .05$).

Below is the excerpt from Dr. O'Day's article in which she describes the adjusted odds ratios for the current analysis. The excerpt refers to Table 12.2, which is identical to Table 12.1 (presented earlier).

Table 12.2

Predicting Exercise Status: Results from Logistic Regression Analysis

Predictor variable	B	SE	Wald χ^2	p	Adjusted odds ratios OR	[95% CI]
1. Health motivation	0.254	0.203	1.554	.213	1.289	[0.865, 1.920]
2. Desire to be thin	0.071	0.147	0.236	.627	1.074	[0.805, 1.433]
3. Dislike of exercise	-0.540	0.149	13.103	<.001	0.583	[0.435, 0.781]
4. Lack of time for exercise	-0.297	0.130	5.216	.022	0.743	[0.576, 0.959]
(Constant)	2.579	1.386	3.462	.063	13.183	

Note. $N = 238$. For the model containing only the constant, $-2LL = 209.017$; For the model containing the constant plus all predictors: $-2LL = 173.889$. Model χ^2 (4, $N = 238$) = 35.128, $p < .001$. Cox & Snell $R^2 = .137$. Nagelkerke $R^2 = .235$. B = Logistic coefficients. SE = Standard errors for logistic coefficients. Wald χ^2 = Test of null hypothesis that $B = 0$ (for each test, $df = 1$). p = Probability value for Wald χ^2. OR = Adjusted odds ratio. 95% CI = 95% Confidence interval for the adjusted odds ratio.

Table 12.2 presents the adjusted odd is ratios for the predictor variables below the heading *OR*. The adjusted odds ratio for dislike of exercise was 0.583, and the adjusted odds ratio for lack of time for exercise was 0.743. For each of these predictor variables, the 95% confidence interval did not include the value of 1.0, indicating that the adjusted odds ratio was statistically significant ($p < .05$). For each predictor, the adjusted odds ratio was less than 1.0, indicating that participants with high scores on these two variables were less likely to be in the exercisers subgroup.

The adjusted odds ratio for health motivation was 1.289, and the adjusted odds ratio for desire to be thin was 1.074. The 95% CI for both statistics included the value of 1.0, indicating that neither adjusted odds ratio was statistically significant ($\alpha = .05$).

Nature of the Relationship: Summary

Table 12.2 shows that the results related to the adjusted odds ratios (*AOR*) were consistent with the results related to the logistic regression coefficients (*B*). Regardless of which statistic was consulted, it is clear that only two of the predictor variables made statistically significant contributions to the logistic regression equation. These were (a) the dislike of exercise, and (b) the lack of time for exercise.

The nature of the coefficients shows that subjects with high scores on dislike of exercise were less likely to be in the exercisers subgroup. They showed that subjects with high scores on lack of time for exercise were also less likely to be in the exercisers subgroup.

How Strong Is the Relationship?

The model χ^2 test described in an earlier section is a null-hypothesis significance test (NHST). It tells us whether we can reject the null hypothesis that all regression coefficients are equal to zero in the population. But it does not tell us anything about the *strength* of the relationship between the criterion variable and the predictor variables, taken as a group. For this, researchers usually turn to the logistic regression counterparts to the R^2 statistic that is used in linear multiple regression. In some cases, they may also perform a classification analysis similar to the procedure used with discriminant analysis.

PSEUDO-R^2 STATISTICS. First the bad news: In logistic regression, there is no single statistic that is universally accepted as the direct counterpart to the R^2 statistic that is used in linear regression. However, there are at least two statistics (sometimes called *pseudo-R^2 statistics*) that are similar to R^2 and are often reported: the Cox & Snell R^2 and the Nagelkerke R^2.

The Cox & Snell R^2 is an index of the strength of the association between the dichotomous criterion variable and the predictor variables, and it is similar to linear-regression R^2 in that lower values (closer to zero) indicate weaker relationships, whereas higher values (closer to 1) indicate stronger relationships. However, the Cox & Snell R^2 has the limitation that it cannot reach the maximum value of 1 that may be reached with the linear regression R^2.

In contrast, the Nagelkerke R^2 (Nagelkerke, 1991) may reach the value of 1. It indicates the percent of variation in the criterion variable that is explained by the logistic regression model (Norusis, 2005).

Be warned, however, both statistics suffer from the weakness of frequently underestimating the values of R^2 that would be obtained with linear multiple regression. This is one of the reasons that they are called *pseudo-R^2* statistics.

Here is the way that Dr. O'Day described these statistics for the current analysis:

> For the full model, the Cox & Snell R^2 = .137, and the Nagelkerke R^2 = .235. These R^2 statistics are appropriate indices of effect size for logistic regression, but are not directly comparable to the R^2 statistics computed in linear multiple

regression, as they typically underestimate the values of R^2 that would be obtained with linear regression.

CLASSIFICATION ANALYSIS. Articles sometimes report a classification analysis in which participants are classified into groups based on their predicted probability scores. The predicted probability scores, in turn, are determined by the participants' scores on the predictor variables. The logic is that this classification table can provide a sense as to how well the model fits the data: If the logistic regression model accurately classifies participants into the correct group under the dichotomous criterion variable, it may be seen as evidence that the predictors are doing a good job of predicting the criterion (but be warned that the results of this classification analysis can also be misleading—as we shall soon see).

Below is the portion of Dr. O'Day's article that describes the classification analysis from her investigation.

Table 12.3

Results of Classification Analysis Using the Logistic Regression Equation

Actual group	Predicted group		% Correctly classified
	Nonexercisers	Exercisers	
Nonexercisers (n = 38)	6	32	16%
Exercisers (n = 200)	3	197	99%

Note. N = 238. Values in table are frequencies unless otherwise indicated. Overall percentage of correct predictions = 85%.

Table 12.3 presents the results of a classification analysis based on the logistic regression equation containing the constant term plus all four predictor variables. A cut value of .5 was used to classify participants into groups. Overall, 85% of the participants were correctly classified. However, this finding is misleadingly high due to the fact that 229 of the 238 participants (96%) were classified as being exercisers—the largest subgroup (exercisers actually constituted 84% of the sample). The model correctly classified 197 of the 200 exercisers (indicating that sensitively was 99%), but only 6 of the 38 nonexercisers (indicating that specificity was only 16%).

With binary logistic regression, the cut score is usually .5, meaning that participants with a predicted probability of .5 (or higher) of being in the "event" group are predicted as being in the "event" group, and those with a probability less than .5 are predicted to be in the "non-event" group. Remember that, in this study, the "event" group consisted of the exercisers, and the "non-event" group consisted of the nonexercisers.

SENSITIVITY VERSUS SPECIFICITY. When participants are classified into groups in a 2 × 2 table as they are in Table 12.3, the article sometimes discusses sensitivity, specificity, and related concepts. In this context, **sensitivity** represents the proportion of participants who (a) have the characteristic of interest and (b) are correctly predicted by the logistic regression equation to have that characteristic. Sensitivity is sometimes called the *correct classification of true positives*. On the other hand, **specificity** represents the proportion of participants who (a) do not have the characteristic, and (b) are correctly predicted to not have that characteristic. Specificity is sometimes called the *correct classification of true negatives*.

Both sensitivity and specificity are measures of classification accuracy, and researchers typically hope to obtain high values for both (Wright, 1995). As you saw in the preceding excerpt, sensitivity in Dr. O'Day's study was high (99%), but specificity was low (16%).

Along with sensitivity and specificity, articles sometimes also report the false positive rate and the false negative rate. The ***false positive rate*** refers to the proportion of participants who do not have the characteristic, but are incorrectly predicted to have it. The ***false negative rate*** is the proportion of participants who do have the characteristic, but are incorrectly predicted to not have it (Norusis, 2005). Researchers typically hope to have a small false positive rate as well as a small false negative rate.

PROBLEMS WITH CLASSIFICATION ANALYSES. In many instances, classification analyses are not terribly useful for evaluating a logistic regression equation. This is especially likely to be true in investigations in which one of the subgroups is much larger than the other. For example, in the current study, 84% of all participants were actually in the exercisers subgroup. This means that a researcher could have achieved an overall correct classification rate of 84% by simply predicting that all subjects were in the "exerciser" group, regardless of their scores on the predictor variables. Notice that the 84% correct classification rate achieved by this very crude technique is not much lower than the 85% correct classification rate that was achieved by using the four-variable logistic regression equation (*ouch!*). For these and related reasons, some experts advise that classification analyses should be performed only if classification had been the initial reason for the research (e.g., King, 2008).

Example 2: Sequential Logistic Regression

With logistic regression, researchers do not necessarily have to enter predictor variables into the equation all at once, as was illustrated with Example 1. It is also possible to add predictors to the equation in a sequence of pre-determined steps. Among other things, this allows researchers to determine whether a set of theoretically important variables (added to the equation at a later step) account for a significant amount of variance in the criterion variable, beyond the variance accounted for by a group of control variables (which had been added at an early step). Analyses of this sort go by a variety of names, including *sequential logistic regression*, *logistic regression with hierarchical entry*, and *logistic regression with block-wise entry*.

Potential Confounding Variables

In the previous section, you read about the analysis in which Dr. O'Day found a significant relationship between exercise status and a logistic regression equation containing four predictor variables. Assume that Dr. O'Day's colleague, Dr. Grey, is concerned that the relationships that she found could be spurious, due to the influence of several underlying variables. The four potential confounding variables are:

- The number of hours per week that the subjects are employed
- Participant age
- Participant sex
- Participant athletic status (i.e., whether the subject is on an official university-sponsored athletic team, such as Women's Track)

In response, Dr. O'Day now wishes to determine whether there is still a relationship between exercise status and the variables of the four-factor model after she has statistically controlled for these four potential confounding variables. She will use sequential logistic regression to do this.

Statistically Controlling for the Confounding Variables

In order to statistically control for the potential confounding variables, Dr. O'Day will add predictor variables to a logistic regression equation in a sequence of two predetermined steps. At Step 1, she will add the four potential confounding variables listed above: hours employed, participant age, participant sex, and athletic status. These four variables will serve as control variables. At Step 2, she will add the four variables of theoretical interest: the variables that constitute her four-factor model of exercise. If adding the four variables of the four-factor model at Step 2 results in a significant improvement in the model's ability to predict exercise status, it will strengthen the argument that the relationship between exercise status and the four-factor model is a real relationship, not a spurious relationship due to the influence of the potential confounding variables.

Results Produced by Adding Predictor Variables in Steps

Assume that Dr. O'Day performs the analysis described above. In her research article, she describes the analysis and results in this way:

> In an additional set of analyses, sets of predictor variables were added to a logistic regression equation in separate steps. The purpose of this analysis was to determine whether the variables that constitute the four-factor model will display a significant relationship with exercise status after statistically controlling for four subject variables.
>
> The $-2LL$ for the logistic regression equation that contained only the constant was $-2LL = 209.02$. At Step 1, four control variables were added to this equation: *hours employed* (the number of hours the participant worked at a paying job per week), *participant age, participant sex* (coded 0 = female, 1 = male), and *athletic status* (coded 0 = not a member of a university athletic team, 1 = a member of a university athletic team). The model chi-square statistic at the bottom of Table 12.4 shows that adding these four control variables to an equation that contained only the constant term did not result in a significant improvement in the model's ability to predict the criterion variable, model χ^2 (4, $N = 238$)= 3.51, $p > .05$. For the model containing just the constant and the four control variables, the Cox & Snell model R^2 was $R^2 = .015$, and the Nagelkerke model R^2 was $R^2 = .025$.

Table 12.4

Results from Sequential Logistic Regression: Predicting Exercise Status from Four Control Variables and the Four-Factor Model

	Adjusted odds ratios (*OR*)	
	Model 1	Model 2
	OR [95% CI]	*OR* [95% CI]

Predictors added at each step:		
Step 1: Creating Model 1		
(Constant, B_0)	2.55	4.92
Hours employed	1.01 [0.98, 1.04]	1.02 [0.99, 1.06]
Participant age	1.03 [0.94, 1.13]	1.04 [0.95, 1.14]
Participant sex	0.92 [0.43, 1.95]	0.98 [0.40, 2.2]
Athletic status	4.42 [0.55, 35.27]	2.02 [0.21, 19.65]
Step 2: Creating Model 2		
Health Motivation		1.29 [0.86, 1.94]
Desire to be thin		1.12 [0.83, 1.52]
Dislike of exercise		0.59 [0.44, 0.79]
Lack of time for exercise		0.69 [0.52, 0.91]
Model summary statistics:		
Model $-2LL$	205.51	169.53
Model χ^2 (*df*)	3.51(4)	39.49(8)***
$\Delta\chi^2$ (*df*)	3.51(4)	35.99(4)***
Cox & Snell Model R^2	.015	.153
Nagelkerke Model R^2	.025	.262

Note. $N = 238$. The $-2LL$ for the logistic regression equation that contained only the constant was 209.017. *OR* = Adjusted odds ratio for predictor variable. CI = 95% Confidence interval for adjusted odds ratios. Model $-2LL$ = -2 Log likelihood for model. Model χ^2 = Chi-square test of null hypothesis that the logistic regression coefficients for all predictor variables included in the model are equal to zero. *df* = Degrees of freedom. $\Delta\chi^2$ = Chi-square test of null hypothesis that the logistic regression coefficients for predictors added to model at most recent step are equal to zero.
* $p < .05$. ** $p < .01$. *** $p < .001$.

UNDERSTANDING INFORMATION ABOUT MODEL 1 IN THE TABLE. Just to make sure that you understand the layout of this information-rich table: The right side of the table consists of two large columns of numbers headed "Model 1" and "Model 2." Below the heading "Model 1" you will find information for the logistic regression equation that contains just the constant plus the four control variables. Below the heading "Model 2" you will find information for the logistic regression equation that contains the constant, the four control variables, and the four variables of the four-factor model of exercise behavior.

At the location where the column headed "Model 1" intersects with the row headed "Model $\chi^2(df)$," you will find the *model χ^2 statistic*. You will recall that the **model χ^2 statistic** tests the null hypothesis that, in the population, the logistic regression coefficient for each of the predictor variables included in the equation is equal to zero. You can see that the model chi-square statistic for Model 1 was $\chi^2 = 3.51$, which is nonsignificant. The fact that this model chi-square statistic is nonsignificant tells us that none of the control variables displayed a logistic regression coefficient that was significantly different from zero.

At the location where the column headed "Model 1" intersects with the row headed "$\Delta\chi^2(df)$," you will find the *model-improvement χ^2 statistic*. You will recall that the **model-improvement χ^2 statistic** determines whether the predictor variables added to an equation at the most recent step results in a significant improvement in the equation's ability to predict the criterion variable. At Step 1 of the current analysis, the four control variables were added to an equation that contained just the constant term. This means that, at Step 1, the model-improvement χ^2 statistic is essentially equivalent to the model χ^2 statistic described above. Sure enough, Table 12.4 shows that $\chi^2 = 3.51$ for both statistics.

Finally, at the bottom of the column for Model 1, we can see that the Cox & Snell model R^2 was very small at $R^2 = .015$, and the Nagelkerke model R^2 was similarly small at $R^2 = .025$. Combined, these results show that the four control variables do not display much of a relationship with exercise status.

ADDING THE PREDICTORS OF THEORETICAL INTEREST. Remember that Dr. O'Day doesn't care much about the four control variables added at Step 1. After all, she is only using them as *control* variables. What she really wishes to know is whether the four variables that constitute her four-factor model of exercise are still related to the criterion variable after she has statistically controlled for these four nuisance variables. Therefore, she adds the four variables of theoretical interest at Step 2. Below, she reports what she found:

> At Step 2, the four variables that constitute the four-factor model of exercise were added to Model 1, resulting in Model 2. Summary statistics at the bottom of Table 12.4 show that adding these four variables resulted in a significant improvement in the model's ability to predict the criterion variable, $\Delta\chi^2$ (4, $N = 238$)= 35.99, $p < .001$. The final Model 2 equation (which included all eight predictor variables) displayed a statistically significant relationship with the criterion variable: The model χ^2 statistic was χ^2 (8, $N = 238$)= 39.49, $p < .001$, the Cox & Snell model R^2 was $R^2 = .153$, and the Nagelkerke model R^2 was $R^2 = .262$.

Success! Dr. O'Day's block of four theoretical variables not only predicts exercise status, but it even predicts exercise status after she has statistically controlled for the four potential confounding variables!

All of this establishes that Dr. O'Day has obtained a statistically significant relationship. But what is the *nature* of that relationship? Are all four theoretical variables related to exercise status, or just a few? And will the direction of these relationships be as she had initially predicted? To find out, we turn to the adjusted odds ratios.

Adjusted Odds Ratios for the Individual Predictors

To further investigate the relationship between exercise status and the four predictor variables of theoretical interest, Dr. O'Day could interpret either the logistic regression coefficients or the adjusted odds ratios. Table 12.4 did not really have enough room to present both indices, so Dr. O'Day included just the adjusted odds ratios. Here is what she said about them in the text:

> The adjusted odds ratios for the eight predictor variables in Model 2 appear in the upper section of Table 12.4. The adjusted odds ratio for a given predictor variable is statistically significant ($p < .05$) if the 95% confidence interval for that adjusted odds ratio does not contain the value of 1.0. Inspection of the confidence intervals under the heading "Model 2" shows that only two predictor variables were statistically significant: dislike of exercise (OR = 0.59) and lack of time for exercise (OR = 0.69). The adjusted odds ratio for both variables was less than 1, indicating that high scores on each variable were associated with a decreased probability that the participant was in the "exerciser" subgroup.

Summary for Example 2

In Example 2, Dr. O'Day found pretty much the same results that she found in Example 1—the inclusion of the four control variables did not affect the substantive interpretation of her findings. The equation containing the four variables that constitute her four-factor model displayed a statistically significant relationship with exercise status regardless of whether she statistically controlled for the four potential confounding variables. Also regardless of the control variables, she found that only two of her predictor variables displayed an *AOR* that was statistically significant. The nature of these *AORs* showed that participants were more likely to be in the exerciser subgroup if they reported that they (a) liked to exercise and (b) had adequate free time to exercise. On the whole, it is fair to say that these results provided mixed support for Dr. O'Day's four-factor model of exercise.

Additional Issues Related to this Procedure

This section wraps things up by describing the differences between binary logistic regression (the topic of the current chapter) versus its first-cousin, multinomial logistic regression. The section ends with a discussion of the assumptions underlying logistic regression.

Binary Versus Multinomial Logistic Regression

This chapter focused on **binary logistic regression**, the version of logistic regression that is appropriate when the criterion variable consists of exactly two categories. In contrast, **multinomial logistic regression** is an alternative procedure that may be used when the criterion variable consists of more than two categories. With this latter procedure, the criterion variable is typically on an ordinal scale of measurement.

For example, in the current chapter, the criterion variable (exercise status) was a dichotomous variable: Participants were either in the exerciser category or were in the nonexerciser category. But what if Dr. O'Day had instead placed each participant in one of *three* possible categories: (a) nonexercisers, (b) infrequent exercisers, and (c) frequent exercisers? The criterion variable now has three possible values, and is on an ordinal scale of measurement. In this case, she may have chosen to analyze the data using multinomial logistic regression.

The good news is that when reading about multinomial logistic regression, you will see many of the same statistics that have already been covered in this chapter (e.g., the $-2LL$, the χ^2 tests, the logistic regression coefficients, and the adjusted odds ratios). Perhaps the biggest difference is the fact that, with multinomial logistic regression, more than one version of the logistic regression equation is estimated and presented in the article. The number of regression equations presented is equal to $M - 1$, where M = the number of categories under the criterion variable (Menard, 1995).

For example, consider the hypothetical study described above in which there are now three categories for exercise status rather than just two. In multinomial logistic regression, Dr. O'Day would choose one of the categories as the *reference category*—the category against which the other two categories would be compared (let's assume that she chooses the *nonexercisers* as her reference category). She would then perform two comparisons: In the first comparison, she would estimate a logistic regression equation in which the criterion variable essentially had just two categories: the *nonexercisers* versus the *infrequent exercisers*. She would compute indices of model fit (similar to the model χ^2 statistics described above) and would estimate the logistic regression coefficients, the adjusted odds ratios, and related statistics for each predictor variable. Combined, these statistics would tell her how well the predictor variables (as a set) predicted membership in the nonexerciser group versus the infrequent exerciser group.

In the second comparison, she would estimate a new logistic regression equation in which the criterion variable would once again have just two categories. For this analysis, however, the two categories would now be the *nonexercisers* versus the *frequent exercisers*. For this logistic regression equation, she would compute the same indices that were computed for the first comparison (the model χ^2, the logistic regression coefficients, the adjusted odds ratios, and so forth).

In her article, Dr. O'Day would present the statistics obtained from the two separate logistic regression models, and would interpret the "big picture" that emerges when the two sets of results are reviewed in aggregate (see Tabachnick & Fidell, 2007, pp. 481-499 for examples). As was stated earlier, once you understand the results generated in a binary logistic regression analysis (as presented in this chapter), understanding the results from a multinomial logistic regression analysis is fairly straightforward.

Assumptions Underlying this Procedure

When attempting to predict a dichotomous criterion variable from multiple predictor variables, researchers are sometimes in a position to choose between using either discriminant analysis (covered in *Chapter 11*) or logistic regression. In these situations, logistic regression is sometimes the more attractive alternative because it has fewer assumptions. The assumptions for logistic regression include the following (from Field, 2009a; Tabachnick & Fidell, 2007; Warner, 2008; Wright, 1995).

DICHOTOMOUS CRITERION VARIABLE. For binary logistic regression (the main focus of the current chapter), the criterion variable should consist of just two categories. Each participant should belong to one category or the other category, and no participant should belong to both.

LINEARITY. There should be a linear relationship between any continuous predictor variable and the logit of Y of the criterion variable (Field, 2009a; Hosmer & Lemeshow, 1989).

INDEPENDENCE OF OBSERVATIONS. Logistic regression assumes that each observation is unrelated to every other observation. In practical terms, this means that the researcher should not obtain repeated measures from the same participant.

ABSENCE OF SPECIFICATION ERRORS. The logistic regression equation should not omit any important predictor variables, and it should not include any irrelevant predictor variables.

ABSENCE OF MULTICOLLINEARITY. Multicollinearity occurs when two or more of the predictor variables display very strong correlations with one another. Multicollinearity may result in misleadingly large standard errors for the parameter estimates, along with other problems.

Chapter 13

MANOVA and ANOVA

This Chapter's *Terra Incognita*

You are reading an article about a study in which the predictor variable is *team & exercise status:* a limited-value variable which consists of three groups: nonexercisers, individual exercisers, and team exercisers. The researcher wishes to determine whether this predictor variable is related to four criterion variables. The *Results* section of the article reports the following:

> Data were analyzed using a one-way multivariate analysis of variance (MANOVA) in which the predictor variable was team & exercise status and the criterion variables were health motivation, desire to be thin, dislike of exercise, and lack of time for exercise. For this analysis, Wilks' lambda was used as the multivariate test statistic. With alpha set at .05, the exact F value for Wilks' lambda was significant, $\Lambda = .81$, $F(8, 464) = 6.48$, $p < .001$.
>
> Four univariate ANOVAs were performed as follow-up procedures, one for each of the four criterion variables. The Bonferroni adjustment was used to maintain an experimentwise Type I error rate of $\alpha = .05$. With this adjustment, an F statistic for an individual ANOVA was considered to be statistically significant only if its p value was less than .0125. Using this criterion, the univariate ANOVA was significant for only three criterion variables: health motivation, dislike of exercise, and lack of time for exercise.
>
> To understand the specific nature of the group differences, the Games-Howell procedure (Games & Howell, 1976) was used as a post-hoc test. These tests revealed that: (a) for health motivation, the team exercisers scored higher than the other two groups; (b) for dislike of exercise, the nonexercisers scored higher than the other two groups; and (c) for lack of time for exercise, the nonexercisers scored higher than the individual exercisers, who in turn scored higher than the team exercisers.

Do not be afraid. All shall be revealed.

The Basics of this Procedure

This chapter describes ***multivariate analysis of variance*** (MANOVA), a statistical procedure that allows researchers to investigate differences between two or more groups with respect to their mean scores on multiple quantitative criterion variables. MANOVA is the multivariate extension of univariate analysis of variance (ANOVA). While univariate ANOVA allows the analysis of just one criterion variable, MANOVA allows the analysis of two or more. This is illustrated in Figure 13.1, which presents the generic analysis model for MANOVA.

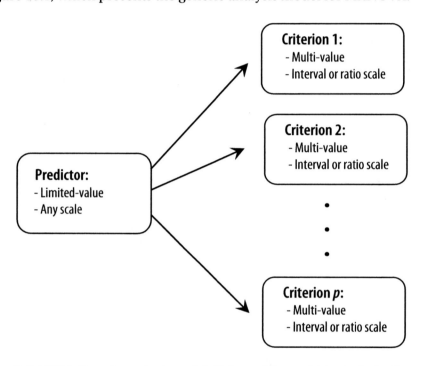

FIGURE 13.1. Generic analysis model: Nature of the variables included in a typical one-way MANOVA.

Figure 13.1 presents the generic analysis model for a ***one-way MANOVA***—an analysis that involves just one predictor variable. An analysis with more than one predictor variables is called a ***factorial MANOVA***. The current chapter focuses on one-way MANOVA with one between-subjects factor. The fact that the predictor variable is a ***between-subjects factor*** indicates that a given participant appears in just one of the conditions under the predictor variable. If the study had instead included a ***within-subjects factor***, it would have indicated that the researcher took repeated measures from each participant. Between-subjects factors and within-subjects factors were discussed in greater detail in *Chapter 2*.

Variables Investigated in MANOVA

Figure 13.1 shows that a one-way MANOVA allows researchers to investigate the relationship between a single limited-value predictor variable and multiple criterion variables, with each criterion assessed on an interval or ratio scale of measurement. MANOVA is a multivariate test of group differences—it allows researchers to determine whether there is a significant difference between at least two of the groups with respect to their mean score on several criterion variables, taken as a set. Theoretically, there is no limit to the number of criterion variables that may be included in the analysis, although Stevens (1980) recommends fewer than 10.

Ideally, the various criterion variables included in a MANOVA should be related to one another in some theoretical sense (Weinfurt, 1995). One way to accomplish this is to select criterion variables that are all measures of the same underlying construct. This approach is illustrated in the "Illustrative Investigation" section, later in this chapter.

Steps Followed in MANOVA

Multivariate analysis of variance is typically conducted in a sequence of steps. First, the researcher computes a multivariate test statistic to determine whether there is a relationship between the criterion variables (taken as a set), and the predictor variable. The researcher typically interprets just one of four popular multivariate test statistics for this purpose: the Pillai-Bartlett trace (V), the Hotelling-Lawley trace (T^2), Wilks' lambda (Λ), or Roy's largest root, (Θ). If the multivariate relationship is significant, the researcher next performs a follow-up procedure to better understand the nature of the relationship. The follow-up procedure typically involves one or more of the following: (a) a separate univariate ANOVA performed on each criterion variable, (b) a discriminant analysis (c) a stepdown analysis, or (d) multivariate planned comparisons. These follow-up procedures are described in greater detail in a later section of this chapter.

Coverage of One-Way ANOVA

Because researchers often report univariate ANOVAs for each criterion variable included in a MANOVA, a good deal of this chapter focuses on the interpretation of one-way ANOVAs with one between-subjects factor. It shows how one-way ANOVAs are often presented in the text, tables, and figures of published articles. It also discusses the various indices of effect size (e.g., η^2) as well as the various multiple-comparison procedures (e.g., the Tukey HSD test) that are often used as part of a one-way ANOVA.

Illustrative Investigation

Assume that the fictitious researcher, Dr. O'Day, continues her program of research in which she investigates attitudes and behaviors related to health and exercise. This chapter illustrates MANOVA by describing a study in which Dr.

O'Day explores the relationship between a categorical predictor variable and four quantitative criterion variables, all of which reflect attitudes toward health and exercise.

PROCEDURE. In this investigation, Dr. O'Day administered the *Questionnaire on Eating and Exercising* (from *Appendix A*) to 260 participants enrolled in psychology courses at a regional state university in the Midwest. She obtained usable responses from 238 students.

VARIABLES INVESTIGATED. The predictor variable in the current investigation was labeled *team & exercise status*. This was a nominal-scale variable that classified each participant as falling into one of three categories:

- The *nonexercisers* were subjects who did not engage in regular aerobic exercise at all ($n = 35$).

- The *individual exercisers* were subjects who engaged in aerobic exercise on a regular basis, but did so on their own—not as a part of any organized sport ($n = 143$).

- The *team exercisers* were subjects who engaged in aerobic exercise on a regular basis, and did so as members of teams that were part of a university athletic program, a city league, or some other entity ($n = 60$).

The study's criterion variables were the four variables that constituted Dr. O'Day's *four-factor model of exercise behavior*. These were four attitudinal variables related to health and exercise. Each variable was assessed using a multiple-item summated rating scale whose scores could range from 1.00 to 7.00, with higher values indicating higher levels of the construct being assessed. The name of each variable is provided below, along with (a) the item numbers for the questionnaire items constituting the scale and (b) one representative item from each scale.

- *Health motivation* was measured by items 27-29 on the *Questionnaire on Eating and Exercising* (from *Appendix B*). Sample item: "It is important to me to be healthy."

- *Desire to be thin* was measured by items 31-33. Sample item: "It is important to me to be thin."

- *Dislike of exercise* was measured by items 47-49. Sample item: "I find aerobic exercise to be unpleasant."

- *Lack of time for exercise* was measured by items 51-53. Sample item: "I am too busy to engage in frequent aerobic exercise."

RESEARCH HYPOTHESES. Imagine that Dr. O'Day began her investigation with the hypothesis that the variable *team & exercise status* has a causal effect on the study's four criterion variables. She predicted that when people engage in the behavior of exercising, this behavior should cause their attitudes in these four areas to become more pro-exercise and pro-health. Specifically, she predicted that, compared to the nonexercisers, the individual exercisers and the team exercisers would:

- display higher scores on health motivation,
- display higher scores on desire to be thin,
- display lower scores on dislike of exercise, and
- display lower scores on lack of time for exercise.

The nature of the hypothesized relationships is illustrated in Figure 13.2.

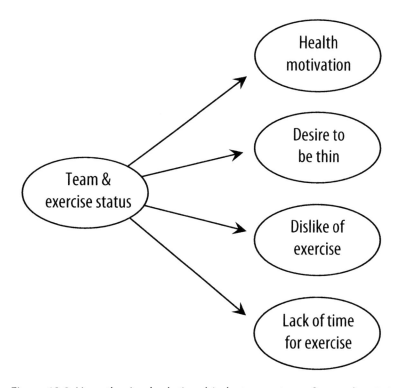

Figure 13.2. Hypothesized relationship between team & exercise status and four criterion variables.

CAVEAT REGARDING THIS STUDY. Although Dr. O'Day hypothesized that team & exercise status has a causal effect on the four criterion variables, most researchers would be concerned that the present investigation is capable of providing only weak evidence related to that hypothesis. This is because the current investigation is a nonexperimental study: She did not manipulate an independent variable by actively assigning participants to conditions. Instead, she merely determined where the participants already stood on naturally occurring variables (e.g., she did not assign some participants to be individual exercisers—she simply determined which participants were already individual exercisers at the time the study was conducted). As you learned in *Chapter 2*, when researchers wish to obtain strong evidence of causal relationships, they should in most cases conduct a true experiment in which participants are randomly assigned to treatment conditions.

Although the study reported here is a nonexperimental investigation, it is still acceptable to analyze the results using MANOVA and ANOVA—these procedures may be used nonexperimental studies as well as with true experiments. Just remember that, even if the results are statistically significant, they will not provide terribly strong evidence that team & exercise status actually had a *causal* effect on the four criterion variables.

Research Questions Addressed

The research questions that may be answered in a one-way MANOVA are determined, in part, by the nature of the follow-up procedures that are used (e.g., discriminant analysis versus stepdown analysis versus the other procedures to be described here). This chapter will illustrate the widely used procedure in which the researcher first verifies that the multivariate test statistic is significant and then performs a series of univariate ANOVAs on the individual criterion variables. The multivariate test statistic allows the researcher to determine whether there is a relationship between the predictor variable and the multiple criterion variables, taken as a set, and also allows the researcher to assess the strength of that relationship. If the multivariate statistic is significant, the researcher then follows up by performing a series of univariate ANOVAs—one for each criterion variable. These univariate ANOVAs allow the researcher to:

- Determine whether there is a significant relationship between a specific criterion variable and the predictor variable.

- Determine which groups displayed relatively high mean scores and which groups displayed relatively low mean scores on the criterion.

- Determine which pairs of groups displayed mean scores that were significantly different according to planned contrasts or post-hoc tests.

This chapter discusses the use of univariate ANOVAs as follow-up tests because this approach is often reported in published articles. Please do not take this as an endorsement of univariate ANOVAs as a follow-up procedure—as you will learn in the next section, there are a number of weaknesses associated with the use of univariate ANOVAs as a follow-up to MANOVA, and few experts recommend it as the best alternative.

Details, Details, Details

An earlier section indicated that MANOVA is the multivariate extension of univariate ANOVA. With simple univariate ANOVA, researchers determine whether there is a difference between conditions with respect to a single criterion variable. With MANOVA, on the other hand, researchers determine whether there is a difference between conditions with respect to a *variate*—an optimally weighted combination of several criterion variables. This section describes the nature of this variate in greater detail, discusses the multivariate significance tests used with MANOVA, and provides an overview of the follow-up procedures that are often a part of the analysis.

Null Hypothesis Tested

When researchers perform a univariate ANOVA, they typically test a null hypothesis that states that there is no difference between the means of any of the conditions in the population. In a study with three treatment conditions, the null hypothesis is stated as follows:

> H_0: $\mu_1 = \mu_2 = \mu_3$. In the population, there is no difference between the three conditions with respect to mean scores on the criterion variable.

In the preceding null hypothesis, the three μ symbols represent the means for the three conditions in the population. To test this null hypothesis, the ANOVA produces an obtained F statistic along with a corresponding probability value (i.e., p value) for the F statistic. If the p value is below the selected criterion for statistical significance (typically $\alpha = .05$), this null hypothesis is rejected and the researcher concludes that there is a difference between at least two of the three conditions.

The preceding shows that, with univariate ANOVA, the three conditions are being compared on just one criterion variable. In contrast, with MANOVA, the conditions are compared on multiple criterion variables (four criterion variables in Dr. O'Day's current study). The set of criterion variables under a specific condition is called a *vector*, and the null hypothesis tested in MANOVA refers to this vector. Dr. O'Day's current investigation involves three conditions. The multivariate statistic that she will compute as part of the MANOVA will therefore test the following null hypothesis:

> H_0: In the population, there is no difference between the three conditions with respect to the vectors of means on the criterion variables.

If the obtained multivariate test statistic produces a p value that is lower than the standard criterion (say, $\alpha = .05$), the researcher rejects this null hypothesis and concludes that there is a difference between at least two of the conditions. This paves the way for a series of follow-up tests that will allow her to understand the specific nature of this difference.

The Discriminant Function Variates

In the section that follows this section, you will learn about the four multivariate test statistics that are used to determine whether a MANOVA is statistically significant. Before you can understand these statistics, however, you must first understand the concept of the ***discriminant function variate*** that is created in a MANOVA. The concepts discussed in this section are based largely on Field (2009a, pp. 598-603).

VARIATE. One key to understanding MANOVA is to understand that the multiple criterion variables being analyzed are actually combined into a single synthetic variable called a *discriminant function variate* (we will simply call it a *variate*). In this context, a **variate** is an optimally weighted linear combination of the criterion variables. In creating this synthetic variable, the various criterion variables are given different weights so as to maximize the differences between the groups being compared (in Dr. O'Day's study, this involves maximizing the differences between the nonexercisers versus the individual exercises versus the team exercisers). It is important to remember that (synthetic or not) this variate is a quantitative variable, and each participant has a score on it (such as 0.15, −0.52, 0.89, and so forth).

The number of variates created in a MANOVA will be equal to either p (the number of criterion variables) or $k − 1$ (where k is the number of conditions), whichever is smaller. In the current study, there are four criterion variables (which means that $p = 4$), and three treatment conditions (which means that $k − 1 = 3 − 1 = 2$). The smaller of these two values is 2, so we know that two variates will be created.

When more than one variate is created in a MANOVA, the first variate is always the most important variate. In this case, "most important" means "this is the variate that does the best job of maximizing the differences between the various groups." In studies in which more than one variate is created, each subsequent variate will (a) be uncorrelated with all previous variates, and (b) will further maximize the differences between the groups (to the extent that this is possible, given the nature of the previous variates).

USING DISCRIMINANT ANALYSIS TO UNDERSTAND THE VARIATES. This section has discussed the concept of a discriminant function variate only because understanding this concept is necessary if you are to understand the differences between the four multivariate test statistics to be presented later. In practice, the nature of the variates created in a given MANOVA is typically not written about in a research article and is typically not even available to researchers who perform MANOVA using statistical software. In these cases, the variates are behind the scenes, invisible to both researcher and reader.

The variates are invisible, that is, unless the researcher performs some type of follow-up procedure that reveals the nature of the variate. Perhaps the best follow-up procedure for this purpose is *discriminant function analysis* (or *discriminant analysis*, for short). A discriminant analysis produces standardized discriminant function coefficients, structure coefficients, and other results that allow researchers to interpret the nature of the variates—to understand how the various criterion variables contributed to the variate.

Chapter 11 of this book covered discriminant analysis in some detail. In fact, the main research example used in that chapter is the very same data set used in the current chapter (i.e., the same criterion variables, the same predictor variable, the same subjects—everything). This means that the discriminant analysis presented in *Chapter 11* serves as one possible follow-up to the MANOVA reported here.

After completing the current chapter, you are encouraged to read (or re-read) *Chapter 11* in order to obtain a better understanding of the variates that were analyzed in the current MANOVA.

Four Multivariate Test Statistics

Most popular statistical applications compute four different multivariate test statistics to test the null hypothesis described earlier. There are conceptual differences between these statistics, and researchers typically report just one of the four in the published article.

As was mentioned earlier, Field (2009a) provides an excellent discussion of the variates that are created in MANOVA and discriminant analysis, and his discussion is relevant to the four multivariate test statistics to be discussed here. Field's discussion helps us to see that each of the four multivariate test statistics have counterparts in univariate ANOVA (i.e., in analysis of variance with just one criterion variable). The four multivariate test statistics are as follows:

PILLAI'S TRACE. *Pillai's trace* (symbol: *V*) is sometimes reported as the *Pillai-Bartlett trace*. It represents the sum of the proportion of explained variance on the variates. This makes Pillai's trace analogous to the R^2 statistic that can be computed in regression or univariate ANOVA (remember that, in a regression analysis, the R^2 statistics is computed as SS_{Model}/SS_{Total}).

HOTELLING'S T^2. *Hotelling's T^2* is sometimes reported as the *Hotelling-Lawley trace*. It represents the ratio of explained variance to unexplained variance, summed for each variate in the MANOVA. In this sense, Hotelling's T^2 is analogous to the F statistic that is computed in univariate ANOVA.

WILKS' LAMBDA. *Wilks' lambda* is represented as Λ, the Greek capital letter *lambda*. It is computed by (a) calculating the unexplained variance for each variate and (b) multiplying these values together. The resulting statistic may range from zero through 1, with smaller values (closer to zero) representing a stronger "effect." Wilks' lambda is one of the relatively few statistics for which smaller values indicate a stronger relationship. It is analogous to the ratio of error variance to total variance in a univariate ANOVA (that is, it is analogous to the ratio: SS_{Error}/SS_{Total}).

ROY'S LARGEST ROOT. *Roy's largest root* is represented as Θ, the Greek capital letter *theta*. It is similar to Hotelling's T^2, in that it is also analogous to the F statistic from univariate ANOVA. Both Hotelling's T^2 and Roy's largest root represent the ratio of explained variance to unexplained variance. However, with Hotelling's T^2, this ratio is summed across all variates, whereas with Roy's largest root, it is computed for the first variate only. In a typical MANOVA, the first variate is the one that best maximizes differences between conditions (groups). Therefore, in many research situations, Roy's largest root displays greater statistical power than the other multivariate statistics described in this section.

WHICH STATISTIC IS BEST? When just two groups are being compared in a MANOVA, the four multivariate test statistics produce the same results. That is, if there are just two groups and one of the multivariate statistics is significant, the other three multivariate statistics will be significant as well. However, when more than two groups are being compared, the four statistics may produce different results, and there is no consensus as to which statistic is always the "preferred" statistic in that situation.

In the typical social science investigation, Roy's largest root is likely to be slightly more powerful, but this statistic is not robust against violations of the assumption of homogeneity of covariance matrices (to be discussed at the end of the chapter). When the analysis involves more than one variate, Pillai's trace may be slightly more powerful than the others (Stevens, 1996). However, many researchers avoid using this statistic because they have no idea how to pronounce the name "Pillai" at conferences.

The illustrative investigation presented in this chapter will report Wilks' lambda as the multivariate test statistic. In part, this is because Wilks' lambda is the oldest and most widely reported multivariate statistic (Weinfurt, 1995). In part, this is also because it is easy to compute η^2 (eta-squared, an index of effect size) from Wilks' lambda. As you no doubt know by now, indices of effect size loom large in this book.

Follow-Up Procedures Used with MANOVA

When a multivariate test statistic (such as Wilks' lambda) is statistically significant, the researcher typically performs a follow-up procedure to better understand the nature of the results. As was mentioned earlier, these follow-up procedures allow the researcher to determine just what a significant multivariate result *means:* Which criterion variable(s) contributed the most to the significant results? What is the general trend of the results?

There is a variety of different follow-up procedures that researchers may use for this purpose, and this section briefly describes the ones that you are most likely to encounter in the published literature. It begins with the approach that is probably the most popular, although not necessarily the most highly recommended: performing a series of univariate ANOVAs.

UNIVARIATE ANOVAS. One popular approach for MANOVA involves (a) checking the multivariate test statistic (such as Wilks' lambda) and, if it is significant, next (b) performing a separate univariate ANOVA for each of the criterion variables in the analysis. In the case of Dr. O'Day's study, this would involve performing four one-way ANOVAs: one in which the criterion variable is health motivation, one in which the criterion variable is desire to be thin, and so forth. In each ANOVA, the predictor variable would continue to be team & exercise status.

Some researchers use univariate ANOVAs as follow-up tests due to a misunderstanding regarding the familywise Type I error rate (sometimes called the experimentwise Type I error rate). As you will recall from *Chapter 5*, a ***Type I error*** occurs whenever researchers reject the null hypothesis when doing so is a mistake. In other words, a Type I error occurs whenever researchers tell the world that their results are statistically significant, but are wrong in arriving at that decision. If a researcher sets alpha at α = .05, performs just one significance test, and satisfies all relevant assumptions, the probability of making a Type I error is equal to .05 (because that is the level of alpha that was selected). However, if the researcher performs multiple significance tests on the same dataset, the probability of making at least one Type I error out of all of those tests becomes larger than .05. This inflated probability of an error is the actual familywise Type I error rate. In general, the *familywise Type I error rate* (symbol: α_{FW}) is the probability of making at least one Type I error out of all of the tests that are performed when multiple significance tests are performed on the same data. Investigators typically try to keep the familywise Type I error rate at a small value such as α_{FW} = .05.

Researchers typically perform MANOVA in order to prevent an inflated familywise Type I error rate. For example, if Dr. O'Day sets alpha at α = .05 and then computes a single multivariate significance test that involves all four criterion variables (say, by computing Wilks' lambda), the probability of making a Type I error in just that test is indeed equal to .05. So far so good.

Problems would emerge, however, if Dr. O'Day went on to perform several univariate ANOVAs as follow-up tests—one for each criterion variable. Some researchers erroneously believe that, if they set alpha at α = .05 for each of these univariate ANOVAs, the familywise Type I error rate will be held at α = .05. They believe that since the multivariate test (Wilks' lambda, in this case) was significant, they do not have to worry about the familywise Type I error rate becoming inflated with the group of univariate ANOVAs that they subsequently perform. But they are mistaken in this belief.

Bray and Maxwell (1982) describe a (probably common) scenario in which a researcher performs MANOVA on a group of criterion variables, some of which are related to the predictor variable in the population, and some of which are not. As one might expect, the researcher finds that the multivariate test statistic (say, Wilks' lambda) is statistically significant. This is okay, because some of the criterion variables are, in fact, related to the predictor in the population. However, if the researcher goes on to perform a separate univariate ANOVA on each criterion variable, the fact that *multiple* analyses are being performed means that the familywise Type I error rate will be inflated for those criterion variables that are not related to the predictor variable in the population. The greater the number of criterion variables, the greater the inflation.

For these and other reasons, some have cautioned against using univariate ANOVAs as a follow up to MANOVA. At a minimum, they have advised researchers to use some type of correction (such as the Bonferroni correction) to protect against an inflated familywise Type I error rate (Tabachnick & Fidell, 2007). And some have recommended that univariate ANOVAs should be supplemented with a multivariate statistical procedure—such as discriminant analysis—as a follow-up to MANOVA (Field, 2009a). That approach is discussed next.

DISCRIMINANT ANALYSIS. Discriminant function analysis is an advanced statistical procedure that allows researchers to investigate the relationship between a single categorical variable (called a *grouping variable*), and multiple quantitative variables (called *attribute variables*). An earlier section of this chapter described how MANOVA involves the creation and analysis of *variates*: optimally weighted linear composites of the criterion variables. Discriminant analysis allows researchers to explore the *nature* of these variates: Which criterion variables are positively correlated with the variate? Which are negatively correlated with it? Which criterion variables display near-zero relationships with it?

Field (2009a) recommends discriminant analysis as a follow-up to MANOVA because discriminant analysis acknowledges the intercorrelated, multivariate nature of the data. Discriminant analysis reduces multiple criterion variables to a smaller number of dimensions, and it allows the researcher to investigate the nature of these dimensions.

As was mentioned earlier, *Chapter 11* of this book provides the results of a discriminant analysis performed on the variables that are analyzed with MANOVA in the current chapter. Interested readers are referred to *Chapter 11* to see how discriminant analysis allows for a richer and more nuanced understanding of multivariate relationships described here.

STEPDOWN ANALYSIS. *Stepdown analysis* is a special application of analysis of covariance, or ANCOVA (ANCOVA is covered in *Chapter 14* of this book). With stepdown analysis, the criterion variables from the MANOVA are added to an ANCOVA equation in a series of predetermined steps (Stevens, 1996; Tabachnick & Fidell, 2007).

A stepdown analysis may be used in situations in which it is possible to order the criterion variables according to practical or theoretical considerations. For example, imagine that, in a study with four criterion variables, they have been rank-ordered by the researcher in terms of their theoretical importance. The abbreviation *CV1* is used to represent the most-important criterion variable, *CV2* is used to represent the second-most important criterion variable, and so forth. The stepdown analysis will also make use of the categorical predictor variable from the MANOVA (remember that *team & exercise status* served as the categorical predictor in Dr. O'Day's study).

Stepdown analysis proceeds in a sequence of steps. In the first step, an ANOVA is performed in which CV1 serves as the criterion, and the categorical predictor variable from the MANOVA serves as the predictor variable. At Step 2, an ANCOVA is performed in which CV2 serves as the criterion, the categorical predictor from the MANOVA again serves as the predictor variable, and CV1 serves as the covariate. The goal of this analysis is to determine whether there is still a significant relationship between CV2 and the categorical predictor variable after statistically controlling for CV1. At Step 3, an ANCOVA is performed in which CV3 is the criterion, the MANOVA's categorical predictor serves as the predictor variable, and CV1 and CV2 both serves as the covariates. The analysis proceeds in this way, with each criterion variable being added to the list of covariates. One purpose is to determine just how much a given criterion variable adds to an equation that already contains higher-priority criterion variables as covariates. If a given criterion variable (by itself) had displayed a significant relationship with the predictor variable, but does not display a significant relationship in an ANCOVA that includes higher-priority criterion variables, the interpretation is that whatever variance this criterion variable shares with the predictor must have already been accounted for by the higher-priority criterion variables.

Confused? Don't worry about it. You will run into a stepdown analysis about as often as you run into a millionaire college professor. As was mentioned above, stepdown analysis is appropriate only in situations in which there is some theoretical justification for ordering the criterion variables. Since this is seldom feasible, relatively few articles report the procedure (okay, if you *do* run into a stepdown analysis, be sure to see Stevens, 1996, and Tabachnick and Fidell, 2007, as they do a much better job of explaining all this).

POST-HOC TESTS. In general, *post-hoc tests* involve making a relatively large number of comparisons between groups (possibly all possible comparisons between groups) in order to identify the specific pairwise-comparisons that are responsible for the significant multivariate effect. These tests are often included as part of the "univariate ANOVA" strategy described above. Post-hoc procedures may include special applications of the Hotelling T^2 statistic, univariate t tests, and the Tukey simultaneous confidence interval (Stevens, 1996). This approach is more appropriate during the early, exploratory stages of a program of research, when the researchers are learning about the nature of the "effects" for the first time. Because so many comparisons are made, post-hoc tests tend to be less powerful than the planned comparisons to be described below.

MULTIVARIATE PLANNED COMPARISONS. When researchers use *multivariate planned comparisons*, they make a limited number of comparisons involving a limited number of variables in order to identify the specific groups and/or criterion variables that have made the largest contributions to the multivariate results. This approach is appropriate when a program of research has already established that an effect exists, and the researchers now wish to better understand just *why* it exists. With planned comparisons, the number of tests performed is typically smaller than with post-hoc tests, and the tests tend to have more power (Stevens, 1996).

WHICH APPROACH IS BEST? There is no single follow-up procedure to MANOVA that will always be the "correct" approach. Many experts lean toward statistical procedures that acknowledge the multivariate nature of the data: discriminant analysis, stepdown analysis, and multivariate planned comparisons. On the other hand, the researchers who actually conduct and publish research have traditionally favored multiple univariate ANOVAs. So that you will be prepared to read and understand the multivariate approach, *Chapter 11* of this book covers discriminant analysis. And so that you will be prepared to read and understand the univariate approach, the next few sections present an example in which individual ANOVAs are used as a follow-up to a significant MANOVA.

Results from the Investigation

Let's assume that Dr. O'Day began by performing some exploratory data analyses. She screened the data for skewness, outliers, and related problems. She verified that it satisfy the assumptions of multivariate normality, equality of covariance matrices, and the other assumptions listed at the end of this chapter. With this done, she proceeded to the analyses that were most relevant to her research hypotheses.

Descriptive Statistics

For statistical procedures that include multiple quantitative criterion variables, it is standard to begin with a table of basic descriptive statistics: means, standard deviations, scale reliabilities, and all possible correlations between the variables. Dr. O'Day does this in the following excerpt:

Table 13.1

Correlations, Descriptive Statistics, and Coefficient Alpha Reliability Estimates (on the Diagonal) for the Criterion Variables, Decimals Omitted

			Correlations			
Variable	M	SD	1	2	3	4
1. Health motivation	5.76	0.98	(83)			
2. Desire to be thin	5.30	1.35	30***	(85)		
3. Dislike of exercise	2.36	1.31	-31***	-15*	(92)	
4. Lack of time for exercise	3.67	1.71	-20**	-08	39***	(95)

Note. $N = 238$.
*$p < .05$; **$p < .01$; ***$p < .001$.

Table 13.1 presents means, standard deviations, and intercorrelations between the four criterion variables. Coefficient alpha reliability estimates for each variable exceeded the criterion of $\alpha \geq +.70$ recommended by Nunnally (1978).

Is the Multivariate Test Statistically Significant?

The central null hypothesis in MANOVA states that, in the population, there is no difference between the various conditions with respect to the vectors of means on the criterion variables. Below, Dr. O'Day describes the analysis and results relevant to this null hypothesis:

> Data were analyzed using a one-way multivariate analysis of variance (MANOVA) in which the predictor variable was team & exercise status and the criterion variables were health motivation, desire to be thin, dislike of exercise, and lack of time for exercise. For this analysis, Wilks' lambda was used as the multivariate test statistic. With alpha set at .05, the exact F value for Wilks' lambda was significant, $\Lambda = .81$, $F(8, 464) = 6.48$, $p < .001$.

The preceding excerpt includes the words "the exact F value for Wilks' lambda." This refers to the fact that Dr. O'Day used a statistical application that converted Wilks' lambda into an obtained F statistic. In this analysis, Wilks' lambda produced an obtained F statistic of $F = 6.48$. With 8 and 464 degrees of freedom, the probability value for this F statistic was $p < .001$. This was smaller than the standard criterion of $\alpha = .05$, so she concluded that the results were statistically significant.

An earlier section indicated that values of Wilks' lambda (Λ) may range from zero through 1, with lower values (closer to zero) indicating a stronger relationship. For the current study, the obtained value was $\Lambda = .81$. A later section will review some guidelines for interpreting the relative strength of the relationship revealed by this statistic.

What Is the Nature of the Relationship?

An earlier section discussed the fact that there are a number of different follow-up procedures that researchers use to understand the nature of the relationship between the predictor variable and the criterion variables in a MANOVA. Many experts argue that multivariate statistics deserve multivariate follow-up procedures such as discriminant analysis. Therefore, *Chapter 11* of this book shows how discriminant analysis may be used for the current investigation.

Despite the virtues of multivariate procedures, published articles often use a series of univariate ANOVAs as a follow-up to a significant MANOVA. So that you will be well-prepared to interpret such articles, the current section illustrates this univariate-ANOVA approach.

RESULTS FROM THE UNIVARIATE ANOVAS. Dr. O'Day introduces the results from the individual ANOVAs in this excerpt:

> As a follow-up procedure, four univariate ANOVAs were performed—one for each criterion variable. In each ANOVA, the predictor variable was team & exercise status. The results of these analyses are presented in Table 13.2.

Table 13.2

Results of Univariate ANOVAs for Four Criterion Variables (in Each ANOVA, the Predictor Variable Is Team & Exercise Status)

Criterion variable	SS	MS	(MSE)	$F(2, 235)$	p	η^2
Health motivation	13.99	7.00	(0.91)	7.69	<.001	.06
Desire to be thin	5.27	2.63	(1.81)	1.46	.236	.01
Dislike of exercise	53.01	26.51	(1.51)	17.55	<.001	.13
Lack of time for exercise	61.92	30.96	(2.70)	11.47	<.001	.09

Note. $N = 238$. SS = Sum of squares; MS = Mean square; MSE = Mean square error; η^2 = eta squared (percent of variance in criterion variable accounted for by team & exercise status).

If an investigation reports an ANOVA performed on just one criterion variable, the researcher typically just presents the results within the text of the article. If, on the other hand, the study reports ANOVAs performed on several criterion variables, the researcher might choose instead to summarize the results in a table. This is what Dr. O'Day does with Table 13.2.

Table 13.2 follows a fairly standard format used with ANOVA summary tables in empirical journals. If you need a quick refresher regarding the contents of ANOVA summary tables (e.g., what, exactly does "*MS*" stand for?), see the section titled "Understanding the Contents of ANOVA Summary Tables" in the "Additional Issues" section near the end of this chapter.

In the following excerpt, Dr. O'Day summarizes the most important findings from Table 13.2.

> Because four analyses are reported in Table 13.2, the Bonferroni adjustment was used to maintain a familywise Type I error rate of $\alpha_{FW} = .05$. With this adjustment, an *F* statistic in Table 13.2 was considered to be statistically significant only if the *p* value for that *F* statistic was less than .0125. Using this criterion, the univariate ANOVA *F* statistic was significant for only three of the four criterion variables: health motivation, $F(2, 235) = 7.69$, $p < .001$, $\eta^2 = .06$; dislike of exercise, $F(2, 235) = 17.55$, $p < .001$, $\eta^2 = .13$; and lack of time for exercise, $F(2, 235) = 11.41$, $p < .001$, $\eta^2 = .09$.

In the preceding excerpt, Dr. O'Day refers to the *Bonferroni adjustment*. The **Bonferroni adjustment** is a procedure in which the researcher adjusts the alpha level that will be used to determine whether a given statistic is significant. Researchers use it when they need to perform more than one significance test on the same data set and do not want the familywise Type I error rate to become inflated. Dr. O'Day used it here because she performed four ANOVAs, one for each criterion variable.

In *Chapter 5*, you learned that the general form for the Bonferroni adjustment is as follows:

$$\alpha_{ADJ} = (\alpha_{FW} / K)$$

With the preceding formula, α_{ADJ} = the adjusted value of alpha that will be used for each individual significance test; α_{FW} = the desired value for the familywise Type I error rate; and K = the number of comparisons to be made (i.e., the number of significance tests to be performed).

Dr. O'Day wanted the familywise Type I error rate to be $\alpha_{FW} = .05$. She planned to perform four significance tests (one for each criterion variable), which meant that $K = 4$. The adjusted level of alpha was therefore computed as:

$$\alpha_{ADJ} = (\alpha_{FW} / K) = (.05 / 4) = .0125$$

The above shows that the adjusted level of alpha was $\alpha_{ADJ} = .0125$. This value meant that Dr. O'Day would view a univariate ANOVA as being statistically significant only if its *p* value were less than .0125.

GROUP MEANS ON THE CRITERION VARIABLES. We now know which criterion variables displayed significant univariate ANOVAs. But are the means in the direction that Dr. O'Day had hypothesized? She addresses this issue next:

Table 13.3

Means [and 95% Confidence Intervals for Means] for Each Criterion Variable as a Function of Team & Exercise Status

	Nonexercisers (n = 35)		Individual exercisers (n = 143)		Team exercisers (n = 60)	
Criterion variable	M	[95% CI]	M	[95% CI]	M	[95% CI]
Health motivation	5.33	[5.02, 5.65]	5.72	[5.56, 5.88]	6.11	[5.87, 6.35]
Desire to be thin	4.95	[4.50, 5.40]	5.38	[5.16, 5.61]	5.32	[4.98, 5.66]
Dislike of exercise	3.45	[3.04, 3.86]	2.28	[2.08, 2.48]	1.94	[1.63, 2.25]
Lack of time for exercise	4.63	[4.08, 5.18]	3.73	[3.46, 4.00]	2.97	[2.55, 3.39]

Note. N = 238. M = Mean; CI = 95% Confidence interval for the mean.

Table 13.3 presents means on the four criterion variables for each condition under team & exercise status. The 95% confidence interval for each mean appears in brackets.

For the three criterion variables that displayed significant F statistics, the Games-Howell post-hoc test was used to identify the pairs of group means that were significantly different (for the Games-Howell procedure, alpha was set at α = .05). The results of these post-hoc tests are summarized in Table 13.4.

The four criterion variables are displayed as rows in Table 13.4. For a given criterion variable, means identified with the same subscript are not significantly different according to the Games-Howell post-hoc test (α = .05). These tests showed that: (a) for health motivation, the team exercisers scored significantly higher than the other two groups; (b) for dislike of exercise, the nonexercisers scored significantly higher than the other two groups; and (c) for lack of time for exercise, the nonexercisers scored significantly higher than the individual exercisers, who in turn scored significantly higher than the team exercisers.

Table 13.4

Group Means and Results from Games-Howell Post-Hoc Tests

Criterion variable	Nonexercisers ($n = 35$) M (SD)	Individual exercisers ($n = 143$) M (SD)	Team exercisers ($n = 60$) M (SD)
Health motivation	5.33_a (0.95)	5.72_a (1.03)	6.11_b (0.74)
Desire to be thin	4.95_a (1.28)	5.38_a (1.38)	5.32_a (1.30)
Dislike of exercise	3.45_a (1.43)	2.28_b (1.21)	1.94_b (1.15)
Lack of time for exercise	4.63_a (1.62)	3.73_b (1.60)	2.97_c (1.76)

Note. $N = 238$. *M* = Mean; *SD* = Standard deviation. Within the same row, means identified with the same subscript are not significantly different according to the Games-Howell post-hoc test, $\alpha = .05$.

When reporting results from multiple-comparison procedures, some articles use a convention in which same-letter subscripts are used to identify means that are not significantly different (this is the convention used in Table 13.4). However, in other articles, same-letter subscripts have the opposite interpretation. In those articles, same-letter subscripts identify pairs of means that *are* significantly different. As a reader of research, you must always check the note at the bottom of the table to determine which convention is being used.

The Games-Howell test described above (Games & Howell, 1976) is an example of a ***multiple-comparison procedure:*** a statistical procedure that allows researchers to determine which specific conditions are significantly different from each other. Multiple-comparison procedures are useful in ANOVA because the *F* statistic from the ANOVA can indicate whether the results are statistically significant in general, but it cannot indicate which specific pairs of conditions are significantly different.

Researchers have developed many different multiple comparison procedures, and different procedures are appropriate in different situations. The "Additional Issues" section near the end of this chapter provides recommendations for choosing appropriate procedures according to the purpose of the analysis and whether the sample sizes and variances are equal across conditions.

DISPLAYING GROUP MEANS IN A FIGURE. Researchers sometimes use bar charts, line graphs, or other types of figures to illustrate mean scores. In Figure 13.3, a bar chart is used to display mean scores on health motivation for the three groups being compared.

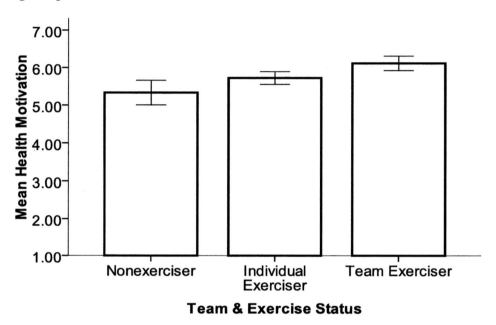

Figure 13.3. Mean scores on health motivation as a function of team & exercise status. Error bars are 95% confidence intervals.

Figure 13.3 shows that the mean score on health motivation was higher for the team exercisers than for other two groups. The error bars at the top of each bar represent the 95% confidence intervals for the means. You can see that these confidence intervals are fairly narrow, indicating that the means are fairly precise estimates.

Remember that Figure 13.3 illustrates just one of the three criterion variables that displayed a significant result in a univariate ANOVA. In practice, most journal editors would discourage Dr. O'Day from including bar charts for an investigation such as this. This is because figures take up a good deal of space, and for an investigation as simple as the current investigation, figures are not really needed to help readers understand the results. With Dr. O'Day's investigation, the group means reported in Table 13.3 and Table 13.4 are probably adequate. Journal editors typically allow the inclusion of figures only when the study is fairly complex, and figures are needed to help readers understand the findings.

How Strong Is the Relationship?

There are at least two ways to investigate the strength of the relationship between the variables investigated in a MANOVA. One involves the value of Wilks' lambda (Λ) computed in the omnibus multivariate analysis that included all four criterion variables. The second involves values of eta squared (η^2) computed in the four separate univariate ANOVAs performed on the individual criterion variables. In this section, Dr. O'Day will report both.

EFFECT SIZE FOR THE MULTIVARIATE ANALYSIS. In an earlier excerpt, Dr. O'Day reported Wilks' lambda (Λ), the index of multivariate association. Part of that excerpt is reproduced here:

> With alpha set at .05, the exact F value for Wilks' lambda was significant, $\Lambda = .81$, $F(8, 464) = 6.48$, $p < .001$.

When researchers use Wilks' lambda in a one-way MANOVA, it is fairly straightforward to compute ***eta squared*** (η^2): the proportion of variance in the criterion variables accounted for by the predictor variable (Huberty & Smith, 1982; Weinfurt, 1995). The formula is:

$$\eta^2 = 1 - \Lambda$$

Where Λ = Wilks' lambda. In Dr. O'Day's study, $\Lambda = .81$, so eta squared is computed as:

$$\eta^2 = 1 - \Lambda = 1 - .81 = .19$$

This indicates that team & exercise status accounts for 19% of the variance in the four criterion variables, taken as a set. In evaluating the relative size of eta squared, the best practice is to compare it to the indices found by other researchers when they studied the same phenomena. If there are no analogous studies in the research literature, the obtained value can be compared against Cohen's (1988) criteria:

> $\eta^2 = .01$ indicates a small effect,
>
> $\eta^2 = .06$ indicates a medium effect, and
>
> $\eta^2 = .14$ indicates a large effect.

According to these criteria, the value of η^2 obtained by Dr. O'Day ($\eta^2 = .19$) represents a large effect. But that was for the *multivariate* analysis. Is it possible that some of her individual criterion variables displayed relatively strong relationships with team & exercise status while other criterion variables displayed weaker relationships?

EFFECT SIZES FOR THE UNIVARIATE ANOVAs. Table 13.2 (presented earlier) displayed the F values, eta squared statistics, and related results for the univariate ANOVAs. When interpreting effect size for the individual criterion variables, these values of η^2 may be compared to the same criteria used for the multivariate η^2 statistic (presented above). For the four individual criterion variables, Dr. O'Day found the following η^2 values:

- For health motivation, $\eta^2 = .06$ (a medium effect).
- For desire to be thin, $\eta^2 = .01$ (a small effect).
- For dislike of exercise, $\eta^2 = .13$ (just short of a large effect).
- For lack of time for exercise, $\eta^2 = .09$ (between a medium effect and a large effect).

This chapter has focused on just one index of effect size used with ANOVA: eta squared. To learn about alternative indices (such as η_p^2, ω^2, ω_p^2 and $\hat{d}$), see the section titled "Other Indices of Effect Size Used with ANOVA" in the "Additional Issues" section at the end of this chapter.

Summary of Findings

Dr. O'Day's results revealed a significant multivariate relationship between team & exercise status and the four criterion variables taken as a set. The *F* statistic derived from Wilks' lambda was statistically significant, and this allowed Dr. O'Day to reject the null hypothesis of equal vectors of means for the various groups. Results from the univariate ANOVAs revealed a significant relationship between the predictor variable and three of the four criterion variables. For these three variables, the Games-Howell post-hoc test revealed the following:

- The team exercisers displayed higher mean scores on health motivation, compared to the other two groups.
- The nonexercisers displayed higher mean scores on dislike of exercise, compared to the other two groups.
- The nonexercisers displayed higher mean scores on lack of time for exercise compared to the individual exercisers. The individual exercisers, in turn, displayed higher mean scores than the team exercisers.

Wilks' lambda for the multivariate relationship between team & exercise status and the four criterion variables was $\Lambda = .81$, and this corresponded to a value of eta squared of $\eta^2 = .19$ (a large effect). Values of eta squared from the univariate ANOVAs ranged from $\eta^2 = .01$ for desire to be thin to $\eta^2 = .13$ for dislike of exercise.

Additional Issues Related to this Procedure

This section provides more details regarding the contents of the typical ANOVA summary table, alternative indices of effect size used with ANOVA, and alternative multiple-comparison procedures. It ends with the usual summary of assumptions underlying the procedure.

Understanding the Contents of ANOVA Summary Tables

Table 13.2 (which appeared earlier in this chapter) presented results from the univariate ANOVAs performed on each criterion variable. This section describes the contents of that table in a bit more detail. This should serve as a refresher if it has been a while since you learned about one-way ANOVA. For convenience, Table 13.2 is reproduced here as Table 13.5.

Table 13.5

Results of Univariate ANOVAs for Four Criterion Variables (in Each ANOVA, the Predictor Variable Is Team & Exercise Status)

Criterion variable	SS	MS	(MSE)	$F(2, 235)$	p	η^2
Health motivation	13.99	7.00	(0.91)	7.69	<.001	.06
Desire to be thin	5.27	2.63	(1.81)	1.46	.236	.01
Dislike of exercise	53.01	26.51	(1.51)	17.55	<.001	.13
Lack of time for exercise	61.92	30.96	(2.70)	11.47	<.001	.09

Note. N = 238. *SS* = Sum of squares; *MS* = Mean square; *MSE* = Mean square error; η^2 = eta squared (percent of variance in criterion variable accounted for by team & exercise status).

The following sections explain the information presented in each of the columns appearing in Table 13.5. Formulas are presented for some of the statistics presented in the table.

THE COLUMN HEADED "CRITERION VARIABLE." This column simply identifies the criterion variable being analyzed. In Table 13.5, each row represents a different criterion variable, and a separate univariate ANOVA was performed on each.

THE COLUMN HEADED "SS." This symbol stands for "Sum of Squares." The ***sum of the squares*** is a relatively crude measure of variability. It is important primarily because it is used in computing other terms in the ANOVA, such as the mean squares. For a given criterion variable, the sum of squares reported in this table is the ***sum of squares between-groups*** (also known as the *sum of squares for the model* or *sum of squares for the effect*). Other things held constant, this term will be large to the extent that the means for the three groups are different from one another.

THE COLUMN HEADED "MS." In Table 13.5, the heading "*MS*" is short for "Mean Square." In ANOVA, a **mean square** is a *variance estimate:* an estimate of the criterion variable's variance in the population, based on some aspect of the variance observed in the sample. Different types of mean squares are computed in ANOVA, and the entries in the column headed "*MS*" in the current table are *mean squares between groups* (symbol: MS_{bn}). The **mean square between groups** reflects (a) error variance plus (b) additional variance due to the treatment effect (if there was any treatment effect). The MS_{bn} will be large to the extent that the predictor variable had a large "effect" on the criterion variable. The MS_{bn} is computed in the following way:

$$MS_{bn} = \frac{SS_{bn}}{df_{bn}}$$

where SS_{bn} = the sum of squares between groups (described earlier), and df_{bn} = the **degrees of freedom between groups**. The df_{bn} is also called the *degrees of freedom for the numerator, degrees of freedom for the model*, or the *degrees of freedom for the effect*. The formula for computing the df_{bn} is:

$$df_{bn} = k - 1$$

Where k = the number of conditions under the predictor variable. In the present study, $k = 3$ (because there were three conditions), and therefore $df_{bn} = 2$.

THE COLUMN HEADED "(MSE)." Under the heading "(*MSE*)," you will find the **mean square error** for each ANOVA. This statistic is also called the *mean square within groups* (MS_{wn}) or the *mean square residual* (MS_{res}). As its name would suggest, the mean square error reflects *error variance*: the pooled average of the variances observed in the various treatment-conditions (Sheng, 2008). Unlike the MS_{bn}, the *MSE* is *not* made larger when the predictor variable has a substantial effect on the criterion variable. The *MSE* is computed as:

$$MSE = \frac{SS_{wn}}{df_{wn}}$$

In the preceding formula, SS_{wn} = the *sum of squares within groups* (not presented in the table) and df_{wn} = the **degrees of freedom within groups** (this is alternatively called the *degrees of freedom for the denominator, degrees of freedom for the error term,* or *degrees of freedom residual*). For a one-way ANOVA with one between-subjects factor, the formula for computing the degrees of freedom within groups is:

$$df_{wn} = N - k$$

Where N = the total number of participants in the study, and k = the number of conditions. In this study, the total number of subjects was 238, and the number of conditions was 3, so the $df_{wn} = 235$.

THE COLUMN HEADED "F." This column presents the obtained F statistic for each ANOVA. The ***F statistic*** is used to determine whether the results of an analysis of variance are statistically significant. It is computed as:

$$F = \frac{MS_{bn}}{MSE}$$

Where MS_{bn} = the mean square between groups and MSE = the mean square error. Values of F may range from zero through infinity. The "no-effect" value for an F statistic is 1.00. In other words, if there is no relationship between the predictor variable and the criterion variable, we expect the obtained F statistic to be approximately equal to 1.00. When there is a relationship between the predictor and the criterion, we expect the F statistic to be greater than 1.00.

The top of the column that presents the F statistic is actually headed "$F(2, 235)$." The values within the parentheses are the degrees of freedom that were used with each univariate ANOVA. The first number always represents the degrees of freedom between groups (df_{bn}), which in this case was 2. The second number always represents the degrees of freedom within groups (df_{wn}), which in this case was 235.

THE COLUMN HEADED "p." This column presents ***probability values*** (i.e., p values) for the obtained F statistics. A given p value estimates the probability that we would have obtained the current sample results if the null hypothesis were true. Normally, the researcher would consider a given ANOVA to be statistically significant if the p value in this column is less than .05. However, in the current analysis Dr. O'Day is using the Bonferroni adjustment, so she will consider the ANOVA for a given criterion variable to be significant only if the obtained p value is less than .0125. Using this criterion, three of the four ANOVAs in this analysis were statistically significant.

THE COLUMN HEADED "η^2." This column presented values of ***eta squared*** (symbol: η^2): an index of effect size that is widely used with ANOVA. Eta squared indicates the proportion of variance in the criterion variable that is accounted for by the predictor variable. Values may range from .00 through 1.00 with higher values representing stronger relationships. The no-effect value is $\eta^2 = .00$. Eta square is computed as:

$$\eta^2 = \frac{SS_{bn}}{SS_{tot}}$$

Where SS_{bn} = the sum of the squares between groups, and SS_{tot} = the sum of the squares total. The *sum of the squares total* is a crude index of the total variability in the sample (it was not reported in the table).

Selecting the Appropriate Multiple-Comparison Procedure

Earlier sections of this chapter discussed a multiple-comparison procedure called the Games-Howell test. ***Multiple-comparison procedures*** are statistical tests that allow researchers to determine which specific conditions are significantly different from one another. They are typically performed in ANOVAs that include three or more conditions. This is because the F statistic from the ANOVA may indicate that the overall results are significant, but it does not indicate which specific *pairs* of conditions are significantly different. Multiple-comparison procedures, on the other hand, were developed for this very purpose.

Researchers and statisticians have developed a very long list of multiple comparison procedures, and most fall into one of two general categories: ***planned contrasts*** (which are appropriate for testing a limited number of a priori hypotheses) and ***post-hoc tests*** (which are appropriate for testing hypotheses generated after the data have been inspected or for comparing all possible pairs of conditions). The selection of a specific multiple comparison procedure for a given investigation is determined by a number of factors, including the following:

- **NATURE OF THE RESEARCH HYPOTHESES.** If the research hypotheses were developed before the researcher obtained and inspected the data, and if these hypotheses deal with just a limited number of comparisons (not all possible comparisons), planned contrasts may be appropriate. However, if the research hypotheses were developed after the data were inspected or if the researcher wishes to investigate all possible pairings of conditions, post-hoc tests may be more appropriate.

- **EQUAL VERSUS UNEQUAL SAMPLE SIZES.** Some multiple comparison procedures will give correct results only if the ns are equal (i.e., if the same number of participants appear in each condition or cell). Other procedures will give correct results with equal ns as well as unequal ns.

- **EQUAL VERSUS UNEQUAL VARIANCES.** Some multiple comparison procedures will give correct results only if the variances are *homogeneous* (i.e., if the amount of variability on the criterion variable is pretty much the same within each treatment condition). Other procedures will give correct results regardless of the homogeneity of the variance.

- **TYPE I ERRORS VERSUS TYPE II ERRORS.** Some multiple comparison procedures are somewhat more likely to result in Type I errors (i.e., telling the world that the results are statistically significant when, unbeknownst to the researcher, this is an incorrect conclusion). Other procedures are more likely to produce Type II errors (i.e., telling the world that the results are nonsignificant when this is an incorrect conclusion).

A number of writers (e.g., Field, 2009a; Kirk, 1995; Toothaker, 1993) have summarized how well the different multiple comparison procedures perform with respect to a variety of outcomes (e.g., making Type I errors; making Type II errors) and under a variety of conditions (e.g., with equal ns versus unequal ns; with equal variances versus unequal variances). This section summarizes some of their recommendations.

PLANNED CONTRASTS FOR TESTING A PRIORI HYPOTHESES. An ***a priori hypothesis*** is a research hypothesis that was formed based on theory and previous research prior to inspecting the data obtained from the current investigation. When circumstances allow researchers to develop specific hypotheses based on theory and previous research (and when these hypotheses deal with only a *subset* of all possible comparisons), it is generally desirable to perform planned contrasts such as Student's multiple *t* test or Dunn's multiple comparison test. ***Planned contrasts*** are statistical procedures that test a priori hypotheses and involve just a limited number of comparisons. Planned contrasts are valued, in part, because they tend to have greater statistical power, compared to post-hoc tests. Kirk (1995) discusses the advantages and disadvantages of various planned-contrast procedures, and Field (2009a) shows how to perform orthogonal contrasts (and other types of contrasts) using SPSS.

POST-HOC TESTS WHEN VARIANCES ARE EQUAL AND SAMPLE SIZES ARE EQUAL. ***Post-hoc tests*** are appropriate when researchers wish to test hypotheses that are formed only after the data have been gathered and inspected. Some post-hoc tests may also be appropriate when the researcher wishes to investigate every possible pairing of conditions.

When the sample variances are equal and the *n*s in the various conditions are also equal, experts often recommend the REGWQ procedure (REGWQ stands for Ryan, Einot, Gabriel, and Welsch Q), and Tukey's HSD test. Dunn's multiple comparison test (also known as the Bonferroni procedure) may be appropriate when making just a few comparisons, but it suffers the disadvantage of being fairly conservative (i.e., of being somewhat prone to Type II errors).

POST-HOC TESTS WHEN VARIANCES ARE EQUAL AND SAMPLE SIZES ARE DIFFERENT. When the variances are equal and the sample sizes are slightly different, the best bet is Gabriel's procedure, although the Dunn test may also be used if making just a few comparisons (once again, the Dunn test is conservative). If the sample sizes are quite different, Hochberg's GT2 is a better choice.

POST-HOC TESTS WHEN VARIANCES ARE DIFFERENT. The Games-Howell test (Games & Howell, 1976) is the best procedure to use when the sample variances are different. It may be used with equal or unequal *n*s.

Other Indices of Effect Size Used with ANOVA

In the univariate ANOVAs reported in this chapter, the primary index of effect size was *eta squared*. ***Eta squared*** (symbol: η^2) represents the proportion of variance in the criterion variable accounted for by the predictor variable(s). Values may range from 0.00 to 1.00 with higher values representing stronger relationships.

The eta squared statistic (η^2) is equivalent to another statistic called *R squared* (symbol: R^2). Both statistics are computed using essentially the same formula, and both statistics indicate the proportion of variance in the criterion variable that is accounted for by the predictor variable(s). Although there are plenty of exceptions, the general rule goes like this: The symbol "η^2" is more likely to be used if the researcher has performed ANOVA, while the symbol "R^2" is more likely to be used if the researcher has performed multiple regression (as covered in *Chapter 9* and *Chapter 10*).

OMEGA SQUARED. Eta squared is a descriptive statistic: It describes the proportion of variance in the criterion variable accounted for by the predictor variable *in the current sample*. In many situations, however, researchers do not want to merely describe the proportion of variance accounted for in this one sample—instead, they want to estimate the proportion of variance actually accounted for in the larger population. This means trouble, because eta squared provides a biased estimate when used for this purpose—it tends to overestimate the actual proportion of variance accounted for. The smaller the sample, the greater the overestimation.

To avoid this bias, some articles report a different index of effect size called ***omega squared*** (symbol: ω^2). Omega squared is similar to eta squared in that both indices deal with the proportion of variance in the criterion variable accounted for by the predictor variable. However, ω^2 provides an *unbiased estimate* of the variance actually accounted for in the population. Omega squared is reported somewhat less frequently in journal articles because it can be difficult to compute with some types of research designs.

When evaluating the value of omega squared obtained in a specific study, Kirk (1996) recommends that $\omega^2 = .01$ be interpreted as a small effect, $\omega^2 = .06$ be interpreted as a medium effect, and $\omega^2 = .14$ be interpreted as a large effect. These are the same criteria used to interpret eta squared.

PARTIAL ETA SQUARED AND PARTIAL OMEGA SQUARED. An ANOVA that includes more than one predictor variable is called a ***factorial ANOVA***. Researchers performing factorial ANOVA often report *partial eta squared* as the index of effect size. ***Partial eta squared*** (symbol: η_p^2) represents the proportion of variance in the criterion accounted for by a given predictor while excluding the other predictor variables (Field, 2009a). The specific way that η_p^2 is computed depends on the statistical procedure being performed. For one-way ANOVA with one between-subjects factor, it is computed as:

$$\eta_p^2 = \frac{SS_{\text{effect}}}{SS_{\text{effect}} + SS_{\text{error}}}$$

Where SS_{effect} = the sum of squares for the effect (i.e., the sum of the squares between groups), and SS_{error} = the sum of squares for the error term (i.e., the sum of squares within-groups).

For a one-way ANOVA with one between-subjects factor, η_p^2 will always be equal to η^2. On the other hand, when the ANOVA is a *factorial* ANOVA—an analysis that includes more than one predictor variable—the two indices will usually be different. In those situations, η_p^2 is typically the preferred index.

Like eta squared, partial eta squared is also a biased descriptive statistic that tends to overestimate the actual proportion of variance accounted for in the population. Researchers who wish to avoid this limitation may instead compute partial omega squared (ω_p^2), an unbiased estimate of variance accounted for in the population.

To interpret the relative size of η_p^2 and ω_p^2, it is best to compare them to the indices found by other researchers who studied the same variables. When this is not possible, they may be evaluated using the same criteria used with η^2 and ω^2. Specifically, .01 is interpreted as a small effect, .06 is interpreted as a medium effect, and .14 is interpreted as a large effect (Huck, 2012).

THE *d* FAMILY OF INDICES. All of the effect-size indices discussed up to this point represent what Rosenthal (1994) calls the *r-family* of indices, as they each represent correlations (or squared correlations) between the predictor variable and the criterion variable. When reading about results from an ANOVA, however, you may also encounter indices from the *d-family:* indices that represent the difference between the means of two conditions, as measured in standard deviations. The symbol for such an index may be *d* or $\hat{d}$.

For example, if a researcher indicates that the effect size for the difference between Condition 1 and Condition 2 is $\hat{d}$ = .80, it tells us that the mean for Condition 1 was .80 standard deviations higher than the mean for Condition 2 (just how the "standard deviation" is computed depends on the nature of the study). Effect indices from the *d* family can be valuable when reporting results from an ANOVA because they are *specific*—they communicate the size of the difference between the means of just two specific conditions. Comparisons such as this may be of greater interest to the reader compared to the more nebulous "overall" effect for the ANOVA as represented by η^2 (or any of the other *r*-family statistics discussed in this section).

As always, researchers who compute $\hat{d}$ as an index of effect size should compare their results against those found in similar investigations. When this is not possible, Cohen (1988) offers these last resort criteria for the *d* statistic: $d \geq \pm 0.20$ represents a small effect, $d \geq \pm 0.50$ represents a medium effect, and $d \geq \pm 0.80$ represents a large effect. For details regarding the *d*-family of indices, see Howell (2008).

Assumptions Underlying this Procedure

Because MANOVA is essentially a multivariate extension of ANOVA, the assumptions underlying MANOVA are generally either identical to the assumptions for ANOVA, or are multivariate extensions of those assumptions. Many of the assumptions listed below are discussed in greater detail in *Chapter 2*. You may view an article as being more credible if it indicates that the data satisfied these assumptions (taken from Field, 2009a; Stevens, 1996; Tabachnick & Fidell, 2007):

RANDOM SAMPLING. The groups should be random samples from the populations of interest.

INTERVAL- OR RATIO-SCALE MEASUREMENT. The criterion variables should be quantitative variables assessed on an interval scale or ratio scale of measurement.

MULTIVARIATE NORMALITY. Within conditions, the criterion variables should display multivariate normality. With sample sizes that are typical for social science research, the univariate and multivariate statistics are robust against modest violations of this assumption, as long as the violations are not due to outliers.

HOMOGENEITY OF THE COVARIANCE MATRICES. The various groups should be sampled from populations with equal variance-covariance matrices. This homogeneity assumption may be tested with Box's M test, although significant results for Box's M are not necessarily a cause for alarm as the test is very sensitive. If the sample sizes are equal, the significance tests for the treatment effects in the MANOVA are robust against modest violations of this assumption.

INDEPENDENCE OF OBSERVATIONS. In part, this assumption means that the researcher should not obtain repeated measures from the same participant.

Chapter 14

Analysis of Covariance (ANCOVA)

This Chapter's *Terra Incognita*

You are reading about an investigation that compared two alternative approaches for teaching about diet and nutrition. The diet-related information was presented to one group of subjects by means of a videotaped presentation and was presented to the second group by means of a 10-page written text. The researcher used a true-false quiz to assess the subjects' knowledge about healthy versus unhealthy diets. This quiz was administered as a pre-test just before the new information was presented to participants and was administered a second time as a post-test just after the new information was presented. The researcher wanted to determine whether greater learning took place in the video condition compared to the written-text condition. The published article reported the following:

> Data were analyzed using a one-way analysis of covariance (ANCOVA) in which the dependent variable was knowledge post-test scores, the between-subjects factor was instruction format (video versus text), and the covariate was knowledge pre-test scores. The ANCOVA revealed a significant treatment effect for instruction format, $F(1, 33) = 14.61$, $MSE = 3.21$, $p < .001$, $\eta_p^2 = .31$.
>
> After adjusting for knowledge pre-test scores, the mean score on the knowledge post-test was higher in the video condition ($M = 43.68$, $SE = 0.44$) than in the text condition ($M = 41.21$, $SE = 0.44$). The difference between the two adjusted post-test means was 2.46 points, 95% CI [1.15, 3.77].

Do not fear. All shall be revealed.

The Basics of this Procedure

Analysis of covariance (abbreviation: **ANCOVA**) is a variation of analysis of variance (ANOVA). ANOVA allows researchers to determine whether there is a relationship between a categorical predictor variable and a continuous quantitative criterion variable. ANCOVA, on the other hand, allows researchers to determine whether there is a relationship between a categorical predictor variable and a continuous quantitative criterion variable after statistically controlling for variance that the criterion variable shares with another variable called a *covariate*. In other words, ANCOVA allows researchers to determine whether there is a difference between the means of two or more treatment conditions after statistically controlling for the covariate.

Generic Analysis Model for ANCOVA

Figure 14.1 presents the generic analysis model for the simplest version of ANCOVA. With this version, scores on a quantitative criterion variable are viewed as being a function of just two variables: (a) a limited-value predictor variable and (b) a quantitative covariate.

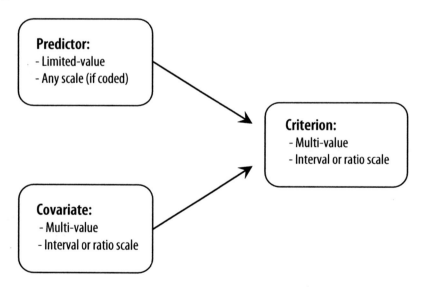

Figure 14.1. Generic analysis model: Nature of the variables included in the simplest version of analysis of covariance (ANCOVA).

The characteristics of the three types of variables included in an ANCOVA (as illustrated in Figure 14.1) are elaborated below:

- **PREDICTOR VARIABLE.** The predictor variable in ANCOVA is typically a ***limited-value variable:*** a variable that contains just two to six categories. This chapter will therefore refer to the predictor variable as a *categorical* predictor variable. This variable may be assessed on any scale of measurement as long as it has been appropriately transformed using dummy-coding or some similar coding scheme (dummy coding was explained in *Chapter 10*). An analysis that includes just one categorical predictor variable is called a ***one-way ANCOVA***, and an analysis that includes more than one categorical predictor are called a ***factorial ANCOVA***. It is unusual for analyses to include more than three or four categorical predictor variables. The analysis illustrated in the current chapter includes just one categorical predictor variable.

- **COVARIATE**. The *covariate* is typically a multi-value quantitative variable measured on an interval scale or ratio scale. The covariate is typically correlated with the study's criterion variable, and is often used to add an element statistical control to the analysis. Later sections will explain just what is meant by "statistical control." Theoretically, there is no limit to the number of covariates that may be included in an analysis, although, in practice, it is unusual to see more than six or so. The analysis illustrated in the current chapter will include just one covariate.

- **CRITERION VARIABLE**. The criterion variable is typically a multi-value quantitative variable assessed on an interval scale or ratio scale. An analysis that includes multiple criterion variables is called a ***multivariate analysis of covariance***, or **MANCOVA**. The analysis presented in this chapter will include just one criterion variable.

Advantages of ANCOVA over ANOVA

Consider the typical situation in which a researcher wishes to investigate the relationship between (a) a categorical predictor variable consisting of two or more conditions, and (b) a continuous quantitative criterion variable. Such an analysis involves computing the mean score on the criterion variable for each condition and then determining whether there is a significant difference between any of these treatment condition means. We generally think of this as the prototypical situation in which it would be appropriate to analyze the data using a one-way ANOVA (as covered in *Chapter 13*).

However, in some situations the researcher may have also obtained scores on an additional variable: a continuous quantitative variable that is correlated with the criterion variable. If certain conditions are satisfied, it may be possible to use this additional variable as a *covariate* in the analysis. Using this variable as a covariate will convert the analysis from a simple ANOVA into a somewhat-more complex analysis called ANCOVA.

In many ways, analyzing the data with ANCOVA is very similar to analyzing it with ANOVA. For example, both procedures allow the researcher to determine whether there is a significant difference between the treatment condition means. However, under certain optimal conditions, analyzing the data with ANCOVA rather than ANOVA may result in two important benefits: (a) it may reduce the size of the error term (thus increasing the power of the test), and (b) it may adjust the mean scores on the criterion variable for the covariate, thus adding an element of statistical control to the investigation. We shall consider these benefits in turn.

ANCOVA REDUCES THE SIZE OF THE ERROR TERM. Including a covariate in an ANCOVA often causes the error term for the analysis (i.e., the MSE or MS_{wn}) to be smaller than it otherwise would be. This, in turn, typically causes the obtained F statistic for the categorical predictor variable to be larger than it otherwise would be. This is usually a good thing, because larger F statistics are more likely to be statistically significant. In other words, including a covariate often increases the *statistical power* of the analysis: It increases the likelihood that the researcher will be able to reject the null hypothesis and tell the world that there are significant differences between the treatment-condition means.

ANCOVA ADJUSTS MEAN SCORES ON THE CRITERION VARIABLE FOR THE COVARIATE. When researchers randomly assign participants to treatment conditions, this random assignment typically produces groups that are equivalent to each other on most subject variables at the very beginning of the study (before the independent variable is manipulated). In this context, "equivalent" means that the various groups will be similar to one another in terms of age, income, gender, and other subject characteristics. This is why random assignment is such an important foundation to doing good research.

Unfortunately, in rare instances random assignment may not work as advertised. In rare instances, researchers may randomly assign subjects to condition but still end up with groups that are not equivalent on all subject variables, just due to chance. For example, the average subject age in Group 1 may be a bit lower (younger) than the average subject age in Group 2 just due to chance. Such a state of affairs is particularly worrisome if the subject variable in question happens to be correlated with the study's criterion variable.

For example, imagine that a researcher hypothesizes that drinking a protein shake every day will cause better muscle tone in adults. He begins with a pool of 40 subjects, randomly assigning 20 subjects to the experimental condition (which drinks a protein shake every day) and the remaining 20 subjects to the control condition (which drinks a no-protein placebo shake every day). At the end of the study, he analyzes the data with a simple ANOVA and finds that the mean score on muscle tone is significantly higher in the experimental group than in the control group. *Success!*

Or is it? When the researcher takes a closer look at his data, he notices that the mean age for the subjects in the experimental group was significantly lower (i.e., younger) than the mean age for the subjects in the control condition. This type of thing should not happen often if subjects are randomly assigned to conditions, but it happened here. The researcher knows that age is negatively correlated with muscle tone: Younger people tend to have better tone than older people. Now the researcher cannot be sure why the experimental group displayed better muscle tone at the end of the study. Was it because the people in the experimental group drank the protein shakes? Or was it simply because they were younger than the subjects in the other group, and therefore had better tone from the start? The researcher cannot be terribly sure, because subject age is now operating as a *confounding variable* (the concept of *confounding variables* was introduced in Chapter 8 of this book).

ANCOVA can be helpful in situations such as this. If the researcher reanalyzes the data with ANCOVA and uses subject age as a covariate, the ANCOVA will produce **adjusted group means**: estimates of the means that the various conditions probably would have displayed if they had displayed equal means on the covariate (i.e., subject age) at the very beginning of the study. If the researcher performs the ANCOVA and finds that there is still a significant difference between the adjusted group means for the two groups (after statistically controlling for subject age), he can be more confident that the differences are actually due to the independent variable (the protein shakes) and are not due to the confounding variable of subject age.

SUMMARY. When ANCOVA is used in the appropriate situation and is used correctly, it can provide results that are superior to the results produced by an ANOVA. In these situations, ANCOVA provides superior results by (a) increasing the statistical power of the analysis and (b) producing mean scores that have been adjusted for differences on the covariate. Later sections will provide more detail regarding the types of situations that are appropriate for ANCOVA and will describe the characteristics that a covariate must have if it is to produce the benefits described here.

Illustrative Investigation

Imagine that the fictitious researcher Dr. O'Day conducts a true experiment as part of her research program dealing with attitudes toward healthy diets. In the current investigation, the criterion variable is *knowledge about healthy diets:* the extent to which participants understand which foods promote good health (e.g., fruits and vegetables) and which foods promote poor health (e.g., trans fats and highly processed foods). She develops a true-false test to measure this variable. With this instrument, scores may range from 10 to 50, and higher scores indicate better knowledge about diet and nutrition.

RESEARCH HYPOTHESIS. Dr. O'Day wants to determine whether some instruction formats are more effective than others when it comes to educating people about healthy eating. She reviews theory and previous research on this topic and is impressed by a theory from the field of social psychology that says that learning is more efficient when the information is presented visually by models that have been videotaped. The theory says that learning tends to be much slower and less efficient when learners have to read the information from printed text documents. Based on her review of the literature, she designs an empirical investigation to test the following research hypothesis:

> Instructional methods in which information about healthy diets is presented by means of videotaped models will be more effective in increasing participant knowledge, compared to instructional methods in which the same information is presented by means of a printed text document.

Figure 14.2 illustrates this hypothesis. The single-headed arrow in the figure represents the predicted causal relationship.

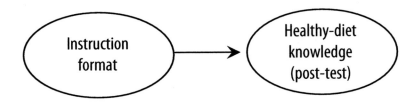

Figure 14.2. Hypothesized relationship between instruction format and healthy-diet knowledge.

METHOD. Dr. O'Day decides to conduct a true experiment to test this hypothesis. She decides to directly compare the two instruction formats by randomly assigning participants to one of two treatment conditions:

- The first treatment condition consists of a *video format*. Participants in this condition observe a 15-minute videotaped presentation in which a model illustrates how college students on limited budgets can select and prepare inexpensive meals that are balanced and nutritious.

- The second treatment condition consists of a *text format*. Participants in this condition are given the same facts and suggestions that were presented in the video condition, but for this group the information is presented in the form of a 10-page document that they must read. This 10-page document is essentially just a transcript of the monolog spoken by the model in the other condition (i.e., the video condition).

Dr. O'Day wishes to determine whether one format is more effective than the other in increasing the participants' knowledge about healthy diets. She therefore administers the true-false knowledge test to the participants at two points in time:

- The first administration is at the very beginning of the experimental session, prior to either seeing the video or reading the text document. Scores obtained at this time will be referred to as the *knowledge pre-test* scores.

- The second administration is immediately after the participants have been exposed to the relevant information. In other words, the second administration occurs immediately after seeing the video (for those in the video condition) or immediately after reading the text document (for those in the text condition). Scores obtained at this time will be referred to as the *knowledge post-test* scores.

The study is conducted using a sample of 36 participants. Half of the participants ($n = 18$) are randomly assigned to the video condition, and the other half ($n = 18$) are randomly assigned to the text condition.

INITIAL ANALYSIS WITH ANOVA. After conducting the experiment, Dr. O'Day initially analyzes the data using ANOVA (*not* ANCOVA). This section illustrates the use of ANOVA so that when she reanalyzes the same data using ANCOVA (later in this chapter) we will be able to directly compare the results produced by the two procedures. Let's assume that, at this point, Dr. O'Day does not see any need to analyze the data using ANCOVA. But this will change.

Table 14.1

Means and Standard Errors for Healthy-Diet Knowledge as a Function of Instruction Method (Text Versus Video) and Time (Pre-test Versus Post-test)

	Means		SE_M	
	Video	Text	Video	Text
Knowledge pre-test	40.39	38.67	0.49	0.54
Knowledge post-test	44.06	40.83	0.41	0.53

Note. $N = 36$. SE_m = standard error of the mean.

Here are the details of the initial analysis: Dr. O'Day analyzes her data using a one-way ANOVA with one between-subjects factor. In this ANOVA, the criterion variable is *knowledge post-test scores.* This is a multi-value quantitative variable assessed on an interval scale of measurement. In the analysis, the predictor variable is *instruction format* (video versus text). This is a dichotomous variable. For the moment, she does not include scores on the knowledge pre-test in this ANOVA in any way.

Results from the ANOVA reveal a statistically significant difference between the video condition versus the text condition with respect to mean scores on the knowledge post-test, $F(1, 34) = 23.12$, $p < .001$, $\eta_p^2 = .41$. Means and standard errors are presented in the second row of Table 14.1. This row shows that the mean knowledge post-test score is higher for the video condition ($M = 44.06$) than for the text condition ($M = 40.83$).

Table 14.1 also reports means and standard errors for the knowledge pre-test, but remember that the knowledge pre-test was not actually included in the ANOVA in any way. Those means are presented merely to set up a mini-drama involving Dr. O'Day and her academic arch nemesis.

Pleased with her findings, Dr. O'Day shares them with her faculty colleague, Dr. Grey. Dr O'Day points at Table 14.1 and says:

> Look—the mean knowledge post-test score for the video condition is 44.06, whereas the mean post-test score for the text condition is lower at 40.83. My ANOVA shows that this difference is statistically significant. So these results support my hypothesis that the video format is superior to the text format for increasing knowledge scores.

"These results prove nothing of the sort," scoffs Dr. Grey. He continues:

> Look at the knowledge *pre-test* scores: At pre-test, the people in the video group already had higher knowledge scores than the people in the text group. The means on the knowledge pre-test were 40.39 for the video group and 38.67 for the text group. And, this difference was displayed at the very *beginning* of the study—well before you manipulated your independent variable. So the fact that the video group is still higher at the end of the study—as illustrated by the mean post-test scores—probably just reflects the fact that they had better knowledge before your experiment even began.

Dr. O'Day stares at the table. The video group *was* higher than the text group at pre-test. She had randomly assigned subjects to treatment conditions, and she knew that random assignment typically results in groups that are fairly equivalent at the beginning of a study. However, she also knew that once in a great while, random sampling can produce groups that are *not* equivalent at the beginning of a study. This is especially possible when the samples are small (as they are in her current study). She did some additional analyses and found that the difference between the two groups at pre-test was large enough to be statistically significant.

Crap!

REANALYSIS WITH ANCOVA. Dr. O'Day considers her options. She knows that one possibility involves reanalyzing her data using analysis of covariance, rather than analysis of variance. If she uses ANCOVA, she could use subjects' scores on the knowledge pre-test as a covariate. Among other things, this would allow her to estimate what the means on the knowledge post-test probably would have been if the two groups had been equal on the knowledge pre-test at the very beginning of the study.

She checks her data, and finds that they satisfy the assumptions required by ANCOVA. She decides to do the analysis. The variables constituting this ANCOVA are illustrated in Figure 14.3.

Figure 14.3 shows that the ANCOVA will include the following variables:

- The categorical predictor variable will be *instruction format* (video versus text). This was the same categorical predictor variable that had been used in the ANOVA.

- The criterion variable will be scores on the *knowledge post-test*. Again, this was the same criterion variable that had been used with the ANOVA.

- The covariate will be scores on the *knowledge pre-test.* You will recall that she had not used the pre-test scores in the original ANOVA at all.

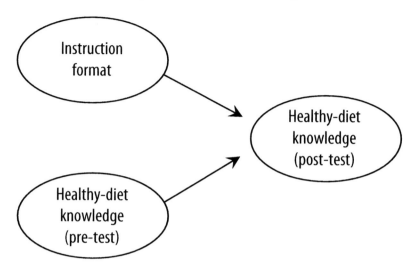

Figure 14.3. Variables to be analyzed in the ANCOVA: Instruction format (the categorical predictor variable), healthy-diet knowledge pre-test scores (the covariate), and healthy-diet knowledge post-test scores (the criterion variable).

The simple ANOVA that Dr. O'Day performed earlier had indicated that there was a significant relationship between instruction format (video versus text) and knowledge post-test scores. The current ANCOVA will indicate whether there is still a relationship between instruction format and knowledge post-test scores after she has statistically controlled for the covariate, knowledge pre-test scores. At least two alternative outcomes are possible:

- It is possible that, after controlling for knowledge pre-test scores, the relationship between instruction format and knowledge post-test scores will become *nonsignificant* (in other words, the difference between the two groups on mean knowledge post-test scores could become *nonsignificant* after she has statistically controlled for the pre-test scores). Such an outcome would be consistent with Dr. Grey's argument that the instruction format independent variable had never really had any effect at all—it would be consistent with his argument that the two groups were different at the time of the post-test simply because they had already been different at the time of the pre-test. This outcome would be bad news for Dr. O'Day.

- On the other hand, it is possible that there will still be a significant relationship between instruction format and knowledge post-test scores even after she has statistically controlled for knowledge pre-test scores. If there is such a significant relationship (and if the means are still in the direction that Dr. O'Day had predicted), it would be consistent with her argument that the independent variable (instruction format) really did have an effect on the dependent variable (knowledge post-test scores). This outcome would be good news for Dr. O'Day.

Which will it be? All is revealed following *Details, Details, Details*.

Details, Details, Details

In the published literature, you are likely to encounter applications of ANCOVA that are much more complicated than the one described in the current chapter. This section discusses the ways that these analyses can be made more complicated by including additional predictor variables, additional criterion variables, or additional covariates. It also discusses characteristics that are desirable in a covariate, so that you can better evaluate an investigation that uses ANCOVA.

Alternative Versions of ANCOVA

To keep things simple, this chapter emphasizes the most elementary version of ANCOVA: an analysis that involves just one criterion variable, just one predictor variable, and just one covariate. Remember, however, that in the research literature you are likely to encounter variations on this basic model. Some examples:

- Analyses sometimes include more than one categorical predictor variable (i.e., more than one factor). Depending on the number of factors and conditions, such an analysis may be called a *factorial ANCOVA*, a *two-way ANCOVA*, a *2 × 3 × 2 ANCOVA*, and so forth. Researchers perform these analyses because they wish to understand the relationship between two or more predictor variables and the criterion variable after statistically controlling for a covariate.

- Analyses sometimes include more than one criterion variable. These analyses are called *multivariate analysis of covariance*, or *MANCOVA*.

- Analyses sometimes include multiple covariates. Performing ANCOVA with multiple covariates allows researchers to understand the relationship between a categorical predictor variable and a continuous criterion variable after statistically controlling for a number of potential confounding variables.

You can see that analysis of covariance is a highly flexible procedure. Theoretically, it allows researchers to include any number of predictor variables, criterion variables, or covariates.

This flexibility, however, does not mean that it is always a good idea to include a large number of variables in an analysis. In particular, researchers should be judicious when selecting the variables that will serve as *covariates* in an ANCOVA, because not every variable displays the properties that are desirable in a covariate. These properties are described next.

Characteristics of a Good Covariate

Dr. O'Day's current investigation is an example of a *pre-test/post-test* experimental design. Much of this chapter focuses on this design because it produces data that are particularly well-suited for ANCOVA. The pre-test scores and the post-test scores are essentially the same variable—a variable that is measured at two points in time. This means that the pre-test scores will very likely be correlated with the post-test scores. This is a good thing, because a variable should typically be used as a covariate only if it displays a substantial correlation with the criterion variable.

However, an investigation does not necessarily have to use a pre-test/post-test experimental design to produce data appropriate for ANCOVA. In many cases, ANCOVA is used to analyze data from studies that use a *post-test-only* experimental design. In these investigations, there are no pre-test scores on the criterion variable, just post-test scores (in other words, scores on the criterion variable obtained at just one point in time—after the independent variable has been manipulated). In these studies, the researcher might use some *subject variable* (i.e., some naturally-occurring characteristic of the participants) as a covariate in the ANCOVA. For example, Dr. O'Day might have used *subject age* as a covariate in the current analysis. Or she might have used *socio-economic status* or *attitudes toward health* or any of a number of other variables.

She should, however, be *picky* in selecting her covariates. Not just any variable can be used if she wants to enjoy the statistical benefits that are associated with ANCOVA—benefits such as increased statistical power and appropriately adjusted mean scores on the criterion variable. In a given research situation, some variables make better covariates than others. This section summarizes the characteristics that a variable should display if a researcher plans to use it as a covariate in ANCOVA (the following is based on Huck, 2004, Keppel & Zedeck, 1989, Pedhazur, 1982, and Tabachnick & Fidell, 2007).

1) **THE COVARIATE MUST BE CORRELATED WITH THE CRITERION VARIABLE.** Huck (2004) recommends that within each treatment condition, the correlation between the covariate and the criterion variable should exceed $r = \pm.20$. When the correlation between the covariate and the criterion variable is less than $\pm.20$, the analysis will typically display greater statistical power if the researcher simply drops the covariate and analyzes the data using ANOVA rather than ANCOVA (Keppel & Zedeck, 1989).

2) **THE RELATIONSHIP BETWEEN THE COVARIATE AND THE CRITERION VARIABLE SHOULD BE LINEAR.** In *Chapter 6*, you learned that, when there is a ***linear relationship*** between X and Y, it means that the Y scores change in only one direction as the X scores change, so that it is possible to draw a best-fitting straight line through the center of the scatterplot. ANCOVA requires that the relationship between the covariate and the criterion variable must be linear.

3) **THE DATA MUST DISPLAY HOMOGENEITY OF REGRESSION.** A dataset displays *homogeneity of regression* if the relationship between the criterion variable and the covariate is essentially the same within each of the conditions (i.e., groups) under the categorical predictor variable. The nature of the relationship between the criterion variable and the covariate may be illustrated by drawing a regression line, and may be represented numerically by the *unstandardized regression coefficient, b* (you learned about these concepts in *Chapter 7*). For example, imagine that Dr. O'Day computed the unstandardized regression coefficient (b), which represented the nature of the relationship between knowledge post-test scores (the criterion variable), and knowledge pre-test scores (treated as the predictor variable in this specific analysis). Further, imagine that she computed one regression coefficient using just the 18 subjects in the video condition (symbol: b_V), and a separate regression coefficient using just the 18 subjects in the text condition (symbol: b_T). Her dataset would display *homogeneity of regression* if there were *not* a significant difference between these two regression coefficients. This would be the case if the two regression coefficients were very similar values (for example, if $b_V = +.43$ and $b_T = +.45$). In contrast, her dataset would display *heterogeneity of regression* if there *were* a significant difference between these two regression coefficients. This would be the case if the two regression coefficients were very different values (for example, if $b_V = +.43$ and $b_T = +.08$). If the data display heterogeneity of regression (also known as a *treatment × covariate interaction*), it is not appropriate to proceed with the ANCOVA. When the data display heterogeneity of regression, the researcher might choose to drop the covariate and analyze the data using ANOVA rather than ANCOVA, might choose to perform a more sophisticated procedure such as the *Johnson-Neyman technique* (Pedhazur, 1982), or might pursue yet some other alternative. This assumption of homogeneity of regression is discussed in greater detail in the "Assumptions" section at the end of this chapter.

4) **THE NUMBER OF COVARIATES SHOULD BE KEPT TO A MINIMAL NUMBER.** In performing the F test for the independent variable, the researcher loses one error-term degree of freedom for each additional covariate included to the analysis. With other factors held constant, this means that, with the addition of each new covariate to the ANCOVA, it becomes a bit more difficult for the independent variable to display statistical significance (Tabachnick & Fidell, 2007). This means that researchers should generally resist the temptation to throw several dozen covariates into the analysis in order to "control" for every possible confounding variable. In most situations, researchers achieve the optimal benefits from ANCOVA by keeping the number of covariates relatively small.

5) **IF MORE THAN ONE COVARIATE IS USED, THE COVARIATES SHOULD DISPLAY LITTLE OR NO CORRELATION WITH ONE ANOTHER.** When two covariates are strongly correlated with one another, the second covariate does not achieve any meaningful adjustment beyond the adjustment achieved by the first covariate.

Therefore, adding the second covariate only has the undesirable effect of decreasing the error-term degrees of freedom, thus decreasing the power of the *F* test for the categorical predictor variable. The implication is this: When multiple covariates are included in an analysis, the researcher will maximize statistical benefits by including covariates that display (a) strong correlations with the criterion variable, but (b) weak (or zero) correlations with one another.

6) **SCORES ON THE COVARIATE MUST BE RELIABLE AND VALID MEASURES OF THE INTENDED CONSTRUCT.** Remember that ANCOVA is first-cousin to multiple regression—the widely used statistical procedure covered in *Chapter 9* and *Chapter 10*. One of the assumptions underlying multiple regression is the requirement that the predictor variables should be measured with perfect reliability (i.e., that they should be free of measurement error). Since ANCOVA may be viewed as a special case of multiple regression, the implication is that the covariates used in ANCOVA should also be measured with perfect reliability. You may be shocked...*shocked*...to learn that some variables studied in the social and behavioral sciences are not measured with perfect reliability. Like just about *all* of them, for example. Tabachnick and Fidell (2007) warn that unreliable covariates can lead to incorrect results from ANCOVA, especially when ANCOVA is used with nonexperimental data. Because it is not always possible to use *perfectly* reliable variables as covariates, Tabachnick and Fidell recommend that researchers do the next best thing—use *highly* reliable variables. Specifically, they recommend that variables should be used as covariates only if their reliability estimates exceed $r_{xx} > .80$.

7) **THE COVARIATE MUST NOT BE AFFECTED BY THE INDEPENDENT VARIABLE.** For example, in Dr. O'Day's investigation, this assumption requires that scores on the knowledge pre-test (the covariate) must not be affected by the instruction-format manipulation (the independent variable). This requirement is most clearly satisfied when the investigation is a true experiment in which the covariate is measured prior to manipulating the independent variable. Dr. O'Day's study is an example of such a situation—she first obtained scores on the knowledge pre-test and then subsequently manipulated the independent variable by either showing the video or asking participants to read the printed document. In a true experiment such as this, researchers can be fairly confident that they have satisfied this assumption. On the other hand, when ANCOVA is used with nonexperimental studies, it is more difficult to know whether the assumption is satisfied. The use of ANCOVA in nonexperimental investigations is discussed in the "Additional Issues" section near the end of this chapter.

Results from the ANCOVA

Dr. O'Day's current study differs from many investigations described in this book in that it is a ***true experiment:*** a study in which the researcher randomly assigns subjects to conditions, actively manipulates the independent variable, and maintains a high level of control over environmental circumstances. Because it is a true experiment, she may use the terms *independent variable* and *dependent variable* rather than the more generic terms *predictor variable* and *criterion variable* when she describes it in a research article.

Reprise: Dr. O'Day's Research Method

To refresh our memory regarding the details of Dr. O'Day's study, an excerpt from the *Method* section of her article is presented here. Make a mental note regarding the names that she assigns to her variables, as we will see these names again in the *Results* section to follow.

> The dependent variable was the participants' knowledge about diet and nutrition, as assessed by a true/false-format knowledge test (typical item: "Baked potato chips are usually lower in saturated fat than fried potato chips."). Test scores could range from 10 to 50, with higher scores indicating greater knowledge.
>
> After providing informed consent, each participant completed the true/false test. Scores from this first administration constituted the *knowledge pre-test* variable.
>
> The study's independent variable was then manipulated. The independent variable was *instructional method:* the medium used to communicate information to participants. This variable was manipulated by randomly assigning half of the participants ($n = 18$) to the video condition, and the other half ($n = 18$) to the text condition. Participants in the video condition watched a 15-minute presentation in which a model discussed the characteristics of a healthy diet and showed how a college student living on a limited budget could prepare inexpensive healthy meals. Participants in the text condition read a 10-page document that presented the same information (the 10-page document was essentially a written version of the script that had been delivered by the model in the video condition).
>
> After receiving this information via one of the two formats, each participant again completed the true/false test. Scores from this second administration constituted the *knowledge post-test* variable. Finally, each participant was debriefed.

Descriptive Statistics

Researchers often begin the *Results* section of an article by reporting descriptive statistics for the study's variables. Dr. O'Day does this in the following excerpt.

> Table 14.2 presents descriptive statistics and correlations for the study's variables. The table shows that the knowledge pre-test scores displayed a relatively strong correlation with knowledge post-test scores, $r = .58$, $p < .01$. It also shows that

the Kuder-Richardson reliability estimate for the knowledge pre-test was $KR20 = .81$, satisfying Tabachnick and Fidell's (2007) recommendation that the reliability of covariates should exceed $r_{xx} = .80$. These findings supported the use of the knowledge pre-test as a covariate in the ANCOVA to be reported below.

Table 14.2

Means, Standard Deviations, and Pearson Correlations (Decimals Omitted) for Study Variables

			Pearson correlations		
Variable	M	SD	1	2	3
1. Knowledge pre-test	39.53	2.33	(81)		
2. Knowledge post-test	42.44	2.57	58**	(84)	
3. Instruction format[a]	0.50	0.51	38*	64**	(--)

Note. $N = 36$. Values in parentheses on diagonal are $KR20$ reliability estimates for the healthy-diet knowledge test.
[a] Dummy coded 0 = text instructions, 1 = video instructions.
*$p < .05$. **$p < .01$.

The diagonal of the correlation matrix in Table 14.2 presents **Kuder-Richardson reliability estimates** (symbol: *KR20*) for the knowledge pre-test and the knowledge post-test. The *KR20* reliability estimate is a special case of the coefficient alpha reliability estimate covered in *Chapter 3*. The *KR20* estimate is often used when a test allows for dichotomous responses (such as *true/false* responses).

Having covered the descriptive statistics, Dr. O'Day moves on to results that are more directly relevant to her research hypothesis: mean scores on the criterion variable for the two groups. This is done in the following excerpt:

Figure 14.4 displays observed scores on the knowledge test as a function of instruction format (video versus text) and time (pre versus post). The figure shows that mean scores on the knowledge post-test were higher for the video condition ($M = 44.06$, $SE = 0.41$) than for the text condition ($M = 40.83$, $SE = 0.53$).

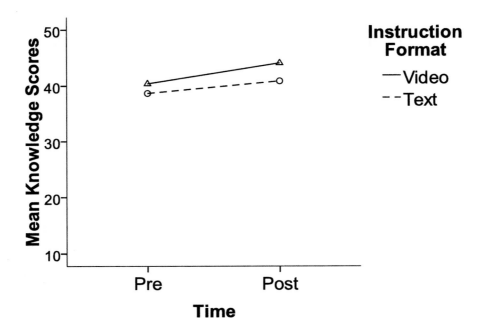

Figure 14.4. Healthy-diet knowledge as a function of instruction format and time.

Dr. O'Day presents mean knowledge scores in Figure 14.4 to communicate the general trend of the findings. The figure shows that the video group scored higher than the text group on the knowledge post-test (which is a good thing), but the video group also scored higher than the text group on the knowledge pre-test (which is a bad thing). A bit later in the article, she will use these differences on the pre-test as one of the justifications for performing an ANCOVA.

Is There a Significant Relationship Between the Variables?

For many researchers, the first step in determining whether there is a relationship between variables involves conducting a null-hypothesis significance test. If the results are statistically significant (i.e., if the null hypothesis is rejected), it is seen as evidence that the variables are related.

THE NULL HYPOTHESIS. Using notation suggested by Warner (2008), the null hypothesis for a one-way ANCOVA may be stated as:

$$H_0: \mu^*_{A1} = \mu^*_{A2} = \ldots \mu^*_{Ak}$$

where μ^*_{Ak} = the adjusted mean score (in the population) for the kth level of Factor A (the predictor variable). The asterisk (*) next to the symbol for the population mean (μ) indicates that these means have been adjusted for the covariate.

Dr. O'Day's predictor variable (instruction format) included just two conditions (video versus text). The null hypothesis for this predictor variable may therefore be stated as:

$H_0: \mu_{A1}^* = \mu_{A2}^*$. In the population, there is no difference between the video condition versus the text condition with respect to the adjusted means on the knowledge post-test.

DR. O'DAY'S ANOVA. Dr. O'Day will actually perform two separate analyses on the knowledge post-test scores: an ANOVA (in which scores on the post-test have not been adjusted for pre-test scores), as well as an ANCOVA (in which the post-test scores have been adjusted for pre-test scores). In an actual published article, most researchers would have presented only the ANCOVA. However, both analyses are presented here so that we can see what happens to the error term in the analysis when the researcher controls for a covariate.

Two analyses were performed. In both analyses, the between-subjects factor was instruction format (video versus text), and the dependent variable was knowledge post-test scores. The first analysis was a traditional ANOVA that did not include a covariate. The second analysis was an ANCOVA in which knowledge pre-test scores were used as a covariate.

The first analysis (the ANOVA with no covariate) showed that the mean post-test score for the video condition was significantly higher than the mean post-test score for the text condition, $F(1, 34) = 23.12$, $MSE = 4.04$, $p < .001$, $\eta_p^2 = .41$. This ANOVA is summarized in the upper section of Table 14.3.

Although the results from this ANOVA are consistent with the prediction that instruction format would have an effect on post-test knowledge scores, Figure 14.4 (presented earlier) shows that there also appears to be a difference between the two groups with respect to their knowledge pre-test scores, with the video condition ($M = 40.39$, $SE = 0.49$) scoring somewhat higher than the text condition ($M = 38.67$, $SE = 0.54$). This raised the possibility that the difference between the two groups that had been observed in the post-test scores was not due to the independent variable (instruction format), but was instead due to differences between the two groups that had existed prior to the manipulation of the independent variable (such differences can occur in situations in which random assignment to conditions does not produce groups of participants that are equal on all relevant variables).

To address this possibility, the data were reanalyzed using analysis of covariance (ANCOVA). This analysis included the same variables that had been included in the ANOVA described above: the dependent variable was once again knowledge post-test scores and the between-subjects factor was once again instruction format. The ANCOVA, however, made use of a covariate: scores on the knowledge pre-test. By including this covariate, the ANCOVA would indicate whether there was still a significant effect for instruction format after statistically controlling for knowledge pre-test scores.

Table 14.3

Knowledge About Healthy-Diets (Post-test Scores) as a Function of Instruction Format (Video Versus Text): Results from ANOVA and ANCOVA

Analysis and Source	df	SS[a]	MS	F	p	η_p^2
ANOVA (no covariate)						
Instruction format	1	93.44	93.44	23.12	<.001	.41
Within-cells error	34	137.44	4.04			
Corrected total	35	230.89				
ANCOVA (covariate: pre-test)						
Knowledge pre-test	1	31.60	31.60	9.85	.004	.23
Instruction format	1	46.86	46.86	14.61	<.001	.31
Within-cells error	33	105.84	3.21			
Corrected total	35	230.89				

Note. $N = 36$.
[a] Type III sums of squares.

DR. O'DAY'S ANCOVA. Having provided the rationale for the ANCOVA, Dr. O'Day moves on to the analysis and results. She begins with a discussion of the assumptions underlying ANCOVA, and then turns to the results themselves.

> Data were screened to verify that variables were normally distributed within each condition and that there was a linear relationship between pre-test scores and post-test scores. The test for homogeneity of regression (Pedhazur, 1982) showed that the regression coefficient representing the relationship between pre-test scores and post-test scores was not significantly different in the two treatment conditions, $F(1, 32) = 0.29$, $MSE = 3.28$, $p = .595$, $\eta_p^2 = .01$. These and additional analyses indicated that all assumptions required for the ANCOVA were satisfied.
>
> The results of the ANCOVA are presented in the lower section of Table 14.3. These results show that, after controlling for knowledge pre-test scores, there was still a significant treatment effect for instruction format, $F(1, 33) = 14.61$, $MSE = 3.21$, $p < .001$, $\eta_p^2 = .31$. According to Cohen's (1988) criteria, the obtained value of partial eta squared ($\eta_p^2 = .31$) represented a large effect.

So the difference between means is still statistically significant, even after adjusting for knowledge pre-test scores. Dr. O'Day may reject the null hypothesis.

SMALLER ERROR TERM WITH THE ANCOVA. An earlier section indicated that including a good covariate in an analysis often means that the error term in the ANCOVA will be smaller than the error term in the corresponding ANOVA. This is generally a good thing, because the error term is the denominator in the formula for the F statistic for the treatment effect (instruction format, in this case), and a smaller error term can increase the likelihood that this F statistic will be statistically significant.

This can be seen in Table 14.3—the table that summarizes the ANOVA and ANCOVA performed by Dr. O'Day. First, look in the upper portion of the table, which presents results from the ANOVA. Where the row headed "Within-cells error" intersects with the column headed "MS," you will find the *mean square within-cells* (symbol: MS_{wn}), also known as the *mean square error* (symbol: MSE). The MSE is the error term used in computing F statistics for the analysis. The table shows that, for the ANOVA (which included no covariate), the MSE is 4.04.

Next, look in the lower portion of the same table, which presents results from the ANCOVA. Where the row headed "Within-cells error" intersects with the column headed "MS," you will find the MSE from the ANCOVA. Here, you can see that the MSE is now 3.21, which is smaller than the value of 4.04 that had been obtained in the ANOVA. This is just one illustration of how researchers can sometimes reduce the size of the error term in their analyses by including an effective covariate.

DEGREES OF FREEDOM FOR THE F STATISTIC. Researchers routinely report the degrees of freedom for their F tests, and understanding how these values are computed will help you understand the nature of their investigation (e.g., the number of treatment conditions, the number of participants). Here we will focus on the F statistic for the categorical predictor variable in the ANCOVA (this may be referred to as the *independent variable*, the *factor*, the *treatment effect*, or the *main effect*). In Dr. O'Day's study, the categorical predictor variable was *instruction format* (video versus text). Below is the excerpt in which she reported the F statistic for this categorical predictor variable:

> These results show that, after controlling for knowledge pre-test scores, there was still a significant treatment effect for instruction format, $F(1, 33) = 14.61$, $MSE = 3.21$, $p < .001$, $\eta_p^2 = .31$.

The numbers within parentheses (i.e., "1, 33") are the degrees of freedom for the F statistic. The first number (the "1," in this case) represents the *degrees of freedom between groups* (symbol: df_{bn}). This is sometimes referred to as the *degrees of freedom for the numerator*. The formula for computing the df_{bn} is:

$$df_{bn} = k - 1$$

Where k = the number of conditions under the categorical predictor variable. In the present study, $k = 2$ (because there were two conditions), and therefore $df_{bn} = 1$.

The second number (the "33," in this case) represents the *degrees of freedom within groups* (symbol: df_{wn}). This is sometimes referred to as the *degrees of freedom for the denominator, degrees of freedom for the error term,* or *degrees of freedom for the residual term*. With ANCOVA, the formula for computing the df_{wn} is:

$$df_{wn} = N - k - c$$

where N = the total number of participants in the study, k = the number of conditions under the categorical predictor variable, and c = the number of covariates. In the present study, $N = 36$, $k = 2$, and $c = 1$ (because there was just one covariate: knowledge pre-test scores). This means that the df_{wn} may be computed as:

$$df_{wn} = N - k - c = 36 - 2 - 1 = 33$$

REVERSE-ENGINEERING THE DEGREES OF FREEDOM. Now that you know the formulas for the degrees of freedom, you can reverse-engineer them in order to determine the number of conditions in a study, as well as the number of participants. This trick is useful for those occasions in which you have skipped the method section and have gone straight to the results, but still wish to understand some of the basic features of the study.

When the data are analyzed using a one-way between-subjects ANCOVA (as is the case with Dr. O'Day's investigation), the number of treatment conditions will be equal to $df_{bn} + 1$. So, if an article says:

- "$F(1, 33) = 14.61$," it means that there were two treatment conditions
- "$F(2, 160) = 1.27$," it means that there were three treatment conditions
- "$F(4, 231) = 18.54$," it means that there were five treatment conditions

With the same type of ANCOVA, the number of subjects included in the analysis will be equal to $df_{bn} + df_{wn} + c + 1$. So, if an ANCOVA included one covariate and the article says:

- "$F(1, 33) = 14.61$," it means that 36 subjects were included in the analysis
- "$F(2, 160) = 1.27$," it means that 164 subjects were included in the analysis
- "$F(4, 231) = 18.54$," it means that 237 subjects were included in the analysis

What is the Nature of the Relationship?

When we discuss the "nature of the relationship" with a procedure such as ANCOVA, we are typically most interested in the ***adjusted means:*** the means of the treatment conditions after they have been adjusted for the covariates. Dr. O'Day discusses these in the following excerpt:

> Table 14.4 provides mean scores for the healthy-diet knowledge variable. The last row of the table displays mean knowledge post-test scores adjusted for

knowledge pre-test scores in the ANCOVA. This row shows that the adjusted mean was higher in the video condition (M = 43.67, SE = 0.44) compared to the text condition (M = 41.21, SE = 0.44). The difference between the two adjusted means was 2.46 points, 95% CI [1.15, 3.77].

Table 14.4

Means and Standard Errors for Healthy-Diet Knowledge as a Function of Instruction Method (Text Versus Video) and Time (Pre-test Versus Post-test)

	Means		SE_M	
	Video	Text	Video	Text
Knowledge pre-test	40.39	38.67	0.49	0.54
Knowledge post-test, unadjusted	44.06	40.83	0.47	0.47
Knowledge post-test, adjusted[a]	43.67	41.21	0.44	0.44

Note. N = 36. SE_m = standard error of the mean.
[a]The knowledge post-test means have been adjusted by using knowledge pre-test scores as a covariate in an ANCOVA.

UNDERSTANDING THE ROWS OF THE TABLE. Table 14.4 is easiest to understand if we focus on the two columns under the heading "Means." These columns present mean scores on the knowledge test for the two treatment conditions.

The top row of the table provides mean scores on the knowledge pre-test. These were the scores obtained prior to manipulation of the independent variable.

The second row provides unadjusted means on the knowledge post-test. These are the means that were observed after the manipulation of the independent variable, not yet adjusted for the covariate. It is generally a good idea to present these means in the article so that the reader can see just how much they change after being adjusted.

The last row of the table provides means on the knowledge post-test after being adjusted for the covariate. Notice that these means are a bit less extreme, compared to the unadjusted means—controlling for the covariate seems to have decreased the size of the difference between them. Specifically:

- For the video condition, the unadjusted post-test mean had been fairly high at $M = 44.06$, but the adjusted mean for the same condition is a bit lower (and therefore less extreme) at $M = 43.67$.

- For the text condition, the unadjusted post-test mean had been fairly low at $M = 40.83$, but the adjusted mean for the same condition is a bit higher (and is therefore less extreme) at $M = 41.21$.

It makes sense that the means of the two conditions should be "pulled in" to be less extreme after controlling for the covariate. Remember the main reason that Dr. O'Day added a covariate to the analysis in the first place: She noticed that the video condition was higher than the text condition at the very beginning of the study. This meant that the observed difference between the two conditions at the end of the study may not have been due to the manipulation of the independent variable—it may have been due merely to the differences that had existed from the beginning. This raised the question:

> Will there still be a significant difference between the two group means if I control for the differences that had existed at the beginning of the study (i.e., the differences displayed on the pre-test)?

Theoretically, the two adjusted post-test means could have been pulled in so much that they would have been equal to one another, no longer displaying a statistically significant difference. Such an outcome would have been consistent with the idea that the independent variable had no effect at all.

But this is not what happened, as can be seen by the fact that there is still a difference between the two adjusted post-test means presented as the last row in Table 14.4. Therefore, the answer to the preceding question appears to be:

> Yes, there is still a difference between the two group means, even after controlling for the differences that had existed at the beginning of the study.

ALTERNATIVE TABLE FOR PRESENTING THE MEANS. Table 14.4 illustrates one approach for presenting means for the treatment conditions. If Dr. O'Day had wanted to present confidence intervals for the knowledge post-test means, she might have instead prepared a table such as Table 14.5. The table shows that the confidence intervals are fairly narrow, indicating that the means are reasonably precise estimates of the population means.

MULTIPLE COMPARISON PROCEDURES USED WITH ANCOVA. In an investigation that includes three or more treatment conditions, the F statistic for the categorical predictor variable might indicate that there is overall statistically significant effect, but it does not identify the specific pairs of means that are significantly different. In those situations, researchers often perform a ***multiple comparison procedure:*** an analysis that determines whether there is a significant difference between specific pairs of treatment conditions.

Since Dr. O'Day's current investigation included just two conditions, multiple comparison procedures were not needed. Nevertheless, this section will briefly review these procedures so that you will be prepared when you encounter them in published articles.

Table 14.5

Healthy-Diet Knowledge Post-test Scores as a Function of Instruction Method (Text Versus Video): Unadjusted Means from the ANOVA and Adjusted Means from the ANCOVA

	M	SE_M	95% CI LL	95% CI UL
Unadjusted means from ANOVA				
Video condition	44.06	0.47	43.09	45.02
Text condition	40.83	0.47	39.87	41.80
Adjusted means from ANCOVA[a]				
Video condition	43.67	0.44	42.78	44.57
Text condition	41.21	0.44	40.32	42.11

Note. N = 36. SE_M = standard error of the mean. LL and UL represent lower limit and upper limit (respectively) for 95% confidence interval for mean.
[a]Means have been adjusted by using knowledge pre-test scores as a covariate in an ANCOVA.

Most multiple comparison procedures fall into one of two general categories. The first category consists of **planned contrasts**: tests that are used when the researcher wishes to test hypotheses that were developed prior to actually conducting the study and inspecting the data. Planned contrasts typically involve just a limited number of comparisons (out of all of the treatment-mean comparisons that are theoretically possible).

The second category consists of **post-hoc tests**: procedures used to either (a) test hypotheses that were developed after the data were inspected or (b) test hypotheses involving every possible pairing of treatment conditions. Post-hoc tests typically display poorer statistical power, compared to planned contrasts.

The "Additional Issues" section of *Chapter 13* provides a more detailed discussion of the planned contrasts and post-hoc tests that are used with simple ANOVA. *Chapter 13* also shows how researchers often present the results of post-hoc tests in the tables and text of published articles.

Popular data-analysis applications such as SAS and SPSS typically offer only a limited number of multiple-comparison procedures that can be used as part of ANCOVA, compared to the relatively large number that are available with ANOVA. The remainder of this section describes these procedures along with the references that show how to use them.

Tabachnick and Fidell (2007) provide syntax for performing orthogonal contrasts in SAS and SPSS. They also provide a formula for using the Scheffé adjustment for controlling the Type I error rate in post-hoc tests performed as part of an ANCOVA.

Stevens (1996) illustrates the Bryant-Paulson simultaneous test, a post-hoc test for detecting significant pairwise differences in ANCOVA. He provides versions of the test that are appropriate for randomized research designs, for non-randomized designs, for studies with just one covariate, and for studies with several covariates.

Field (2009a) shows how to use SPSS to perform simple contrasts and post-hoc tests as part of an ANCOVA. Field and Miles (2010) show how to use SAS to perform similar procedures. Warner (2008) shows how to use SPSS to perform contrasts and also provides guidance on how the results can be summarized in tables and text.

How Strong Is the Relationship?

An ***index of effect size*** is a value that represents the strength of the relationship between a predictor variable and a criterion variable. An index that is particularly appropriate for use with ANCOVA is *partial eta squared* (symbol: η_p^2). ***Partial eta squared*** represents the proportion of variance in a criterion variable explained by a predictor variable while excluding the other predictor variables (Field, 2009a).

The specific way that η_p^2 is computed depends on the statistical procedure being performed. For one-way ANCOVA with one between-subjects factor, it is computed as:

$$\eta_p^2 = \frac{SS'_{effect}}{SS'_{effect} + SS'_{error}}$$

Where:

SS'_{effect} = the adjusted sum of squares for the effect of interest. The term "adjusted" means that this is the sum of squares from the ANCOVA, not the ANOVA. In a one-way ANCOVA, the "effect of interest" is typically the categorical predictor variable (which, in this case, is *instruction format*).

SS'_{error} = the adjusted sum of squares for the within-cells error. Again, the term "adjusted" means that this is the sum of squares from the ANCOVA, not the ANOVA.

The ANCOVA summary table for the current study (presented as Table 14.3) showed that the sum of squares for the effect (instruction format) was 46.86, and the sum of the squares for the error term was 105.84. Therefore, η_p^2 is computed as:

$$\eta_p^2 = \frac{SS'_{effect}}{SS'_{effect} + SS'_{error}} = \frac{46.86}{46.86 + 105.84} = \frac{46.86}{152.70} = .307 = .31$$

So partial eta squared for the *instruction format* independent variable was $\eta_p^2 = .31$. This is the same value that had previously been reported in Table 14.3. It indicates that instruction format accounts for 31% of the variance in knowledge post-test scores, after adjusting for knowledge pre-test scores.

As always, it is best to interpret the effect size for a given study against effect sizes found in other investigations of the same variables. When this is not possible, the obtained value of partial eta squared (η_p^2) may be interpreted using the same criteria that Cohen (1988) provides for eta squared (η^2):

$\eta_p^2 = .01$ is interpreted as a small effect

$\eta_p^2 = .06$ is interpreted as a medium effect

$\eta_p^2 = .14$ is interpreted as a large effect

The value of $\eta_p^2 = .31$ that was obtained in the current study may therefore be interpreted as a large effect. Please remember that these results are fictitious, however.

Summary of Findings

These analyses revealed a statistically significant relationship between instruction format and healthy-diet knowledge. The ANCOVA showed that there was a difference between the mean knowledge post-test score displayed by the video group versus the text group, and that this difference was statistically significant even after controlling for scores on the knowledge pre-test, $F(1, 33) = 14.61$, $p < .001$.

The results showed that the adjusted mean on the knowledge post-test was higher in the video condition than in the text condition. For the video condition the adjusted mean was 43.67, 95% CI [42.78, 44.57], whereas for the text condition the adjusted mean was 41.21, 95% CI [40.32, 42.11].

For the independent variable in the ANCOVA (instruction format), the index of effect size was $\eta_p^2 = .31$. This represents a large effect, according to Cohen's (1988) guidelines.

Additional Issues Related to this Procedure

Statistics textbooks (e.g., Huck, 2004; Keppel & Zedeck, 1989; Pedhazur, 1982) warn that ANCOVA is a statistical procedure that is especially prone to being misused. To help you identify such misuse, this section describes some of the problems that might occur when ANCOVA is used to analyze data from nonexperimental studies. The section ends by summarizing the assumptions underlying ANCOVA.

Using ANCOVA in Nonexperimental Research

Most of this chapter has illustrated how ANCOVA can be used to analyze data from true experiments. Most experts agree that this is typically a good application of the procedure, because using ANCOVA to analyze data from experiments utilizing a pre-test/post-test design can increase the statistical power of the analysis and can also adjust the post-test means for small differences on the pre-test (Tabachnick & Fidell, 2007; Warner, 2008).

ANCOVA is also used to analyze data from nonexperimental research, and this use of the procedure can be more controversial. An example is when ANCOVA is used to analyze data from *intact groups*—groups that are not created through random assignment to conditions. In these cases, the subjects typically assign themselves to "conditions" long before the study is conducted. As a result, the groups usually differ on a large number of confounding variables from the outset. Researchers may then attempt to use ANCOVA to statistically control for these confounding variables, in an effort to "equate" the groups—to make them equivalent on the confounding variables. Authorities (e.g., Huck, 2004) advise that readers should be very cautious when interpreting the results from such an analysis.

EXAMPLE. Imagine that Dr. O'Day's faculty colleague, Dr. Grey, has a hypothesis that students in a *PSYC 100 General Psychology* course learn more when lectures are delivered using PowerPoint slides, compared to when they are delivered without PowerPoint slides (*PowerPoint®* is the computer application that instructors often use as a visual aid during lectures). Dr. Grey happens to be teaching two sections of *General Psychology* this semester, so he conducts an investigation to test his hypothesis:

- One of his sections of *General Psychology* meets on Wednesdays at 2:00 PM. Dr. Grey decides that this section will be the *PowerPoint condition*. He delivers his lectures to this group using PowerPoint slides as visual aids.

- One of his sections of *General Psychology* meets on Friday mornings at 8:00 AM. Dr. Grey decides that this section will be the *no-PowerPoint condition*, so he delivers his lectures to this group without the slides.

Remember that this is *not* a true experiment. Dr. Grey did not begin with a pool of potential participants and then randomly assign half of them to the 2:00 PM PowerPoint condition and the other half to the 8:00 AM no-PowerPoint condition. Instead, the students assigned themselves to conditions by either signing up for the 2:00 PM class or the 8:00 AM class. From the outset, this means that the two groups may have differed on important variables before the semester even began (we shall revisit these potential differences in a moment).

INITIAL ANALYSIS WITH ANOVA. Dr. Grey conducts his study according to plan. At the end of the semester, he prepares a final exam to assess the knowledge and skills that the students have developed during the semester. Scores on this final exam may range from 0 to 100, with higher scores indicating better learning. He administers the test to both sections of the course and analyzes the results using a simple ANOVA (*not* an ANCOVA). In this analysis, the criterion variable is *final exam scores* and the categorical predictor variable is *lecture format* (PowerPoint versus no PowerPoint). The results show that the PowerPoint group scored significantly higher than the no-PowerPoint group. The mean score for the PowerPoint group was $M = 80$, and the mean score for the no-PowerPoint group was just $M = 70$. *Success!*

Or was it?

Some of Dr. Grey's colleagues point out that, although there was a significant difference between the two means, this difference may not have been due to the PowerPoint manipulation at all. Maybe the two means were different because the typical individual who signs up for a 2:00 PM class tends to be a very different type of student, compared to the typical individual who signs up for an 8:00 AM class. For example, maybe the students who sign up for a 2:00 class tend to have higher levels of academic aptitude. If this were true, then the difference between the two means could have been due to differences in academic aptitude, not the PowerPoint manipulation. Skeptics who argue this are arguing that academic aptitude may be operating as a *confounding variable* (as defined in *Chapter 2*).

RE-ANALYSIS WITH ANCOVA. Deeply offended by his colleague's skepticism, Dr. Grey is eager to show that his results really were due to the PowerPoint manipulation, not the subjects' differing levels of academic aptitude. To prove this, he decides to repeat the analysis, this time using ANCOVA to statistically control for the potential confounding variable. In the analysis, the criterion variable will once again be *final exam scores*, and the categorical predictor variable will once again be *lecture format*. In the ANCOVA, the covariate will be student scores on the *Scholastic Aptitude Test* (i.e., the *SAT*). Dr. Grey knows that his colleagues will agree that the SAT is a good measure of academic aptitude.

The results from the ANCOVA showed that, after controlling for SAT scores, the adjusted mean score on the final exam were now $M = 78$ for the PowerPoint group and $M = 72$ for the no-PowerPoint group. The two means were not quite as far apart as they had been in the first analysis, but the difference was still large enough to be statistically significant. *Success!*

Or was it?

PROBLEMS WITH THIS USE OF ANCOVA. Colleagues point out that the observed difference between the two groups still may not have been entirely due to the PowerPoint manipulation. Maybe the students in the PowerPoint group scored higher because they were older students. Or maybe they scored higher because they were more conscientious. Or maybe they scored higher because they were not as sleepy as the other students. Or maybe because they were Aquarians.

Eventually, Dr. Grey will see that his study suffers from dozens of potential confounding variables. He cannot statistically control for all of these variables because he did not *measure* all of them (and you cannot statistically control for variables that you did not measure). For these and related reasons, experts generally recommend that readers should be cautious when interpreting results from nonexperimental studies in which ANCOVA is used in an effort to equate nonequivalent groups (Huck, 2004; Pedhazur, 1982). The experts generally caution that, when investigating hypothesized causal relationships, the use of ANCOVA cannot serve as an effective substitute for random assignment to conditions (Tabachnick & Fidell, 2007).

None of this is to say that ANCOVA must never be used with data from nonexperimental investigations. There are many situations in which it is either impossible or unethical to conduct a true experiment, and in those situations ANCOVA can be a useful tool for data analysis. Cook and Campbell's (1979) book on quasi-experimentation provides guidance regarding research design and data analysis for those situations for which true experiments are not feasible.

Also remember that, in some investigations, researchers are not testing causal hypotheses at all—they merely wish to investigate the relationship between a categorical predictor variable and a quantitative criterion variable after statistically controlling for one or more covariates. If the relevant assumptions have been met, ANCOVA may be appropriate in those situations.

CONCLUSION. In summary:

- Articles sometimes report studies that, on the surface, look like true experiments (because the study includes two or more groups of participants).
- We (the readers of the article), should check closely to determine whether the participants were randomly assigned to treatment conditions, whether the researcher actively manipulated an independent variable, and whether the researcher exercised high levels of experimental control. If not, we should seriously consider the possibility that the results obtained at the end of the study may have been due to differences that existed from the very beginning (due to the use of intact groups, the presence of confounding variables, or other indications of weak research design).
- If the researcher attempts to control for these pre-existing differences using ANCOVA, we should seriously consider the possibility that the ANCOVA is not adequately controlling for all of the relevant confounding variables.

Assumptions Underlying this Procedure

As a statistical procedure, ANCOVA combines elements of ANOVA with elements of linear regression. Because of this, it comes as no surprise that that ANCOVA has

(a) one set of assumptions related to ANOVA, along with (b) a separate set of assumptions related to linear regression. Both sets of assumptions are briefly summarized here. Some of the following assumptions are described in greater detail in *Chapter 2*.

The first four assumptions listed below are assumptions underlying one-way ANOVA with one between-subjects factor. These assumptions are similar to those listed in *Chapter 13*.

- **INTERVAL- OR RATIO-SCALE MEASUREMENT.** The criterion variable should be a quantitative variable that is assessed on an interval scale or a ratio scale of measurement (Field, 2009a).

- **NORMALLY DISTRIBUTED DATA.** In the population, scores on the criterion variable should follow an approximate normal distribution within each condition. Violations of the normality assumption are less problematic if the departure from normality is not extreme or if the samples are large (Aron et al., 2006; Gravetter & Wallnau, 2000).

- **HOMOGENEITY OF VARIANCE.** Scores on the criterion variable should display equal variances in the population (Aron et al., 2006; Field, 2009a; Gravetter & Wallnau, 2000). The F test from ANOVA is robust against small to moderate violations of this assumption as long as the sample sizes are equal (Field, 2009a), but it is not robust against major violations of the homogeneity assumption, particularly when the sample sizes are small or unequal (Sheng, 2008).

- **INDEPENDENCE OF OBSERVATIONS.** Within each condition, the observations should be independent (Field, 2009a; Gravetter & Wallnau, 2000). The independence assumption is met if no observation or score (on the criterion variable) is influenced by any other observation or score.

Along with the preceding requirements, Stevens (1996) also provides the following three assumptions for ANCOVA. Notice that these are similar to some of the assumptions underlying multiple regression (covered in *Chapter 9* and *Chapter 10* of this book).

- **LINEAR RELATIONSHIP.** In ANCOVA, the relationship between the criterion variable and each of the covariates must be linear (Pedhazur, 1982). When there is a ***linear relationship*** between X and Y, it means that the Y scores change in only one direction as the X scores change, so that it is possible to draw a best-fitting straight line through the center of the scatterplot.

    If the relationship between the criterion variable and a covariate is nonlinear (i.e., if there is a curvilinear relationship between the criterion variable and the covariate), the adjustment of the treatment means for that covariate will be incorrect. In these situations, Stevens (1996) recommends either (a) transforming the variables so as to create a linear relationship, or (b) fitting a polynomial analysis-of-covariance model to the data. Pedhazur (1982) shows how multiple regression may be used to investigate curvilinear relationships in general, and Huitema (1980) discusses nonlinear ANCOVA.

- **ABSENCE OF MEASUREMENT ERROR ON THE COVARIATE.** Technically, the covariate used in ANCOVA should be perfectly reliable—it should be measured without error (Tabachnick & Fidell, 2007). This assumption should sound familiar to you, because "absence of measurement error" was one of the assumptions for the predictor variables used in multiple regression (covered in *Chapter 9* of this book), and ANCOVA is a close relative of multiple regression.

    This assumption is more important for nonexperimental studies than for true experiments. In a true experiment, when the covariates display less-than-perfect reliability the ANCOVA may have relatively poor statistical power. This is not terrific, but is not a catastrophe.

    On the other hand, in a nonexperimental study the stakes are much higher. When the covariates display less-than-perfect reliability in this type of investigation the ANCOVA may produce adjusted group means that are biased and misleading. This is a much more serious problem. Stevens (1996, p. 323), for example, presents a hypothetical ANCOVA in which the presence of measurement error in a covariate causes the adjusted mean for Group 2 to be significantly higher than the adjusted mean for Group 1, whereas the use of a covariate free of measurement error would have caused the relationship between the two adjusted means to be reversed.

    The implication is that readers should consider the reliability of the covariate when interpreting results from an ANCOVA. Tabachnick and Fidell (2007) suggest that, although there are a small number of demographic variables (such as sex and age) that may be measured with perfect reliability, variables that are measured "psychometrically" (e.g., by using multiple-item summated rating scales) typically fall short of perfect reliability. To protect against the biased results described above, they recommend that only variables with reliability estimates exceeding $r_{xx} > .80$ should be used as covariates in nonexperimental research.

- **HOMOGENEITY OF REGRESSION.** This assumption requires that the slope (i.e., the regression coefficient, symbol: *b*) that represents the relationship between the criterion variable and the covariate must be the same within each condition under the categorical predictor variable. With factorial designs (i.e., studies that contain more than one categorical predictor variable), the corresponding assumption is that the criterion-covariate slope must be the same within each *cell* (Pedhazur, 1982). When this assumption is met, the slopes are said to be *homogeneous*, and the data are said to display **homogeneity of regression**. When this assumption is not met, the slopes are said to be *heterogeneous*, and the data are said to display **heterogeneity of regression**.

    To test this assumption, the researcher typically creates one or more terms that represent the interaction between the categorical predictor variable and the quantitative covariate (Field, 2009a). If this interaction term is statistically significant, it means that the homogeneity of regression assumption has been violated (i.e., that the relationship between the criterion variable and the covariate is different in different conditions under the categorical predictor variable). Such an outcome is sometimes referred to as a significant **covariate × treatments interaction** (Keppel & Zedeck, 1989). In such situations, it is inappropriate to use the covariate to adjust the treatment means, as is usually done in ANCOVA. In such situations, the researcher may choose to just drop the covariate and analyze the data using ANOVA rather than ANCOVA.

However, simply dropping the covariate is not the only alternative that is available to the researcher when the homogeneity of regression assumption has been violated. In some cases, the researcher may choose to investigate the nature of this covariate × treatments interaction in order to better understand it. For example, imagine that Dr. O'Day modified her study so that the covariate was *IQ scores* rather than *knowledge pre-test scores*. This would mean that the criterion variable was still knowledge post-test scores, the categorical predictor variable was still instruction format (video versus text), and the covariate is now IQ. Imagine that Dr. O'Day found a significant covariate × treatments interaction that showed that (a) there was a strong positive correlation between IQ and knowledge post-test scores for subjects in the text condition, but that (b) there was essentially no correlation between IQ and knowledge post-test scores for subjects in the video condition. On the negative side, this result means that she has violated the homogeneity of regression assumption, and it therefore will not be appropriate to adjust mean scores based on the IQ covariate. On the positive side, however, Dr. O'Day may decide that this unexpected interaction is of theoretical importance, and is worthy of additional investigation. To better understand this result, she may choose to analyze the data using some advanced procedure appropriate for such interactions, such as the Johnson-Neyman technique (Pedhazur, 1982). Since she is no longer adjusting group means for the covariate, it is no longer appropriate to call the analysis an ANCOVA (Warner, 2008), but this does not mean that she must abandon her analyses or the dataset.

Chapter 15

Exploratory Factor Analysis

This Chapter's *Terra Incognita*

You are reading a journal article. It describes a study in which the researcher has analyzed data from a 12-item questionnaire assessing attitudes toward health and exercise. She reports the following in the *Results* section:

> An exploratory factor analysis was performed to investigate the factor structure underlying responses to the 12-item questionnaire. In this analysis, principal axis factoring was used to extract the factors, and squared multiple correlations were used as prior communality estimates. A scree test suggested four meaningful factors, so a four-factor solution was rotated using a promax (oblique) rotation. A questionnaire item was said to load on a factor if the factor loading for that item was ≥ ±.40. Using this criterion, the factor pattern matrix displayed simple structure and was interpretable. The factor pattern indicated that (a) Factor 1 was measured by items 51, 52, and 53; given the nature of these items, this construct was labeled the *lack of time for exercise* factor; (b) Factor 2 was measured by items 47, 48, and 49 and was labeled the *dislike of exercise* factor; (c) Factor 3 was measured by items 31, 32, and 33 and was labeled the *desire to be thin* factor; and (d) Factor 4 was measured by items 27, 28, and 29 and was labeled the *health motivation* factor. Results from the analysis were used to create estimated factor scores for each participant on each of the four factors.

Do not be afraid. All shall be revealed.

The Basics of this Procedure

Exploratory factor analysis (EFA) is a multivariate procedure used to determine the number and nature of the common factors that underlie a set of manifest variables. ***Manifest variables*** are observed variables—the real-world variables that the researcher actually measures in a direct way. Manifest variables are also called *observed variables, measured variables,* and *indicator variables*. The ***common factors***, on the other hand, are hypothetical constructs: unseen variables that the researcher does not measure in a direct way, but are assumed to be present because of the effect that they have on the manifest variables. Common factors are also called *latent variables* and *latent factors*. In this context, the word *latent* means *underlying*.

This procedure is often used in investigations in which participants have responded to a large number of Likert-type rating items on a questionnaire. The researcher then performs an exploratory factor analysis of the subjects' responses to the individual rating items. The researcher uses EFA to determine: (a) how many latent factors underlie responses to the questionnaire items, (b) which questionnaire items are measuring each factor, (c) what substantive label should be given to each factor, and (d) the nature of the correlations between the factors.

Exploratory factor analysis is often called a ***dimension-reduction procedure***. This label acknowledges the fact that the researcher typically begins with a relatively large number of dimensions (e.g., responses to the relatively large number of rating items on the questionnaire) and uses EFA to reduce these to a smaller number of dimensions (e.g., the smaller number of latent factors identified in the course of the analysis).

Illustrative Investigation

This chapter illustrates exploratory factor analysis using a scenario involving the fictitious researcher, Dr. O'Day. Imagine that, not long after being employed by her current university, Dr. O'Day tells Dr. Silver (a faculty colleague) about her interest in conducting research on exercise, nutrition, and health. Dr. Silver tells Dr. O'Day that he is interested in the same issues, and in fact has developed a 12-item questionnaire that measures a number of health-related attitudes. He has already administered the questionnaire to a large sample of participants, although he never got around to actually analyzing the data.

The items constituting Dr. Silver's 12-item questionnaire appear in Table 15.1. Subjects responded to each item using a Likert-type response format in which 1 = *Extremely Disagree* and 7 = *Extremely Agree*. Note that the items in this table also appear in the *Questionnaire on Eating and Exercising*, which is located in *Appendix A* of this book. The somewhat unusual item numbers appearing in Table 15.1 (i.e., "27," 28," etc.) are the original item numbers that these statements were assigned in the much-longer questionnaire in *Appendix A*. This is done so that it will be easy for you to find these items in the questionnaire in the appendix, should you wish to do so.

Table 15.1

The 12-Item Questionnaire To Be Analyzed in the Current Investigation

---

Questionnaire Item

---

27. It is important to me to be healthy.
28. I am willing to sacrifice in order to stay healthy.
29. I am willing to work hard in order to avoid medical problems.
31. It is important to me to be thin.
32. I very much want to have a fairly low body weight.
33. It is important to me to have a lean physical appearance.
47. I **hate** to engage in aerobic exercise.
48. I find aerobic exercise to be unpleasant.
49. I find aerobic exercise to be punishing.
51. I am too busy to engage in frequent aerobic exercise.
52. I am under too much time pressure to engage in frequent aerobic exercise.
53. With my busy schedule, there is very little time for frequent aerobic exercise.

Note. Item numbers are the original item numbers from *Questionnaire on Eating and Exercising*, which appears in *Appendix A*.

Although Dr. Silver's questionnaire consists of 12 items, Dr. O'Day does not believe that it is actually measuring 12 different constructs. For example, items 27, 28, and 29 all seem to be measuring a single construct that might be labeled *heath motivation*. Other groups of items might be measuring other constructs.

With Dr. Silver's encouragement, Dr. O'Day analyzes the data using exploratory factor analysis. She knows that exploratory factor analysis will allow her to gain a better understanding of just what is really being measured by this 12-item questionnaire. EFA will help her determine the number of latent factors that are being measured by the 12 individual rating items, will help her assign meaningful names to these factors, will allow her to determine the pattern of correlations between the factors, and will allow her to investigate the relationship between these factors and other variables (such as *subject age*).

Types of Variables That May be Analyzed

Exploratory factor analysis is typically performed on quantitative variables that are assessed on an interval scale or ratio scale of measurement. Additional assumptions are summarized at the end of this chapter.

The manifest variables analyzed in an exploratory factor analysis do not have to be responses to questionnaire items—they can be any quantitative variable that is assessed on an interval or ratio scale (provided that the assumptions are met). The manifest variables may be scores on ability tests, achievement tests, personality inventories, physical measurements, or any other metric variable. In the current example, the manifest variables are responses to individual questionnaire items. This scenario is illustrated because it appears so often in the published literature—not necessarily because it is a best practice. In fact, some authors argue that it is a very questionable practice.

Some experts (e.g., Wirth & Edwards, 2007) have argued that Likert-type rating items (such as the 12 questionnaire items in Dr. O'Day's study) produce variables that are actually measured on an ordinal scale of measurement, not an interval scale. Because of this, it follows that responses from Likert-type items should not be analyzed using traditional EFA. Theirs is a well-reasoned argument, and if it comes to be generally accepted by the research community, the use of EFA with Likert-type rating items may disappear from the published literature.

But that has not happened just yet (for example, see Warner, 2008, for a description of the lively debate surrounding this scale-of-measurement controversy). At this time, most journals continue to publish articles that describe the use of EFA with Likert-type rating scales. In keeping with this popular practice, Dr. O'Day's current investigation will include an exploratory factor analysis performed on just this type of data.

There are limits, however. The EFA procedures described here are not appropriate for manifest variables that allow for mere dichotomous responses (e.g., questionnaire items with *Yes/No* responses). It is possible to perform dimension-reduction procedures on dichotomous variables, but the researcher typically needs to compute special correlation coefficients and use specialized software and procedures (see Field, 2009a, p. 650 for an overview).

Research Questions Addressed

Exploratory factor analysis allows researchers to investigate the factor structure that underlies a set of manifest variables. This ***factor structure*** refers to (a) the number and nature of the latent factors being measured, (b) the relationship between these latent factors and the manifest variables that measure them, and (c) the correlations among the latent factors themselves. An EFA allows researchers to investigate questions such as the following:

- How many factors are measured by this set of manifest variables? In other words, what is the optimal number of factors I should retain in order to account for most of the common variance, while still keeping my solution parsimonious? (Remember that *parsimonious* means *simple*).

- What is the nature of Factor 1? In other words, what name should be given to this factor? What about Factor 2? What about each of the remaining factors?

- What is the pattern of intercorrelations between the factors in my factor solution?

- What is the relationship between (a) the factors that I created in my EFA and (b) other variables that are important in my program of research? Are there significant and substantial relationships between these factors and other variables that are theoretically important?

Details, Details, Details

This section goes into more detail in describing what is meant by a *factor structure*. It also sharpens your understanding of the differences between exploratory factor analysis versus confirmatory factor analysis, as well as the differences between orthogonal factors versus oblique factors.

Example of an Underlying Factor Structure

Prior to performing the exploratory factor analysis, Dr. O'Day does not know what factor structure might underlie responses to the 12-item questionnaire. She will perform the EFA in an effort to discover that factor structure. So that you (the reader) might better understand the *concept* of factor structure, this section will peek into the future to provide a preview of the results that she will eventually obtain by the end of her analyses. This preview will make it easier to grasp some of the new terms and concepts to be introduced in this chapter (terms such as *factor loadings* and *inter-factor correlations*).

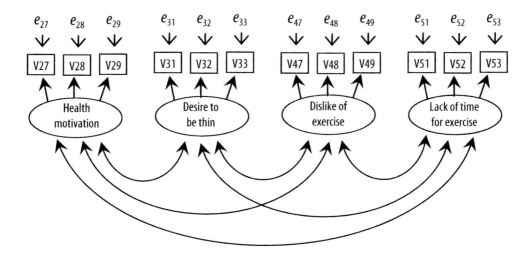

Figure 15.1. Path diagram illustrating the factor structure underlying the 12-item questionnaire.

Figure 15.1 presents a path diagram that illustrates the factor structure underlying the 12-item questionnaire. The figure uses symbols and conventions that are often used by researchers performing exploratory factor analysis, confirmatory factor analysis, and structural equation modeling. The remainder of this section explains the meaning of these symbols and conventions as they pertain to Dr. O'Day's investigation.

MANIFEST VARIABLES. First, the squares in the figure represent ***manifest variables***. These are the real-world variables that are actually provided by the participants and analyzed by the researcher. In the current analysis, each square represents one of the items on the 12-item questionnaire that was presented in Table 15.1. For example, the square labeled "V27" represents responses to questionnaire item 27: "It is important to me to be healthy." The square labeled "V28" represents responses to item 28: "I am willing to sacrifice in order to stay healthy." The remaining squares may be interpreted in the same way. The "V" in each name stands for "Variable."

LATENT FACTORS. Next, the ovals in the figure represent ***latent factors***: hypothetical constructs that are not observed in a direct manner, but which are assumed to exist because of the apparent effects that they exert on the manifest variables. The "effect" that a latent factor exerts on a manifest variable is represented by the straight singled-headed arrow that goes from an oval to a square.

In the figure, the first oval represents the *health motivation* latent factor. Imagine that there is some invisible construct within the participants—some hidden disposition that reflects the importance that individuals place on staying healthy and avoiding illness. If there were such a disposition, it would probably exert a strong influence on the way that people respond to questionnaire items such as "It is important to me to be healthy." If researchers could identify such an invisible construct, they would probably give it a name like *health motivation*.

That has been done in Figure 15.1. You can see that the first oval represents the health motivation latent factor. Straight single-headed-arrows go from this oval to the squares for Item 27, Item 28, and Item 29. These arrows reflect the prediction that responses to these three items are influenced by the relative standing of participants on the underlying factor of health motivation.

The second oval in Figure 15.1 represents a latent factor labeled *the desire to be thin*. Imagine that there is some separate invisible construct within the participants, and this construct influences the way that they respond to items such as "It is important to me to be thin." (Item 31 on the questionnaire). That is why there is an arrow going from the second oval to the squares labeled "V31," "V32," and "V33."

The remaining ovals may be interpreted in the same way. There are four ovals in Figure 15.1, reflecting the idea that responses to the 12-item questionnaire are influenced by four underlying factors. Those of you with good memories have noticed that these are the four constructs that constitute Dr. O'Day's *four-factor model of exercise behavior*—a fictitious mini-theory featured in many of the other chapters in this book. In the current chapter, we travel back in time to the investigation in which she first developed this four-factor model.

Some additional notes about the conventions used in Figure 15.1:

- Straight, single-headed arrows go from a given oval to just three squares. This reflects two simultaneous predictions: (a) each latent factor is measured by three items on the questionnaire, and (b) no questionnaire item is influenced by more than one common latent factor. This is a relatively uncomplicated factor model, and researchers typically hope to find relatively simple models such as this. In practice, however, actual factor models can certainly be more complex. An example would be a factor model in which some or all of the manifest variables are influenced by more than one latent factor. Such a model could still be analyzed using EFA—it would just be a bit more challenging to interpret.

- The latent factors in an EFA are sometimes called **_common factors_**. This term reflects the fact that a group of manifest variables usually have some latent factor in common. For example, Figure 15.1 shows that V27, V28, and V29 have the health motivation factor in common; V31, V32, and V33 have the desire to be thin factor in common, and so forth.

- Each square in the figure is actually affected by two sources. First, an arrow goes from an oval to a given square, representing the effect that a common factor has on that manifest variable. In addition, a smaller arrow goes from an *e* term to a corresponding manifest variable. For example, an arrow goes from e_{27} to V27 and a separate arrow goes from e_{28} to V28. These *e* terms represent **_unique factors_**: constructs and sources of error that are specific to a given manifest variable. Think about it this way: Scores on V27 (the questionnaire item "It is important to me to be healthy.") are influenced, to some extent, by participants' standing on the latent factor *health motivation* (represented by the oval). However, not all of the variance in V27 is accounted for by this common factor. The term e_{27} represents the other sources of variability (including error) that account for all of the remaining variance in V27.

- The curved two-headed arrows that connect each oval to every other oval in Figure 15.1 represent the **_inter-factor correlations_**: correlations between the latent factors. Certain types of EFA allow researchers to determine the sign and size of the correlations between the latent factors.

DISCOVERING THE UNDERLYING FACTOR STRUCTURE. Remember that, prior to conducting the EFA, Dr. O'Day *does not know* that responses to the 12-item questionnaire are due to the underlying factor structure displayed in Figure 15.1. Exploratory factor analysis allows her to *discover* this underlying factor structure. In other words, it allows her to discover that there are four factors underlying the questionnaire; that one of the factors influences responses to items 27, 28, and 29; that a second factor influences responses to items 31, 32, and 33; and so forth.

Exploratory Versus Confirmatory Factor Analysis

Researchers sometimes use exploratory factor analysis in situations in which they already have a clear hypothesis about the factor structure underlying a dataset. In these situations, they may hope to use EFA to "confirm" their hypothesis. Some experts (e.g., Obsborne, Costello, & Kellow, 2008) argue that using exploratory factor analysis for this purpose is not a great idea.

These authors argue that exploratory factor analysis should be used in *exploratory* situations. These are situations in which the researcher does not have a strong hypothesis about the number of factors that underlie the dataset, which questionnaire items are measuring which factors, and so forth. They argue that EFA is well-suited for *discovering* a factor structure, not confirming it.

In those situations in which the researchers already have clear hypotheses regarding the factor structure underlying a dataset, experts typically argue that the data should be analyzed using *confirmatory factor analysis* (CFA), not exploratory factor analysis. Confirmatory factor analysis produces a large number of ***fit indices:*** statistics that represent the goodness of fit between the theoretical factor model and the relationships that are actually observed in the sample data. These indices go by abbreviations such as the NFI, the SRMR, the CFI, and the RMSEA. Fit indices tell the researcher whether the theoretical model appears to be more or less "correct," given the actual relationships observed in the sample data. Most approaches to exploratory factor analysis do not produce fit indices of this sort.

This book covers confirmatory factor analysis in *Chapter 17*. There, you will learn about the fit indices listed above as well as many other aspects of CFA that make it more appropriate (compared to EFA) for testing hypotheses about factor structure.

Results from the Investigation

To recap: Although the questionnaire presented in Table 15.1 consists of 12 items, Dr. O'Day believed that it was actually measuring a smaller number of underlying factors. To discover this factor structure, she performed an exploratory factor analysis and prepared an article in which she summarized her analyses and results. The remainder of this chapter presents excerpts from this article, interspersed with relevant explanations.

Descriptive Statistics for the Questionnaire Items

When researchers perform an exploratory factor analysis, the *Results* section usually begins with a description of the manifest variables being analyzed. For the current article, Dr. O'Day would probably provide the actual questionnaire items as they were presented in Table 15.1.

Articles should typically include a table that presents means and standard deviations for the manifest variables. If space permits, this table should also include all possible correlations between the manifest variables, as this will make it possible for other researchers to re-analyze the data, possibly using different statistical procedures. If space does not permit, the correlation matrix (or raw data) should be made available to researchers in some other way, possibly at the journal's web site.

To save space, this chapter will omit the article excerpt in which Dr. O'Day presents the means, standard deviations, and correlations for her 12 manifest variables. It will also omit the excerpt in which she explains how she inspected her data to verify that it satisfied the assumptions underlying factor analysis (these assumptions are listed in the "Additional Issues" section at the end of the chapter). For guidance on how the assumptions for EFA can be assessed using SPSS, see Field (2009a). Instead, the current chapter will begin where the factor analysis itself begins: with the initial extraction of the factors.

Initial Extraction of the Factors

The first step in an exploratory factor analysis is called the *initial extraction of the factors*. In this context, the word *extraction* means *creation*.

Understanding this process of extraction will be easier if we first understand the definition of a factor. There are two conceptual ways of thinking about the factors that are studied in an exploratory factor analysis:

1) Earlier, you learned that a factor is an *unseen latent variable* that influences responses to the questionnaire items. From this perspective, factors are the underlying constructs represented as ovals in Figure 15.1.

2) For purposes of actually performing the factor analysis, a factor can also be viewed as a *variate*—a *synthetic variable* created by the statistical application.

For the next little while, we will focus on the second of these two definitions: We will think of a factor as being a synthetic variable created by the statistical application. In *Chapter 7: Bivariate Regression*, you learned that a **synthetic variable** (i.e., a **variate**) is an artificial variable, typically created by applying optimal weights to one or more manifest variables. Synthetic or not, a variate created in EFA is a quantitative variable, and this means that each participant has a score on it such as +0.19 or -1.20. The next few sections describe the nature of the variates (i.e., factors) created in EFA, and provide a conceptual explanation of just how they are created.

EXTRACTION OF THE FIRST FACTOR. Exploratory factor analysis is actually performed by a statistical application using matrix algebra. It is useful to think of this application as creating the factors one at a time (technically, this is not exactly the way that it is done, but thinking of it in this way will make the major concepts easier to grasp).

In creating the first factor (Factor 1), the matrix-algebra formula used by the application attempts to create a synthetic variable that will account for as much of the common variance in the dataset's manifest variables as is possible. The manifest variables, in this case, are responses to the 12 questionnaire items. In other words, the matrix-algebra formula is designed to create a factor that will be strongly correlated with responses to some subset of items on the questionnaire.

In the current analysis, the formula accomplishes this by creating a factor that is strongly correlated with the following three questionnaire items:

```
51. I am too busy to engage in frequent aerobic exercise.
52. I am under too much time pressure to engage in frequent aerobic exercise.
53. With my busy schedule, there is very little time for frequent aerobic
    exercise.
```

Notice that all three of these items deal with the issue of *time pressure*—people who agree with these items believe that they are under too much time pressure to engage in aerobic exercise. In order to account for the variance in these items, the statistical application created a factor that we might label the *lack of time for exercise* factor. This factor is a synthetic variable, and each participant will have a quantitative score on it. Scores on this synthetic variable will display strong correlations with responses to the preceding three items. Later in the chapter, we will see the formula that the statistical application used to estimate participants' scores on this factor.

EXTRACTION OF THE SECOND FACTOR. The statistical application is now ready to create Factor 2 for Dr. O'Day's analysis. Factor 2 will be created so as to satisfy two criteria:

- Factor 2 must be uncorrelated with Factor 1.

- Factor 2 must account for as much of the common variance in the questionnaire items as is possible (as long as it remains uncorrelated with Factor 1).

To achieve this, there is no point in creating a factor that is strongly correlated with the three questionnaire items that deal with lack of time for exercise—Factor 1 has already accounted for that variance. Instead, the statistical application creates Factor 2 so that it will be strongly correlated with a different subset of questionnaire items—the subset presented below:

```
47. I hate to engage in aerobic exercise.
48. I find aerobic exercise to be unpleasant.
49. I find aerobic exercise to be punishing.
```

These three items all deal with the negative emotional association that many people have with physical exercise. Therefore, it might make sense to call Factor 2 the *dislike of exercise* factor.

Remember that Factor 2 is also a synthetic variable, and every participant will have a quantitative score on it. Scores on Factor 2 will:

- be uncorrelated with scores on Factor 1, and

- display strong correlations with questionnaire items 47, 48, and 49 (presented above).

TWO CHARACTERISTICS OF THE FACTORS INITIALLY EXTRACTED. And so it goes. The statistical application continues extracting factors in this way. During the initial extraction, each factor will display the following two characteristics:

- Each factor will be uncorrelated with the factors that preceded it.

- Each factor will account for a maximal amount of common variance in the data set that has not already been accounted for by the factors that preceded it.

NUMBER OF FACTORS INITIALLY EXTRACTED. So far this section has discussed the first two factors created in the course of Dr. O'Day's EFA. But the statistical application does not stop at just two factors. It continues extracting factors—with each new factor displaying the two characteristics described above—until it reaches its limit, as described below:

> The number of factors initially extracted in an exploratory factor analysis is always equal to the number of manifest variables being analyzed.

Dr. O'Day is performing a factor analysis of 12 questionnaire items. This means that 12 factors will be created in the initial extraction. If she were performing a factor analysis of a 50-item questionnaire, then 50 factors would be initially extracted.

This fact should surprise you. After all, one purpose of an EFA is to reduce a large number of manifest variables to a smaller number of factors. What good is it if the number of factors extracted is equal to the number of manifest variables that you started with?

Not to worry. By the end of the analysis, the number of factors retained in the analysis will be smaller than the number of manifest variables being analyzed. This is because, in the typical EFA, only the first few factors will prove to be meaningful factors. Therefore, only these first few factors will be retained and interpreted. The later (unimportant) factors will be disregarded. In one factor analysis, the researcher might conclude that the first six factors are meaningful. In another factor analysis, the researcher might determine that only one factor is meaningful.

Researchers use a variety of strategies for determining the number of factors that should be retained and interpreted. Those criteria will be described later in the chapter.

METHODS FOR EXTRACTING FACTORS. Popular statistical applications such as SAS and SPSS typically offer a variety of methods that can be used to extract (create) the factors in a factor analysis. These methods have names such as *unweighted least squares, alpha-factoring*, and *image-factoring*.

The *maximum likelihood* method (abbreviation: *ML*) is a generally well-regarded procedure for extracting factors (Osbourne et al., 2008). In part, this is because ML typically produces accurate estimates for factor loadings and other statistics, and in part it is because ML produces a significance test that can help researchers decide how many factors to retain and interpret. A potential downside to the ML method is the fact that it requires that somewhat more restrictive assumptions be met, such as multivariate normality in the manifest variables. When data do not satisfy this assumption, Osbourne et al. (2008) recommends *principal axis factoring* (abbreviation: *PAF*) as a possible alternative.

Many researchers use a procedure called the *principal components* method of extraction. You should be aware that this method results in a *principal components analysis* (PCA), not an exploratory factor analysis (EFA). Although PCA is fine for reducing a large number of manifest variables into a smaller number of synthetic variables called *principal components*, it is not capable of discovering the factor structure that underlies a dataset, and therefore should really not be called *factor analysis*. The "Additional Issues" section near the end of this chapter explains the reasons for this, and discusses the similarities and differences between the two procedures.

EXTRACTION METHOD USED BY DR. O'DAY. In the following excerpt, Dr. O'Day describes the method used to extract factors in her EFA. This excerpt would have appeared early in the *Results* section, shortly after the descriptive statistics.

> **Initial extraction of factors.** Exploratory factor analysis was performed on responses to the 12 questionnaire items listed in Table 15.1. In the analysis, squared multiple correlations served as prior communality estimates, the principal axis factoring method was used to extract the factors, and all analyses were performed on the correlation matrix rather than the covariance matrix.

COMMUNALITY. In the preceding paragraph, Dr. O'Day says that "squared multiple correlations served as prior communality estimates." **Communality** (symbol: h^2) is a characteristic of a manifest variable: it is the proportion of variance in that variable that is accounted for by the common factors in the analysis. In general, the communality for a given variable should range from .00 to 1.00. For example, if the communality for Item 27 from Dr. O'Day's study were $h^2 = .60$, it would mean that 60% of the variance in responses to this item was accounted for by the underlying factors in the analysis. A manifest variable with a large communality is typically viewed as being an important variable in the analysis—the large communality is evidence that the manifest variable must be measuring at least one of the underlying factors.

The actual communality for a given manifest variable in the population is unknown, and therefore must be estimated from sample data. In fact, we estimate the communality of a given item in two different ways at two different points in an exploratory factor analysis. At the very beginning of a factor analysis, we compute *prior communality estimates*, and at the end we compute *final communality estimates*.

First, a ***prior communality estimate*** is a tentative estimate of the manifest variable's communality—an estimate that is computed before the factor analysis is actually performed. Prior communality estimates are used to resolve a dilemma that researchers find themselves in at the very beginning of the factor analysis:

- They cannot estimate communalities very accurately until they have completed the factor analysis.
- But they cannot *perform* the factor analysis unless they have communality estimates.

The way that most researchers deal with this problem is by using *squared multiple correlations* (*SMCs*) as prior communality estimates. This involves performing a multiple regression in which a specific questionnaire item serves as the criterion variable, and the remaining questionnaire items serve as the predictor variables. The resulting R^2 statistic (or SMC) is the prior communality estimate for that questionnaire item. The process is then repeated for the remaining questionnaire items (with each item taking its turn as the criterion variable in the regression analysis). The resulting SMCs serve as the prior communality estimates for the manifest variables, and the researcher may now perform the factor analysis.

At the end of the analysis, a different type of communality estimate is computed. A ***final communality estimate*** is the proportion of variance in a manifest variable that is accounted for by the retained common factors. Final communality estimates are computed as the sum of that item's squared factor loadings for the retained factors on the original unrotated factor pattern (you will learn about factor loadings and factor patterns later in this chapter). The final communality estimate is the researcher's best estimate for the manifest variable's actual communality in the population and is often reported in the published article.

Determining the Number of Factors To Retain

An earlier section indicated that the number of factors initially extracted in a factor analysis is always equal to the number of manifest variables being analyzed. Researchers typically find that the first few factors are important (i.e., they account for a meaningful amount of common variance), whereas the latter factors are unimportant (i.e., they appear to account for little more than error variance). Researchers usually want to retain and interpret only those factors that are important and meaningful. Therefore, they need criteria for deciding how many factors to retain.

Statisticians and researchers have developed a variety of criteria for this purpose. No single method has been universally embraced as the one that always results in a "correct" decision; instead, most researchers use a combination of methods. This section discusses some of the more widely-used alternatives.

EIGENVALUES IN FACTOR ANALYSIS. Most of the criteria used to answer the number-of-factors question involve the use of *eigenvalues*. An ***eigenvalue*** is a measure of the variance that is accounted for by an individual factor. The size of the eigenvalue for a given factor tells us how important that factor is. Factors with large eigenvalues are accounting for a lot of the variance in a dataset and are therefore "important."

Table 15.2 presents the eigenvalues for each factor extracted in Dr. O'Day's analysis. The table shows that Factor 1 displayed an eigenvalue of 4.06, Factor 2 displayed an eigenvalue of 2.10, and so forth. The table shows that the first few factors have large eigenvalues, and the eigenvalues for each succeeding factor become progressively smaller.

Table 15.2

Initial Eigenvalues from Exploratory Factor Analysis of 12 Questionnaire Items

Factor	Eigenvalue
1	4.06
2	2.10
3	1.38
4	1.07
5	0.02
6	-0.01
7	-0.04
8	-0.06
9	-0.09
10	-0.11
11	-0.14
12	-0.20

Note. N = 238.

WHY SOME EIGENVALUES ARE NEGATIVE. Some of the factors in Table 15.2 displayed negative eigenvalues (i.e., Factors 6-12). You may be wondering how an eigenvalue could assume a negative value. After all, eigenvalues are supposed to represent *variance*, and we generally think of variances as assuming only positive values.

The answer involves the fact that the prior communality estimates included in the factor analysis were not the true communalities for the items, but were instead estimates of the communalities—estimates that contained some amount of error. If the prior communality estimates had been the true values for the population communalities, then the first few factors in Dr. O'Day's factor analysis would have displayed relatively large positive eigenvalues, and the remaining factors would have displayed eigenvalues of zero. Dr. O'Day would have retained and interpreted the first factors, and would have disregarded the factors with eigenvalues of zero. The fact that the prior communality estimates contained error resulted in some counter-intuitive statistical results such as the negative eigenvalues observed in the table. A more detailed explanation of this phenomenon is provided in Hatcher (1994, p. 85).

THE EIGENVALUE-1 CRITERION. Back to the issue at hand: deciding how many factors to retain. Many researchers follow the strategy of retaining and interpreting any factor that displays an eigenvalue equal to or greater than 1.00. This approach is called the *eigenvalue-1 criterion*, the *Kaiser criterion*, the *Kaiser rule*, *Kaiser-Guttman* rule, or just the *K1 criterion* (Kaiser, 1970; Rummel, 1970). The eigenvalue-1 criterion is discussed here because it is popular, not because it is a best practice. In fact, it is a pretty absurd practice to use as part of an exploratory factor analysis. Technically, the eigenvalue-1 criterion was developed for use with *principal component analysis* (*PCA*)—a statistical procedure that many people confuse with factor analysis. The differences between PCA and EFA are briefly discussed in the "Additional Issues" section at the end of this chapter.

The ***eigenvalue-1 criterion*** says that the researcher should retain and interpret any factor (or principal component) with an eigenvalue of 1.00 or higher. Table 15.2 shows that Factors 1 through 4 display eigenvalues greater than 1.00; therefore, if Dr. O'Day were to apply the eigenvalue-1 criterion, she would retain and interpret the first four factors.

But Dr. O'Day should not apply the eigenvalue-1 criterion in this analysis. There are two reasons:

1) She is performing an exploratory factor analysis, and the eigenvalue-1 criterion is appropriate for principal component analyses, not EFA.

2) If she were conducting a PCA, she *still* should not use the eigenvalue-1 criterion. Studies that have systematically compared the eigenvalue-1 criterion against other procedures typically conclude that the eigenvalue-1 criterion is unreliable, often results in retaining the wrong number of components, and generally performs worse than alternative methods (Cortina, 2002; Velicer, Eaton, & Fava, 2000; Zwick & Velicer, 1986).

Fortunately, there are better approaches for solving the number-of-factors problem in an EFA. One of these is the *scree test*, and it is discussed next.

THE SCREE TEST. A *scree test* involves visually inspecting a plot of eigenvalues so as to locate the "break" between (a) the factors with large and meaningful eigenvalues versus (b) those with relatively small and trivial eigenvalues. The factors that appear to the left of the break are then retained and interpreted.

A scree test begins by creating a ***scree plot***: a figure that presents *factor numbers* on the horizontal axis and the *eigenvalues* for the factors on the vertical axis. This is done in Figure 15.2, which displays the same eigenvalues that were presented earlier in Table 15.2. The plot shows that the eigenvalue for Factor 1 was 4.06, the eigenvalue for Factor 2 was 2.10, and so forth for the remaining factors.

Figure 15.2 shows that the eigenvalue for Factor 1 was relatively large, indicating that Factor 1 accounted for a large and meaningful amount of common variance. The eigenvalue for Factor 2 was not quite as large, but was probably still large enough to be retained and interpreted. As the line moves to the right, it drops lower and lower, indicating that each successive factor accounts for less and less variance. On the right side of the graph, the line becomes essentially flat, indicating that those factors account for very little variance (probably just error variance).

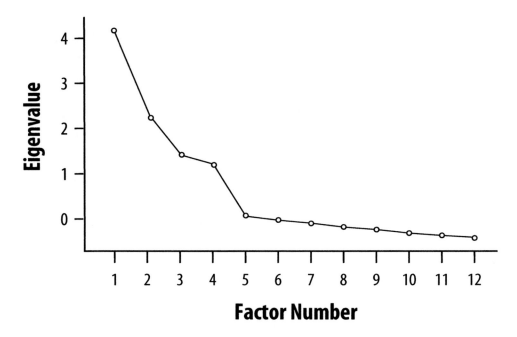

Figure 15.2. Scree plot from factor analysis of 12-item questionnaire.

To perform a ***scree test***, the researcher:

- Looks for the last big "break" in the scree plot—the last large gap before the line levels off and becomes relatively flat.
- The researcher then retains and interprets those factors whose eigenvalues came *before* this break; the researcher disregards the factors whose eigenvalues came after the break.

Figure 15.2 shows that Factors 1-4 display fairly large eigenvalues, and that there is a relatively large break between Factor 4 and Factor 5. The precise values in Table 15.2 show that the eigenvalue for Factor 4 was relatively large at 1.07, but the eigenvalue for Factor 5 was substantially smaller at 0.02. So there was a fairly large gap between Factor 4 and Factor 5.

On the other hand, there was not much of a gap between the eigenvalue for Factor 5 (0.02) versus the eigenvalue for Factor 6 (–0.01). This tells us that Factor 5 is part of the relatively flat part of the scree plot. Factors that are part of the "flat line" are viewed as being trivial factors that should not be retained.

In short, the scree test tells Dr. O'Day that she should retain Factors 1-4. These are the "meaningful" factors that she will rotate and interpret.

As subjective as all this might sound, it actually appears to work. Simulation studies show that the scree test performs fairly well when compared to alternative procedures for solving the number-of-factors problem (e.g., Zwick & Velicer, 1986).

Dr. O'Day describes her scree test below:

> A scree test (Cattell, 1966; Rummel, 1970) was performed by creating a plot of eigenvalues for the 12 factors. This scree plot is presented as Figure 15.2. The plot reveals a large break between Factor 4 and Factor 5, with the eigenvalues leveling off beyond the break. This was taken as evidence that Factors 1-4 were meaningful factors, and Factors 5-12 were measuring little other than error variance.

In case you were wondering: The word *scree* is an ancient Scandinavian term for the loose rocks and rubble that collect at the bottom of a cliff. You can see why it is a pretty good name for the graph in Figure 15.2. The break between Factor 4 and Factor 5 represent the edge of the cliff, and relatively small eigenvalues for Factors 5 through 12 represent the *scree*—the unimportant rubble that accumulate below.

THE INTERPRETABILITY CRITERIA. Perhaps the most important criterion that Dr. O'Day will use in making her number-of-factors decision is the *interpretability criterion*. When researchers use the ***interpretability criterion***, they determine whether the factor solution makes sense—whether the manifest variables appear to cluster together in ways that seem logical and reasonable, given constructs that are being measured.

For example, imagine that (in a totally different fictitious investigation) a researcher creates a 10-item instrument to obtain student evaluations of college teachers. The instrument contains items such as "The teacher treated students with respect," and students responded to each item using a 7-point response format that ranged from 1 = *Disagree Strongly* to 7 = *Agree Strongly*. The researcher obtained evaluations of a large number of college teachers, and analyzed the results using exploratory factor analysis.

Below is a description of one hypothetical outcome for this study. The following is a best-case scenario for the analysis: If the researcher obtained the results described here, most readers would agree that the factor solution scored high on the interpretability criterion:

1) *Items measuring the same factor obviously share the same conceptual theme.* Imagine that the analysis suggests that the 10 items on the questionnaire are actually measuring just two factors, with questionnaire items 1-5 measuring Factor 1 and questionnaire items 6-10 measuring Factor 2. If the results of the factor analysis indicate that questionnaire items 1-5 are all measuring the same factor, then it should be obvious to most readers that these five items have something in common. When people read the questionnaire items, they should see a clear conceptual theme running through them, tying them together. For example, imagine that Factor 1 consists of items such as "The teacher treated students with respect" and "The teacher was kind to students." These two items share the same conceptual theme—a theme that we might label *teacher's respect*

for students. If all five of the items measuring Factor 1 obviously share this same theme, then Factor 1 is said to be *interpretable*.

2) *Groups of items measuring different factors are obviously different in terms of conceptual theme.* Again, imagine that questionnaire items 1-5 are measuring Factor 1 and items 6-10 are measuring Factor 2. If this were the case, then it should be obvious to anyone reading the items that there are clear thematic differences between the two sets of items. In the preceding paragraph, we established that questionnaire items 1-5 all share a theme that we might label *teacher's respect for students*. In contrast, imagine that items 6-10 (which measure Factor 2) consist of statements such as "The teacher was knowledgeable regarding the topics being taught" and "The teacher was familiar with current research in the area." These two items share a conceptual theme that we might call *teacher's knowledge regarding course content*. This conceptual theme for Factor 2 (knowledge regarding course content) is obviously distinct from the conceptual theme for Factor 1 (respect for students). These findings further support the argument that this two-factor solution scores high on the interpretability criterion.

With the above paragraphs, we are getting ahead of ourselves a bit, as we have not yet discussed the procedure that researchers follow when they "interpret a solution." That procedure is explained in upcoming sections that show how researchers interpret the factor loadings that appear in a pattern matrix and a structure matrix. Nevertheless, you should still get the major point: A factor solution satisfies the interpretability criterion if the results seem reasonable and logical in terms of what is already known about the variables being studied.

Consider how Dr. O'Day might apply the interpretability criterion in the current analysis. Up until this point, her analyses have suggested that she should probably retain and interpret a four-factor solution (i.e., Factors 1-4). To be sure that this is the best solution, Dr. O'Day will compare it to two alternatives. She will print out the complete results for a three-factor solution, a four-factor solution, and a five-factor solution. She will carefully review each of these, determining which solution makes the most sense in terms of what she already knows about the constructs being measured. In the end, her "preferred solution" will be the one that is the most interpretable.

Factor Loadings, Factor Matrices, and Factor Rotations

Over the next few sections, you will see how Dr. O'Day actually interprets a factor solution. The process involves:

- Making a tentative decision regarding the number of factors to retain. In this example, Dr. O'Day will tentatively decide on a four-factor solution.

- Reviewing the questionnaire items that have relatively large factor loadings for a given factor to see what these items have in common. This makes it possible for Dr. O'Day to come up with meaningful names for each of the factors.

- Stepping back to simultaneously take in all of the major results from analysis: the factor pattern, the factor structure, and the inter-factor correlations. This provides the "big picture" regarding the relationships between the constructs being measured.

FACTOR LOADINGS. The basic building block of this process is a *factor loading*: a statistic that represents the nature of the relationship between a factor and a manifest variable (such as a questionnaire item). Factor loadings are actually different statistics in different matrices. The factor loadings in factor pattern matrices are multiple regression coefficients, and the factor loadings in factor structure matrices are bivariate (Pearson) correlation coefficients. Factor loadings generally range in size from -1.00 through 0.00 through $+1.00$ (although there are exceptions to this rule).

Factor loadings are important because they are used to determine which manifest variables are measuring which factors. For a given factor, if the loading for a manifest variable is relatively large in absolute value, that item is assumed to be measuring that factor. When a manifest variable is measuring a factor, we say that it is *loading* on that factor.

How large does a factor loading have to be to be considered meaningful? Popular criteria are in the range of $\pm.30$, $\pm.35$, and $\pm.40$ (Obsborne et al., 2008). For example, in a given investigation, the researcher might decide that, if the factor loading for a given questionnaire item exceeds $\pm.40$ (in absolute value), it will be taken as evidence that this item is in fact loading on this factor. In the social and behavioral sciences, the value of $\pm.40$ may be the most widely-used criterion.

FACTOR MATRICES. An important step in a factor analysis is the interpretation of the factor matrices. A *factor matrix* is a table that presents factor loadings for the relationships between factors and manifest variables. Eventually, Dr. O'Day will review a factor matrix to determine which questionnaire items are measuring Factor 1, which items are measuring Factor 2, and which items are measuring each of the remaining factors. Through this process, she will arrive at meaningful labels for each factor.

TWO TYPES OF FACTOR ROTATION. A number of different factor matrices are created in the course of a factor analysis. The first of these—the *initial factor matrix*—is seldom interpreted. This is because the initial factor matrix is typically difficult to interpret—it typically makes it hard for the researcher to see which items that are actually measuring a given factor. To address this problem, researchers almost always perform a mathematical transformation called a *factor rotation* as part of an EFA.

A *factor rotation* is a linear transformation performed on a factor matrix for the purpose of achieving simple structure. For the moment, we can informally define *simple structure* as *easy to interpret* (we will see a more comprehensive definition a bit later). When a factor matrix displays simple structure, the relative size of the factor loadings make it easy to determine which items are measuring Factor 1, which items are measuring Factor 2, and so forth. When researchers perform a rotation, it transforms the factor loadings so that (in most cases), the large loadings remain large and the small loadings become smaller. Because of this, it becomes easier to discriminate between the items that are measuring a factor versus those that are not.

Statistical applications offer a wide variety of rotations to be used with factor analysis, and these rotations have exotic names like *direct oblimin* and *varimax*. As a reader of research, you do not need to be terribly familiar with the mathematical differences between these methods. You should, however, know that almost all of these procedures fall into one of two categories: orthogonal rotation methods versus oblique rotation methods. And, yes, you should understand the conceptual difference between these two categories.

First, **orthogonal rotations** result in **orthogonal factors**: factors that are uncorrelated with one another (in statistics, the word *orthogonal* means *uncorrelated*). If you perform an orthogonal rotation and then compute Pearson correlations between the resulting factors, each correlation will be equal to zero. Popular orthogonal rotation methods have names such as equimax, parsimax, quartimax, and varimax. The most widely-used approach is probably the varimax rotation.

Second, **oblique rotations** result in **oblique factors:** factors that are allowed to be correlated with one another. An oblique rotation does not *force* factors to be correlated. However, if the factors represent constructs that are naturally correlated with one another in the real world, an oblique rotation *allows* the corresponding factors from the EFA to be correlated as well. Oblique rotation methods include the following: biquartimin, covarimin, direct oblimin, direct quartimin, maxplan, promax, oblimax, and procrustes. In published research, direct oblimin and promax rotations seem to be especially popular.

The researcher must decide whether to use an orthogonal versus an oblique rotation, and the choice will have important implications for the results of the analysis. Here are some of the advantages and disadvantages of the two approaches:

- Orthogonal rotations are typically easier to interpret because they produce just one factor matrix. However, orthogonal rotations (by definition) produce factors that are forced to be uncorrelated with one another. This means that, in many situations, orthogonal factor solutions will provide a poor fit to the data, especially when the corresponding constructs in the real world are actually correlated. In the social and behavioral sciences, it is unusual to find naturally occurring constructs that are completely uncorrelated, so there is a danger that orthogonal rotations may place unrealistic constraints on the analysis, possibly distorting the results.

- Oblique rotations allow factors to be correlated, so most experts view them as being more realistic. If the constructs being investigated in a factor analysis are correlated with one another in the real world, the corresponding factors produced by an oblique rotation will be correlated as well. In those situations, an oblique rotation typically results in a better fit to the real-world data, compared to an orthogonal rotation. But this improved fit comes at a price: Oblique rotations can be a bit more complex to interpret. To interpret an oblique rotation, the researcher must interpret two different matrices of factor loadings, whereas an orthogonal rotation requires just one.

DR. O'DAY'S FACTOR ROTATION. For what it is worth, I always advise people to bite the bullet and go with the oblique rotation. And since she almost always takes my advice, that is what Dr. O'Day chooses to do. She describes this rotation in the following excerpt:

> **Rotated factor matrices.** A promax (oblique) rotation was used to achieve simple structure in the matrices of factor loadings. A given variable (questionnaire item) was said to load on a factor if the factor loading was ≥ ±.40 for that variable. In the table, asterisks identify loadings that are equal to or greater than ±.40. In the pattern matrix, three items loaded on each of the four retained factors, and none of the 12 items loaded on more than one factor.

Interpreting the Factor Pattern and Factor Structure

Soon we will see the procedure that Dr. O'Day follows when interpreting a matrix of factor loadings. First, however, this section will explain the differences between the two matrices to be reviewed: the factor pattern matrix versus the factor structure matrix.

THE FACTOR PATTERN MATRIX. In a *factor pattern matrix*, the factor loadings are standardized multiple regression coefficients, similar to the beta weights (i.e., β weights) obtained in multiple regression. More specifically, the factor loadings for a given manifest variable in the factor pattern matrix represent the standardized multiple regression coefficients that would be obtained from a multiple regression analysis in which the manifest variable (such as a questionnaire item) serves as the criterion variable, and the retained factors serve as the predictor variables. If the factor loading for a given factor is relatively large, that factor is viewed as being an important predictor of the manifest variable.

When performing an oblique rotation, the primary interpretation of the solution should be based on this factor pattern matrix. In part, this is because the pattern matrix is usually the easiest matrix to understand—it usually provides the clearest picture as to which manifest variable is measuring which latent factor (Rummel, 1970). Table 15.3 presents the pattern matrix for Dr. O'Day's current factor analysis.

Table 15.3

Promax-Rotated Factor Pattern for the 12 Questionnaire Items, Decimals Omitted (Factor Loadings Are Standardized Multiple Regression Coefficients)

	Factor				
1	2	3	4		Questionnaire item
03	03	-02	79*	27.	It is important to me to be healthy.
-05	02	-01	81*	28.	I am willing to sacrifice in order to stay healthy.
03	-09	04	72*	29.	I am willing to work hard in order to avoid medical problems.
04	-03	85*	-01	31.	It is important to me to be thin.
04	02	82*	-06	32.	I very much want to have a fairly low body weight.
-10	02	68*	11	33.	It is important to me to have a lean physical appearance.
04	85*	-03	-04	47.	I *hate* to engage in aerobic exercise.
00	94*	00	02	48.	I find aerobic exercise to be unpleasant.
00	83*	03	00	49.	I find aerobic exercise to be punishing.
92*	01	-01	-02	51.	I am too busy to engage in frequent aerobic exercise.
88*	08	03	03	52.	I am under too much time pressure to engage in frequent aerobic exercise.
95*	-05	-02	00	53.	With my busy schedule, there is very little time for frequent aerobic exercise.

Note. N = 238. An asterisk (*) flags any loading $\geq .\pm 40$.

Under the general heading "Factor," Table 15.3 provides the factor loadings for Factors 1, 2, 3, and 4. Consider just the first row running horizontally at the top of the table. This row provides factor loadings for questionnaire Item 27: "It is important to me to be healthy." The factor loadings in this row are the standardized multiple regression coefficients obtained from a multiple regression analysis in which Item 27 (serving as the criterion variable) is regressed on Factors 1-4 (serving as the predictor variables). The multiple regression equation might be presented in this way:

$$V27 = .03(F1) + .03(F2) + -.02(F3) + .79(F4)$$

In the preceding equation, "V27" represents scores on questionnaire Item 27, "F1" represents scores on Factor 1, "F2" represents scores on Factor 2, and so forth.

The numbers next to the "F" symbols are standardized multiple regression coefficients. The preceding multiple regression equation shows that, in order to compute a given participant's observed score on questionnaire item 27 (in standardized form), we must:

- Multiply the participant's standardized score on Factor 1 by .03,
- multiply the participant's standardized score on Factor 2 by .03,
- multiply the participant's standardized score on Factor 3 by –.02,
- multiply the participant's standardized score on Factor 4 by .79,
- and then sum these products.

Notice that Factor 4 was given the most weight in determining a given participant's score on questionnaire Item 27. This tells us that questionnaire item 27 is measuring Factor 4.

The preceding indicated that the factor loadings in the first row of Table 15.3 may be interpreted as multiple regression coefficients from an analysis in which questionnaire Item 27 serves as the criterion variable, and Factors 1, 2, 3, and 4 serve as the predictor variables. The factor loadings in each of the remaining rows may be interpreted in a similar way: In each of the remaining rows, a different questionnaire item serves as the criterion in a multiple regression analysis, and the four factors serve as predictors.

You will recall that a multiple regression coefficient is a *partial* regression coefficient—it represents the nature of the relationship between one predictor variable and the criterion variable while statistically controlling for the other predictor variables in the equation. This notion of statistical control is what separates the factor loadings in the pattern matrix (presented above) from the factor loadings in the structure matrix (to be presented a bit later).

INTERPRETING THE PATTERN MATRIX. We are finally ready to interpret the factor solution (look alive now—this step is the main reason we *did* the EFA). In general, this involves coming up with a meaningful name for each of the retained factors. In doing this, we will focus primarily on the factor pattern matrix.

To interpret a given factor, we must identify the questionnaire items that load on that factor (i.e., the questionnaire items that display large loadings for that factor). In a previous excerpt, Dr. O'Day indicated that she would use the criterion of ±.40: She would conclude that a given item loads on a factor if its factor loading was equal to or greater than .40 in absolute value. This is a fairly typical criterion for assessing factor loadings.

Step by step, this is how we will interpret the factor pattern:

- First, focus on just the column of factor loadings for Factor 1 (the first vertical column of factor loadings, running up and down). Which questionnaire items have high loadings for this factor (i.e., loadings ≥±.40)? To make this task easier, Dr. O'Day has already flagged these loadings with asterisks.

- Now read the questionnaire items themselves (the questionnaire items that you have just identified as loading on Factor 1). What do these items have in common? What common construct do they all seem to measure? What would be a good name for this construct? Use that name as your label for Factor 1.

- Now go to Factor 2, and repeat this process.

- And then repeat the process for each of the remaining factors in the table.

When finished, you will have a tentative name for each of the four factors in the table. It is *tentative* at this point because you have not yet reviewed the structure matrix (to be covered below).

In the following excerpt from her article, Dr. O'Day describes how she interpreted the solution, based on this factor pattern matrix.

> **Interpretation of the factors.** Table 15.3 shows that items 51, 52, and 53 (near the bottom of the table) displayed loadings greater than or equal to ±.40 on Factor 1. Given the nature of these questionnaire items, Factor 1 was labeled *lack of time for exercise*. Items 47, 48, and 49 displayed large loadings for Factor 2. Given the nature of these items, Factor 2 was labeled *dislike of exercise*. Items 31, 32, and 33 loaded on Factor 3, which was labeled *desire to be thin*. Finally, items 27, 28, and 29 loaded on Factor 4, which was labeled *health motivation*.

SIMPLE STRUCTURE. An earlier section indicated that researchers perform factor rotations in the hope of achieving simple structure in the matrix of factor loadings. At that time, *simple structure* was informally defined *easy to interpret*. More specifically, when a matrix of factor loadings displays **simple structure**, it means that the matrix displays two characteristics: *simple columns* and *simple rows*. Table 15.3 displays simple structure, so these two characteristics can be explained in greater detail below by referring to this table:

- *Simple columns.* To achieve simple structure, the columns for a given factor (running vertically) should display high loadings for some of the manifest variables, but near-zero loadings for the remaining variables. Notice that this is true for Table 15.3: Factor 1 has high loadings for items 51, 52, and 53, but near-zero loadings for the remaining items. In the same way, the columns for the remaining factors display high loadings for three questionnaire items each, but near-zero loadings for the other items.

- *Simple rows.* To achieve simple structure, the row for a given manifest variable (running horizontally) should display a high loading for one factor but near-zero loadings for the remaining factors. Notice that this is true for Table 15.3: Questionnaire item 27 displays a high loading for Factor 4, but near-zero loadings for the remaining three factors. In the same way, the remaining 11 items display high loadings for one factor each, but near-zero loadings for the remaining factors.

Simple structure makes it easy to interpret the solution and label the factors. Because the factor pattern in Table 15.3 displays simple structure, it is easy to see that items 51, 52, and 53 measure Factor 1: items 47, 48, and 49 measure Factor 2, and so forth.

Researchers often use simple structure as one of their criteria for deciding how many factors to retain. For example, imagine that the scree plot created by Dr. O'Day for the current analysis was ambiguous—imagine that it did not clearly indicate whether she should retain three factors, four factors, or five factors. In such a situation, Dr. O'Day would probably print out the complete results (including factor patterns) for all three solutions. She would interpret the factor patterns and other results using the procedures described here. If only one of the solutions produced a factor pattern that displayed simple structure (and if only that solution satisfied the interpretability criteria), Dr. O'Day would probably adopt it as her final preferred solution.

CORRELATIONS BETWEEN LATENT FACTORS. Dr. O'Day has not finished interpreting the factor solution. Most important is the fact that she has not yet reviewed the factor structure matrix. However, that matrix will make more sense if she first reviews the ***inter-factor correlations:*** a matrix that contains all possible Pearson correlations between the latent factors extracted in the factor analysis.

In Figure 15.1 (presented much earlier in this chapter), four ovals represented the four latent factors underlying scores on the 12 questionnaire items. You may recall that these four ovals were connected to each other by a set of curved double-headed arrows. These curved double-headed arrows represent the inter-factor correlations. During the current step of the analysis, Dr. O'Day will review the sign and size of each of these correlations.

To put these correlations in context: All exploratory factor analyses begin with orthogonal factors (factors that are uncorrelated with one another). If the researcher requests an *orthogonal* rotation (such as a varimax rotation), the factors remain orthogonal throughout the analysis. In fact, when researchers request orthogonal rotations, most statistical applications do not produce a matrix of inter-factor correlations. Such a matrix is not needed because it is understood that all of the inter-factor correlations are equal to zero.

However, if the researcher requests an *oblique* rotation (such as the promax rotation that Dr. O'Day requested), the application typically does produce a matrix of inter-factor correlations. In most cases, these correlations will be some value other than zero.

An oblique rotation is actually conducted in a series of steps. At the beginning of the analysis, the factors that are created in the initial extraction are orthogonal (uncorrelated). However, if the researcher has requested an oblique rotation, the application then relaxes the orthogonality of the factors during the rotation itself. This means that the application allows the factors to become correlated if the corresponding constructs are actually correlated in the real world. By the end of the analysis, the rotated factors will display a pattern of correlations with one another that mirrors the correlations displayed by the real-world constructs.

Table 15.4 presents the inter-factor correlations obtained from Dr. O'Day's EFA. Each factor is identified using the labels that were produced during the interpretation of the factor pattern (e.g., F_1 is labeled as the "lack of time for exercise" factor).

Table 15.4

Inter-Factor Correlations Obtained from Factor Analysis (with Promax Rotation) of 12-Item Questionnaire

	Correlations			
Factor	F_1	F_2	F_3	F_4
F_1. Lack of time for exercise	--			
F_2. Dislike of exercise	.40	--		
F_3. Desire to be thin	-.09	-.17	--	
F_4. Health motivation	-.23	-.35	.35	--

Note. $N = 238$.

It is always a good idea to review the inter-factor correlations to see if the pattern of relationships makes sense, given what is known about the variables. If they do not make sense, it may indicate that errors were made in the analysis or interpretation.

As an illustration, consider the correlations reported in the column headed "F_1" in Table 15.4. These correlations show that F_1 (the *lack of time for exercise* factor) displays:

- A moderate positive correlation of +.40 with F_2, the *dislike of exercise* factor,
- a near-zero correlation of −.09 with F_3, the *desire to be thin* factor, and
- a small negative correlation of −.23 with F_4, the *health motivation* factor.

The sign and size of these correlations (and the other correlations in the table) seem reasonable. Therefore, the correlations themselves do not constitute any reason to suspect that errors were made in the EFA.

THE FACTOR STRUCTURE MATRIX. We are now ready for the second of the two matrices. Table 15.5 presents the structure matrix for the current factor analysis. The factor loadings that appear in the ***factor structure matrix*** are simple bivariate correlation coefficients similar to the Pearson *r*. A given factor loading in this matrix represents the simple bivariate correlation between a manifest variable and a factor. These are *not* partial correlations—nothing is statistically controlled or partialed out.

Table 15.5

Promax-Rotated Factor Structure and Final Communality Estimates for the 12 Questionnaire Items, Decimals Omitted (Factor Loadings Are Bivariate Correlations)

Factor					
1	2	3	4	h^2	Questionnaire item
-13	-22	25	76*	58	27. It is important to me to be healthy.
-22	-27	28	81*	65	28. I am willing to sacrifice in order to stay healthy.
-17	-33	31	76*	58	29. I am willing to work hard in order to avoid medical problems.
-05	-16	85*	30	72	31. It is important to me to be thin.
-01	-08	79*	21	64	32. I very much want to have a fairly low body weight.
-18	-18	73*	37	55	33. It is important to me to have a lean physical appearance.
39	88*	-20	-35	78	47. I **hate** to engage in aerobic exercise.
37	93*	-15	-30	87	48. I find aerobic exercise to be unpleasant.
33	82*	-11	-27	68	49. I find aerobic exercise to be punishing.
93*	38	-10	-24	87	51. I am too busy to engage in frequent aerobic exercise.
90*	41	-05	-19	82	52. I am under too much time pressure to engage in frequent aerobic exercise.
93*	34	-09	-21	87	53. With my busy schedule, there is very little time for frequent aerobic exercise.

Note. N = 238. An asterisk (*) flags any loading $\geq .\pm 40$. h^2 = Final communality estimates.

The general trend of factor loadings in Table 15.5 is similar to the trend seen in Table 15.3: For Factor 1, the highest loadings are still for questionnaire items 51, 52, and 53; for Factor 2, the highest loadings are still for questionnaire items 47, 48, and 49. The same trend appears for Factors 3 and 4.

However, the one obvious difference between the two matrices is this: Table 15.5 (the structure matrix) does not display the clean simple structure that had been displayed by Table 15.3 (the pattern matrix). In a pattern matrix, it is common for the high loadings to be very high, and the low loadings to be very low (often near zero). In a structure matrix, on the other hand, the high loadings may be just as high, but the low loadings are typically not quite as low.

For example, consider questionnaire item 47: "I *hate* to engage in exercise." In Table 15.3 (the pattern matrix), the factor loading for this item on Factor 1 was only .04—close to zero. However, in Table 15.5 (the structure matrix) the loading for the same item on the same factor was substantially larger at .39. Why is the loading so much larger in the structure matrix? The answer goes like this:

- Item 47 ("I hate to engage in aerobic exercise") measures Factor 2 (the *dislike of exercise* factor).
- Factor 2 (*dislike of exercise*) is positively correlated with Factor 1 (*lack of time for exercise*). In fact, Table 15.4 shows a correlation of $r = .40$ between Factor 1 and Factor 2.

Because of the moderately strong correlation between Factor 1 and Factor 2, any questionnaire item that displays a high structure loading for Factor 2 will just about have to display non-zero structure loadings for Factor 1. This is a mathematical certainty when factor loadings are simple bivariate correlation coefficients, and that is exactly what the factor loadings are in a structure matrix.

Things are different in a factor pattern matrix, however. The factor loadings in a factor pattern matrix are standardized multiple regression coefficients obtained when a questionnaire item is the criterion variable and the factors are the predictor variables. These loadings are *partial regression coefficients*, which take into account the correlations between the factors. As a result, it is possible for a given questionnaire item to display a very high loading on one factor, but near-zero loadings for the others.

The upshot is this: When interpreting an oblique rotation, the primary interpretation is placed on the pattern matrix. This is because the pattern matrix is more likely to display simple structure and is therefore easier to interpret.

Once this is accomplished, however, the researcher should still review the structure matrix. This is because the factor loadings in the structure matrix are simple bivariate correlations that reveal the actual relationships between the manifest variables and the factors, without any partialing. While an item might display a near-zero loading for a given factor in the pattern matrix (thus misleading you into believing that it is unrelated to that factor), the same item might display a loading of .39 in the structure matrix (revealing that the item is, in fact, related to the factor).

In short, researchers must review both matrices if they hope to fully understand the results from an oblique rotation.

WHAT IF AN ARTICLE PRESENTS JUST ONE TABLE OF FACTOR LOADINGS? You will often encounter articles that report just one table of loadings, possibly with a vague title such as "Factor Loadings" or "Factor Matrix." Why did the article not present the two separate tables (i.e. the factor pattern and the factor structure) discussed in this chapter? There are at least two possibilities:

1) *The EFA included an orthogonal rotation.* An earlier section indicated that an *orthogonal rotation* results in factors that are uncorrelated with one another. An orthogonal rotation does not produce a separate factor pattern matrix and factor structure matrix—it produces just one matrix of rotated factor loadings (Field, 2009a). It is technically correct to view the factor loadings in this table as being the standardized multiple regression coefficients that appear in the pattern matrix, and at the same time as being the simple bivariate correlation coefficients that appear in the structure matrix (Rummel, 1970, p. 399). This is possible because the factors in an orthogonal rotation are uncorrelated with one another. This means that your task (as the reader) is easy: You need to interpret just one matrix of rotated factor loadings. How will you know that the researchers have used an orthogonal rotation? Your tip-off will usually be some mention of one of the following orthogonal rotation methods: equimax, parsimax, quartimax, or varimax.

2) *The EFA included an oblique rotation.* An earlier section indicated that an *oblique rotation* allows the resulting factors to be correlated with one another. When researchers perform an oblique rotation, they really should include two separate tables in the published article: the factor pattern matrix as well as the factor structure matrix. If they have included just one of these two tables, it will probably be the factor pattern matrix, since this is the preferred table for interpreting the meaning of the factors. But you (as the reader) should interpret it with caution, as you have been denied the information that you would have obtained from the structure coefficients. How will you know that the researchers have used an oblique rotation? Your clue will probably be some mention of one of the following oblique rotation methods: biquartimin, covarimin, direct oblimin, direct quartimin, maxplan, promax, oblimax, or procrustes.

FINAL COMMUNALITY ESTIMATES. Along with the factor loadings, Table 15.5 also has a column headed h^2. You may recall that h^2 is the symbol for *communalities*, and this column contains the final communality estimates from the factor analysis.

An earlier section indicated that a ***final communality estimate*** represents the proportion of variance in a manifest variable that is accounted for by the retained factors. The communalities are often used to assess the relative importance of the manifest variables in the analysis: If an item has a large communality, that item must have been an important measure of at least one of the study's latent factors. Table 15.5 shows that the final communality estimates for the 12 questionnaire items in the current study ranged from $h^2 = .55$ (for Item 33) to $h^2 = .87$ (for Items 48, 51, and 53).

Dr. O'Day's Decision Regarding the Number of Factors

It has been a long road, but Dr. O'Day is finally ready to make a decision regarding the number of factors to retain and interpret. As you might have guessed, she decides that her final preferred solution will be the four-factor solution. Although the questionnaire contained 12 rating items, the factor analysis showed that it was not really measuring 12 separate dimensions—it was most likely measuring just four.

This chapter has been lengthy because it had to cover a long list of criteria that researchers use in making this number-of-factors decision. Dr. O'Day relied on multiple methods to arrive at her decision, and the use of multiple methods is almost always a best practice in factor analysis. In the following excerpt, she summarizes her decision:

> The four-factor solution was selected as the investigation's final preferred solution. This decision was based on the fact that (a) the scree plot displayed a large break after Factor 4, after which the eigenvalues leveled off, and (b) the four-factor solution proved to be interpretable.

It has been a long road, but she is not home just yet. Depending on the purpose of her investigation, Dr. O'Day might use the results from the analysis to compute *estimated factors scores*. The following section explains just what these scores are and why she might need them.

Estimated Factor Scores

Interpreting the factor matrices is not necessarily the final step in a factor analysis. Once researchers have determined how many factors are being measured by a questionnaire, they often use the results to create ***estimated factor scores***: synthetic variables that represent our best guess as to where the subjects stand on the latent constructs being studied. The researcher may then explore the relationship between these estimated factor scores and other variables assessed in the investigation.

Notice that these new variables are called *estimated* factor scores, not factor scores. This is because estimated factor scores are merely our best guesses as to where the participants probably stand on the latent constructs being studied. In a moment, you will see that the statistical application creates estimated factor scores by adding together optimally weighted subject scores on the manifest variables. Because these manifest variables contain error variance (as well as "true score" variance), it is not possible for the computer to arrive at a pure factor score, free of error. Therefore, we call these values *estimated* factor scores rather than factor scores.

EQUATIONS USED IN CREATING THE SCORES. For each retained factor, the statistical application creates an equation that applies optimal weights to each manifest variable included in the EFA. A given subject's responses to the manifest variables are converted into z scores, the z scores are inserted into the equation, and the equation then computes the subject's estimated factor score.

As an illustration, below is the equation for computing estimated factor scores for Factor 1 from the current factor analysis:

$$F_1 = -.001(V27) + .000(V28) + -.010(V29) + .002(V31) + .004(V32) + -.001(V33)$$
$$+ .016(V47) + -.015(V48) + -005(V49) + \mathbf{.350(V51)} + \mathbf{.246(V52)} + \mathbf{.404(V53)}$$

The preceding equation shows that, in computing a given participant's estimated factor score on Factor 1, relatively large weights are given to the last three items (51, 52, and 53), and very small weights are given to the remaining nine items. That is because Factor 1 is the *lack of time for exercise* factor, and items 51, 52, and 53 are the items that measure the *lack of time for exercise* construct.

Different weights are applied to the questionnaire items in computing estimated factor scores on Factor 2. That equation is presented below:

$$F_2 = -.006(V27) + .002(V28) + -.028(V29) + -.004(V31) + -.002(V32) + .000(V33)$$
$$+ \mathbf{.286(V47)} + \mathbf{.511(V48)} + \mathbf{.192(V49)} + .006(V51) + .030(V52) + -.002(V53)$$

Factor 2 reflects *dislike of exercise*. Notice how the preceding equation gave a lot of weight to items 47, 48, and 49, but little weight to the other items. That is because items 47, 48, and 49 measure dislike of exercise.

A different set of weights are applied to the manifest variables to create estimated factor scores for Factor 3 and for Factor 4. Those equations are presented below. You can see that for each factor, the item numbers for items given the greatest amount of weight are presented in boldface.

$$F_3 = .010(V27) + .017(V28) + .037(V29) + \mathbf{.432(V31)} + \mathbf{.331(V32)} + \mathbf{.232(V33)}$$
$$+ -.028(V47) + -.001(V48) + .014(V49) + -.009(V51) + .038(V52) + -.033(V53)$$

$$F_4 = \mathbf{.297(V27)} + \mathbf{.384(V28)} + \mathbf{.286(V29)} + .032(V31) + -.018(V32) + .063(V33)$$
$$+ -.040(V47) + -.003(V48) + -.018(V49) + -.041(V51) + .023(V52) + -.002(V53)$$

COMPUTING CORRELATIONS INVOLVING THE ESTIMATED FACTOR SCORES. Once the researcher has created these estimated-factor-score variables, she may use them in the same way that she would use any other quantitative variable in a study. The following section describes how Dr. O'Day performed some correlational analyses that made use of these new variables.

Correlation between factor scores and other variables. For each participant, estimated factor scores were created for each of the four retained factors. These estimated-factor-score variables appear as variables 1-4 in Table 15.6. The table displays Pearson correlations between these four estimated-factor-score variables and a separate variable called *exercise sessions*. The exercise sessions variable indicates the number of times each week that a participant engaged in aerobic exercise that lasted at least 15 minutes.

Table 15.6

Means, Standard Deviations, and Pearson Correlations (Decimals Omitted) for the Four Estimated Factor-Score Variables (F_i) and Exercise Sessions

Variable	M	SD	1	2	3	4	5
1. F_1 Lack of time for exercise	0.00	0.967	--				
2. F_2 Dislike of exercise	0.00	0.954	43***	--			
3. F_3 Desire to be thin	0.00	0.907	-10	-20**	--		
4. F_4 Health motivation	0.00	0.899	-26***	-40***	42***	--	
5. Exercise sessions	2.93	2.313	-46***	-39***	17*	33***	--

Note. $N = 238$.
*$p < .05$; **$p < .01$; ***$p < .001$.

In Table 15.6, variables 1-4 represent the estimated-factor-score variables created as part of the EFA. If you compare the correlations in Table 15.6 to the inter-factor correlations that were presented earlier in this chapter (in Table 15.4), you will see that the pattern of the correlations is very similar, but that the actual values of the correlations are not identical. They are not identical because the inter-factor correlations presented in Table 15.4 represented the correlations between the *actual factors*, whereas the correlations in Table 15.6 represent the correlations between *estimated-factor-score variables*. An earlier section explained that "actual factors" and "estimated factor-score variables" are not exactly the same thing, so we cannot expect the inter-factor correlations to be identical for the two sets of variables.

Finally, the row headed "5. Exercise sessions" in the table displays the correlations between *exercise sessions* and the four estimated-factor-score variables. The table shows that exercise sessions display the following correlations with the four estimated-factor-score variables, ordered from the strongest relationship to the weakest:

- The strongest predictor of exercise sessions was F_1, lack of time for exercise ($r = -.46, p < .001$).

- The next-strongest predictor of exercise sessions was F_2, dislike of exercise ($r = -.39, p < .001$).
- The next-strongest predictor of exercise sessions was F_4, health motivation ($r = .33, p < .001$).
- The weakest predictor of exercise sessions was F_3, desire to be thin ($r = .17, p < .01$).

If we simultaneously consider the size as well as the sign of these four correlations, we can construct a rough "profile" of the type of individual who engages in a large number of exercise sessions each week. People who engage in many exercise sessions each week:

- Say that they have plenty of time to exercise,
- say that they like to exercise,
- are motivated to stay healthy, and
- want to have a thin physical appearance.

Remember that some of these predictors are more important than others. For example, the correlation between desire to be thin and exercise sessions was so weak ($r = .17$) that some researchers would view it as an unimportant predictor.

Putting it All Together: Writing Up the Factor Analysis

One purpose of this book is to show how specific statistical analyses might be summarized in a journal article. Although this chapter has shown how specific parts of Dr. O'Day's factor analysis might have been written up for publication, these excerpts have been presented in an isolated and somewhat disjointed fashion. This section attempts to make amends by showing how these various pieces could be consolidated into one unified excerpt from a journal article:

> An exploratory factor analysis was performed on responses to the 12 questionnaire items presented in Table 15.1. Principal axis factoring was used to extract the factors, the analysis was performed on the matrix of correlation coefficients, and squared multiple correlations served as prior communality estimates.
>
> A scree test revealed a large break between Factor 4 and Factor 5, suggesting that four factors should be retained. Following a promax (oblique) rotation, the factor pattern matrix for the four-factor solution displayed simple structure and proved to be interpretable. The rotated factor pattern appears in Table 15.3.
>
> An item was said to load on a given factor if the if the relevant loading in the rotated pattern matrix was $\geq \pm.40$. Using this criterion, items 51, 52, and 53 loaded on Factor 1 (subsequently labeled *lack of time for exercise*), items 47, 48, and 49 loaded on Factor 2 (labeled *dislike of exercise*), items 31, 32, and 33 loaded on Factor 3 (labeled *desire to be thin*), and items 27, 28, and 29 loaded on Factor 4 (labeled *health motivation*).

The inter-factor correlations (presented in Table 15.4) ranged from $r = -.09$ to $r = .40$. The factor structure matrix is presented in Table 15.5.

Results from the analysis were used to create optimally weighted estimated factor scores for each of the four factors. These estimated-factor-score variables were then correlated with *exercise sessions:* the number of times that the participant engaged in aerobic exercise each week. Exercise sessions displayed the following correlations with the estimated-factor-score variables: for F_1, lack of time for exercise: $r = -.46$, $p < .001$; for F_2, dislike of exercise, $r = -.39$, $p < .001$; for F_3, desire to be thin: $r = .17$, $p < .01$; and for F_4, health motivation: $r = .33$, $p < .001$.

Additional Issues Related to this Procedure

This final section briefly discusses principal component analysis (PCA), a multivariate statistical procedure that in some ways is similar to exploratory factor analysis. The section ends with a list of assumptions underlying EFA.

EFA Versus Principal Component Analysis

If you read much research, sooner or later you will encounter an article in which the researchers say that they have analyzed their data using *exploratory factor analysis (EFA)* when, upon closer inspection, it turns out that they had actually used a separate procedure called *principal component analysis (PCA)*. The confusion arises because EFA and PCA appear to be very similar procedures (on the surface, at least). Both EFA and PCA are dimension-reduction procedures. Both follow the same sequence of steps: initial extraction of the factors (or components), rotation of the matrix of loadings, and so forth.

Despite the apparent similarities, PCA has a great limitation compared to EFA: Only exploratory factor analysis allows researchers to discover the factor structure underlying a set of manifest variables. Only EFA allows the researcher to make statements such as "The 12-item questionnaire is measuring four latent factors. One factor is the *lack of time for exercise* factor." Only EFA allows researchers to draw path models (similar to Figure 15.1) in which the ovals represent unseen latent factors that have an effect on observed variables.

Principal component analysis *does* allow researchers to reduce a large number of questionnaire items to a smaller number of components. On the surface, these components may appear to be identical to the latent factors that are studied in EFA. But they are not conceptually equivalent, and it is technically incorrect to refer to them as "factors." To keep things brief, this section will not explain the technical reasons for this, but will instead focus on how you can determine whether an article is describing EFA versus PCA (readers who are interested in the technical differences between EFA and PCA are referred to Hatcher, 1994, *Chapter 2*, as well as Rummel, 1970).

How do you determine that a given analysis was actually a principal component analysis rather than an exploratory factor analysis? Some clues are listed below.

Some of these clues, when present in an article, indicate that the analysis was *definitely* a PCA, whereas other clues merely indicate that it *might* have been a PCA. But any of the following should be seen as a potential red flag:

- If the researchers describe it as a *principal component analysis*, or indicate that they used the *principal components* method of extraction.

- If the researchers refer to the synthetic variables that they have extracted as *principal components* or just *components*.

- If the researchers indicate that they used 1s (ones) as prior communality estimates, or indicate that they analyzed a correlation matrix with 1s on the diagonal.

- If the researchers say that they used the *Kaiser criterion* or the *eigenvalue-1 criterion* to determine the number of components or factors to retain. Actually, this is not a terribly good clue. It *should* be a good clue, because the eigenvalue-1 criterion is only statistically appropriate for PCA (it is just about never appropriate for EFA). Unfortunately, many researchers do not know this, and use the eigenvalue-1 criterion with both procedures.

None of this is to say that principal component analysis is a bad statistical procedure. It is useful as a dimension-reduction technique, and it produces principal components that are perfectly appropriate for a wide variety of uses in research. But principal component analysis is *not* factor analysis: It does not allow researchers to discover the factor structure underlying a set of variables.

Assumptions Underlying this Procedure

My earlier book, which shows how to use SAS to perform structural equation modeling (Hatcher, 1994), also includes a chapter on exploratory factor analysis. It lists the following assumptions for an exploratory factor analysis that uses principal axis factoring for the initial extraction of factors. Some of the following assumptions were explained in greater detail in *Chapter 2*.

- **INTERVAL OR RATIO SCALE VARIABLES.** Variables should be assessed on an interval or ratio scale. Some researchers argue that Likert-type questionnaire items produce data that are on an ordinal scale and are therefore inappropriate for EFA, but many other researchers disagree and argue that EFA is appropriate for this type of data (see Warner, 2008, for a general overview of the debate regarding this scale-of-measurement issue).

- **RANDOM SAMPLES.** Scores should be randomly sampled from the population of interest.

- **LINEAR RELATIONSHIPS.** The relationships between the manifest variables should be linear relationships.

- **NORMAL DISTRIBUTIONS.** Manifest variables should display approximately normal distributions.

- **BIVARIATE NORMALITY.** Ideally, pairs of variables should display bivariate normality, although mild to moderate departures may not be a problem if the sample is large.

The maximum likelihood (ML) method for extracting factors produces a significance test that researchers sometimes use to help determine the correct number of factors to retain. This significance test requires that the following assumption be met (in addition to the preceding assumptions):

- **MULTIVARIATE NORMALITY.** Scores on the manifest variables should follow an approximate multivariate normal distribution.

Field (2009a) shows how to use SPSS to investigate a number of additional assumptions for factor analysis. These include:

- **SAMPLING ADEQUACY.** Field shows how to interpret the Kaiser-Meyer-Olkin measure of sampling adequacy.

- **ADEQUACY OF THE CORRELATION MATRIX.** Field also showed how to use Bartlett's test, the determinant of the R-matrix, and other information to verify that the correlations in the initial correlation matrix are neither too small nor too large for purposes of performing exploratory factor analysis.

Chapter 16

SEM I: Path Analysis with Manifest Variables

This Chapter's *Terra Incognita*

You are reading a journal article. It was written by a researcher who has developed a theoretical path model that hypothesizes causal relationships between a number of attitudinal and behavioral variables related to health and physical exercise. The *Results* section reports the following:

> Data were analyzed using path analysis with manifest variables. The maximum likelihood method was used to estimate parameters, and all analyses were performed on the variance-covariance matrix.
>
> Of the three theoretical models being compared, only Model 3 displayed an acceptable fit to the data. Only Model 3 displayed a model chi-square statistic that was nonsignificant, $\chi^2(4, N = 238) = 5.85$, $p = .211$. It has been recommended (e.g., Hu & Bentler, 1999; Mueller & Hancock, 2008) that a model be viewed as displaying an acceptable fit if the SRMR is $\leq .08$, the RMSEA is $\leq .06$, and the CFI and NNFI are $\geq .95$. Again, only Model 3 satisfied all of these criteria.
>
> The signs of the path coefficients in the final preferred model were consistent with the signs in the initial theoretical model. Each path coefficient was significantly different from zero, $\alpha = .05$. The three variables having the largest total effects on exercise sessions were lack of time for exercise ($EC = -.43$), dislike of exercise ($EC = -.22$), and health motivation ($EC = +.17$).

Take a deep breath. All shall be revealed.

The Basics of this Statistical Procedure

Path analysis is a highly flexible statistical procedure that allows researchers to test causal models by analyzing correlational data. It requires that the researcher be very familiar with previous research and use that familiarity to develop a theoretical ***path model***: a simplified representation of the hypothesized causal relationships between the relevant variables. The researcher then analyzes data from a sample of participants and determines whether the actual relationships between the variables are consistent with the relationships that were predicted by the theoretical path model.

The type of analysis described in this chapter is most frequently called *path analysis*, but may also be called *path analysis with manifest variables* or *path analysis with manifest indicators*. It belongs to a larger family of statistical procedures called *structural equation modeling (SEM)* or *covariance structural modeling (CSM)*.

A path analysis produces an extensive set of statistical results. These include *fit indices* (which indicate the extent to which the model fits the data), *path coefficients* (which indicate the size of the direct effects that antecedent variables have on consequent variables), and R^2 *statistics* (that indicate the percent of variance in the consequent variable that is accounted for by its antecedent variables). The results may also include *model modification indices*–statistics that help the researcher modify the model in order to achieve a better fit to the data.

Because it is based on correlational research designs, path analysis typically does not provide evidence of cause and effect that is as persuasive as the evidence provided by true experiments. Nevertheless, for researchers working in areas where it is difficult to actually manipulate independent variables, path analysis provides a flexible and informative procedure for investigating hypothesized relationships between naturally occurring variables.

Types of Variables That May Be Analyzed

Many researchers use a maximum likelihood (ML) estimation procedure to perform path analysis, and this procedure requires that the variables be assessed on an interval or ratio scale (Klem, 1995) and that they follow a multivariate normal distribution (McDonald & Ho, 2002). Although these assumptions imply that all constructs must be continuous quantitative variables, in practice it is not unusual to see path models that include dichotomous exogenous variables that have been dummy-coded (a later section of this chapter will explain what an *exogenous variable* is; for a refresher on dummy-coding, see *Chapter 10*).

The maximum likelihood method is not the only estimation procedure available in path analysis. Alternative estimation procedures have been developed to handle nonnormal data as well as categorical variables and ordinal-scale variables (e.g., Browne, 1984; Muthén, 1984).

There are a number of additional assumptions that are required by path analysis with manifest variables. These are summarized at the end of the chapter.

Path Analysis with Manifest Variables Versus Latent Factors

This chapter focuses on **path analysis with manifest variables**. In this type of path analysis, the variables that appear in the causal model are all **manifest variables**: observed variables that are measured in a direct way. For example, the *health motivation* variable that has been discussed in many previous chapters of this book is an example of a manifest variable. Scores on this manifest variable were the real-world scores obtained from a multiple-item summated rating scale.

In contrast, **latent factors** are *hypothetical constructs*: invisible, underlying factors that cannot be observed in a direct way, but are instead inferred to be present based on the results of factor-analytic statistical procedures. *Chapter 18* of this book will cover a more sophisticated procedure called **path analysis with latent factors**. With this latter procedure, the variables that constitute the main part of the causal model are latent factors, rather than the somewhat more primitive manifest variables to be covered in the current chapter. A path analysis with latent factors is sometimes called a *LISREL analysis*, because LISREL (Jöreskog & Sörbom, 1989) was the first widely available computer program that performed analyses of this sort. Path analysis with latent factors has some advantages over path analysis with manifest variables, as we shall see in *Chapter 18*.

Illustrative Example

This chapter illustrates path analysis with manifest variables by estimating three competing causal models related to health and physical exercise. In each model, the ultimate criterion variable is *exercise sessions*: the number of times that the participants engage in aerobic exercise each week. The researcher (Dr. O'Day, of course) wants to identify the variables that cause people to engage in exercise. Based on a review of theory and previous research, she believes that she has identified the five variables that affect exercise behavior. What's more, she believes that she has determined which variables probably have *direct effects* on exercise behavior, and which probably have only *indirect effects* (an *indirect effect* is an effect that is mediated by an intervening variable—more on this later in the chapter).

METHOD. To investigate her proposed causal model, Dr. O'Day administered the *Questionnaire on Eating and Exercising* (which appears in *Appendix A*) to 260 college students enrolled in psychology courses at a regional state university in the Midwest. She obtained usable responses from 238 subjects.

PRIMARY THEORETICAL MODEL. Figure 16.1 presents Dr. O'Day's hypothesized causal model. It will serve as the primary theoretical model to be tested in this chapter.

The primary criterion variable in Dr. O'Day's causal model is *exercise sessions*: the number of times that the participants engage in aerobic exercise each week. This variable was measured using item 76 on the *Questionnaire*.

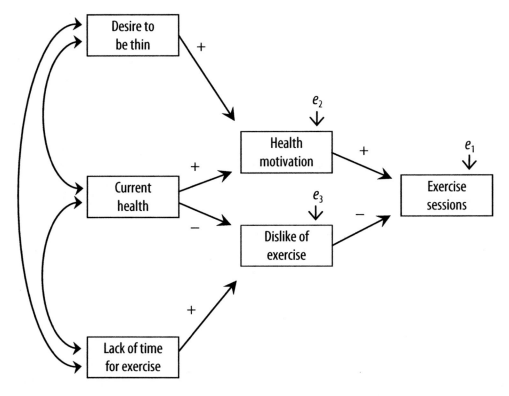

Figure 16.1. Model 1: Path model describing the relationship between exercise sessions and the variables that are hypothesized to have direct or indirect effects on it.

Each of the remaining variables were measured using a multiple-item summated rating scale. Each scale consisted of three to four Likert-type rating items. For each item, possible responses ranged from 1 = *Extremely Disagree* to 7 = *Extremely Agree*. Scores on the resulting variables could range from 1.00 (meaning that the participant had less of the attribute) to 7.00 (meaning that the subject had more of the attribute). Below, the name of each scale is provided along with (a) the item numbers for the items used to create that scale, and (b) a sample item from the scale:

- *Health motivation*, items 27-29; "It is important to me to be healthy."
- *Dislike of exercise*, items 47-49; "I find aerobic exercise to be unpleasant."
- *Desire to be thin*, items 31-33; "It is important to me to be thin."
- *Current health*, items 55-58; "In general, my health is good."
- *Lack of time for exercise*, items 51-53; "I am too busy to engage in frequent aerobic exercise."

ANTECEDENT VERSUS CONSEQUENT VARIABLES. Each of the variables in Figure 16.1 is either an antecedent variable, a consequent variable, or both. An ***antecedent variable*** is a variable that precedes another variable in the causal chain. For example, you can see that the desire to be thin is an antecedent to health motivation. On the other hand, a ***consequent variable*** is a variable that comes after another variable in the causal chain. You can see that exercise sessions is a consequent variable to health motivation. Path models are typically drawn so that causation flows from left to right (i.e., from antecedent variables on the left to the consequent variables on the right).

The straight, single-headed arrows in the model represent ***direct effects***: the prediction that an antecedent variable exerts causal influence on a consequent variable in a direct fashion (i.e., in a fashion not mediated by any "go-between" variables). The signs on the arrows indicate whether the antecedent variable is hypothesized to have a positive effect (+) or a negative effect (−). These signs are similar to the signs of Pearson *r* correlation coefficients that were covered in *Chapter 6:* A positive sign means that higher values on the antecedent variable are predicted to be associated with higher values on the consequent variable, and a negative sign means that higher values on the antecedent variable are predicted to be associated with lower values on the consequent variable.

RESIDUAL TERMS. The symbols e_1, e_2, and e_3 in the figure represent *residual terms*. The *residual term* for a given variable represents the variance in that variable that is not accounted for by the other variables included in the model. The meaning of these residual terms will be explained in more detail later in this chapter.

Research Questions Addressed

Performing path analysis on the data obtained in this investigation would allow Dr. O'Day to answer research questions such as those presented below. Definitions for some of the following terms will be provided in later sections.

- Overall, does my theoretical path model provide a good fit to the data?
- Does adding a new path to this model result in a significant improvement in the model's fit to the data?
- If my theoretical model hypothesizes that one variable has a direct effect on another variable, is the path coefficient for that relationship significant, substantial, and in the predicted direction?
- Which antecedent variables appear to have the strongest total effects on the consequent variables?

Details, Details, Details

Path analysis is a bottomless pit of terms, concepts, statistics, indices, and counter-intuitive ideas. With this section, we cautiously lean forward and take our first tentative peek into the abyss below.

Can Path Analysis Prove Cause-and-Effect Relationships?

No, it cannot. A cautious scientist would never claim than some statistical procedure has "proven" cause and effect. Not even an advanced, hard-to-understand procedure like path analysis.

It's like this: Path analysis is a powerful tool for analyzing data from correlational studies. In the right situations, it can expand our understanding, increase our knowledge base, help us disconfirm bad theories, and help us refine good theories. In short, it is a terrific tool for doing good science.

But it is just a statistical procedure. There's only so much it can do.

For example, imagine that a researcher has worked on some specific research topic for years and has ultimately arrived at a theoretical path model that makes specific predictions regarding the causal relationships between variables. She then measures the relevant variables in a very large sample and analyzes the resulting data using path analysis. When the analysis is complete, the results might provide support for her original theoretical causal model. In a best-case scenario, this would mean that (a) the model provided a good overall fit to the data and (b) each of the paths in the theoretical path model displayed path coefficients that were significant, substantial, and in the predicted direction. These results are consistent with the hypothesized model, and this means that the researcher's model has survived an attempt at disconfirmation. This is good news for the researcher—it gives the model credibility as a potentially useful theory. But few statisticians would conclude that these results have "proven" that the theoretical model was "correct."

Here is one of the most important reasons that we should use caution when interpreting results from a path analysis: Even if the researcher's model provides a good fit to the data, it almost certainly is not the *only* theoretical path model that provides a good fit. It is likely that another researcher could rearrange the same variables to create a new path model (e.g., perhaps with causation flowing in a different direction), perform a path analysis on this new model, and still obtain results indicating that there is a good fit between model and data.

Ouch.

The implication is simple: Path analysis is a terrific statistical procedure, but its results (like the results of any statistical procedure) should be taken with a grain of salt. When path analysis shows that there is a good fit between the researcher's theoretical model and the data, the researcher should rejoice—the model has survived an attempt at disconfirmation, and that means that this model *might* be a reasonable representation of reality. But be skeptical of researchers who take such results as conclusive proof that their model is the one "correct" model. When reading about a path analysis, the best advice is *caveat emptor*.

The Theoretical Models To Be Compared

When using path analysis, some researchers focus on just one theoretical model: They crunch the numbers to determine whether this one model provides a good fit to the data. But they could do better than that.

Researchers who try to follow best practice use path analysis to directly compare two or more competing theoretical models. In such investigations, the competing models might differ in terms of their complexity, the number of causal paths they contain, the nature of their predictions, or in other ways (Thompson, 2000). This approach allows researchers to identify the model that provides the best fit to the data while still remaining parsimonious (in this context, *parsimonious* means *relatively simple*).

The current chapter illustrates this strategy of using path analysis to compare alternative models. The next few sections introduce the three theoretical models that Dr. O'Day plans to compare.

DR. O'DAY'S REVIEW OF THE LITERATURE. Prior to conducting the current study, Dr. O'Day completed a comprehensive review of previous theory and research related to aerobic exercise behavior. Imagine that the top experts in this area disagree as to which variables are the most important determinants of exercise behavior. Therefore she developed three alternative models, with each model reflecting a somewhat different theoretical perspective (as always, remember that all theoretical models presented here are works of fiction, sort of like the maps of middle earth that appear in *Lord of the Rings*).

THE THREE ALTERNATIVE MODELS. In the *Introduction* section of her article, Dr. O'Day summarized the relevant research literature and presented the three alternative models to be investigated. Here's an excerpt:

> One purpose of this study was to compare three alternative models of the determinants of exercise behavior. Figure 16.2 illustrates these alternative models.
>
> Model 1 is the least complex of the three models being compared—it consists of the causal paths represented by arrows with solid lines. Plus signs (+) and minus signs (–) appear on the arrows to represent the hypothesis that an antecedent variable is expected to have either a positive effect (+) or a negative effect (–) on the consequent variable.
>
> Model 2 is the second theoretical model to be investigated. Model 2 is identical to Model 1, except that Model 2 includes an additional path represented by the dashed-line arrow in Figure 16.2. This dashed-line arrow represents the prediction that lack of time for exercise will have a negative direct effect on exercise sessions.
>
> Finally, Model 3 is identical to Model 2, except that Model 3 includes an additional path represented by the dotted-line arrow. This dotted-line arrow represents the prediction that dislike of exercise will have a negative direct effect on health motivation.

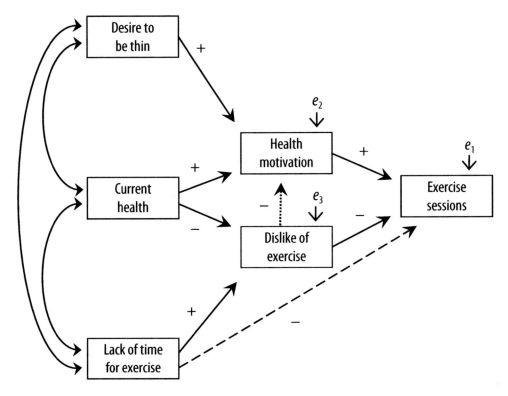

Figure 16.2. Composite model illustrating the three models to be compared. Model 1 consists of the paths represented by solid-line arrows; Model 2 is identical to Model 1, except that Model 2 contains an additional path from lack of time for exercise to exercise sessions (represented by the dashed-line arrow); Model 3 is identical to Model 2, except that Model 3 contains an additional path from dislike of exercise to health motivation (represented by the dotted-line arrow).

Basic Terms and Concepts

Put on your thinking caps—it's time for some new terms. These terms will be relevant not only to the analysis covered in the current chapter, but also to the confirmatory factor analysis to be covered in *Chapter 17*, as well as to the path analysis with latent factors to be covered in *Chapter 18*.

ENDOGENOUS VARIABLES. In path analysis, an ***endogenous variable*** is a variable that is causally determined by other variables in the model. If a variable has a straight, single-headed arrow pointing at it, it is an endogenous variable. In Figure 16.2, the three endogenous variables are exercise sessions, health motivation, and dislike of exercise.

One of the assumptions of path analysis is as follows: If a variable is an *endogenous variable*, then all important determinants of that endogenous variable must be included in the hypothetical path model to be tested. If any of the nontrivial causes of an endogenous variable have been omitted, this constitutes a type of *specification error*, and it may cause parameter estimates (such as path coefficients) to be biased and misleading. To avoid such a specification error, researchers must perform a thorough review of the relevant research literature and develop theoretical models that include all important determinants of the endogenous variables.

EXOGENOUS VARIABLES. In path analysis, an ***exogenous variable*** is a variable that is not causally determined by the other variables in the model. If a variable does not have any straight, single-headed arrows pointing at it, it is an exogenous variable. If a given variable appears as an exogenous variable in a model, it means that the researcher is not attempting to identify the variables that causally determine that variable. In Figure 16.2, the exogenous variables are desire to be thin, current health, and lack of time for exercise. The three "*e*" terms that have been added to this model are also exogenous variables—their meaning will be described below.

CURVED DOUBLE-HEADED ARROWS. A curved double-headed arrow represents a simple correlation between two variables. If two variables are joined by a curved arrow, it means that the model acknowledges that there may be a correlation between the two variables, but the model does not attempt to explain the causal relationship (if any) between them.

You can see that curved two-headed arrows connect the three exogenous manifest variables that appear on the left side of the path model: the desire to be thin, current health, and lack of time for exercise. In path analysis, it is a fairly standard convention to connect the exogenous manifest variables with curved double-headed arrows in this way.

STRAIGHT SINGLE-HEADED ARROWS AND DIRECT EFFECTS. A straight single-headed arrow represents the model's prediction that the antecedent variable (where the arrow begins) has a direct effect on the consequent variable (where the arrow ends). The ***direct effect*** of an antecedent variable on a consequent variable is that part of the effect that is not mediated by another variable (Pedhazur, 1982, p. 181). In Figure 16.2, you can see that health motivation is predicted to have a direct effect on exercise sessions, and dislike of exercise is also predicted to have a direct effect on exercise sessions. The model predicts other direct effects—each is represented by a straight single-headed arrow.

PATH COEFFICIENTS. The term ***path coefficient*** may be defined in a number of different ways. In simple terms, it is a number that indicates the size of the direct effect of one variable on another variable (Klem, 1995). When path analysis is performed using multiple regression, the path coefficients are essentially multiple regression coefficients: The path coefficient for a given antecedent variable is a value that estimates the amount of change in a consequent variable that is associated with a one-unit change in that antecedent variable, while statistically holding constant the remaining antecedent variables.

The sign of a path coefficient is interpreted in the same way that we interpret the sign of a multiple regression coefficient. When a path coefficient is positive (+), it means that an increase in the antecedent variable is associated with an increase in the consequent variable. When the coefficient is negative, it means that an increase in the antecedent variable is associated with a decrease in the consequent variable.

Path coefficients may be in *raw-score form* or they may be in *standardized form*. When a path coefficient is standardized, it is interpreted in the same way that a beta weight (β) from multiple regression is interpreted: It represents the amount of change in the consequent variable (in standard deviations) that is associated with a one-standard deviation increase in the antecedent variable, while statistically holding constant the remaining antecedent variables. Under typical circumstances, standardized path coefficients range in size from –1.00 through 0.00 through +1.00.

RESIDUAL TERMS FOR ENDOGENOUS VARIABLES. In a path model, the *residual term* (sometimes called the *error term* or the *disturbance term*) is typically represented by the symbol e (for "error," as in Figure 16.2) or the symbol d (for "disturbance"). A separate residual term is estimated for each endogenous variable in the model. In Figure 16.2, the symbol e_1 represents the residual term for the first endogenous variable (exercise sessions). A causal path goes from this error term to exercise sessions, representing the direct effect that the residual term has on it. The residual terms for the remaining two endogenous variables in the model may be interpreted in the same way.

The **residual term** for a given variable represents the variance in that variable that is not accounted for by the other variables included in the model. It represents causal effects on the endogenous variable due to antecedent variables that have been left out of the model, to random shocks, or to specification errors (James, Mulaik, & Brett, 1982).

Initial Theoretical Model Versus Final Preferred Model

An earlier section indicated that it is good practice to conduct a study that compares multiple competing causal models. One of these may be the researcher's favored model, and the others may be alternative models proposed by other researchers.

INITIAL AND ALTERNATIVE THEORETICAL MODELS. In a path analysis, the researcher may refer to one causal model as the *initial* or *primary theoretical model*. This represents the researcher's best guess regarding the relationships that should exist (based on his or her interpretation of theory and previous research) prior to analyzing data from the current investigation. In the current study, Model 1 (from Figure 16.2) is Dr. O'Day's primary theoretical model.

In addition, it is ideal if the researcher also proposes one or more *alternative theoretical models* that specify relationships that are different from the relationships hypothesized by the primary model. In the current investigation, Model 2 and Model 3 (from Figure 16.2) serve as the alternative theoretical models.

FINAL PREFERRED MODEL. By the end of the analysis, the researcher will arrive at a *final preferred model*. This is the investigator's best guess regarding the relationships between the variables, based on the results obtained from the path analysis. In a best-case scenario, either the initial theoretical model or one of the alternative theoretical models will provide a good fit to the data, and will be selected as the final model.

DATA-DRIVEN MODEL MODIFICATIONS. But that doesn't always happen. If none of the theoretical models provide a good fit, the researcher may choose to perform a series of ***data-driven model modifications*** to arrive at final model (these are also called *post-hoc model modifications* and the process is sometimes called a *specification search*). In making these modifications, the researcher looks for differences between the initial theoretical model versus the relationships that actually exist in the sample data, and then changes the theoretical model in order to achieve a better fit to the data. Most path analysis applications provide *modification indices:* statistics that "suggest" the changes that should be made in order to achieve a better fit. When making changes in this manner, the researchers will typically reassure the reader that they are only making changes that can be justified on theoretical grounds.

Sounds great, doesn't it? You just sort of put the computer on autopilot and let it tell you how to draw your theoretical model. And it gets better: By making these data-driven model modifications, researchers can just about always arrive at a final model that displays a good fit to the data.

But there is trouble in paradise. Most experts warn that readers should exercise great caution (perhaps even skepticism) when reading about data-driven model modifications. One of the problems is the fact that this process often results in a model that will not generalize to other samples or to the larger population. In making these post-hoc model modifications, researchers often capitalized on chance correlations that are specific to the current sample. As a result, they may arrive at a model that provides a terrific fit in the current sample, but provides a lousy fit when applied to a different sample from the same population. (MacCallum, Roznowski, & Necowitz, 1992).

Data-driven model modifications are somewhat more respectable if they are performed as part of a cross-validation. In this context, a ***cross-validation*** is a procedure in which the researcher divides the sample into two groups: a calibration sample and a validation sample. The researcher uses the calibration sample to developed a good-fitting model (using the data-driven modification indices, described above). The researcher then uses the validation sample to see if the new model still fits the data, when applied this new sample. If yes, readers can have a bit more confidence in the new model.

So don't be surprised when you read about data-driven model modifications in a path analysis. If the researchers go to the effort of conducting a cross-validation, you can have some confidence in their newly-revised model. But if they used data-driven model modifications without cross-validation, you should view their revised model in the same way you view those maps of middle earth in *Lord of the Rings*.

Results from the Investigation

In an article that describes a path analysis, the *Results* section typically begins with a table that provides means, standard deviations, intercorrelations, and coefficient alpha reliability estimates for the variables included in the analysis (similar to the tables presented in *Chapter 6*). Including such a table is a good idea with just about any investigation, but is particularly important when reporting an investigation that makes use of structural equation modeling. By including a table of correlation coefficients and standard deviations, the researchers make it fairly easy for others to replicate and extend their analyses. This is because many applications that perform structural equation modeling allow the data to be input in the form of a table of correlations and standard deviations.

Having stressed the importance of including such a table, this section will violate its own advice and skip the table in order to save space. Think of this chapter as being sort of like your mother: Do as I say, not as I do.

Estimating the Parameters of the Model

After the table of correlations, the *Results* section typically continues by providing a description of the estimation procedure used to perform the path analysis. The **estimation procedure** is the statistical algorithm used to compute **parameter estimates:** the path coefficients, error terms, and other statistics that represent various characteristics of the path model. Dr. O'Day describes the estimation procedure in the following excerpt from her article:

> The SAS PROC CALIS procedure was used to perform path analysis with manifest variables. The maximum likelihood method was used to estimate parameters, and all analyses were performed on the variance-covariance matrix.

The *maximum likelihood procedure* (abbreviation: ML) is probably the most popular estimation procedure used with path analysis. In *Chapter 12*, you learned that it is called **maximum likelihood estimation** because the final parameter estimates are the ones that maximize the likelihood of obtaining the sample data that were actually obtained (Tabachnick & Fidell, 2007).

In the preceding excerpt, Dr. O'Day indicated that she used PROC CALIS (a procedure available with the SAS® application) to perform her statistical analysis. Researchers seldom name their statistical application when performing the older, better-established statistical tests such as ANOVA or multiple regression. However, they are fairly likely to name the application used to perform structural equation modeling (SEM). This is because SEM is a newer procedure that is still being developed, and some applications have capabilities that others do not. Popular applications for performing SEM include LISREL®, EQS®, AMOS®, and the SAS® System's PROC CALIS (full disclosure: Yes, I am author or co-author of several books that show how to use the SAS System, including a book that covers SEM).

The preceding excerpt from Dr. O'Day's article indicated that "all analyses were performed on the variance-covariance matrix." Researchers typically have a choice of analyzing either the variance-covariance matrix or the correlation matrix. The variance-covariance matrix is generally preferred for SEM, as it is more likely to result in correct estimates for the standard errors.

Comparing the Three Path Models

One of the most important things that the researcher must do in a path analysis involves determining whether the theoretical model displays an acceptable fit to the data. Conceptually, this means determining whether the causal relationships hypothesized by the theoretical model are feasible, given the correlational relationships that were actually observed in the real-world sample.

A variety of different statistical results will guide the researcher in making this decision, and the more important ones are described here. This section begins with a discussion of the *matrix of residuals*, a set of results that is especially important because it is used to derive a number of fit indices (such as the chi-square test). This section then discusses *global fit indices:* single values that provide an overall evaluation of fit. Finally, it reviews the *chi-square difference test:* an inferential procedure that evaluates the effects of adding new causal paths to an existing model.

The Actual, Predicted, and Residual Correlations

When performing a path analysis, the computer program analyzes either a matrix of correlations (the correlations that were actually observed between the variables in the data set) or a matrix of covariances (the covariances that were actually observed). Correlations and covariances are two different statistics that reflect the nature of the relationship between pairs of variables.

With path analysis, researchers are more likely to analyze covariances than correlations. However, since readers tend to be more familiar with correlation coefficients than with covariances, this section will describe a situation in which the analysis has been conducted on correlations.

ACTUAL VERSUS PREDICTED CORRELATIONS. In a path analysis, the researcher works with two different matrices of correlation coefficients. First, the *matrix of actual correlations* contains the correlations that were actually observed between the study's manifest variables. These are the real-world correlations that the researcher must analyze in order to perform the path analysis.

Based on this analysis, the statistical application then estimates path coefficients for the paths contained in the researcher's theoretical causal model. Each **path coefficient** is a statistic that estimates the size of the effect that an antecedent variable has on a consequent variable. Path coefficients are analogous to the multiple regression coefficients produced in multiple regression.

Next, the application creates a new matrix of correlation coefficients called the ***predicted model matrix***. Each correlation in this matrix is hypothetical in nature: It is the correlation that we should have observed in the matrix of actual correlations if the researcher's theoretical model were true. In other words, the correlations in the predicted model matrix are the correlations that were *implied* by the researcher's theoretical model. This predicted model matrix is a type of barometer—a barometer which reveals the "correctness" of the researcher's theoretical model. It works like this:

- If the researcher's theoretical model provides a good fit to the data, then the correlations in this predicted model matrix will be identical (or almost identical) to the correlations contained in the matrix of actual correlations. This would be good news for the researcher's theoretical model.

- On the other hand, if the researcher's theoretical model provides a poor fit to the data, then the correlations in this predicted model matrix will be very different from correlations contained the matrix of actual correlations. This would be bad news for the researcher's theoretical model.

RESIDUALS. From the above, it is clear that the researcher needs to know how similar the predicted model matrix is to the matrix of actual correlations. To make this comparison easier, the computer program subtracts each correlation in the predicted model matrix from its counterpart in the matrix of actual correlations. The result is called the ***raw residual matrix*** (or simply the *residual matrix*). The individual values in this matrix are called *residuals* or *discrepancies*.

If the researcher's theoretical model provides a good fit to the data, then each value in this raw residual matrix should be zero (or close to zero)—the larger the residuals, the poorer the fit between model and data. Some experts (e.g., McDonald & Ho, 2002) recommend that researchers should carefully review this residual matrix in order to determine how well the model fits the data, and to identify the specific relationships that are not well-accounted for by the theoretical model.

Global Fit Indices: Basic Concepts

Although the residual matrix is important for determining whether a given model displays a good fit to the data, it is often large and unwieldy, containing dozens and dozens of values. As a result, the residual matrix itself can be difficult to interpret.

GLOBAL FIT INDICES. Wouldn't it be great if we could compute just one number, and this one number would tell us how well the theoretical model fits the data? Yes it would be great, and we would call such a statistic a ***global fit index***: a single value that summarizes the overall fit between the theoretical path model and the data.

There is just one catch: Researchers have developed dozens of different ways of defining "fit of the model," and this has led to the development of dozens of different global fit indices—indices that may differ radically in their interpretation. Some indices (such as the NFI) range from 0 to 1, while others (such as the model χ^2 statistic) may display values much larger than 1. With some of the indices that range from 0 to 1 (such as the CFI and the GFI), higher values indicate a better fit. With other indices that range from 0 to 1 (such as the RMSEA and the SRMR), lower values may indicate a better fit.

It is not feasible to describe all of these global fit indices in this chapter. Instead, the next few sections will describe the most widely-used indices, will explain the differences between them, and will provide some guidelines for their interpretation.

PARSIMONIOUS FIT INDICES. Some of the following indices merely reflect the goodness of fit displayed by the theoretical model. Other indices reflect goodness of fit, but also reward models for being more parsimonious. A *parsimonious* model is a *simpler* model. Researchers know that they can almost always obtain a good fit to the data by adding more and more paths to their theoretical models. However, most experts agree that the best models are the parsimonious models: the ones that do a good job of accounting for the observed relationships in the data, but do so with a relatively small number of paths. ***Parsimonious fit indices*** reflect the goodness of fit between model and data, but also take into account the complexity of the model and reward models for being less complex. Examples of parsimonious indices include the *root mean square error of approximation* (RMSEA; MacCallum, Browne, & Cai, 2006) and the *Akaike information criterion* (AIC; Akaike, 1987). With the RMSEA and the AIC, smaller values indicate a better fit.

THE PARSIMONY RATIO. James, Mulaik, and Brett (1982) propose a ***parsimony ratio*** (symbol: PR): a statistic that merely reflects the parsimony (not the fit) of a structural equation model. Values of the PR may range from .00 (representing the least parsimonious model possible) to 1.00 (representing the most parsimonious model possible).

It is computed as follows:

$$PR = \frac{df_t}{df_n}$$

Where:

PR = Parsimony ratio

df_t = degrees of freedom for the theoretical model of interest

df_n = degrees of freedom for the null model

In the formula for the parsimony ratio, df represents degrees of freedom for a model chi-square statistic. More specifically, df_t represents the degrees of freedom for the theoretical model that the researcher is evaluating (for Dr. O'Day, this would be Model 1, Model 2, or Model 3). Similarly, df_n represents the degrees of freedom for the *null model* (also called the *uncorrelated-variables model* or the *independence model*). The null model is a model in which all correlations between all manifest variables are constrained to be equal to zero. Most applications routinely report the χ^2 statistic and df for the null model.

The parsimony ratio is important for two reasons. First, it is often reported in published articles so that readers can compare the parsimony of competing theoretical models. Second, it is sometimes multiplied by a global fit index to create a parsimonious fit index. For example, The NFI (a global fit index) is multiplied by the PR to create the ***parsimonious normed-fit index*** (symbol: PNFI). The PNFI reflects a given model's fit to the data, but also rewards models for being more parsimonious (James et al., 1982).

Widely Used Global Fit Indices

This section lists the global fit indices that you are likely to encounter when reading about studies that make use of path analysis or other forms of structural equation modeling (Jackson, Gillaspy, & Purc-Stephenson, 2009; McDonald & Ho, 2002; O'Boyle & Williams, 2011). For some indices, this section also provides criteria that have been recommended for determining whether a given model displays an acceptable fit to the data.

Until the late 1990s, most researchers evaluated their models using criteria that, by today's standards, were not terribly stringent. However, in an influential article, Hu and Bentler (1999) provided a more stringent set of criteria, and those criteria appear to have been widely adopted by researchers who perform structural equation modeling. Where possible, this section provides these more stringent criteria (Hu & Bentler, 1999; Mueller & Hancock, 2008).

THE MODEL CHI-SQUARE STATISTIC. Articles reporting path analyses almost always include the *model chi-square statistic* (i.e., the *model* χ^2). This is an inferential procedure that tests the null hypothesis that the model fits the data. Theoretically, when the researcher's model provides a good fit to the data, the model chi-square statistic itself should be a small value (i.e., the χ^2 should be closer to zero), and it should be nonsignificant (i.e., the *p* value for the chi-square statistic should be larger than the standard criterion of .05).

In practice, however, the chi-square statistic tends to be overly sensitive—it is often statistically significant even when the model provides a fairly good fit to the data. Because of this, researchers seldom place much weight on the idea that the *p* value must be nonsignificant for a model to be acceptable. Instead, researchers are more likely to focus on a *ratio*: the ratio of the obtained chi-square statistic to the degrees of freedom for the analysis (i.e., the ratio of χ^2/df). With this approach, the model is seen as providing a good fit to the data when this χ^2/df ratio is lower than some targeted value, perhaps 2 or 3 or 5 or so. Klem (2000) points out that there does not appear to be a widely accepted rule of thumb as to how small the ratio must be.

When preparing an article that describes a path analysis, it is very important to report the model chi-square with its degrees of freedom for each model that you estimate. As was explained above, few researchers pay much attention to the statistical significance of this statistic. However, the model chi-square statistic is important for other purposes, as will be explained a bit later in this chapter.

THE RMSEA. The abbreviation *RMSEA* stands for *root mean square error of approximation* (MacCallum et al., 2006). The RMSEA is very widely reported, possibly because it is a parsimonious fit index. With the RMSEA, smaller values (closer to 0) indicate a better fit between model and data. Traditionally, the fit was viewed as being acceptable if RMSEA was < .08, although more recent criteria suggest that the fit is acceptable if the RMSEA is ≤ .06 (Hu & Bentler, 1999; McDonald & Ho, 2002, p.72; Mueller & Hancock, 2008).

THE AIC. The abbreviation *AIC* stands for *Akaike information criterion* (Akaike, 1987). Like the RMSEA, the AIC is also a parsimonious fit index. With the AIC, smaller values indicate a better fit, although there do not appear to be any widely accepted standards indicating just how small it must be to be acceptable. The AIC is valued, in part, because it allows researchers to compare the fit of competing models that are not nested (the concept of models being *nested* will be explained in the section titled "The Chi-Square Difference Test," later in this chapter). When non-nested models are being compared, the one with the lower value for the AIC is usually preferred (Keith, 2006).

THE SRMR. The abbreviation *SRMR* stands for *standardized root mean-square residual* (Bentler, 1995). With this index, smaller values (closer to 0) indicate a better fit between model and data. A model's fit may be acceptable if the SRMR is ≤ .08 (Hu & Bentler, 1999; Mueller & Hancock, 2008).

THE NFI AND PNFI. Values of the *normed-fit index* (abbreviation: *NFI*; Bentler & Bonett, 1980) may range from 0.00 to 1.00, with higher values (closer to 1.00) indicating better fit. The *parsimonious normed-fit index* (abbreviation: *PNFI*; James et al., 1982) reflects fit, but also rewards models for being more parsimonious. Because the NFI often displays biased results in small samples, its use has been largely supplanted by the comparative fit index (CFI), which is discussed next.

THE CFI AND PCFI. The abbreviation *CFI* stands for *comparative fit index* (Bentler, 1990). Compared to the NFI, the CFI tends to be less biased in small samples. Values of the CFI may range from 0.00 to 1.00, with higher values indicating a better fit. Traditionally, the fit was viewed as being acceptable if the CFI was > .90. More recent criteria require that the CFI should be ≥ .95 (Hu & Bentler, 1999; Mueller & Hancock, 2008). The *parsimony CFI*, or *PCFI* reflects fit, but also rewards models for being more parsimonious (Keith, 2006).

THE NNFI OR TLI. The abbreviation *NNFI* stands for the *nonnormed fit index* (Bentler & Bonett, 1980). The same statistic is also called the *Tucker-Lewis Index*, or *TLI* (Tucker & Lewis, 1973). The NNFI is similar to the CFI, except that its values may in some situations be slightly lower than .00 or slightly higher than 1.00. With the NNFI, higher values indicate a better fit. Traditionally, a model was viewed as being acceptable if the NNFI > .90. More recent criteria require that the NNFI be ≥ .95.

THE GFI AND AGFI. The *goodness-of-fit index* (abbreviation: *GFI*) reflects the overall discrepancy between the actual correlation or covariance matrix versus the predicted correlation or covariance matrix (Joreskog & Sorbom,1984; Mueller &

Hancock, 2008). Values may range from 0 to 1, with higher values indicating a better-fitting model (Bryant & Yarnold, 1995; Huck, 2012). The *adjusted goodness-of-fit index* (abbreviation: *AGFI*) is a parsimonious fit index that reflects this discrepancy but also rewards models for being less complex. With both statistics, values exceeding .90 indicate acceptable fit (Mueller & Hancock, 2008).

Summary of Criteria for Evaluating Global Fit Indices

Is your head swimming with all of these indices? It should be. For better or worse, articles describing a path analysis often report a multitude of global fit indices. Regarding this practice, McDonald and Ho (2002) observed:

> It is sometimes suggested that we should report a large number of these indices, apparently because we do not know how to use any of them. (p.72)

This large number of indices means that there is a lot of information that you (the reader) must critically evaluate. To make this task more manageable, Table 16.1 summarizes most of the global fit indices just described. For each index, the table presents the criterion for an acceptable model fit, as recommended by Hu and Bentler (1999) and Mueller and Hancock (2008). The table omits global fit indices for which these authors did not suggest a specific criterion.

Table 16.1

Widely Used Global-Fit Indices and Criteria for an Acceptable Model Fit

Criterion	Global fit index
≤ 0.06	*RMSEA (Root mean-square error of approximation)* [a]
≤ 0.08	*SRMR (Standardized root mean-square residual)* [a]
≥ 0.90	NFI (Normed-fit index)
≥ 0.95	*CFI (Comparative-fit index)* [a]
≥ 0.95	NNFI (Nonnormed-fit index); TLI (Tucker-Lewis Index)
≥ 0.90	GFI (Goodness-of-fit index)
≥ 0.90	AGFI (Adjusted goodness-of-fit index)

Note. Criteria are based on Hu and Bentler (1999) and Mueller and Hancock (2008).
[a] Indices presented in italics are recommended by Mueller and Hancock (2008).

Global Fit Indices Obtained in the Current Analysis

Enough with the background—it is time to see the fit indices obtained by Dr. O'Day when she compared her three theoretical models. Imagine that, in the *Results* section of her article, she reports the following:

Table 16.2

Path Analysis Results: Fit Indices for a Null Model and Three Theoretical Path Models

Model	\multicolumn{4}{c}{Model χ^2}					
	χ^2	df	p	χ^2/df	SRMR	RMSEA	NFI	CFI	NNFI
Null model	207.54	15	<.001	13.84					
Model 1	51.34	6	<.001	8.56	.10	.18	.75	.76	.41
Model 2	18.90	5	.002	3.78	.07	.11	.91	.93	.78
Model 3	5.85	4	.211	1.46	.03	.04	.97	.99	.96

Note. $N = 238$. SRMR = Standardized root-mean-square residual; RMSEA = root-mean-square error of approximation; NFI = normed-fit index; CFI = comparative fit index; NNFI = nonnormed fit index. Model 1 was the path model consisting of solid-line arrows in Figure 16.2. Model 2 was identical to Model 1, except that Model 2 included an additional path from lack of time for exercise to exercise sessions. Model 3 was identical to Model 2, except that Model 3 included an additional path from dislike of exercise to health motivation.

Table 16.2 presents goodness-of-fit indices obtained from the path analysis. It presents these results for four different models. First, the "null model" is a model in which all variables are uncorrelated (this model serves as a baseline against which the three theoretical models may be compared). The first theoretical model is Model 1, the path model represented by the solid-line arrows in Figure 16.2. Model 2 is identical to Model 1, except that it contains an additional path from lack of time for exercise to exercise sessions (represented by the dashed-line arrow in Figure 16.2). Model 3 is identical to Model 2, except that it contains an additional path from dislike of exercise to health motivation (represented by the dotted-line arrow).

Table 16.2 shows that, of the three theoretical models being compared, only Model 3 displays an acceptable fit to the data. Only Model 3 displays a model chi-square statistic that is nonsignificant, $\chi^2(4, N = 238) = 5.85$, $p = .211$. It has been recommended (e.g., Hu & Bentler, 1999; Mueller & Hancock, 2008) that a model be viewed as displaying an acceptable fit if the SRMR is ≤ .08, the RMSEA is ≤ .06, and the CFI and NNFI are ≥ .95. Again, Table 16.2 shows that only Model 3 satisfies all of these criteria.

The Chi-Square Difference Test

An earlier section indicated that, when feasible, it is a best practice to compare alternative models that reflect different theoretical perspectives. It is ideal if these models are *nested*, because the fit of nested models can be directly compared by using a special application of the chi-square test.

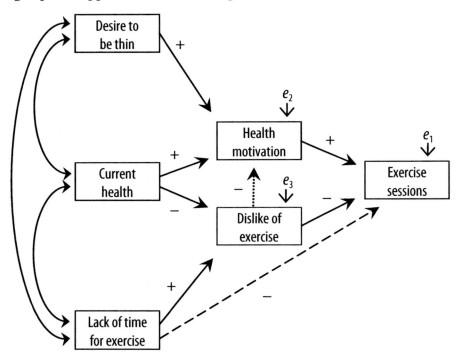

Figure 16.3. Composite model illustrating the three models to be compared. Model 1 consists of the paths represented by solid-line arrows; Model 2 is identical to Model 1, except that Model 2 contains an additional path from lack of time for exercise to exercise sessions (represented by the dashed-line arrow); Model 3 is identical to Model 2, except that Model 3 contains an additional path from dislike of exercise to health motivation (represented by the dotted-line arrow).

NESTED MODELS. A simpler model is said to be *nested* within a more complex model if deleting one or more paths (or other parameters) from the more complex model results in the simpler model. Given this definition, you can see from Figure 16.3 that Dr. O'Day's Model 1 is nested within Model 2. You know that the models are nested because Model 2 is identical to Model 1, except that it contains one additional path—the dashed-line arrow from lack of time for exercise to exercise sessions. Model 1 is essentially just a simpler version of Model 2, and that means that it is *nested within* Model 2. In the same way, the figure also shows that Dr. O'Day's Model 2 is nested within Model 3.

THE LOGIC OF THE TEST. Adding a new path to a simpler model almost always improves that model's fit to the data. However, in many cases, the improvement in fit is only trivial in size. Making matters worse is the fact that adding the new path will certainly make the model less parsimonious, and a less parsimonious model is usually less desirable. Therefore, researchers typically want to add a new path to a model only if adding that path results in a substantial improvement in the model's fit to the data.

Because of these considerations, researchers sometimes perform a significance test to help determine whether adding the new path to an existing model results in a statistically significant improvement in the model's fit. This significance test is the ***chi-square difference test***, and it is based on the ***chi-square difference statistic*** (symbol: $\Delta\chi^2$).

OBTAINING MODEL χ^2 FOR THE TWO MODELS. Before the researcher can perform this significance test, she must first obtain the model chi-square statistic for two path models: the simpler model that does not contain the new path of interest (Model 1, in this case), and the more complex model that does contain the new path of interest (Model 2, in this case). In fact, these χ^2 statistics for Model 1 and Model 2 already appear in Table 16.2, under the heading "χ^2."

THE χ^2 DIFFERENCE TEST. To compute the difference statistic, the researcher begins with the model chi-square statistic for the simpler model, and subtracts from it the model chi-square statistic for the more complex model The result is the observed *chi-square difference statistic* (symbol: $\Delta\chi^2$).

To determine whether this $\Delta\chi^2$ is statistically significant, the researcher looks up the critical value in a table of chi-square in the back of a statistics textbook. The degrees of freedom for this test are equal to the number of paths in the more complex model minus the number of paths in the simpler model. If the observed $\Delta\chi^2$ statistic is larger than this critical value of χ^2, this means that the addition of the new path to the model resulted in a statistically significant improvement in the model's fit to the data. In other words, if the results are statistically significant, it means that the complex model provides a fit that is significantly better than the fit displayed by the simpler model.

COMPARING MODEL 1 TO MODEL 2. Table 16.2 shows that the model chi-square statistic for Model 1 is 51.34, and the model chi-square statistic for Model 2 is 18.90. Subtracting the latter from the former results in a chi-square difference statistic of $\Delta\chi^2 = 32.44$. Model 2 has one more causal path than Model 1, and this means that the chi-square difference test has 1 degree of freedom. A table of chi-square indicates that, with alpha set at $\alpha = .05$, the critical value of chi-square with 1 *df* is $\chi^2_{\text{crit}} = 3.84$. Because the observed chi-square difference score (32.44) is larger than this critical value (3.84), Dr. O'Day concludes that adding the dashed-line path to Model 1 resulted in a significant improvement in the model's fit to the data. In other words, Model 2 provides a significantly better fit than Model 1.

COMPARING MODEL 2 TO MODEL 3. In Figure 16.3, Model 3 is identical to Model 2, except that Model 3 contains one additional path: The dotted-line path that goes from dislike of exercise to health motivation. To determine whether the addition of this path results in another significant improvement in the model's fit, Dr. O'Day performs another chi-square difference test.

Table 16.2 shows that the model chi-square statistic for Model 2 is 18.90, and the model chi-square statistic for Model 3 is 5.85. Subtracting the latter from the former results in a chi-square difference statistic of $\Delta\chi^2 = 13.05$. Model 3 has one more causal path than Model 2, which means that the chi-square difference test again has 1 degree of freedom. This, in turn, means that the critical value of chi-square is once again $\chi^2_{crit} = 3.84$. The observed chi-square difference score (13.05) is larger than this critical value (3.84), so Dr. O'Day concludes that adding the dotted-line path from dislike of exercise to health motivation resulted in a significant improvement in the model's fit to the data. In other words, Model 3 provides a significantly better fit than Model 2.

R^2 Statistics for the Competing Models

Table 16.3 presents R^2 statistics for the three models being compared in the current study. In the section headed "R^2 Statistics," you will find the three endogenous variables included in the study: *dislike of exercise, health motivation*, and *exercise sessions*. Under the heading "Model," you will find the names of the three models being compared.

An R^2 statistic indicates the percent of variance in a given endogenous variable that is accounted for by the variables hypothesized to have an effect on it. This statistic is interpreted in the same way that the R^2 statistic from multiple regression is interpreted: Its values may range from .00 to 1.00, and higher values indicate that the antecedent variables are doing a better job of predicting the endogenous variable in question.

In Table 16.3, you can see that the variables included in Model 1 accounted for 17% of the variance in dislike of exercise, 19% of the variance in health motivation, and 15% of the variance in exercise sessions. The R^2 statistic for Model 2 and Model 3 may be interpreted in the same way.

For a given endogenous variable, comparing the R^2 statistic for different models shows what happens to that statistic when a new path is added to a model. For example, with Model 1, there was no path from lack of time for exercise to exercise sessions, and the two variables that did have a direct effect on exercise sessions accounted for only 15% of the variance in it. With Model 2, a direct path was added from lack of time for exercise to exercise sessions, and this change resulted in a model that accounted for 26% of the variance in exercise sessions. The R^2 statistics for the other variables and other models may be interpreted in the same way.

Table 16.3

Path Analysis Results: R^2 Statistics for Three Endogenous Variables Under Models 1, 2, and 3

	R^2 Statistics		
Model	Dislike of exercise	Health motivation	Exercise sessions
Model 1	.17	.19	.15
Model 2	.17	.19	.26
Model 3	.17	.22	.28

Note. $N = 238$. Model 1 was the path model consisting of solid-line arrows in Figure 16.3. Model 2 was identical to Model 1, except that Model 2 included an additional path from lack of time for exercise to exercise sessions. Model 3 was identical to Model 2, except that Model 3 included an additional path from dislike of exercise to health motivation.

The results presented in Table 16.3 are generally supportive of Dr. O'Day's Model 3. The R^2 statistics show that the two paths that appear in Model 3 (but not in Model 1) result in a substantial improvement in the model's ability to account for variance in health motivation and exercise sessions.

Selecting the Final Preferred Model

Up to this point, this chapter has gone into excruciating detail in describing the various criteria and statistics that researchers use when evaluating structural equation models. It is now time *(finally!)* for Dr. O'Day to put these ideas together. Notice how she does this in the following article excerpt, systematically moving from model to model, narrowing it down to her final preferred model.

Table 16.4

Path Analysis Results: Fit Indices for a Null Model and Three Theoretical Path Models

Model	Model χ^2				SRMR	RMSEA	NFI	CFI	NNFI
	χ^2	df	p	χ^2/df					
Null model	207.54	15	<.001	13.84					
Model 1	51.34	6	<.001	8.56	.10	.18	.75	.76	.41
Model 2	18.90	5	.002	3.78	.07	.11	.91	.93	.78
Model 3	5.85	4	.211	1.46	.03	.04	.97	.99	.96

Note. N = 238. SRMR = Standardized root-mean-square residual; RMSEA = root-mean-square error of approximation; NFI = normed-fit index; CFI = comparative fit index; NNFI = nonnormed fit index. Model 1 was the path model consisting of solid-line arrows in Figure 16.3. Model 2 was identical to Model 1, except that Model 2 included an additional path from lack of time for exercise to exercise sessions. Model 3 was identical to Model 2, except that Model 3 included an additional path from dislike of exercise to health motivation.

One purpose of this investigation was to compare three theoretical models (Model 1, Model 2, and Model 3) to determine which provides the best fit to the data. These comparisons began by reviewing the global fit indices in Table 16.4.

It has been suggested (e.g., Hu & Bentler, 1999; Mueller & Hancock, 2008) that a structural equation model displays an acceptable fit to the data when the SRMR is $\leq$ .08, the RMSEA is $\leq$ .06, and the CFI and the NNFI are $\geq$ .95. Table 16.4 shows that only Model 3 meets all of these criteria.

A series of chi-square difference tests were performed to determine whether adding additional paths to Model 1 resulted in a significant improvement in the model's fit to the data. These chi-square difference tests were based on the model chi-square statistics that appear in Table 16.4. They test the three theoretical models represented in Figure 16.4.

Figure 16.4 presents a composite of the three theoretical models being compared in this study. Model 1 is the simplest of the three models—it consists of the solid-line causal arrows. Adding a new path from lack of time for exercise to exercise sessions (the dashed-line arrow) produces Model 2. The addition of this new path caused a significant reduction in the model chi-square statistic, $\Delta\chi^2 =$ 32.44, $p < .05$. This indicated that Model 2 provided a fit to the data that was significantly better than the fit provided by Model 1. The addition of this path also caused the R^2 statistic for exercise sessions to increase from .15 to .26.

To create Model 3, a new path was added to Model 2. This new path went from dislike of exercise to health motivation (as represented by the dotted-line arrow). This addition caused another significant reduction in the model chi-square statistic, $\Delta\chi^2 = 13.05$, $p < .05$, indicating that Model 3 provided a significantly better fit than Model 2. The addition of this path caused the R^2 statistic for health motivation to increase from .19 to .22.

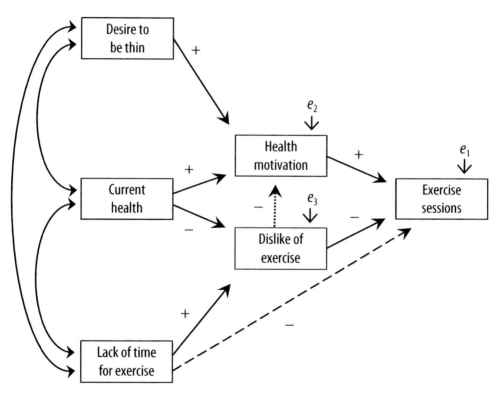

Figure 16.4. Composite of three models being compared.

Table 16.5 presents the observed correlations between the six manifest variables in the current analysis, along with the raw residual matrix from the analysis of Model 3 (observed correlations are below the diagonal, and residuals are above the diagonal). The table shows that most residuals were near-zero in magnitude. None of the asymptotically standardized residuals (not presented here) were larger than ±1.80.

Table 16.5

Means, Standard Deviations, Observed Correlations (Below the Diagonal) and Residuals from the Raw Residual Matrix (Above the Diagonal) for Model 3.

Variable	M	SD	1	2	3	4	5	6
1. Exercise sessions	2.93	2.31	--	.02	-.00	.06	.07	-.01
2. Health motivation	5.76	0.98	.29	--	-.02	.02	.00	-.04
3. Dislike of exercise	2.36	1.31	-.37	-.31	--	-.10	.00	.00
4. Desire to be thin	5.30	1.35	.14	.30	-.15	--	.00	.00
5. Current health	5.46	0.94	.25	.35	-.22	.16	--	.00
6. Lack of time for exercise	3.67	1.71	-.46	-.20	.39	-.08	-.22	--

Note. $N = 238$.

Based on these analyses, Model 3 was selected as the final preferred model. This selection was based on the fact that (a) all of the causal paths in Model 3 were derived from theory and previous research, (b) only Model 3 displayed global fit indices that met all criteria suggested by Hu and Bentler (1999) and Mueller and Hancock (2008); (c) chi-square difference tests showed that adding two new causal paths to Model 1 (and thus creating Model 3) each resulted in a significant improvement in the model's fit to the data; (d) R^2 values for the three endogenous variables in Model 3 were of meaningful size; (e) values in the residual matrix for Model 3 were near-zero, and (f) path coefficients for all causal paths hypothesized by Model 3 were statistically significant ($\alpha = .05$).

Figure 16.5 presents Model 3. The straight, single-headed arrows represent direct effects, and are identified with standardized path coefficients. The curved, double-headed arrows represent Pearson correlations between exogenous variables.

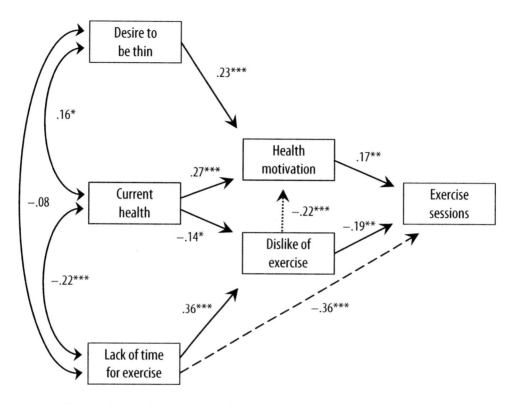

Figure 16.5. Model 3 with standardized path coefficients for causal paths (represented by straight, single-headed arrows) and Pearson *r* coefficients for correlations between exogenous variables (represented by curved, two-headed arrows). Coefficients flagged with asterisks are significantly different from zero, $*p < .05$; $**p < .01$; $***p < .001$.

Path Coefficients in the Final Preferred Model

The results presented in Figure 16.5 are of particular importance to the researcher, as these results reflect the direct effects that the antecedent variables had on consequent variables. The ***direct effect*** of an antecedent variable on a consequent variable is that part of the effect that is not mediated by another variable (Pedhazur, 1982). We make a distinction between *direct effects* (which are covered in this section) versus *indirect effects* (which will be covered a bit later).

STANDARDIZED PATH COEFFICIENTS. In Figure 16.5, direct effects are represented by the straight, single-headed arrows. The numbers on these arrows are standardized path coefficients—statistics which represent the size of the effects. These path coefficients are similar to the standardized multiple regression coefficients (i.e., beta weights, symbol: β) that are obtained in multiple regression. In fact, researchers sometimes use the symbol β to represent standardized path coefficients (e.g., Judge, Hurst, & Simon, 2009).

A ***standardized path coefficient*** indicates the amount of change in a consequent variable (as measured in standard deviations) that is associated with a one-standard deviation increase in the antecedent variable. Path coefficients may be either positive or negative. A statistically significant positive path coefficient is typically interpreted as evidence that an increase in scores on the antecedent variable brings about an increase in scores on the consequent variable. For example, the arrow connecting health motivation to exercise sessions in Figure 16.5 is identified with a standardized path coefficient of .17 (a positive value). The size of the coefficient indicates that, for every increase of one standard deviation in health motivation, there is an increase of 0.17 standard deviations in exercise sessions.

Conversely, a statistically significant negative path coefficient means that an increase in scores on the antecedent variable brings about an *decrease* in scores on the consequent variable. For example, the arrow connecting lack of time for exercise to exercise sessions is identified with a standardized path coefficient of −.36. The size of the coefficient indicates that, for every increase of one standard deviation in lack of time for exercise, there is a *decrease* of 0.36 standard deviations in exercise sessions.

Theoretically, standardized path coefficients are the statistics that would be obtained if all variables had been standardized to have a mean of zero and a standard deviation of one. Standardized path coefficients typically range in value from −1.00 through 0.00 through +1.00. When a standardized path coefficient is equal to zero, it means that the antecedent variable did not have any direct effect on the consequent variable. Larger coefficients (in absolute value) represent stronger direct effects.

Articles often use asterisks to indicate whether specific path coefficients are statistically significant. In this case, "statistically significant" means "significantly different from zero."

UNSTANDARDIZED PATH COEFFICIENTS. When research articles report path coefficients, they typically report the standardized path coefficients described above. In rare instances, they may instead report *unstandardized path coefficients*. An ***unstandardized path coefficient*** represents the amount of change in a consequent variable that is associated with a one-unit increase in the antecedent variable, but is computed using variables that have *not* been standardized to each have a mean of zero and a standard deviation of one. Because the variables have not be standardized, the resulting path coefficients are not constrained to range from −1.00 through 0.00 through +1.00. This, in turn, means that the relative size of unstandardized path coefficients cannot be used to determine which effects were relatively strong versus weak.

CRITERIA FOR INCLUDING A HYPOTHETICAL CAUSAL PATH. The inclusion of a path in the final preferred path model can be justified only if a number of criteria have been met. One obvious criterion is that the model as a whole should display an acceptable fit to the data. Much of the current chapter has focused on this issue of model fit.

A second criterion is that the path coefficient for the proposed causal relationship should be significantly different from zero. All of the path coefficients in Figure 16.5 are identified with one or more asterisks, indicating that all are significantly different from zero ($p < .05$ or smaller).

A final criterion is that the standardized path coefficient should be nontrivial in size. Just how large a coefficient must be to be considered "nontrivial in size" is determined, in part, by the size of the coefficients found by other researchers who have studied the same variables. One general-purpose criterion suggested by Billings and Wroten (1978) is that a standardized path coefficient should exceed ±.05 (in absolute value) to be considered meaningful. All of the path coefficients in Figure 16.5 satisfy this criterion as well.

SYMBOLS USED TO REPRESENT PATH COEFFICIENTS. Researchers use a variety of different symbols to represent path coefficients. First, many authors use the symbol *p* to represent path coefficients. This can cause confusion of course, since the symbol *p* also represents the *p* value, or probability value, from a significance test. If the variables in a causal model have been identified with numbers, then specific path coefficients can be represented using the convention p_{YX}, where *Y* represents the consequent variable where the arrow ends, and *X* represents the antecedent variable where the arrow begins (Pedhazur, 1982). For example, the symbol p_{21} might represent the path coefficient for the direct effect of Variable 1 on Variable 2. A similar convention is used when variables have been identified with letters. For example, if the variable *exercise sessions* is identified with the letter "E," and *health motivation* is identified with the letter "H," then the path coefficient for the effect of health motivation on exercise sessions might be identified using the symbol p_{EH} (Klem, 1995).

Some authors use Greek letters to represent path coefficients (Klem, 2000). With this convention, the lower-case Greek letter gamma (γ) represents the effect of an exogenous variable on an endogenous variable, and the lower-case Greek letter beta (β) represents the effect of one endogenous variable on another endogenous variable. In some texts, subscripts are once again used to identify the specific variables included in the causal path (Long, 1983b). For example, the symbol β_{EH} might represent the effect of health motivation on exercise sessions, and γ_{EL} might represent the effect of lack of time for exercise on exercise sessions.

FIGURES THAT PRESENT *e* TERMS IN A PATH MODEL. Earlier, you learned that the symbol *e* is often used to represent residual terms for the endogenous variables in a causal model. The **residual term** (also called the *error term* or *disturbance term*) represents the variance in the endogenous variable that was not accounted for by its antecedent variables. The size of a residual term is inversely related to the size of the R^2 statistic for a given endogenous variable: As R^2 increases in size, the *e* term decreases.

The following formula is used to compute the residual term for an endogenous variable:

$$e = \sqrt{(1 - R^2)}$$

Where:

e = the residual term for the endogenous variable

R^2 = the variance in that endogenous variable that is accounted for by antecedent variables in the theoretical path model

For example, in Dr. ODay's Model 3, the R^2 value for the exercise sessions was R^2 = .28 (this value was presented in Table 16.3, earlier). Inserting this value in the preceding formula allows us to compute the residual term for exercise sessions:

$$e = \sqrt{(1 - R^2)}$$

$$e = \sqrt{(1 - .28)}$$

$$e = \sqrt{(.72)}$$

$$e = .848$$

$$e = .85$$

So the residual term for exercise sessions is e = .85. We can use the same formula to compute the residual terms for the study's other endogenous variables. Table 16.3 shows that (for Model 3), The R^2 value for health motivation was R^2 = .22; this produces a residual term of e = .88. The R^2 value for dislike of exercise was R^2 = .17; this produces a residual term of e = .91.

Published articles sometimes include the e term for each endogenous variable as part of the figure that illustrates a causal model, such as Figure 16.5. The e term is typically presented simply as a number (such as ".85") with a single-headed causal arrow going from this number to the relevant endogenous variable. The e terms were omitted from Figure 16.5 as the figure was already fairly well cluttered with path coefficients.

Direct Effects, Indirect Effects, and Total Effects

Once researchers have selected a final preferred model, they may present a table of direct, indirect, and total effects. These effects are computed for each endogenous variable in the analysis. The results for Dr. O'Day's Model 3 are presented in Table 16.6.

DIRECT EFFECTS. As was explained above, the ***direct effect*** (symbol: *DE*) of an antecedent variable on a consequent variable is that part of the effect that is not mediated by another variable (Pedhazur, 1982). The numbers on the straight, single-headed arrows in Figure 16.5 are standardized path coefficients, and these path coefficients may also be interpreted as direct effects. They are listed in Table 16.6 under the heading "Direct."

The most important consequent variable in Dr. O'Day's study was exercise behavior. Table 16.6 shows that the three variables predicted to have direct effects on exercise behavior did, in fact, display fairly large direct effect coefficients: $DE = +.17$ for health motivation, $DE = -.19$ for dislike of exercise, and $DE = -.36$ for lack of time for exercise.

Table 16.6

Path Analysis Results: Direct Effects, Indirect Effects, and Total Effects of Antecedent Variables on Endogenous Variables (for Model 3)

Antecedent variable	Exercise sessions			Health motivation			Dislike of Exercise		
	Direct	Indirect	Total	Direct	Indirect	Total	Direct	Indirect	Total
Health motivation	.17		.17						
Dislike of exercise	-.19	-.04	-.22	-.22		-.22			
Desire to be thin		.04	.04	.23		.23			
Current health		.08	.08	.27	.03	.30	-.14		-.14
Lack of time for exercise	-.36	-.08	-.43		-.08	-.08	.36		.36

Note. $N = 238$.

INDIRECT EFFECTS. In contrast to a direct effect, the ***indirect effect*** (symbol: *IE*) of one variable on a second variable represents the part of the effect that is transmitted by one or more mediating variables in the model (Pedhazur, 1982). A ***mediating variable*** is a variable that intervenes between an antecedent variable and a consequent variable and transmits the effect from the antecedent to the consequent.

An antecedent variable may have just one indirect effect on a consequent variable, or it may have multiple indirect effects. For example, Figure 16.5 shows that lack of time for exercise has two indirect effects on exercise sessions:

- With one indirect effect, lack of time for exercise has a positive effect on dislike of exercise, which in turn has a negative effect on exercise sessions.
- With the second indirect effect, lack of time for exercise has a positive effect on dislike of exercise, which in turn has a negative effect on health motivation, which in turn has a positive effect on exercise sessions.

In Table 16.6, the indirect effects of antecedent variables on consequent variables appear under the heading "Indirect." The table shows that the indirect effects of antecedent variables on exercise sessions were relatively small, ranging (in absolute value) from .04 to .08.

TOTAL EFFECTS. Finally, the statistics that may be of greatest interest in Table 16.6 are the *total effects*, which are sometimes called ***effect coefficients*** (symbol: *EC*). The ***total effect*** of an antecedent variable on a consequent variable refers to the sum of its direct effect plus its indirect effects on that consequent variable (Klem, 1995; Pedhazur, 1982). For example, Table 16.6 shows that:

- The direct effect of lack of time for exercise on exercise sessions is *DE* = −.36.

- The indirect effect of lack of time for exercise on exercise sessions is *IE* = −.08.

- Therefore the total effect of lack of time for exercise on exercise sessions is equal to the sum of the preceding two statistics: (−.36) + (−.08) = −.44. The fact that the total effect coefficient reported in the table (*EC* = −.43) is not exactly equal to the sum computed above (−.44) is merely due to rounding error.

Researchers often review the total effects from a path analysis in order to identify the antecedent variables that may be the most important overall determinants of the consequent variables. This is one of the ways in which path analysis is superior to multiple regression: Multiple regression is only capable of identifying the variables with the most important direct effects on a criterion variable (that is, if we think of the relationships studied in multiple regression as being *causal* relationships). However, path analysis allows the computation of total effects, which reflect direct as well as indirect effects. In this way, path analysis provides a more nuanced and comprehensive understanding of the relationships between the variables.

In Table 16.6, total effect coefficients appear below the heading "Total." The table shows that the three variables having the largest total effects on exercise sessions were lack of time for exercise (*EC* = −.43), dislike of exercise (*EC* = −.22), and health motivation (*EC* = +.17).

Summary of Findings

Of the three theoretical path models investigated, Model 3 displayed the best fit to the data—a fit that was significantly better than that displayed by Model 1 or Model 2. Model 3 displayed a nonsignificant model chi-squared statistic of $\chi^2(4, N = 238) = 5.85, p = .211$. For Model 3, the SRMR = .03 (lower than the recommended value of $\leq .08$), the RMSEA = .04 (lower than the recommended value of $\leq .06$), the CFI = .99 and the NNFI = .96 (both higher than the recommended values of $\geq .95$). For the three endogenous variables, R^2 statistics were respectable, ranging from .17 to .28.

For Model 3, the signs of the standardized path coefficients (as presented in Figure 16.5) were consistent with those initially proposed by Dr. O'Day. Each path coefficient was nontrivial in size according to criteria provided by Billings and Wroten (1978). For the various relationships between variables, total effect coefficients ranged from .04 to −.43.

Additional Issues Related to this Procedure

Here, we tie up some loose ends. This section describes (a) how modern SEM software now allows researchers to model the measurement error in the observed variables, thus satisfying one of the oft-violated assumptions underlying path analysis, (b) the difference between recursive versus nonrecursive path models, and (c) the popular (although somewhat questionable) practice of making data-driven model modifications. The section ends with a list of assumptions underlying path analysis.

Modeling Measurement Error in the Observed Variables

One of the assumptions underlying path analysis (to be summarized at the end of this chapter) is the requirement that the exogenous manifest variables included in the analysis must be perfectly reliable—that is, they must be measured without error. When this assumption is not satisfied, the path coefficients obtained from the analysis can be biased and misleading.

Shockingly, this assumption of perfectly reliable measurement has not always been met in path-analytic investigations. As a result, the published research literature almost certainly contains path analyses whose results should be taken with a grain of salt.

Fortunately, however, everything changed in the 1970s. That decade saw the development and spread of computer applications that made it possible for researchers to analyze nonexperimental data using a family of procedures called *structural equation modeling*. At the beginning, the best-known of these applications was *LISREL*, short for *linear structural relations* (Jöreskog & Sörbom, 1989).

Structural equation modeling has provided two approaches for addressing with the requirement that exogenous manifest variables must be measured without error. First, it is now possible to perform *path analysis with latent factors* (this procedure will be covered in *Chapter 18* of this book). The latent factors included in these analyses are measured without error (theoretically, at least), and this satisfies the assumption of perfectly reliable variables.

When performing a path analysis with latent factors is not possible, however, there is still a second alternative, and this second alternative is somewhat simpler. It involves performing path analysis with manifest variables (as described in the current chapter), and incorporating the known reliability of the variables as part of the model. In simple terms, this means that the researcher supplies the computer application with a reliability estimate for each manifest variable (along with some additional information), and the computer application then takes this reliability into account when estimating the path coefficients. As a result, the path coefficients tend to be less biased, compared to the situation in which the reliability of the measures is not taken into account at all.

For a more detailed discussion of these issues, see Klem (2000). For an example of a study in which a path model was estimated using this approach, see Netemeyer, Johnston, and Burton (1990).

Recursive Versus Nonrecursive Models

All of the path models tested in this chapter were *recursive models*. **Recursive models** are causal models in which causation flows in only one direction. For example, in Dr. O'Day's Model 1 (which was first presented in Figure 16.1), causation flowed from the antecedent variables on the left side of the figure to the consequent variables on the right side.

It is also possible to perform path analysis using ***nonrecursive models***: models in which causation flows in more than one direction. A path model may be a nonrecursive path model because it contains a feedback loop, or possibly because there is reciprocal causation between variables (e.g., Variable A affects Variable B, and at the same time Variable B affects Variable A).

Years ago, it was difficult to estimate and test nonrecursive models. However, with the development of structural equation modeling software (such as LISREL, AMOS, EQS, and PROC CALIS), this has become much easier. As a consequence, today you are much more likely to encounter articles that report path models with feedback loops, correlated error terms, and other characteristics of nonrecursive models.

Assumptions Underlying Path Analysis with Manifest Variables

Maximum likelihood (ML) estimation is probably the most popular procedure used to estimate parameters in a path analysis, and this procedure requires that the variables be assessed on an interval or ratio scale of measurement (Klem, 1995) and that they follow a multivariate normal distribution (McDonald & Ho, 2002). Although these assumptions imply that all variables must be continuous quantitative variables, in practice it is not unusual to see dichotomous exogenous variables that have been dummy-coded. Some computer applications (such as *Mplus*) have been developed specifically to handle nonnormal data, categorical variables, and ordinal-scale variables (e.g., Browne, 1984; Muthén, 1984; Muthén & Muthén., 2001).

The assumptions underlying path analysis with manifest variables include the following (taken from Hatcher, 1994; Keith, 2006; Kenny, 1979; Klem, 1995). Some of these assumptions are also discussed in greater detail in *Chapter 2*.

INDEPENDENT OBSERVATIONS. Each observation should be drawn independently from the population of interest. Among other things, this means that the researcher should not take repeated measures from the same subject (unless special procedures have been used to accommodate repeated measures).

MULTIVARIATE NORMALITY. Many of the statistical tests that are computed with a path analysis (such as the model chi-square test and the significance tests for path coefficients) assume that the data follow a multivariate normal distribution. Fortunately, the maximum likelihood estimation procedure appears to be fairly robust against mild to moderate violations of this assumption.

ABSENCE OF SPECIFICATION ERRORS. In part, "absence of specification errors" means that (a) all important determinants of the model's endogenous variables are included in the path model, and (b) no unimportant determinants have been included.

ABSENCE OF MEASUREMENT ERROR. All exogenous variables must be measured with perfect reliability.

CAUSAL MODEL MUST BE THEORETICALLY CORRECT. This means that the correct variables must be included in the model and the causal ordering of these variables must be correct. For guidance in developing theoretically correct causal models, see James et al. (1982).

LINEAR AND ADDITIVE RELATIONSHIPS. The relationships between all variables in the path model must be linear and additive. In other words, the relationships should not be curvilinear, and there should not be any interactive relationships between variables (again, unless special procedures have been used to model such curvilinear or interactive relationships).

ABSENCE OF MULTICOLLINEARITY. The variables should not demonstrate extremely strong correlations with one another. One popular rule of thumb says that there may be a multicollinearity problem if the Pearson correlation between two variables exceeds $r = \pm.80$.

IDENTIFICATION. The path model should be identified. The concept of *identification* is complex; see Bryant and Yarnold (1995, pp. 117-118) or Tabachnick and Fidell (2007, pp. 709-712) for an introduction.

LARGE SAMPLES. In most cases, path analysis should be based on large samples—Klem (1995) recommends 200 to 300 observations for most analyses. Some experts suggest a bare minimum of 5 to 10 participants for each parameter to be estimated. These parameters include all path coefficients, variances, and covariances contained in the path model (Bentler & Chou, 1988).

Chapter 17

SEM II: Confirmatory Factor Analysis

This Chapter's *Terra Incognita*

You are reading about an empirical investigation in which the researcher has administered an 18-item questionnaire to over 200 college students. She has a detailed set of hypotheses regarding the factor structure that underlies responses to these items and is using a procedure called *confirmatory factor analysis* to test these hypotheses. This is what she reports:

> Global fit indices showed that the six-factor measurement model provided an acceptable fit to the data: CFI = .97; NNFI = .96; RMSEA = .05; SRMR = .05. All of these indices met or exceeded the criteria for an acceptable fit recommended by Hu and Bentler (1999) and Mueller and Hancock (2008).
>
> For all six latent factors, composite reliabilities satisfied the criterion of $\alpha \geq .70$ recommended for coefficient alpha (Nunnally, 1978). Variance-extracted estimates for the six factors ranged from .54 to .87, satisfying the criterion of $\geq .50$ recommended by Fornell and Larcker (1981).
>
> Inter-factor correlations showed that the *exercise behavior* factor displayed the strongest correlation with the *lack of time for exercise* factor ($r = -.52$, $p < .001$). It displayed the weakest correlation with the *desire to be thin* factor ($r = .16$, $p < .05$).

Show a little backbone, will you? All shall be revealed.

The Basics of this Procedure

Confirmatory factor analysis (CFA) allows researchers to investigate the factor structure underlying a set of manifest variables. In this context, a ***factor*** is a hypothetical construct that is not directly observed but is assumed to be present because of the apparent effect that is has on the study's manifest variables. The ***manifest variables*** are the real-world variables that are directly observed—they may be test scores, responses to rating items on questionnaires, or other variables. A ***factor structure*** is a type of causal model that hypothesizes the number of factors being measured by a set of manifest variables, which specific manifest variables are measuring each factor, and related issues.

Confirmatory factor analysis produces indices that reflect the goodness of fit between the hypothesized factor structure and the relationships that are actually observed in the sample. It also produces factor loadings, inter-factor correlations, and other results related to the psychometric properties of the model.

CFA belongs to a larger family of procedures called *structural equation modeling* (*SEM*) or *covariance structure modeling* (*CSM*). **Structural equation modeling** is a highly flexible set of procedures that allows researchers to estimate and test theory-derived models that hypothesize causal relationships between variables (Klem, 2000; Mueller & Hancock, 2008).

Many of the concepts discussed here build on the more elementary concepts that were introduced in the previous chapter (*Chapter 16*). If you are new to structural equation modeling, you may wish to read that chapter before continuing.

Illustrative Example: A Six-Factor Model of Exercise

Many of the earlier chapters of this book discussed something called the *four-factor model of exercise behavior*. This is a theoretical model developed by the fictitious researcher Dr. O'Day, and it has been used to illustrate statistical procedures such as multiple regression and discriminant analysis.

THE SIX-FACTOR MODEL OF EXERCISE BEHAVIOR. Imagine that Dr. O'Day now revises this theoretical framework, expanding it into something that we will call the *six-factor model of exercise behavior*. This revised model now includes the following six constructs:

1. Exercise behavior
2. Health motivation
3. Dislike of exercise
4. Desire to be thin
5. Current health
6. Lack of time for exercise

This revised model contains the four variables that had been included in the old four-factor model, along with two new variables. The four variables that had constituted the old model are *health motivation, desire to be thin, dislike of exercise,* and *lack of time for exercise*. One of the new variables is *current health*: the extent to which the participant feels good physically and is relatively free of medical problems. The second new variable is *exercise behavior:* a measure that reflects how frequently the participant exercises as well as the total minutes of exercise that the participant gets each week.

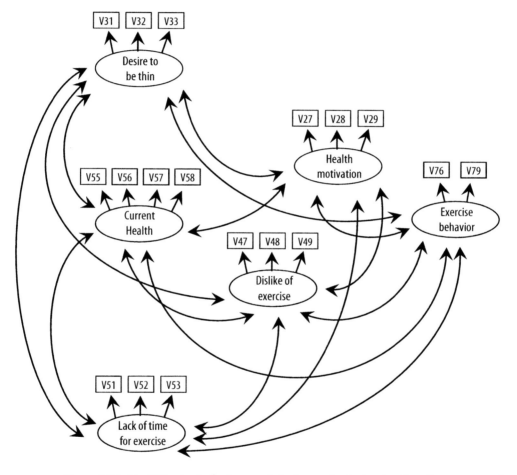

Figure 17.1. Dr. O'Day's six-factor model of exercise behavior (i.e., the proposed measurement model to be analyzed in this chapter).

FIGURE ILLUSTRATING THE SIX-FACTOR MODEL. In the current chapter, the six preceding variables (along with the questionnaire items believed to measure them) will be organized into a *measurement model*, also called a *factor model*. A **measurement model** is a theoretical path model that makes specific predictions about the number of latent factors that are being measured, which manifest variables are measuring each factor, and other characteristics of the data set. Once the variables have been organized into such a factor model, Dr. O'Day will use confirmatory factor analysis to investigate it—to determine whether the theoretical model provides a good fit to the data. Figure 17.1 illustrates the measurement model that she will explore in this chapter.

Conventions Used to Illustrate CFA Models

Looks like a big mess, doesn't it? For better or worse, the ovals, squares, arrows, and other features appearing in Figure 17.1 are fairly typical of the conventions that you will encounter when you read about confirmatory factor analyses in published articles. This section describes the meaning of these conventions.

OVALS FOR LATENT FACTORS. The ovals in Figure 17.1 represent **latent factors**: hypothetical constructs that are not observed in a direct manner, but are assumed to be present based on the way that they influence the manifest indicator variables that *are* observed in a direct manner. In this case, each oval represents one of the factors in Dr. O'Day's 6-factor model of exercise: The oval at the top left of the figure represents the *desire to be thin* factor, the oval below it represents the *current health* factor, and the remaining ovals may be interpreted in the same way. Latent factors are sometimes called *latent variables, unobserved variables, unmeasured variables*, or just *factors*.

SQUARES FOR MANIFEST VARIABLES. In Figure 17.1, the squares represent **manifest variables**: the real-world variables that are actually obtained and measured in a direct way by the researcher. Manifest variables are sometimes called *indicators, indicator variables, observed variables,* or *measured variables*. In the current analysis, each square represents one of the items on the *Questionnaire on Eating and Exercising*, which appears in *Appendix A*. For example, you can see that V31 is one of the indicator variables that measures the *desire to be thin* factor. The "V31" stands for "Variable 31" (in other words, "Item 31" on the *Questionnaire*). Item 31 contains the statement "It is important to me to be thin." You will remember that participants responded to this item by circling any number from 1 (*Extremely Disagree*) to 7 (*Extremely Agree*). Figure 17.1 shows that questionnaire items 31, 32, and 33 all measure the *desire to be thin* factor; items 55, 56, 57, and 58 all measure the *current health* factor, and so forth. In the investigation, Dr. O'Day analyzes these manifest variables, in part, to see if there is any evidence that they really are measuring the latent factors in the way that Figure 17.1 predicts.

In the current study, the manifest variables are individual items on a questionnaire, but the manifest variables in a CFA do not necessarily have to be responses to individual questionnaire items. Sometimes the manifest variables are physical measurements (such as subject weight or height) and sometimes they are summative scores from a multiple-item test or subtest. For example, imagine that a group of researchers perform a CFA, and one of the latent factors (i.e., ovals) in their factor model is "general intelligence." One of the manifest variables (i.e., squares) measuring this factor might be full-test IQ scores from the *Wechsler Adult Intelligence Scale*. A second manifest variable measuring the same factor might be IQ scores from the *Stanford-Binet Intelligence Scales*. A third manifest variable might be scores from the *Wonderlic Cognitive Ability Test* (a test of general intelligence sometimes used in business organizations).

The point is this: Manifest variables do not have to be responses to individual items from a questionnaire. They can be just about any sort of observed variable, as long as the data meet certain statistical assumptions (to be discussed a bit later).

STRAIGHT ARROWS FOR FACTOR LOADINGS. Back to the factor model illustrated in Figure 17.1: With CFA, the straight, single-headed arrows represent the *causal effects* that underlying factors have on the manifest variables. For example, the arrows that go from the oval labeled *desire to be thin* to the three squares above it represent the causal effect that this latent factor is predicted to have on those three manifest variables (questionnaire items 31, 32, and 33). This means that, in CFA, these arrows may be viewed as representing *factor loadings*. You first learned about factor loadings in *Chapter 15 Exploratory Factor Analysis*. In the current chapter, you will see that the concept of *factor loadings* has a similar interpretation in CFA.

CURVED ARROWS FOR INTER-FACTOR CORRELATIONS. In Figure 17.1, the curved, double-headed arrows represent the correlations between the latent factors. If you look closely, you will see that every oval is connected to every other oval by a curved, two-headed arrow. These arrows reflect Dr. O'Day's prediction that every latent factor is correlated with every other latent factor. With CFA, it is fairly standard that all underlying factors are allowed to freely correlate with one another in this way. When the CFA is complete, the software that performs the analysis will compute the value for each of these correlations (which are essentially just Pearson product-moment correlations, symbol: r). Dr. O'Day will then know the size, sign, and statistical significance for every possible correlation between her six factors.

SUMMARY. And that is Dr. O'Day's measurement model. She has made very specific predictions about how many factors are present and which questionnaire items are measuring each factor. The results from the confirmatory factor analysis will allow her to decide whether the proposed model illustrated in Figure 17.1 provides an acceptable fit to the data—i.e., whether the hypothetical factor model provides a reasonably good representation of "reality."

Research Method for the Current Investigation

An earlier section explained that the squares in Figure 17.1 represent individual items from the *Questionnaire on Eating and Exercising* that appears in *Appendix A* of this book. The square labeled "V31" represents item 31 from the questionnaire; the square labeled "V32" represents item 32, and so forth. Participants responded to most of these items by circling a number on a 7-point response scale which ranged from 1 (*Extremely Disagree*) to 7 (*Extremely Agree*).

The two questionnaire items that assessed the *exercise behavior* factor allowed open-ended responses, rather than using this 7-point rating format. Item 76 asked "How many times each week do you engage in a session of aerobic exercise that lasts at least 15 minutes?" Item 79 asked "How many minutes per week (total) do you engage in aerobic exercise?"

Questionnaires were administered to 260 undergraduate students at a regional state university in the Midwest. Students completed questionnaires in exchange for course credit. Completed questionnaires were obtained from 238 participants.

Typical Research Questions Addressed

Confirmatory factor analysis allows researchers to answer questions such as the following. Some of the terms introduced below will be explained in later sections.

- If there a good fit between my hypothesized factor model and the data?
- What is the reliability estimate for each manifest indicator variable?
- Do the indicators that are hypothesized to measure the same underlying factor display convergent validity?
- What is the composite reliability estimate for each factor?
- What are the correlations between the latent factors?

Details, Details, Details

This section describes the types of variables that are usually analyzed in a CFA and explains how confirmatory factor analysis differs from exploratory factor analysis. It ends with a discussion of how CFA is often used as a first step toward a relatively sophisticated procedure called *path analysis with latent factors*.

Types of Variables That May Be Analyzed

Confirmatory factor analysis is often performed using maximum likelihood (ML) estimation. ML estimation requires that the variables be assessed on an interval or ratio scale (Klem, 1995) and that they follow a multivariate normal distribution (McDonald & Ho, 2002).

In the social and behavioral sciences, the manifest variables that are analyzed in a factor analysis are very often responses to rating items on questionnaires. This practice is somewhat controversial, as many researchers argue that rating items on questionnaires typically produces variables that are on an ordinal scale and are, therefore, inappropriate for procedures that require interval-level measurement (e.g., Wirth & Edwards, 2007). At the present time, however, journals continue to publish studies in which CFA was performed on rating-scale items similar to the items presented in this chapter, and it is not clear how this controversy will be resolved. When the manifest variables being analyzed are obviously assessed on a nominal scale or ordinal scale, it may still be possible to perform CFA by using statistical procedures that differ from the procedures described here (see, for example, Muthén, 1984; Muthén & Muthén, 2001).

Exploratory Versus Confirmatory Factor Analysis

Statisticians make a distinction between exploratory factor analysis (EFA) versus confirmatory factor analysis (CFA). *Exploratory factor analysis* is a dimension-reduction procedure. Researchers use it when they wish to reduce the number of manifest variables they are working with to a smaller number of latent factors. In contrast to CFA, exploratory factor analysis is appropriate when researchers do not have a clear, well-developed hypothesis regarding the factor structure that underlies their manifest variables (Osborne, Costello, & Kellow, 2008). They perform EFA to discover that factor structure. Exploratory factor analysis was covered in *Chapter 15* of this book.

In contrast, researchers perform *confirmatory factor analysis* when they have a clear, well-developed hypothesis regarding the factor structure underlying a set of measures (MacCallum, 2009). They perform CFA to test that hypothesis: They assess how well the proposed factor model fits the data, assess the factor loadings for size and statistical significance, assess the reliability of the latent factors, and investigate various other psychometric properties of the factors and indicators. When the results are not supportive of the proposed factor model, they may modify it so has to achieve a better fit.

CFA as a Prelude to Path Analysis with Latent Factors

There are at least two situations in which researchers use confirmatory factor analysis. As we shall see, one of these scenarios is much simpler than the other.

SITUATION 1: USING CFA AS AN END UNTO ITSELF. In Situation 1, researchers perform CFA simply for the purpose of investigating the factor structure that underlies a set of manifest variables. An example would be a scenario in which a researcher has developed a new questionnaire for measuring job satisfaction in employees. He begins with a theoretical model which hypothesizes that there are six different dimensions of job satisfaction. He then develops a questionnaire to assess these dimensions, with Items 1-5 measuring Dimension 1, Items 6-10 measuring Dimension 2, and so forth. He obtains completed questionnaires from a sample of 700 employees, and uses CFA to determine whether there is a good fit between his theoretical model and the data. If he finds that the fit is good (and he obtains additional results that are supportive of his model), he may publish his findings in a journal article. The basic thrust of the article would be something like this: (a) "I developed a theoretical model regarding the factor structure of job satisfaction," (b) "I used this theoretical model to develop an instrument for measuring job satisfaction," (c) "Results from a confirmatory factor analysis were generally supportive of my theoretical model as well as the instrument," (d) "Interested parties may purchase copies of my job satisfaction questionnaire at $10 per copy for use in their own consulting work", and (e) "Yes, I aggressively litigate against anyone who has the temerity to use my questionnaire without paying."

And *viola*: end of story. Once the CFA was completed, the study was over.

SITUATION 2: USING CFA AS STEP 1 IN A TWO-STEP PATH ANALYSIS. Situation 2 involves a more complex scenario. With Situation 2, CFA is used as the first step of a 2-step procedure for testing complex theoretical path models. When they wish to test theoretical models that propose causal relationships between latent variables, many researchers now follow a highly-recommended two-step approach suggested by Anderson and Gerbing (1988). In Step 1, they use confirmatory factor analysis to verify that they have an acceptable measurement model. As you learned earlier, this *measurement model* is a factor-analytic model (similar to Figure 17.1) consisting of a number of latent factors, each of which is measured by two or more manifest variables. With this measurement model, each latent factor is usually allowed to correlate with every other latent factor.

Once an acceptable measurement model has been developed, the study moves to Step 2, in which the researchers perform a relatively sophisticated procedure called *path analysis with latent factors*. In this step, they usually replace some of the curved, two-headed arrows in the measurement model with straight, single-headed arrows and delete some of the remaining curved arrows entirely. The result is a *latent-factor path model*: a causal model that predicts that some of the latent factors have causal effects on other latent factors. These models are sometimes called *LISREL models*, because LISREL (Jöreskog & Sörbom, 1989) was the first widely available software that made this sort of analysis possible. The estimation of such causal models is currently the cutting-edge approach to performing path analysis.

You will not be surprised to learn that Dr. O'Day plans to estimate one of these latent-factor path models (she will do this in *Chapter 18*). As a first step toward this goal, she will first use CFA to verify that she has an acceptable measurement model. She will do this in the current chapter.

Results from the Investigation

When performing CFA, researchers typically investigate two questions: (a) How good is my model's fit to the data?, and (b) How good are the psychometric properties of my model? The way that researchers answer the question about fit is very similar to the way that they assess fit when performing path analysis with manifest variables (as described in *Chapter 16*). The way that researchers investigate the psychometric properties of the solution (e.g., reliability and validity) is new material.

Descriptive Statistics and Matrix of Correlations

In any article dealing with confirmatory factor analysis, it is highly desirable for the authors to present the correlation matrix or variance-covariance matrix that was analyzed. Many researchers present the correlation matrix along with the means and standard deviations for each variable, as it is always possible to convert a correlation matrix into a variance-covariance matrix as long as the standard deviations are provided. When it is not feasible to present either matrix, authors may instead make the matrix (or the raw data) available to readers through the journal's website. To save space, this chapter will omit Dr. O'Day's table of means, standard deviations, and correlations.

Estimating the Parameters of the Model

After the descriptive statistics, the published article typically describes the method used to compute parameter estimates for the measurement model. The parameter estimates in a CFA are the factor loadings, the inter-factor correlations, and related statistics. Below is the excerpt from Dr. O'Day's article in which she describes this step:

> Confirmatory factor analysis (CFA) was performed using the SAS® System's CALIS procedure. The maximum likelihood method was used to estimate parameters, and all analyses were performed on the variance-covariance matrix. The proposed factor model being tested was the factor model illustrated in Figure 17.1.

Results Related to the Fit of the Model

Below, Dr. O'Day describes results related to the fit of her proposed model. She begins by describing the two models included in the table.

Table 17.1

Fit Indices for the Proposed Measurement Model Illustrated in Figure 17.1

Model	Model χ^2				SRMR	RMSEA	NFI	CFI	NNFI
	χ^2	df	p	χ^2/df					
Null model	2706.60	153	<.001	17.69					
Measurement model	191.74	120	<.001	1.60	.05	.05	.93	.97	.96

Note. $N = 238$. SRMR = Standardized root mean-square residual; RMSEA = root mean square error of approximation; NFI = normed fit index; CFI = comparative fit index; NNFI = nonnormed fit index.

Table 17.1 presents global fit indices for two models. The row headed "Null model" presents results for a model in which all manifest variables are uncorrelated with one another. The null model serves as a baseline against which the other model may be compared. The row headed "Measurement model" presents results for the six-factor model of exercise behavior (i.e., the measurement model illustrated in Figure 17.1).

Table 17.1 shows that the model χ^2 for the proposed measurement model was statistically significant, $\chi^2(120, N = 238) = 191.74, p < .001$. Although a significant χ^2 statistic is sometimes interpreted as evidence of a poor fit, it is known that this test becomes so sensitive with large samples that it can result in the rejection of an otherwise well-fitting model (Mulaik, James, Alstine, Bennett, Lind, & Stilwell, 1989). The same tables shows that the ratio of χ^2/df for the current measurement model is 1.60, which is below the informal criterion of 3 used by some researchers (Kline, 1998). This relatively low ratio is indicative of a good fit between model and data.

In the preceding excerpt, Dr. O'Day refers to the common practice of computing the ratio of model χ^2 to df. With this ratio, lower values reflect a better fit. Although there are no widely accepted criteria as to how low this ratio must be to be considered "acceptable," some researchers have noted that ratios in the range of 2, 3, or perhaps even 5 are desirable (Bollen, 1989; Klem, 2000).

As you learned in *Chapter 16*, researchers have developed a large number of global fit indices for structural equation models. Below, Dr. O'Day summarizes some of the fit indices that were obtained for her measurement model:

Table 17.1 provides a number of additional global fit indices, such as the RMSEA and the CFI. In evaluating whether the current model displayed an acceptable fit to the data, each index was compared to the relatively stringent criteria recommended by Hu and Bentler (1999) and Mueller and Hancock (2008).

Table 17.1 shows that for the current measurement model, the comparative fit index (CFI) was .97, and the nonnormed fit index (NNFI) was .96. Both values satisfied the recommended criterion of .95 (with the CFI and the NNFI, higher values indicate a better fit). The observed value for the root mean square error of approximation (RMSEA) was .05, which was lower than the recommended criterion of .06, and the observed value for the standardized root mean-square residual (SRMR) was .05, which was lower than the recommended criterion of .08 (with the RMSEA and the SRMR, lower values indicate a better fit).

In summary, the global fit indices in Table 17.1 indicated that the factor model illustrated in Figure 17.1 provided an acceptable fit to the data. Although the model χ^2 statistic was significant, all of the remaining fit indices satisfied the criteria recommended by Hu and Bentler (1999) and Mueller and Hancock (2008).

All of the fit indices described by Dr. O'Day were discussed in more detail in *Chapter 16*. To refresh your memory:

- The **standardized root mean-square residual** (SRMR; Bentler, 1995) represents the overall discrepancy between the actual covariance matrix versus the matrix implied by the researcher's proposed factor model. With the SRMR, smaller values (closer to zero) indicate a better fit between the proposed factor model and the data.

- The ***root mean square error of approximation*** (RMSEA; MacCallum et al., 2006) is similar to the SRMR, in that it reflects the overall discrepancy between the actual covariance matrix versus the predicted model matrix. However, the RMSEA also rewards theoretical models for being more parsimonious (i.e., for being simpler). Smaller values (closer to zero) indicate a better fit.

- The ***normed fit index*** (NFI; Bentler & Bonett, 1980), the ***comparative fit index*** (CFI, Bentler, 1990), and the ***nonnormed fit index*** (NNFI; Bentler & Bonett, 1980) are incremental indices of fit. They reflect the fit of the present model as compared to a *null model*: A model that displays no relationships at all between any of the observed variables. The NFI and the CFI may range from .00 to 1.00, with higher values (closer to 1.00) indicating a better fit. The NNFI is similar to the CFI, except that its values may in some situations be slightly lower than .00 or slightly higher than 1.00. The NNFI is also called the Tucker-Lewis Index, or TLI (Tucker & Lewis, 1973).

In summary, the overall fit between the data and Dr. O'Day's proposed measurement model appears to be satisfactory. This means that she may now dig a bit deeper by reviewing the psychometric properties of her manifest variables and latent factors.

Reliability and Other Psychometric Properties of the Model

When researchers discuss ***psychometric properties***, they are usually referring to the concepts of reliability and validity. ***Reliability*** is the extent to which a set of scores is consistent or stable. ***Validity*** is the extent to which a set of scores actually measures the construct that it was intended to measure. One of the benefits of the SEM approach to confirmatory factor analysis is the fact that it provides researchers with procedures to assess specific types of reliability and validity, as displayed by the measurement model.

RESULTS RELATED TO THE INDICATOR VARIABLES. Dr. O'Day cannot be confident that her model is acceptable until she has taken a close look at its manifest variables and the latent factors that they are intended to measure. She refers to them in the following excerpt:

> Table 17.2 reports information related to the psychometric properties of the measurement model investigated in the current study. The column on the left side of the table provides the names of each of the six latent factors that appear in the proposed measurement model. Directly below the name of a given factor are the names of the manifest variables that are hypothesized to measure that factor. For example, the Table shows that *exercise behavior* is the first latent factor, and it is measured by manifest variables *V76* and *V79*. The numbers in the names for the manifest variables represent item numbers on the *Questionnaire on Eating and Exercising*. For example, manifest variables V76 and V79 represent questionnaire items 76 and 79, respectively.

Table 17.2

Reliability, Validity, and Other Psychometric Properties of the Latent Factors and Manifest Indicator Variables that Constitute the Measurement Model

Factor and indicators	Results for indicators (items)			Results for composites	
	Standardized loading	t	Item reliability	Composite reliability	Variance extracted
Exercise behavior				.76	.63
V76	.90	12.63*	.81		
V79	.66	9.67*	.44		
Health motivation				.84	.64
V27	.78	13.16*	.61		
V28	.83	14.27*	.69		
V29	.78	13.27*	.61		
Dislike of exercise				.92	.79
V47	.88	17.01*	.77		
V48	.96	19.57*	.92		
V49	.82	15.27*	.67		
Desire to be thin				.85	.66
V31	.91	16.35*	.83		
V32	.80	13.64*	.64		
V33	.72	12.00*	.52		
Current health				.82	.54
V55	.82	14.08*	.67		
V56	.63	10.13*	.40		
V57	.84	14.63*	.71		
V58	.61	9.65*	.37		
Lack of time for exercise				.95	.87
V51	.95	19.31*	.90		
V52	.91	17.96*	.83		
V53	.94	19.07*	.88		

Note. N = 238. The numbers for the manifest indicators (such as "V76") correspond to the item numbers for questionnaire items on the *Questionnaire on Eating and Exercising* in *Appendix A*.
*p < .001

Psychometric properties of the manifest indicator variables. Table 17.2 presents standardized factor loadings for each indicator variable under the heading "Standardized loading." The table shows that these loadings range from .61 to .96 (*Mdn* = 82.5). The column headed "*t*" contains *t* statistics. The *t* statistic for a given indicator variable tests the null hypothesis that the factor loading for that indicator is equal to zero in the population. The table shows that the *t* statistic for each loading is significant ($p < .001$).

In CFA, the *reliability* of an indicator variable refers to the percent of variance in the indicator variable that is accounted for by the underlying factor on which it loads. The reliability of an indicator is estimated by squaring the standardized factor loading (Long, 1983a; Long, 1983b). Reliability estimates for the current analysis appear under the heading "Item reliability" in Table 17.2. The table shows that item reliabilities range from .37 to .92 ($Mdn = .68$).

In CFA, a group of manifest variables displays *convergent validity* if there is evidence that each indicator variable within that group is measuring the same underlying factor (i.e., the same underlying factor that it is hypothesized to measure). In practice, a group of indicator variables is said to display convergent validity if the *t* statistic for each indicator variable is significant (Anderson & Gerbing, 1988). As was reported above, the *t* statistic for each factor loading in Table 17.2 is, in fact, significant ($p < .001$). This finding supports the convergent validity of the indicator variables.

The preceding section of Dr. O'Day's article focused on the manifest variables, also called *indicator variables*. Much of this section dealt with factor loadings. There are different ways to define this concept, but within the context of confirmatory factor analysis, it is useful to think of **factor loadings** as *path coefficients* that represent the size of the effect that a latent factor has on a manifest indicator variable. Standardized factor loadings generally range from −1.00 through 0.00 through +1.00, with values closer to ±1.00 indicating a stronger effect. The indicators with the larger factor loadings (in absolute value) are generally viewed as being the more important measures of the underlying factor.

One of the advantages of confirmatory factor analysis is the fact that it allows researchers to assess the reliability of the manifest indicator variables. The preceding excerpt points out that the reliability of a given indicator variable can be computed by squaring the standardized factor loading for that indicator.

The final paragraph in the preceding excerpt refers to convergent validity. In general, a given instrument (such as a given item on a questionnaire) displays **convergent validity** if it displays substantial correlations with other instruments known to measure the construct of interest. Within the context of a CFA, the convergent validity of a group of manifest indicator variables is demonstrated if each indicator displays a factor loading that is significantly different from zero (i.e, if the *t* statistic or *z* statistic for each indicator's factor loading is significant). The idea is that, if all of the items are statistically significant, that tells us that all of the items must all be measuring the same underlying construct. In other words, the items are converging, or "coming together," in measuring the same thing. This is not a particularly strong test of validity, but it is typically worth reporting.

RESULTS RELATED TO THE LATENT FACTORS. Dr. O'Day now turns her attention to the latent factors (i.e., the hypothetical constructs represented by ovals in Figure 17.1). When she refers to a **composite** in the following section, she is referring to the group of indicators (i.e., questionnaire items) that were intended to measure the same underlying factor.

> **Psychometric properties of the latent factors.** When performing CFA, it is possible to compute a *composite reliability index* for each of the hypothesized latent factors. This index is similar to coefficient alpha (Cronbach, 1951), and reflects the internal consistency of the manifest indicators that are hypothesized to measure a given factor (Fornell & Larcker, 1981). It has been recommended that, to be acceptable, these indices should exceed .70 or possibly .80 (Mueller & Hancock, 2008; Nunnally, 1978).
>
> Table 17.2 presents these reliability estimates under the heading "Composite reliability." For all six factors, composite reliabilities exceeded .70, and for five of the six factors, they exceeded .80.
>
> When performing CFA, the *variance extracted estimate* for a given factor indicates the amount of variance that is captured by that factor in relation to the amount of variance that is due to measurement error (Fornell & Larcker, 1981). Table 17.2 shows that the variance extracted estimates for the six factors in the current analysis ranged from .54 to .87. Each of these estimates exceeded the criterion of .50 recommended by Fornell and Larcker (1981).

The preceding section shows that the latent factors (represented by the ovals in Figure 17.1) display acceptable psychometric properties. These results, when combined with findings related to global fit and the properties of the indicator variables, tell us that Dr. O'Day's model satisfies a variety of criteria used to evaluate measurement models. In other words, the questionnaire items appearing in Figure 17.1 are apparently doing a fairly good job of measuring the six constructs that they were intended to measure.

FORMULAS FOR COMPOSITE RELIABILITIES AND VARIANCE EXTRACTED ESTIMATES. If you perform confirmatory factor analysis as part of your own research, you may find that your software does not automatically compute composite reliabilities or variance extracted estimates for the latent factors. Not to worry, as these indices are easy to compute by hand. For formulas and guidance, see Fornell and Larcker (1981, pp. 45-46), Hatcher (1994, pp. 325-331), and Netemeyer et al. (1990, pp. 151-152).

CORRELATIONS BETWEEN THE FACTORS. Having shown that the measurement model is acceptable, Dr. O'Day now turns her attention to the results that are of greatest interest to the reader: the correlations between the latent factors. Remember that one important objective of her current program of research is to identify variables that are predictive of exercise behavior. One of the latent factors in Dr. O'Day's measurement model is an *exercise behavior* factor. By reviewing the inter-factor correlations from the CFA, she is able to identify the latent factors that display strong correlations with exercise behavior, as well as the latent factors that display weak correlations. She discusses this in the following excerpt:

> **Correlations between the latent factors.** Table 17.3 presents the correlations between the six factors investigated in the current study. Each factor in the table is a latent factor as estimated in the confirmatory factor analysis.

Table 17.3

Correlations Among Latent Factors Estimated in the CFA, Decimals Omitted

Factor	1	2	3	4	5	6
1. Exercise behavior	--					
2. Health motivation	35***	--				
3. Dislike of exercise	-43***	-33***	--			
4. Desire to be thin	16*	33***	-17*	--		
5. Current health	34***	41***	-24***	18*	--	
6. Lack of time for exercise	-52***	-23**	40***	-07	-23***	--

Note. $N = 238$.
$*p < .05$; $**p < .01$; $***p < .001$.

One objective of the current study was to identify the factors that display the strongest correlations with the *exercise behavior* factor (the first latent variable listed in Table 17.3). The table shows that the exercise behavior factor displayed the strongest correlations with the *lack of time for exercise* factor ($r = -.52$, $p < .001$), the *dislike of exercise* factor ($r = -.43$, $p < .001$), and the *health motivation* factor ($r = .35$, $p < .001$). It displayed the weakest correlation with the *desire to be thin* factor ($r = .16$, $p < .05$).

Summary of Findings

The measurement model illustrated in Figure 17.1 (presented much earlier in this chapter) provided an acceptable fit to the data. The comparative fit index (CFI) was .97, the nonnormed fit index (NNFI) was .96, the root mean square error of approximation (RMSEA) was .05, and the standardized root mean-square residual (SRMR) was also .05. All of these indices satisfied the criteria for an acceptable fit recommended by Hu and Bentler (1999) and Mueller and Hancock (2008).

Standardized factor loadings for the manifest variables ranged from .61 to .96 ($Mdn = 82.5$). The t statistic for each factor loading was significant ($p < .001$), and this finding supported the convergent validity of the indicator variables as measures of the underlying factors.

For the six latent factors, composite reliability indices ranged from .76 to 95, exceeding the standard criterion of .70 used with coefficient alpha (Nunnally, 1978). Variance-extracted estimates ranged from .54 to .87, exceeding the criterion of .50 recommended by Fornell and Larcker (1981).

Inter-factor correlations showed that the *exercise behavior* factor displayed the strongest correlation with the *lack of time for exercise* factor ($r = -.52$, $p < .001$). It displayed the weakest correlation with the *desire to be thin* factor ($r = .16$, $p < .05$).

In short, results from the CFA suggest that Dr. O'Day's six-factor measurement model displays a good fit to the data and acceptable psychometric properties. This means that she has completed Step 1 of Anderson and Gerbing's (1988) proposed two-step procedure for testing causal models with latent variables. In Step 2, she will modify the factor model presented in Figure 17.1 so that it proposes causal relationships between some of the latent factors. She will then perform path analysis with latent factors to assess the fit of the resulting causal model. These analyses will be presented in *Chapter 18*.

Additional Issues Related to this Procedure

Given the large amount of information that may be produced in a CFA, this section attempts to lighten your cognitive load by providing a checklist of the most-important characteristics of a well-fitting model. The checklist is followed by another reminder concerning the dangers of data-driven model modifications. It ends with a listing of the assumptions underlying CFA.

Checklist for Evaluating Results from a CFA

In my book on structural equation modeling (Hatcher, 1994), I summarized the characteristics of an ideal solution from a confirmatory factor analysis. That list was designed for researchers who are actually performing CFA.

For the present book, I have instead created a list for those who are *reading* about a confirmatory factor analysis. This list contains some new recommendations that have become widely adopted in the years since the publication of my SEM book in 1994 (e.g., the recommendations from Hu & Bentler, 1999).

Below are the characteristics of the *ideal* measurement model: one that displays a good fit to the data as well as strong psychometric characteristics:

- Ideally, the p value for the model χ^2 statistic should be greater than .05 (that is, the model χ^2 should be nonsignificant). Higher p values indicate better fit. However, remember that this criterion is seldom met, even when there is an otherwise good fit between the model and the data. Because of this, most experts advise that this χ^2 test should be taken with a grain of salt (e.g., Mulaik et al., 1989).

- The ratio of χ^2/df should be a relatively small number, perhaps in the range of 2 or 3 or 5 (Bollen, 1989; Klem, 2000; Kline, 1998).

- According to the relatively stringent criteria for global fit indices suggested by Hu and Bentler (1999) and Mueller and Hancock (2008), the RMSEA should be ≤0.06, the SRMR should be ≤0.08, and the CFI should be ≥0.95.

- Significance tests should show that each factor loading is significantly different from zero. The standardized loadings themselves should be larger than ±.05 in absolute value, since standardized loadings that are less than ±.05 are viewed as being trivial in size.

- For the latent factors, composite reliabilities should be ≥.70, and variance extracted estimates should be ≥.50.

Data-Driven Model Modifications

In this investigation, Dr. O'Day's original measurement model provided a relatively good fit to the data. If it had not provided a good fit, she may have considered making *data-driven model modifications* (also called *post-hoc model modifications*). As you learned in *Chapter 16*, this process typically involves reviewing modification indices and then changing the theoretical model so that it provides a better fit to the sample data.

Chapter 16 indicated that, when researchers make data-driven model modifications, you should view their final preferred model with caution, even if it displays a good fit. This is because data-driven model modifications often capitalize on chance characteristics of the sample data being analyzed—they sometimes result in models that will not generalize to the intended population, or to other samples.

On the other hand, you may have more confidence in a modified model if the researcher created it by using a cross-validation procedure. In *Chapter 16*, you learned that this involves (a) starting with a really large sample; (b) randomly dividing this sample into two groups: a *calibration sample* and a *validation sample*; (c) using the calibration sample to perform the data-driven model modifications; and (d) determining whether the resulting revised model (created with the calibration sample) still provides an acceptable fit when applied to the validation sample. If yes, you can have more faith in the revised measurement model—you can have more faith that it will in fact generalize to new samples.

Assumptions Underlying this Procedure

Because they are both special cases of structural equation modeling, the assumptions underlying confirmatory factor analysis are similar to the assumptions underlying path analysis with manifest variables (covered in *Chapter 16*). Some of the assumptions listed below are explained in greater detail in *Chapter 2* of this book. The assumptions underlying CFA include the following (from Bryant & Yarnold, 1995; Hatcher, 1994; Tabachnick & Fidell, 2007).

INTERVAL-LEVEL MEASUREMENT. Many estimation procedures used with confirmatory factor analysis require that all manifest variables be continuous quantitative variables assessed on an interval scale or ratio scale. Some SEM applications (such as EQS and LISREL) can accommodate ordinal-scale or categorical variables if special steps are taken (see Tabachnick & Fidell, 2007, pp. 729-730). In addition, the application called *Mplus* was developed to accommodate categorical variables (Muthén & Muthén, 2001).

INDEPENDENT OBSERVATIONS. Each participant should be drawn independently from the population of interest. Among other things, this means that the researcher should not take repeated measures from the same subject (unless special procedures have been used to accommodate repeated measures).

LARGE SAMPLES. In most cases, the analysis should be based on large samples. Bryant and Yarnold (1995) recommend that the minimal number of participants should be 5 to 10 times the number of manifest indicators being analyzed (for Dr. O'Day's study, this would translate into a minimum of 90 to 180 participants). MacCallum et al. (1996) provide power tables for determining the sample sizes that are necessary to achieve adequate power for a variety of global fit indices.

MULTIVARIATE NORMALITY. Many of the statistical tests that are computed as part of a CFA assume that the data follow a multivariate normal distribution.

LINEAR AND ADDITIVE RELATIONSHIPS. The relationships between all variables in the model must be linear and additive.

ABSENCE OF MULTICOLLINEARITY. The variables should not demonstrate extremely strong correlations with one another.

IDENTIFICATION. The measurement model should be identified. The concept of *identification* is complex; see Bryant and Yarnold (1995, pp. 117-118) or Tabachnick and Fidell (2007, pp. 709-712) for an introduction.

Chapter 18

SEM III: Path Analysis with Latent Factors

This Chapter's *Terra Incognita*

You are reading an article about three complex causal models. Each model offers a slightly different theoretical perspective regarding the determinants of *exercise behavior*. The researcher has used path analysis with latent factors to investigate these models. An excerpt from her *Results* section appears below:

> The nomological validity of Model 3 was investigated by subtracting the model χ^2 statistic for the measurement model from the model χ^2 statistic for Model 3. With alpha set at $\alpha = .05$, the resulting χ^2 value was not statistically significant, $\Delta\chi^2(4, N = 238) = 7.64$, *ns*. This finding showed that Model 3 provided a fit to the data that was not significantly worse than the fit provided by the measurement model. Of the three models, only Model 3 passed this nomological validity test.
>
> Next, the RMSEA-P was reviewed in order to assess fit in just the path portion of each model, independent of the measurement model. The RMSEA-P for Model 3 was .062, satisfying the recommended criterion of RMSEA-P < .08. For Model 1, the RMSEA-P was .175, indicating a poor fit. For Model 2, it was .103, also indicating a poor fit.

Don't be afraid. All shall be revealed.

The Basics of this Procedure

Path analysis with latent factors allows researchers to investigate complex theoretical path models which predict causal relationships between variables, with some of these variables being latent factors. These analyses are sometimes called *LISREL analyses*, because LISREL (Jöreskog & Sörbom, 1989) was the first widely available computer application that made them possible. Path analysis with latent factors is currently the preferred approach for testing theoretical path models using correlational data.

Path analysis with latent factors represents a coming together of two separate statistical procedures: (a) traditional path analysis (as covered in *Chapter 16*), and (b) confirmatory factor analysis, or CFA (as covered in *Chapter 17*). With traditional path analysis, the researcher investigates a theoretical model that proposes specific causal relationships between a number of variables. All of the variables appearing in this path model are ***manifest variables***: real-world, tangible variables that are obtained and measured in a direct way. Manifest variables are sometimes called *observed variables, measured variables, indicator variables*, or just *indicators*.

In contrast, when researchers perform path analysis with latent factors, the variables that appear in the path model are ***latent factors***: hypothetical constructs that are not measured directly, but whose presence is inferred based on the analysis of the manifest variables. Researchers typically use two or more manifest variables to measure each latent factor, in the same way that multiple observed variables are used to measure each factor in CFA. In these analyses, the latent factors are sometimes called *constructs, latent variables, unobserved variables*, or *unmeasured variables*.

By combining the two statistical procedures, path analysis with latent factors capitalizes on each procedure's strength. It allows researchers to verify that they are doing a good job of measuring the hypothetical constructs of interest (one of the strengths of CFA). It also allows them to determine whether there is empirical support for the causal relationships predicted by their theoretical path model (one of the strengths of traditional path analysis).

Questions Addressed

Path analysis with latent factors is a sophisticated procedure that involves several steps, a variety of statistical analyses, and a great deal of output. Researchers usually begin with a theoretical path model that illustrates hypothesized causal relationships between a number of latent constructs (perhaps 4-12 latent constructs in the typical investigation). Researchers following best practice will usually begin with two or three competing theoretical path models with each model illustrating a different theoretical perspective. By the end of the analysis, they will answer questions such as the following:

- Are my manifest variables doing an acceptable job of measuring the latent factors?

- Did my initial theoretical path model provide an acceptable fit to the data?

- If several competing path models were offered, did some models display a significantly better fit than others?

- If my theoretical path model predicted that one latent factor had a causal effect on another latent factor, did the results support this hypothesis? For example, was the relevant path coefficient large and statistically significant?

- Did the R^2 statistics show that my model accounts for a substantial amount of variance in the endogenous variables?

- Which antecedent variables had the largest direct effects on the consequent variables? Which had the largest indirect effects? Which had the largest total effects?

Why this Procedure is Important

Path analysis with latent factors is an *uber-important* data analysis procedure in the social and behavioral sciences. There are at least two reasons for this.

First, it allows researchers to investigate hypothesized causal relationships using nonexperimental (correlational) data. The evidence that it provides is typically not as persuasive as the evidence that would be obtained from a true experiment, but it is generally viewed as being the next best thing, especially in those situations in which a true experiment is just not possible.

Second, path analysis with latent factors is important because (as its name would suggest), it allows some of the variables in the theoretical path model to be **latent factors**: hypothetical constructs similar to the factors studied in confirmatory factor analysis. This latter characteristic is one of the features that distinguishes the current procedure from *path analysis with manifest variables* (which was covered in *Chapter 16*). Theoretically, the latent factors that are investigated using the current statistical procedure are free of measurement error (Ullman, 2007). This means that the results from path analysis with latent factors are less likely to be biased or misleading, and are more likely to help us understand the true relationships between the constructs of interest.

Alternative Names

Path analysis with latent factors belongs to a family of procedures called *structural equation modeling* (*SEM*). **Structural equation modeling** is a highly flexible set of procedures that allows researchers to estimate and test theory-derived models that hypothesize causal relationships between variables, some of which may be latent variables (Klem, 2000; Mueller & Hancock, 2008). Some researchers refer to this family of procedures as *factor-analytic structural equation modeling* (*FASEM*), and some refer to it as *covariance structural modeling* (*CSM*).

Earlier, it was noted that some researchers refer to these analyses as *LISREL analyses* because LISREL was the first widely-available software that made them possible. Today, path analysis with latent variables may be performed with a variety of different computer applications, including LISREL, AMOS, EQS, MPlus, and the SAS System's PROC CALIS.

Illustrative Investigation

This chapter describes a study very similar to the study described in *Chapter 16*. In that chapter, Dr. O'Day developed three causal models that identified the determinants of *exercise sessions* (the frequency with which people exercised). She then investigated each model using path analysis with manifest variables. With that type of path analysis, each variable in the causal model is a ***manifest variable***: a set of real-world scores that are directly obtained from the participants.

In the current study, Dr. O'Day will investigate the same basic causal models that she investigated in the previous study, but in this case, the variables that constitute her causal models will be ***latent factors:*** hypothetical constructs that are not measured in a direct way. Each subject will have a score on each of the six latent factors that appear in the causal models, but these scores will be factor scores computed through confirmatory factor analysis. Theoretically, these factor scores will be perfectly reliable (i.e., are measured without error), thus satisfying one of the assumptions of path analysis—an assumption that often is not met when performing path analysis with manifest variables.

In summary, the theoretical models that Dr. O'Day will investigate here are essentially the same models that she investigated in *Chapter 16*. However, in the current chapter she will analyze her data using path analysis with latent factors, rather than path analysis with manifest variables.

Method

In this study, the instrument used was once again the *Questionnaire on Eating and Exercising* that appears in *Appendix A*. This questionnaire was administered to 260 undergraduate students at a regional state university in the Midwest, and completed questionnaires were obtained from 238 participants. The students completed the questionnaire in exchange for course credit.

The manifest indicator variables included in Dr. O'Day's causal models are actually individual items from the *Questionnaire on Eating and Exercising*. Most of these were Likert-type rating items. For example, one of the latent factors appearing in her models is called *the desire to be thin*. This factor was measured by three questionnaire items such as "It is important to me to be thin." Participants responded to most of these items by circling a number on a 7-point response scale that ranged from 1 (*Extremely Disagree*) to 7 (*Extremely Agree*).

The two questionnaire items that assessed the *exercise behavior* factor did not use this 7-point response format; instead, they allowed open-ended responses. On the questionnaire, Item 76 asked "How many times each week do you engage in a session of aerobic exercise that lasts at least 15 minutes?" Item 79 asked "How many minutes per week (total) do you engage in aerobic exercise?"

Example of a Theoretical Path Model

Figure 18.1 presents one of the causal models investigated in this study. It uses many of the same graphic conventions that had been used to illustrate the confirmatory factor analysis (CFA) models that appeared in *Chapter 17*.

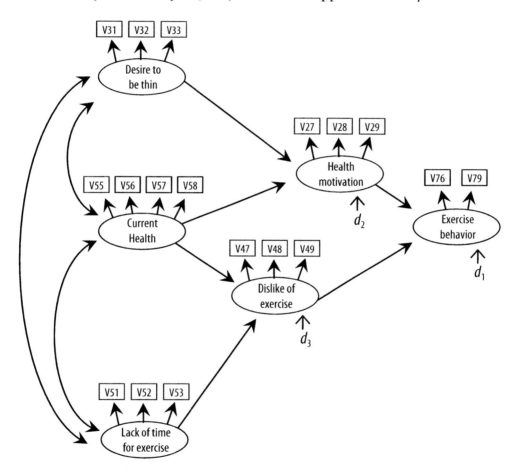

Figure 18.1. Dr. O'Day's Model 1, to be investigated using path analysis with latent variables.

In Figure 18.1, the ovals represent latent factors. The name of a specific latent factor appears inside its oval.

In the figure, the squares represent the manifest indicator variables that measured these latent factors. The numbers inside the squares correspond to the item numbers from the *Questionnaire on Eating and Exercising*. For example, the figure shows that the latent factor called *the desire to be thin* was measured by three manifest variables: Items 31, 32, and 33 on the *Questionnaire on Eating and Exercising*. The remaining latent factors and manifest variables may be interpreted in the same way.

CAUSAL RELATIONSHIPS PREDICTED BY THE THEORETICAL MODEL. In many ways, the causal model illustrated in Figure 18.1 is similar to the confirmatory factor analysis (CFA) model that served as Dr. O'Day's final preferred model in *Chapter 17*. For example, the present model includes the same latent factors, and these factors are measured by the same manifest variables.

However, the current model also differs in some ways from the final preferred CFA model in *Chapter 17*. In the chapter on CFA, all latent factors were allowed to freely correlate with one another. This was represented by the fact that each oval was connected to every other oval by a curved, two-headed arrow.

In contrast, in Figure 18.1, some of these curved, two-headed arrows have been replaced by straight, single-headed arrows. This reflects the fact that Figure 18.1 now predicts that some of the latent factors have causal effects on other latent factors. These arrows reflect predictions that:

- exercise behavior is caused by health motivation and dislike of exercise,
- health motivation is caused by desire to be thin and current health, and
- dislike of exercise is caused by current health and lack of time for exercise.

The causal model illustrated by Figure 18.1 is identical to one of the causal models that Dr. O'Day investigated in *Chapter 16* (the chapter on path analysis with manifest variables). The main difference is that, in *Chapter 16*, all of the variables that constituted the main causal model were manifest variables, whereas in the current chapter, all of the variables that constitute the main causal model are latent factors.

DISTURBANCE TERMS FOR ENDOGENOUS LATENT FACTORS. A latent factor is an ***endogenous latent factor*** if there are any straight, single-headed arrows pointing at it. In Figure 18.1, the endogenous latent factors are exercise behavior, health motivation, and dislike of exercise. The figure includes a *disturbance term* for each of these endogenous factors. These disturbance terms are represented by the symbol *d,* and a small single-headed arrow goes from *d* to a given endogenous latent factor.

The ***disturbance term*** for a given endogenous latent factor reflects things that affect variability in that factor, but were not included as antecedent variables in the theoretical model itself. These "things" include antecedent variables that should have been included (but were erroneously left out by the researcher), random shocks, and specification errors in the structural equation model (James et al., 1982). The *disturbance* terms (symbol: *d*) that are used in latent-factor path analysis correspond to the *residual*, or *error* terms (symbol: *e*) that are used in manifest-variable path analysis.

At the end of the analysis, it is a standard convention for researchers to report R^2 statistics for each endogenous latent factor. These R^2 statistics represent the percent of variance in a given endogenous latent factor that was accounted for by the antecedent variables included in the model. A given R^2 statistic may range from 0.00 to 1.00, with higher values representing a greater percentage of variance accounted for by the antecedent variables.

The presence of a disturbance term in a causal model acknowledges the fact that the antecedent factors almost never predict a consequent factor with perfect accuracy—the disturbance term represents the variance in the consequent factor that was not accounted for by the antecedent factors (Ullman, 2007). The disturbance term for a given endogenous variable is inversely related to the R^2 statistic for that variable. The value for a disturbance term may be computed as follows:

$$d = \sqrt{(1 - R^2)}$$

Where:

d = the disturbance term for an endogenous latent factor, and

R^2 = the variance in that endogenous latent factor that is accounted for by antecedent variables included in the theoretical model.

MEASUREMENT MODEL, PATH MODEL, AND COMBINED MODEL. When performing path analysis with latent factors, it is useful to think of causal models such as the one depicted in Figure 18.1 as consisting of two components: a *measurement model* and a separate *path model*. When we put the two together, the result is a *combined model*. The differences between these three different ways of viewing the model are as follows:

- The ***measurement portion of the model*** (also called the *measurement model*) consists of the latent factors and the indicator variables that measure them. If you evaluated how well the indicator variables (represented by the squares in Figure 18.1) were measuring the latent factors (represented by the ovals), you would be evaluating the measurement model. Dr. O'Day used confirmatory factor analysis to evaluate her measurement model in *Chapter 17*.

- The ***path portion of the model*** (also called the *path model;* sometimes called the *structural model*) consists of the proposed causal relationships between the just latent factors. These causal relationships are represented by the straight, single-headed arrows that connect some ovals to other ovals in Figure 18.1.

- The ***combined model*** (also called the ***composite model*** or the ***full structural model***) is the full structural equation model obtained when the measurement model is combined with the path model. Figure 18.1—when taken as a whole—is an example of a combined model.

When researchers perform path analysis with latent factors, they are typically more interested in the path portion of the model than the measurement portion, since the path portion reflects the causal relationships that are of greatest theoretical importance. But there is a problem: Most global fit indices (such as the

RMSEA, the SRMR, and CFI) reflect the fit displayed by the combined model—the measurement portion mixed in with the path portion. Even worse, these indices tend to be heavily influenced by the fit displayed by the measurement portion of the model, and only minimally influenced by the path portion. This means that when a study reports an acceptable global fit index (such as a CFI that exceeds .95), it is possible that this index is acceptable merely because the measurement portion of the model displays a good fit to the data, and not because the path portion displays a good fit. In fact, reviews of the published literature have shown that articles often report "acceptable" values for global fit indices even when the fit displayed by the path portion of the model is poor (e.g., McDonald & Ho, 2002; O'Boyle & Williams, 2011).

Because of these problems, some experts recommend that researchers should report separate fit indices for the measurement portion of the model, the path portion of the model, and the combined model (e.g., O'Boyle & Williams, 2011). Consistent with these recommendations, a later section of the current chapter describes two indices (the RMSEA-P and the RNFI) that represent the goodness of fit displayed by just the path portion of a model, independent of the measurement portion.

Three Causal Models to be Compared

Figure 18.1 (presented earlier) illustrated just one of the three causal models that Dr. O'Day will investigate in the current chapter. In contrast, Figure 18.2 is a composite that illustrates all three models.

The three theoretical models illustrated in Figure 18.2 differ only with respect to the number of straight, single-headed arrows that connect one latent factor to another latent factor. These arrows represent proposed causal effects.

Model 1 is the simplest of the three models to be investigated. Model 1 consists of the causal paths drawn with solid-line arrows. In other words, Model 1 does not contain the causal paths represented by dotted-line arrows or dashed-line arrows. As explained earlier, Model 1 predicts that (a) exercise behavior is caused by health motivation and dislike of exercise, (b) health motivation is caused by desire to be thin and current health, and (c) dislike of exercise is caused by current health and lack of time for exercise. The plus and minus signs in the figure indicate whether a given antecedent variable is predicted to have a positive effect (+) versus a negative effect (−) on an antecedent variable.

In this investigation, theoretical Model 2 is identical to theoretical Model 1, except that Model 2 includes one additional path: the direct path from lack of time for exercise to exercise behavior. This path is represented by the dashed-line arrow in Figure 18.2.

Finally, theoretical Model 3 is identical to theoretical Model 2, except that Model 3 includes one additional arrow: the path from dislike of exercise to health motivation. This path is represented by the dotted-line arrow in Figure 18.2.

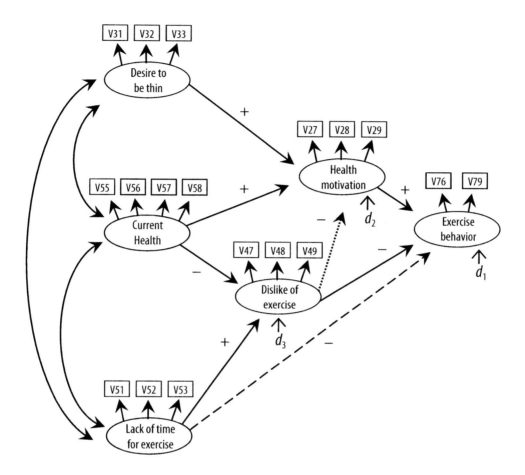

Figure 18.2. Composite illustrating the three theoretical models to be compared.

Research Questions Addressed

The "procedure" described in this chapter is actually an amalgamation of several related procedures. Combined, they allow researchers to investigate a variety of research questions related to the proposed causal models. These questions include the following:

- Is there a good fit between the theoretical model and the data? (i.e., is there a good fit between the relationships that were predicted by the theoretical model and the relationships that actually exist in this sample?).
- Does Model 2 provide a fit to the data that is significantly better than the fit provided by Model 1? Does Model 3 provide a fit that is significantly better than the fit provided by Model 2?
- Regarding the path coefficient that represents a specific causal effect (e.g., the effect of health motivation on exercise behavior): Is this path

coefficient statistically significant? Is it substantial in size? What about the path coefficients for the other effects?

- What are the direct effects, indirect effects, and total effects for the relationships predicted in the model?

Details, Details, Details

This section describes the types of variable that may be used as indicators in path analysis with latent variables, and refreshes your memory regarding the labels that researchers sometimes apply to the models that they estimate (e.g., the *primary theoretical model* and so forth). Finally, it explains why these *"LISREL analyses"* created such a big stir in the scientific community when they first became popular in the 1970s and 1980s.

Types of Variables Analyzed

Maximum likelihood (ML) estimation is probably the most popular method for estimating path models with latent factors, and ML estimation requires that the manifest indicator variables should be on an interval or ratio scale (Klem, 1995) and follow a multivariate normal distribution (McDonald & Ho, 2002). When researchers wish to analyze manifest variables that are nominal-scale variables or ordinal-scale variables, the modeling procedures described here may still be performed, although it may be necessary to conduct the analysis using special estimation methods and, in some cases, special statistical applications. For example, the application *Mplus* (Muthén, 1984; Muthén & Muthén, 2001) was specifically designed for dichotomous and ordinal-scale indicator variables.

Initial Theoretical Models Versus Final Preferred Models

When performing path analysis with latent factors, it is a best practice to estimate and compare more than one causal model. Ideally, the different models should reflect different theoretical positions.

When researchers compare more than one model, they sometimes use standard, boilerplate names to refer to them. For example, when researchers refer to a **primary theoretical model**, this is most likely their "best guess" regarding the relationships that should exist, based on their interpretation of theory and previous research. The researchers' first step will involve determining whether there is a good fit between the data and this initial model. Model 1 (from Figure 18.2) represents Dr. O'Day's primary theoretical model for the current investigation.

Next, the **alternative theoretical models** are models that specify relationships that are different from the relationships hypothesized by the primary model. Ideally, these alternative models should be inspired by competing theories related to the topic. In Figure 18.2, Model 2 and Model 3 are the alternative theoretical models.

The ***final preferred model*** is the winning model, selected at the end of the study. It represents the researcher's best guess regarding the actual relationships between the variables, based on what was found in the current investigation. This final model may be the researcher's primary theoretical model, one of the alternative models, or even a modified version of one of these models.

Does this Statistical Procedure Prove Cause and Effect?

No, it does not.

Path analysis with latent factors is almost always performed on data from nonexperimental investigations. In *Chapter 2* of this book, you learned that nonexperimental research just about never provides strong evidence of cause and effect. Not even when it is analyzed using sophisticated, difficult-to-understand, steroid-injected statistical procedures like this one.

When researchers use path analysis with latent factors to develop a path model that provides a good fit to the data, they may be tempted to tell the world that they have discovered the one true model that correctly depicts the causal relationships between the variables of interest. You, as the reader, should be skeptical of such claims. There are a number of reasons for this, but none is more important than the following: With a given data set, it is usually possible to rearrange the causal ordering of the variables within the theoretical model, and come up with a different causal model that still fits the data (Hershberger, 2006).

Ouch!

This does not mean that path analysis with latent factors is a worthless procedure. I personally believe that it is one of the two most important developments in statistics over the past 60 years (with the other being *meta-analysis*, to be covered in *Chapter 19*).

It's like this: Sometimes we have to do nonexperimental research. And path analysis with latent factors is the *23-speed Osterizer* of nonexperimental research. Because it is first-cousin to confirmatory factor analysis, it allows us to measure our constructs in a way that is far superior to the way that they are measured in path analysis with manifest variables (more on this later).

Latent-factor path analysis allows us to analyze correlational data for the purpose of investigating competing causal models. If we do it right, it can help us make major contributions to a subject area despite the fact that that we are analyzing correlational data. Doing it right involves the following:

- We become excruciatingly familiar with theory and previous research on our topic.
- We develop multiple theoretical path models that describe causal relationships between the constructs of interest, with different models representing different theoretical positions.
- We conduct a program of research on the topic—a program that extends over several years and includes many individual investigations.

- We follow Mueller and Hancock's (2008) recommendation that we should interpret the results from a specific analysis from a largely disconfirmatory perspective. From this perspective: A *bad* fit between model and data may be used to *disconfirm* a specific theoretical model; however, a *good* fit between model and data should not be used to *confirm* a specific theoretical model (i.e., it should not be taken as strong evidence that the theoretical model is true).

If we conduct multiple investigations over an extended period of time, path analysis with latent factors can help us identify the path models that do not work, eventually narrowing the field to a small number of models that do work—a small number of theoretical models that provide a decent fit to the data. We will never conduct a study that proves cause and effect, but will increase the knowledge base, improve our understanding of the constructs, and build better theories. And that is good for science.

What's So Great About this Procedure?

The previous section indicated that—in this author's humble opinion—latent factor path analysis is one of the two most important developments in statistics over the past 60 years. In part, this is because this procedure allows us to perform path analysis while taking into account the *measurement error* that is inherent in our manifest variables. When this measurement error is mathematically modeled by the statistical application, the analysis produces path coefficients (and related statistics) that are much less likely to be biased. In other words, we obtain a better understanding of the actual relationships between the constructs of interest (Thompson, 2000).

For many decades, researchers have known that the manifest variables used in traditional path analysis typically display at least some measurement error. They have also known that this measurement error causes the resulting path coefficients to be biased (Keith, 2006). Here, "traditional" path analysis is path analysis with manifest variables, as was covered in *Chapter 16*.

However, when we perform the special type of path analysis described in the current chapter (i.e., path analysis with latent factors), we address this problem by using *multiple* manifest variables to measure each latent factor. The resulting latent factors reflect the systematic variance that the manifest variables share in common. If all goes well, this means that the latent factors come much closer to representing the true constructs that we really wish to study. Because the measurement error has been estimated and removed, path analysis with latent factors allows us to investigate relationships between latent factors that are free of measurement error (Ullman, 2007). This is a major improvement over the error-laden manifest variables that are used in traditional path analysis.

In summary, the procedures described in this chapter are all about *measurement*: They allow us—for the first time—to model the measurement error displayed by our manifest variables, removing it from the analysis. As a result, these procedures allow us to come much closer to understanding the actual relationships between the constructs of interest.

Results from the Investigation: Overview

Most of this chapter presents excerpts from Dr. O'Day's article in which she describes her analysis and results. To perform these analyses, she follows a two-step protocol recommended by Anderson and Gerbing (1988). This section provides an overview of this two-step approach, along with some initial findings.

Preview of the Analyses to Follow

In Step 1 of Anderson and Gerbing's (1988) two-step approach, Dr. O'Day uses confirmatory factor analysis (CFA) to develop an acceptable measurement model. In this context, a ***measurement model*** is a confirmatory factor analysis model in which each factor is allowed to covary with every other factor. In Step 2, Dr. O'Day modifies this measurement model so that it becomes a ***latent-variable path model***: a model that predicts cause-and-effect relationships between some of the latent factors. In fact, she actually estimates and compares *three* latent-variable path models that reflect three different theoretical perspectives.

Dr. O'Day performs a variety of analyses to compare and understand these three theoretical models. She computes global fit indices (such as the CFI) to evaluate the fit of the combined model (the measurement portion of the model plus the path portion). She reviews other results to evaluate fit in just the path portion of the model (e.g., fit indices such as the RNFI, significance tests for the path coefficients, and R^2 values for the endogenous variables). In the end, she selects one model as her final preferred model, and computes total effect indices for this model.

Estimation Method and Fit Criteria

In the first part of the *Results* section, Dr. O'Day indicates how she will estimate parameters, and describes the cut scores that she will use for evaluating the global goodness-of-fit indices. You can see that she is setting the bar high by using the relatively stringent criteria recommended by Hu and Bentler (1999) and Mueller and Hancock (2008).

> **Data analysis**. The SAS® System's CALIS procedure (SAS Institute, Inc., 2004) was used to estimate and compare four covariance structure models (one measurement model and three path models). Each model consisted of six latent factors with multiple indicators for each factor. All analyses were performed on the variance-covariance matrix, and the maximum likelihood (ML) method was used to estimate parameters.
>
> **Criteria for evaluating goodness of fit**. In evaluating whether a given model displayed an acceptable fit to the data, this investigation used criteria recommended by Hu and Bentler (1999) and Mueller and Hancock (2008). Specifically, a model's fit to the data was viewed as being acceptable if:
> - The standardized root mean-square residual (SRMR) was ≤.08
> - The root mean square error of approximation (RMSEA) was ≤.06
> - The comparative fit index (CFI) was ≥.95

Step 1: Developing the Measurement Model

There is no point in analyzing a proposed causal model if you are doing a bad job of measuring the constructs that appear in it. That is why it is a best practice to develop an acceptable measurement model in Step 1 of the analysis before moving on to the main event: the path analysis with latent factors in Step 2 (Anderson & Gerbing, 1988; McDonald & Ho, 2002; Mueller & Hancock, 2008; O'Boyle & Williams, 2011).

In the following excerpt, Dr. O'Day introduces her measurement model. She then summarizes the statistical results obtained when she analyzed it using confirmatory factor analysis (e.g., the global fit indices, factors loadings, and composite reliability estimates).

> **The measurement model.** This investigation followed the two-step approach recommended by Anderson and Gerbing (1988). At Step 1 of this approach, confirmatory factor analysis (CFA) was used to develop an acceptable measurement model.
>
> The measurement model investigated in Step 1 of the current study is illustrated in Figure 18.3. In the figure, ovals represent the six latent factors that constitute the six-factor model of exercise behavior. Each latent factor was measured by two to four indicator variables (represented by rectangles), and these indicator variables were individual items on the *Questionnaire on Eating and Exercising*. The curved, double-headed arrows in the figure represent correlations—they show that the model's six latent factors were allowed to freely covary with one another in the measurement model.
>
> Global fit indices showed that the six-factor measurement model provided an acceptable fit to the data. The comparative fit index (CFI) was .97, the nonnormed fit index (NNFI) was .96, the root mean square error of approximation (RMSEA) was .05, and the standardized root mean-square residual (SRMR) was .05. All of these indices satisfied the criteria for an acceptable fit recommended by Hu and Bentler (1999) and Mueller and Hancock (2008).
>
> Significance tests showed that each factor loading (represented by unidirectional arrows from ovals to rectangles in Figure 18.3) was significantly different from zero, $p < .001$. The size of each standardized factor loading was substantial, ranging from .61 to .96, Mdn = 82.5. Reliability estimates for the manifest indicators ranged from .37 to .92, Mdn = .68 (Long, 1983a; Long, 1983b).
>
> For the six latent factors, composite reliability indices ranged from .76 to .95, exceeding the recommended criterion of .70 (Fornell & Larcker, 1981; Nunnally, 1978). Variance-extracted estimates for the six factors ranged from .54 to .87, satisfying the criterion of .50 recommended by Fornell and Larcker (1981).
>
> In summary, results from the confirmatory factor analysis indicated that the measurement model in Figure 18.3 displayed a good fit to the data and acceptable psychometric properties. This completed Step 1 of Anderson and Gerbing's (1988) proposed two-step procedure for testing causal models with latent variables.

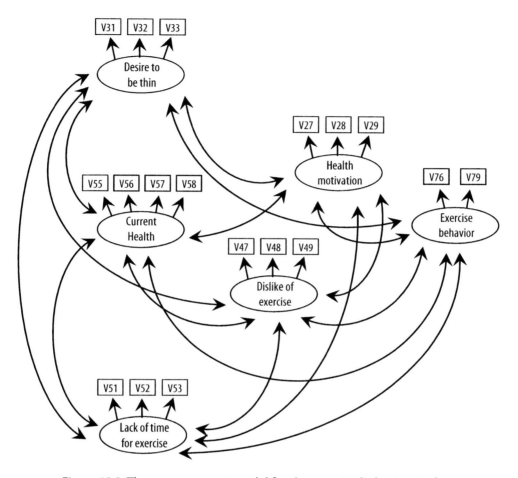

Figure 18.3. The measurement model for the exercise behavior study.

Did the preceding excerpt produce a few seconds of déjà vu? It should, because it was mostly just a retelling of the confirmatory factor analysis that Dr. O'Day had performed in *Chapter 17*.

When you are reading about a path analysis with latent factors, it is not essential that the article reports all of the specific indices listed above (e.g., the CFI, the variance-extracted estimates). But it is important that the researchers provide evidence that their measurement model is a good measurement model with an acceptable fit to the data. And they should provide this evidence before they move on to the path analysis with latent factors. If the researchers did not first verify that their measurement model was acceptable, then you (the reader) should be skeptical about any other results that they present.

Step 2: Comparing Three Latent-Factor Path Models

In the measurement model from Step 1, each latent factor was connected to every other latent factor by a curved, two-headed arrow. These double-headed arrows represented simple correlations. However, in the upcoming analyses, some of these curved, two-headed arrows will be eliminated from the model, and some will be replaced with straight, single-headed arrows. The single-headed arrows represent the causal effects that some latent factors are predicted to have on others.

The Three Theoretical Models

If she were so inclined, Dr. O'Day could have estimated and evaluated just one causal model. However, in studies such as this, it is usually best to estimate more than one model, with each model representing a different theoretical perspective. This allows researchers to determine which perspective is best supported by the sample data.

Dr. O'Day's three competing models. In a published article, Dr. O'Day would almost certainly introduce the three models that she will compare in the *Introduction* section. So let's imagine that the following excerpt is from the *Introduction* section of her article. It is presented at this spot in the current chapter (a bit out of sequence) so that the three theoretical models will be fresh in your mind when you read about the results of her analyses.

> **Path models to be compared**. Figure 18.4 presents a composite of the three theoretical causal models that were investigated in this study. The three models differed only with respect to the number of straight, single-headed arrows connecting one latent factor to another. The latent factors are represented by ovals in the figure, and the single-headed arrows represent the causal effects that one factor is predicted to have on another. Positive signs (+) and negative signs (−) on the arrows represent predicted positive effects and negative effects, respectively.
>
> Model 1 was the simplest of the three models. The only causal paths contained in Model 1 were those represented by solid-line arrows. In other words, Model 1 did not contain the causal paths represented by dotted-line arrows or dashed-line arrows. Model 1 predicted that (a) exercise behavior is caused by health motivation and dislike of exercise, (b) health motivation is caused by desire to be thin and current health, and (c) dislike of exercise is caused by current health and lack of time for exercise.
>
> Model 2 was identical to Model 1, except that it included one additional path: the path from lack of time for exercise to exercise behavior (represented by the dashed-line arrow). Model 3 was identical to Model 2, except that it included one additional path: the path from dislike of exercise to health motivation (represented by the dotted-line arrow).
>
> The three models to be compared in this investigation were *nested models*. Model 1 was nested within Model 2, and Model 2 was nested within Model 3.

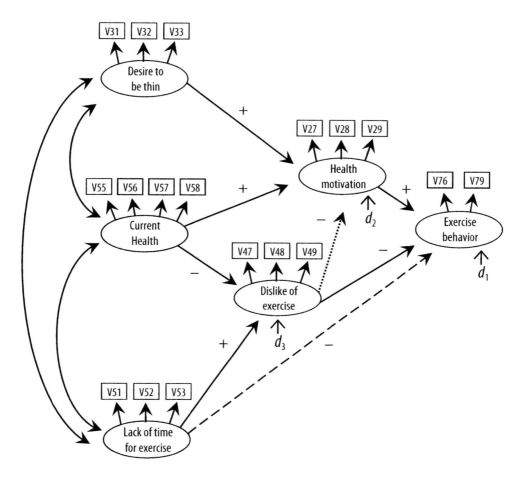

Figure 18.4. Composite illustrating the three theoretical models to be compared.

COMPARING NESTED MODELS. In general, when a researcher says that Model 1 is *nested* within Model 2, it means that Model 1 is a simpler version of Model 2—it means that one or more of the causal paths that appear in Model 2 have been deleted from Model 1. For example, you know that Model 1 is nested within Model 2 in the current study because the dashed-line arrow from lack of time for exercise to exercise behavior appears in Model 2, but not in Model 1.

When models are nested in this way, it is possible to perform a *model-comparison χ^2 test* to determine whether one of the models displays a fit to the data that is significantly better than the fit displayed by another model. This is a very useful test that Dr. O'Day will report later in her article.

COMPARING NON-NESTED MODELS. When the competing theoretical models are not nested, it may still be possible to compare them by using different procedures. One global fit index that is well-suited for comparing non-nested models is the

Akaike information criterion (Akaike, 1987), typically abbreviated as the AIC (Keith, 2006).

Estimating the Three Theoretical Models

The preliminaries are over and it is (*finally!*) time to see the results. In the following excerpt from the *Results* section, Dr. O'Day presents global fit indices for a number of models. Most of these global fit indices were introduced and described in greater detail in *Chapter 16*.

Table 18.1

Fit and Parsimony Indices for Models of Exercise Behavior

Model	χ^2	df	χ^2/df	Combined model					Path model	
				SRMR	RMSEA	CFI	NNFI	PR	RMSEA-P	RNFI
Null model	2706.60	153	17.69							
Uncorrelated factors	397.58	135	2.95	.20	.09	.90	.88	.88	.232	.00
Model 1	241.48	126	1.92	.09	.06	.95	.95	.82	.175	.78
Model 2	209.32	125	1.67	.07	.05	.97	.96	.82	.103	.94
Model 3	199.34	124	1.61	.06	.05	.97	.96	.81	.062	.98
Measurement model	191.70	120	1.60	.05	.05	.97	.96	.78	.000	1.00

Note. N = 238. SRMR = Standardized root mean-square residual; RMSEA = root mean square error of approximation; CFI = comparative fit index; NNFI = nonnormed fit index; PR = parsimony ratio; RMSEA-P = root mean square error of approximation for path portion of model; RNFI = relative normed fit index.

Estimation of the path models. Table 18.1 presents fit indices and parsimony indices for three theoretical models. The entries are ordered from the simpler models (near the top of the table) to the more complex models (toward the bottom).

With the *null model* (sometimes called the *independence model* or *uncorrelated-variables model*), no variable is allowed to correlate with any other variable. This serves as a baseline against which the other models may be compared.

The *uncorrelated factors model* (the second entry in the table) is identical to the measurement model, except that the correlations between the latent factors are all constrained to be equal to zero. Information from the uncorrelated factors model is used to compute some of the fit indices to be discussed later in this article.

> *Model 1, Model 2,* and *Model 3* are the three theoretical causal models that are the central focus of this investigation. They were illustrated in Figure 18.4.
>
> The *measurement model* (the last entry in the table) is a confirmatory factor analysis model in which every latent factor is allowed to covary with every other latent factor. It was illustrated by Figure 18.3 (presented earlier).

Table 18.1 presents a lot of information. But much of the information is merely included for the sake of completeness, and is not terribly central to the research questions at hand (e.g., the *null model* at the top of the table). So disregard the table for the moment, and focus instead on the *text* of the article—which statistics are important enough for Dr. O'Day to actually discuss in the text of her article? Just relax and let her do the heavy lifting for now.

In the following excerpt, Dr. O'Day explains how the columns of the table are organized. She also explains the different types of indices that it presents.

> The left side of Table 18.1 (labeled "Combined model") provides information about the *composite model*—the full structural equation that contains the measurement model combined with the path model. The right side of the table (labeled "Path model") provides information about just the path portion of the model—the proposed causal relationships between the six latent factors.
>
> The table presents different types of statistics. For example, some statistics (such as the standardized root mean-square residual, or SRMR) indicate the extent to which a theoretical model provides a good fit to the data (i.e., the extent to which a theoretical model is an accurate representation of the relationships that actually exist in the sample). Other statistics (such as the parsimony ratio, or PR) merely reflect the parsimony, or simplicity, of the model. Other factors held constant, a model that has fewer causal arrows is a more parsimonious model and will have a higher score on the parsimony ratio. Finally, other statistics (such as the root mean square error of approximation, or RMSEA) simultaneously reflect both the fit and parsimony of the model.

Evaluating the Combined Versions of the Models

To evaluate the goodness of fit displayed by Model 1, Dr. O'Day will focus on a few of the global fit indices presented in the table, along with something called the "nomological validity test." Remember that she has three models to compare, so she will start with the simplest model and will work her way to the most complex.

> **Fit and nomological validity of the combined models.** In Table 18.1, the row headed "Model 1" provides information about the simplest of the three theoretical causal models: the one consisting of solid-line causal arrows in Figure 18.4. For this model, the ratio of χ^2/df was 1.92, which was below the informal criterion of 2 or 3 used by some researchers (Bollen, 1989; Hatcher, 1994; Klem, 2000). Although this finding was supportive of Model 1, the remaining indices were mixed: The SRMR was .09 (higher than the recommended value of $\leq .08$); the RMSEA was .06 (just meeting the recommended value of $\leq .06$); the CFI was .95, and the NNFI was also .95 (both just meeting the recommended value of $\geq .95$).

More troubling were the results from the nomological-validity χ^2 test (Anderson & Gerbing, 1988). With this test, the model χ^2 statistic for the measurement model is subtracted from the model χ^2 statistic for the theoretical model of interest (Model 1, in this case). If the resulting χ^2 difference is statistically significant, it means that Model 1 (in which constraints were placed on some of the relationships between the latent factors) is providing a fit to the data that is significantly worse than the fit provided by the measurement model (in which all latent factors were free to covary). In the current study, when Model 1 was compared to the measurement model, the resulting χ^2 difference was statistically significant, $\Delta\chi^2(6, N = 238) = 49.78, p < .001$. This indicated that Model 1 was not adequately accounting for the actual relationships between the latent factors.

Because of this problem with Model 1, attention was turned to theoretical Model 2. Model 2 was identical to Model 1 except that one additional path appears in Model 2: the path from lack of time for exercise to exercise behavior. A model-comparison χ^2 test showed that adding this new path resulted in a significant improvement in the model's fit to the data, $\Delta\chi^2(1, N = 238) = 32.16, p < .001$ (this test was performed by subtracting the model χ^2 statistic for Model 2 from the model χ^2 statistic for Model 1). The row labeled "Model 2" in Table 18.1 shows that the global fit indices for Model 2 were all better than the global fit indices for Model 1.

The nomological validity of Model 2 was then investigated by subtracting the model χ^2 statistic for the measurement model from the model χ^2 statistic for Model 2. The resulting χ^2 difference was again statistically significant, $\Delta\chi^2(5, N = 238) = 17.62, p < .01$, indicating that Model 2 also provided a fit that was significantly worse than the fit provided by the measurement model. This outcome meant that Model 2 had also failed the nomological validity test.

In the previous paragraphs, Dr. O'Day performed three different tasks:

- She evaluated the overall fit of a given model by reviewing global fit indices such as the SRMR and RMSEA.

- She performed ***model-comparison χ^2 tests***. With these tests, she compares a simpler theoretical model (such as Model 1) to a more complex theoretical model (such as Model 2). If this test is statistically significant, it is good news for the more complex model—it shows that the more complex model (which contains one or more additional paths) displays a fit that is significantly better than the fit displayed by the simpler model.

- She also performed ***nomological-validity χ^2 tests***. This is a χ^2 difference test in which the current theoretical model is compared to the measurement model. If this test is statistically significant, it is bad news for the current theoretical model—it shows that the current theoretical model displays a fit that is significantly worse than the fit displayed by the measurement model.

In the preceding excerpt, Dr. O'Day found that Model 2 displayed a fit that was significantly better than the fit displayed by Model 1. However, she found that neither model passed the nomological validity test. This means that she has one last model to assess: Model 3.

> Model 3 was identical to Model 2 except that one additional path appears in Model 3: the path from dislike of exercise to health motivation. A model-comparison χ^2 test showed that adding this new path resulted in a significant improvement in the model's fit to the data, $\Delta\chi^2(1, N = 238) = 9.98$, $p < .01$. The row for "Model 3" in Table 18.1 shows that the global fit indices for Model 3 were all equal to or better than the fit indices for Model 2. In addition, all of the global fit indices in the table satisfied the criteria for an acceptable fit recommended by Hu and Bentler (1999) and Mueller and Hancock (2008).
> Next, the nomological validity of Model 3 was investigated by subtracting the model χ^2 statistic for the measurement model from the model χ^2 statistic for Model 3. With alpha set at $\alpha = .05$, the resulting χ^2 value was not statistically significant, $\Delta\chi^2(4, N = 238) = 7.64$, ns. This finding showed that Model 3 provided a fit to the data that was not significantly worse than the fit provided by the measurement model. Of the three models, only Model 3 passed this nomological validity test.

Evaluating the Path Portion of the Models

The fit indices reported in the previous excerpt focused on the combined version of Model 3. You will recall that a **combined model** (also called the *composite model* or *full structural model*) refers to the measurement portion of the model combined with the path portion. The indices described above show that the combined version of Model 3 displays a good fit to the data. But, as consumers of research, we should also be interested in the **path portion** of the model: the causal relationships between the latent factors. The path component of the model is represented by the straight, single-headed arrows that connect some ovals to other ovals in Figure 18.4.

FIT INDICES FOR THE PATH PORTION OF THE MODELS. In the following excerpt, Dr. O'Day introduces two new indices that can be used to evaluate fit in just the path portion of a structural equation model.

> **Fit displayed in the path portion of the models**. The right section of Table 18.1 is headed "Path model." This section provides two indices—the RMSEA-P and the RNFI—which reflect fit in just the path portion of each model, independent of the measurement model.
> The abbreviation RMSEA-P represents *root mean square error of approximation of the path component*. With the RMSEA-P, smaller values (closer to zero) indicate a better fit. In their review of 43 published investigations, O'Boyle and Williams (2011) viewed the fit in the path portion of a model as being acceptable if the RMSEA-P was less than .08.

In the current investigation, Table 18.1 shows that the RMSEA-P for Model 3 was .062, satisfying this criterion. For Model 1, the RMSEA-P was .175, indicating a poor fit. For Model 2 it was .103, also indicating a poor fit.

The last column in Table 18.1 is headed "RNFI," which represents *relative normed-fit index*. Values of the RNFI typically range from 0.00 to 1.00, although in some cases they may exceed these bounds. With this index, higher values indicate a better fit (Mulaik et al., 1989). There does not appear to be any widely adopted criterion for evaluating the RNFI, although the traditional criterion for a related statistic, the *normed-fit index* (NFI), has been NFI ≥ 0.90 (Mueller & Hancock, 2008). Table 18.1 shows that the RNFI was equal to .78, .94, and .98 for Model 1, Model 2, and Model 3, respectively.

Some articles fail to report the RMSEA-P, the RNFI, or any other index that reflects fit in just the path portion of a model. In those situations, it may be possible for readers to compute the RMSEA-P and RNFI by making use of other information that appears in the article (such as the χ^2 statistic for various models). Formulas for the RMSEA-P and RNFI are provided near the end of this chapter, in the section titled "Additional Issues Related to this Procedure."

REVIEWING THE R^2 STATISTICS. The R^2 ***statistic*** indicates the percent of variance in an endogenous variable that is accounted for by its antecedent variables. With other things held constant, higher values of R^2 are generally desirable. In the following excerpt, Dr. O'Day discusses the values obtained in the current analysis.

Table 18.2

R^2 Statistics for Three Endogenous Variables Under Models 1, 2, and 3

	R^2 Statistics		
Model	Dislike of exercise	Health motivation	Exercise behavior
Model 1	.19	.25	.22
Model 2	.19	.25	.34
Model 3	.18	.28	.35

Note. N = 238.

R^2 **for the endogenous variables.** Table 18.2 provides R^2 statistics for this investigation's three endogenous variables. The table shows that, for *exercise behavior*, the R^2 statistic for Model 2 (R^2 = .34) was substantially higher than the R^2 statistic for Model 1 (R^2 = .22). This increase in R^2 was due to the fact that Model 2 had one new path not contained in Model 1: The path going from *lack of time for exercise* to *exercise behavior*.

The table also shows that, for *health motivation*, the R^2 statistic for Model 3 (R^2 = .28) was somewhat higher than the R^2 value for Model 2 (R^2 = .25). This increase in R^2 was due to the fact that Model 3 had one new path not contained in Model 2: The path going from *dislike of exercise* to *health motivation*.

EVALUATING THE PATH COEFFICIENTS. Dr. O'Day now comes to one of the most important results: the path coefficients that represent the "effects" that the latent factors have on one another. Think about the straight, single-headed arrows that connect the ovals (i.e., the latent factors) in Figure 18.4 (presented earlier). There is one path coefficient for each of these arrows. The sign of the path coefficient is interpreted in the same way that we interpret the sign of a correlation coefficient: Positive path coefficients indicate that higher scores on the antecedent variable are associated with higher scores on the consequent variable, and negative path coefficients indicate that higher scores on the antecedent are associated with lower scores on the consequent.

There are actually two types of path coefficients. First, the ***unstandardized path coefficients*** are analogous to the unstandardized multiple regression coefficients obtained in a multiple regression analysis (i.e., the raw-score regression coefficients, or *b* weights). These are the path coefficients that are typically reviewed to determine whether they are statistically significant, because most SEM applications report significance tests for the unstandardized path coefficients, but not for the standardized coefficients. However, the unstandardized path coefficients are typically not reviewed to compare the relative size of the effects in a latent-variable path analysis. Because they are not standardized, the size of these coefficients may simply reflect the metric used to measure the variables rather than the magnitude of any causal relationships.

In contrast, the ***standardized path coefficients*** are analogous to standardized multiple regression coefficients (i.e., beta weights, or β weights) from multiple regression. Standardized path coefficients typically range in size from −1.00 through 0.00 through +1.00, and may be directly compared to evaluate the relative magnitude of causal effects: The larger the coefficient in absolute value, the larger the "effect."

In the following excerpts, Dr. O'Day discusses the path coefficients obtained in the analysis of Model 3. She presents the unstandardized coefficients in a table, and the standardized coefficients in a figure.

Path coefficients. Table 18.3 presents the unstandardized path coefficients from the path portion of Model 3, along with their standard errors and *t* values. With alpha set at α = .05, the table shows that all path coefficients were all statistically significant. This finding is supportive of Model 3.

Table 18.3

Unstandardized Path Coefficients from the Path Portion of Model 3

Consequent variable / Antecedent variable	Unstandardized path coefficient	SE	t
Exercise behavior			
Health motivation	.41	.14	2.90**
Dislike of exercise	-.30	.11	-2.80**
Lack of time for exercise	-.50	.09	-5.82***
Health motivation			
Dislike of exercise	-.16	.05	-3.20**
Desire to be thin	.17	.05	3.45***
Current health	.31	.07	4.34***
Dislike of exercise			
Current health	-.22	.09	-2.38*
Lack of time for exercise	.30	.05	5.50***

Note. N = 238.
*p < .05; **p < .01; ***p < .001.

Figure 18.5 presents the standardized path coefficients for the path portion of Model 3. These path coefficients appear on the straight, single-headed arrows that connect one oval to another oval. The figure shows that all coefficients were nontrivial in size—they all exceeded the criterion of ≥±.05 suggested by Billings and Wroten (1978). The values on the curved arrows represent correlations between the exogenous latent factors.

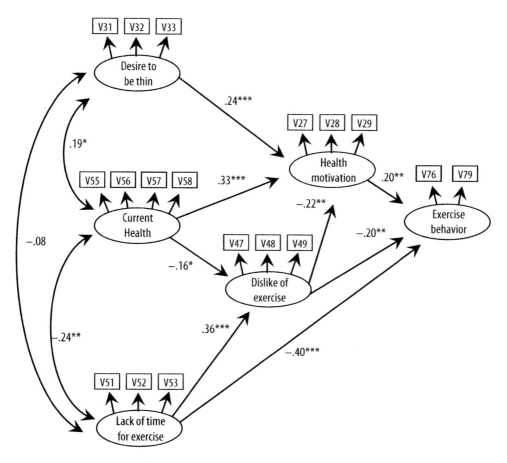

Figure 18.5. Preferred model (Model 3) with standardized path coefficients coefficients (* $p < .05$; ** $p < .01$; *** $p < .001$).

SYMBOLS USED TO REPRESENT PATH COEFFICIENTS IN ARTICLES. Researchers sometimes use Greek letters to represent path coefficients in a latent-factor path analysis. Different Greek letters are used for this purpose, and the choice is determined by the nature of the variables included in the causal path. Specifically:

- Researchers typically use the lower-case Greek letter gamma (γ) to represent the direct effect of an *exogenous* latent factor on an *endogenous* latent factor.

- They typically use the lower-case Greek letter beta (β) to represent the direct effect of one *endogenous* latent factor on another *endogenous* latent factor.

As an illustration, the following excerpt shows how Dr. O'Day could use these conventions to report standardized path coefficients in the text of her article. The standardized path coefficients reported below are taken from Figure 18.5.

Exercise behavior was positively influenced by health motivation (β = .20), and was negatively influenced by dislike of exercise (β = −.20) and lack of time for exercise (γ = −.40). The antecedent variables accounted for 35% of the variance in exercise behavior. Health motivation was negatively influenced by dislike of exercise (β = −.22) and was positively influenced by desire to be thin (γ = .24) and current health (γ = .33). The antecedent variables accounted for 28% of the variance in health motivation. Dislike of exercise was negatively influenced by current health (γ = −.16) and was positively influenced by lack of time for exercise (γ = .36). The antecedent variables accounted for 18% of the variance in dislike of exercise.

In addition to using γ and β to represent path coefficients, some researchers use other Greek letters to represent factor loadings and correlations in causal models. These conventions are described in greater detail in the *Additional Issues* section located toward the end of this chapter.

FIGURES THAT PRESENT *d* TERMS IN A PATH MODEL. Earlier, you learned that the symbol *d* is often used to represent disturbance terms for the endogenous latent factors in a causal model. The ***disturbance term*** represents the variance in the latent factor that was not accounted for by the factor's antecedent variables. The size of a disturbance term is inversely related to the size of the R^2 statistic for a given endogenous factor: As R^2 increases in size, the *d* term decreases. It is computed using the following formula:

$$d = \sqrt{(1 - R^2)}$$

For Dr. O'Day's Model 3, the disturbance term for the exercise behavior factor was $d = .81$. For the health motivation factor, it was $d = .85$, and for the dislike of exercise factor, it was $d = .91$.

Published articles sometimes include the *d* term for each endogenous latent factor as part of the figure that illustrates a causal model, such as Figure 18.5. The *d* term is typically presented simply as a number (such as ".81") with a single-headed causal arrow going from this number to the relevant endogenous variable. The *d* terms were omitted from Figure 18.5 as the figure was already fairly well cluttered with just the path coefficients.

Selecting the Final Preferred Model

For each of the three theoretical models presented earlier in this chapter, Dr. O'Day evaluated the fit of the combined model. She took an even closer look at Model 3, evaluating the fit of the path portion of the model, inspecting path coefficients, and reviewing other results. In the following excerpt, she reports her final decision:

> **Selection of the preferred model.** Based on the results reported here, Model 3 was selected as this study's final preferred model. The selection of Model 3 was based on the following findings: (a) the global fit indices for Model 3 (combined model) satisfied the criteria recommended by Hu and Bentler (1999) and Mueller

and Hancock (2008); (b) for two of the three endogenous variables, the R^2 values found for Model 3 were larger than the R^2 values found in the other two models; (c) although the path portion of Model 3 was less parsimonious than the path portion of the other two models, the RMSEA-P and RNFI showed that it displayed a much better fit to the data; and (d) only Model 3 passed the nomological validity test recommended by Anderson and Gerbing (1988).

Direct Effects, Indirect Effects, and Total Effects

Now that she has selected Model 3 as her "preferred" model, there is one last set of statistics that Dr. O'Day may wish to share with the reader: Direct, indirect, and total effect coefficients. As you may recall from the chapter on path analysis with manifest variables, the ***direct effect*** of an antecedent variable on a consequent variable is an effect that is not mediated by any intervening variables (Cudeck & du Toit, 2009). In Figure 18.5, for example, the arrow that goes directly from dislike of exercise to exercise behavior represents a direct effect.

In contrast, the ***indirect effect*** of an antecedent variable on a consequent variable is the effect that is mediated by one or more intervening variables. For example, Figure 18.5 displays a causal chain in which dislike of exercise affects health motivation and health motivation, in turn, affects exercise behavior. This causal chain represents an indirect effect of dislike of exercise on exercise behavior.

It is possible for an antecedent variable to have more than one indirect effect on a consequent variable. This occurs when the antecedent variable is connected to the consequent variable through more than one causal chain of intervening variables.

Finally, a ***total effect*** is the sum of the direct effects and indirect effects that a given antecedent variable has on a consequent variable. The total effect is often the effect of greatest interest because it provides the *big picture* regarding the relative importance of predictor variables. The statistic that represents the size of a total effect is called an ***effect coefficient***, and is sometimes represented using the abbreviation *EC* (e.g., Pedhazur, 1982).

In the following excerpt, Dr. O'Day introduces a table of effect statistics. The table provides a lot of information; notice how she attempts to simplify things by focusing on just those antecedent variables with the largest total effect coefficients.

> **Direct, indirect, and total effects**. Table 18.4 presents direct effects, indirect effects, and total effects obtained for Model 3. Results in the "Total" effect columns show that the variable that had the strongest total effect on exercise behavior was lack of time for exercise. The effect coefficient for this relationship was *EC* = −.48. The variable that had the strongest total effect on health motivation was current health (*EC* = .36). Finally, the variable that had the strongest total effect on dislike of exercise was lack of time for exercise (*EC* = .36).

Table 18.4

Direct Effects, Indirect Effects, and Total Effects of Antecedent Variables on Consequent Variables for Model 3 (All Estimates Are Standardized).

Antecedent variable	Exercise behavior			Health motivation			Dislike of exercise		
	Direct	Indirect	Total	Direct	Indirect	Total	Direct	Indirect	Total
Health motivation	.20		.20						
Dislike of exercise	-.20	-.04	-.24	-.22		-.22			
Desire to be thin		.05	.05	.24		.24			
Current health		.11	.11	.33	.03	.36	-.16		-.16
Lack of time for exercise	-.40	-.08	-.48		-.08	-.08	.36		.36

Note: $N = 238$.

Summary of Findings

Model 1 and Model 2 did not display a terribly good fit to the data. For both models, the nomological-validity χ^2 test (Anderson & Gerbing, 1988) was statistically significant, indicating that both models displayed a fit to the data that was significantly worse than the fit displayed by the measurement model.

In contrast, Model 3 displayed a fairly good fit to the data. For Model 3, the SRMR = .06, the RMSEA = .05, and the CFI = .97. These results satisfied criteria suggested by Hu and Bentler (1999) and Mueller and Hancock (2008). The χ^2 statistic for the nomological validity test was nonsignificant, indicating that the fit displayed by Model 3 was not significantly worse than the fit displayed by the measurement model.

Although the path portion of Model 3 was less parsimonious than the path portion of the other two models, the RMSEA-P and RNFI showed that it displayed a much better fit to the data. Further, the RMSEA-P satisfied criteria recommended by O'Boyle and Williams (2011).

Each path coefficient in Model 3 was significantly different from zero ($p < .05$), and displayed a sign consistent with the sign that had originally been predicted based on theory and previous research. Each standardized path coefficient exceeded ±.05 in absolute value, indicating that no path coefficient was trivial in size (Billings & Wroten, 1978).

The central criterion variable in the current study was exercise behavior. Results for Model 3 showed that the following constructs had the strongest total effects on exercise behavior: lack of time for exercise ($EC = -.48$), dislike of exercise ($EC = -.24$), and health motivation ($EC = .20$).

Additional Issues Related to this Procedure

This section goes into a bit more detail in explaining the Greek letters (i.e., γ or ϕ) that you will sometimes encounter when reading about a structural equation model. The section also shows how to compute the RNFI and RMSEA-P, provides a checklist for evaluating SEM studies in general, and ends with a list of assumptions underlying path analysis with latent factors.

Greek Letter Department

When reading about a study that used structural equation modeling, you may encounter Greek letters such as λ or γ with no real explanation as to what they represent. The current section provides a pocket translation guide for such situations.

GREEK LETTERS AND LISREL. When you encounter Greek letters in an article that describes a latent-factor path analysis, the letters typically represent specific statistics (such as path coefficients) that were estimated in the analysis. The use of these Greek letters can be traced back to the early popularity of LISREL, the first widely available computer application that performed SEM (Jöreskog & Sörbom, 1989). With LISREL, Greek letters represented specific variables, parameters, and other components of the analysis. Back in the 1970s, a researcher might use up to 13 different Greek letters to represent the different parts of a LISREL analysis (Klem, 2000).

THE MOST IMPORTANT GREEK LETTERS. If you find these Greek letters confusing, you can take heart in some good news. First, not many researchers continue to use the long list of Greek letters that was once required by LISREL. This means that you are now likely to read about an SEM study and not encounter any Greek letters at all—especially if it is a more recent investigation.

Second, there are really only four results from an SEM analysis (that might be represented by a Greek letter) that are important enough to include in the typical article. If you can memorize these four Greek letters, you will be good to go. Here they are:

- *Lambda.* The lower-case Greek letter lambda (λ) represents *factor loadings:* the effects of the latent factors on the observed indicator variables that measure them.

- *Gamma.* The lower-case Greek letter gamma (γ) represents one type of *path coefficient*: a path coefficient that represents the effect of an exogenous latent factor on an endogenous latent factor.

- *Beta.* The lower-case Greek letter beta (β) represents a different type of *path coefficient*: a path coefficient that represents the effect of one endogenous latent factor on another endogenous latent factor.

- *Phi.* The lower-case Greek letter phi (ϕ) represents the *correlation* (or covariance) between two exogenous latent factors. In a Figure (such as Figure 18.5), these correlations or covariances are often represented by curved, two-headed arrows.

Computing the RMSEA-P and RNFI

Previously, this chapter discussed the *root mean square error of approximation of the path component*, or RMSEA-P (O'Boyle & Williams, 2011) and the *relative normed-fit index*, or RNFI (Mulaik et al., 1989). These statistics reflect goodness of fit in just the path portion of the model, independent of the measurement portion. The *path portion* of the model refers to the proposed causal relationships between just the latent factors (e.g., the constructs represented by ovals in Figure 18.5). In most cases, this is the most important part of the model, since it represents the causal relationships that are of greatest theoretical interest.

FAILURE TO REPORT THE RMSEA-P AND RNFI IN PUBLISHED RESEARCH. Unfortunately, relatively few articles dealing with structural equation modeling report the RMSEA-P, the RNFI, or other indices that reflect fit in just the path portion of the model, independent of the measurement model. The global fit indices that are typically reported (such as the RMSEA, the SRMR, and the CFI) reflect fit in the combined model—the full structural equation model that includes the measurement model as well as the path model. Because these global fit indices tend to be heavily influenced by the measurement model, articles often report that a combined model displays an acceptable fit to the data despite having a poor fit in the path portion. This is true even of articles published in prestigious journals (McDonald & Ho, 2002; O'Boyle & Williams, 2011). This is a big problem.

Fortunately, it is sometimes possible for you (the consumer of research) to compute the RNFI and the RMSEA-P from information that appears in the published article. The following sections provide the formula for each index.

COMPUTING THE RNFI. Below is the formula for the RNFI (Mulaik et al., 1989; formula adapted from Hatcher, 1994).

$$\text{RNFI} = \frac{\chi_u^2 - \chi_t^2}{\chi_u^2 - \chi_m^2 - (df_t - df_m)}$$

Where:

RFNI = Relative normed-fit index

χ_u^2 = Model chi-square statistic for the *uncorrelated-factors model*. This is the model in which the correlations between each of the latent factors (represented by ovals in Figure 18.5) are constrained to be equal to zero. Do not confuse this with the *uncorrelated-variables* model (i.e., the *null model* or the *independence model*): the model in which the correlations between the manifest variables are constrained to be equal to zero. Be warned that many articles do not report χ^2 for the uncorrelated-factors model.

χ_t^2 = Model chi-square statistic for the *theoretical model of interest*. For example, in the current investigation, this would be the model chi-square statistic for Model 1, for Model 2, or for Model 3. This should be the *combined model* (i.e., the composite model that contains the measurement portion as well as the path portion of the structural equation model).

χ_m^2 = Model chi-square for the *measurement model*. The measurement model is the confirmatory factor analysis model in which each latent factor is allowed to freely covary with every other latent factor. This model was the main focus of *Chapter 17* in this book.

df_t = Degrees of freedom for the *theoretical model of interest*.

df_m = Degrees of freedom for the *measurement model*.

The preceding formula shows that, in order to compute the RNFI, you must have access to the model chi-square statistic for three models: the uncorrelated factors model, the theoretical model of interest, and the measurement model. Unfortunately (as was mentioned above), many published articles do not report the chi-square statistic for the uncorrelated factors model. In those cases, you may choose to instead compute a different fit index: the RMSEA-P. As you shall see in the following section, the RMSEA-P requires the model chi-square and *df* for just two models: the theoretical model of interest and the measurement model.

COMPUTING THE RMSEA-P. All formulas related to the RMSEA-P are from O'Boyle and Williams (2011, p. 4). They, in turn, give credit to McDonald and Ho (2002).

Prior to computing the RMSEA-P, you must first compute χ_p^2 (the chi-square statistic for just the path portion of the model) and df_p (the degrees of freedom for just the path portion of the model). Below are the formulas for computing these items:

$$\chi_p^2 = \chi_t^2 - \chi_m^2$$
$$df_p = df_t - df_m$$

The definitions for χ_t^2, χ_m^2, df_t, and df_m are the same as those provided in the section on the RNFI, above.

Below is the formula for computing the RMSEA-P. In the following formula, the symbol *n* represents the number of participants included in the sample.

$$RMSEA\text{-}P = \sqrt{\frac{(\chi_p^2 - df_p)}{\left(df_p \times (n-1)\right)}}$$

O'Boyle and Williams (2011, *Appendix B*) also show how to compute 90% confidence interval for the RMSEA-P. In their review of 43 published articles dealing with SEM, they used the two following criteria for determining whether there was an acceptable fit in the path portion of the model: (a) if the point estimate for the RMSEA-P was < .08; and (b) if the lower bound of the 90% CI for the RMSEA-P was < .05 and the upper bound was <.10.

Those of you with alarmingly good memories may have noticed that the current chapter's criterion for the RMSEA-P (<.08) is similar but not identical to its criterion for the RMSEA (≤.06). You will recall that the RMSEA is a *global fit index*—a statistic that reflects fit in the combined model.

Checklist for Evaluating the Study and Results

In my book on structural equation modeling (Hatcher, 1994), I summarized the characteristics of an ideal solution that may be obtained when performing path analysis with latent factors. That list was designed for researchers who are actually performing SEM.

For the present book, however, I have created a list of just those characteristics that you, the reader of the research article, would be in a position to evaluate. In other words, I have omitted the criteria that could only be evaluated by the researcher who was actually conducting the study. For the current list, I have also added some new recommendations that have become popular since the publication of my book on SEM in 1994 (e.g., recommendations from Hu & Bentler, 1999; Mueller & Hancock, 2008; O'Boyle & Williams, 2011).

Below are the characteristics of an *ideal* solution from a path analysis with latent factors:

- The researchers should follow Anderson and Gerbing's (1988) two-step procedure in which they develop an acceptable measurement model during Step 1 and then modify it so that it becomes the theoretical causal model of interest during Step 2.

- The measurement portion of the model should satisfy all of the criteria for an acceptable measurement model that were described in *Chapter 17*. In part, these criteria include the following: (a) the global fit indices such as the CFI should be acceptable; (b) the factor loadings should be significantly different from zero (α = .05); (c) the standardized factor loadings themselves should exceed ±.05 in absolute value; (d) composite reliabilities of the latent factors should exceed .70 and their variance extracted estimates should exceed .50.

- In evaluating the fit of the theoretical causal model, it is ideal if the *p* value for the model χ^2 statistic is greater than .05 (i.e., is nonsignificant). Higher *p* values indicate better fit. But remember that this criterion is seldom achieved, even in models that otherwise display a good fit to the data. So maybe I should not have even brought this up.

- For the theoretical causal model, the ratio of χ^2/df should be less than 2 or 3. But remember that this is an informal benchmark—there are no universally accepted criteria for the χ^2/df ratio (Klem, 2000).

- For the theoretical causal model, the global fit indices should indicate an acceptable fit between model and data. Specifically, the CFI should be ≥.95, the standardized root mean-square residual (SRMR) should be ≤.08, and the root mean square error of approximation (RMSEA) should be ≤.06 (Hu & Bentler, 1999; Mueller & Hancock, 2008).

- The theoretical causal model should demonstrate nomological validity. This means that a χ^2 difference test should reveal a nonsignificant difference between the fit of the theoretical causal model versus the fit of the measurement model.

- The RMSEA-P and RNFI should show that the path portion of the model demonstrates a good fit to the data. O'Boyle and Williams (2011) discussed the following criteria: (a) the point estimate of the RMSEA-P should be <.08; and (b) the lower bound of the 90% confidence interval for the RMSEA-P should be <.05; and (c) the upper bound should be <.10.

- Regarding the path portion of the model: The path coefficients should be significantly different from zero. If the article presents the z statistics or t statistics for the path coefficients, these statistics should exceed ±1.96 for the path coefficients to be statistically significant with α = .05 (assuming that the study was conducted using the fairly large samples that are typically used with SEM).

- Also regarding the path portion of the model: The standardized path coefficients themselves should be nontrivial in size. One criterion suggests that, to be meaningful, path coefficients should exceed ±.05 (Billings & Wroten, 1978).

- The R^2 values for the endogenous variables should be large relative to what has typically been found when researchers have investigated the variables in question.

Assumptions Underlying this Procedure

This section summarizes the assumptions underlying path analysis with latent factors using maximum likelihood estimation (based on Hatcher, 1994; Keith, 2006; Kenny, 1979; Klem, 1995). These assumptions are similar to the assumptions underlying path analysis with manifest variables (covered in *Chapter 16*) and confirmatory factor analysis (covered in *Chapter 17*). Some of these assumptions are discussed in greater detail in *Chapter 2*.

INTERVAL-LEVEL MEASUREMENT. In general, the analysis requires that all manifest variables be continuous quantitative variables assessed on an interval scale or ratio scale. Some SEM applications (such as EQS and LISREL) can accommodate ordinal-scale or categorical variables if special steps are taken (see Tabachnick &

Fidell, 2007, pp. 729-730). In addition, the application called *Mplus* is particularly well-suited to accommodate categorical variables (Muthén & Muthén, 2001).

INDEPENDENT OBSERVATIONS. Each participant should be drawn independently from the population of interest. Among other things, this means that the researcher should not take repeated measures from the same subject unless special procedures have been used to accommodate repeated measures.

MULTIVARIATE NORMALITY. Many of the statistical tests that are computed with a path analysis (such as the model chi-square test and the significance tests for path coefficients) assume that the data follow a multivariate normal distribution. The maximum likelihood estimation procedure appears to be fairly robust against mild to moderate violations of this assumption.

ABSENCE OF SPECIFICATION ERRORS. In part, this means that (a) all important determinants of the model's endogenous variables are included in the path model, and (b) no unimportant determinants are included.

ABSENCE OF MEASUREMENT ERROR. All exogenous variables must be measured with perfect reliability.

CAUSAL MODEL MUST BE THEORETICALLY CORRECT. Among other things, this means that the correct variables must be included in the model, and the causal ordering of these variables must be correct. For guidance on developing theoretically correct causal models, see James et al. (1982).

LINEAR AND ADDITIVE RELATIONSHIPS. The relationships between all variables in the path model must be linear and additive. In other words, the relationships should not be curvilinear, and there should not be any interactive relationships between variables (again, unless special procedures have been used to model such relationships).

ABSENCE OF MULTICOLLINEARITY. The variables should not demonstrate extremely strong correlations with one another. One popular rule of thumb says that there may be a multicollinearity problem if the correlation between two variables exceeds $r = \pm.80$.

MULTIPLE INDICATORS. It is generally best if there are three to five indicator variables measuring each latent factor, although it is possible to perform the analysis with fewer.

IDENTIFICATION. The causal model should be identified. The concept of *identification* is complex; see Bryant and Yarnold (1995, pp. 117-118) or Tabachnick and Fidell (2007, pp. 709-712) for an introduction.

LARGE SAMPLES. In most cases, the analysis should be based on large samples. MacCallum et al. (1996) provide power tables for determining the necessary sample sizes to achieve adequate power for a variety of global fit indices.

Chapter 19

Meta-Analysis

This Chapter's *Terra Incognita*

You are reading a journal article. It was written by a researcher who wanted to determine whether people who frequently play violent video games are more likely to engage in aggressive behavior in real life. To answer this question, the researcher located reports on 70 separate studies dealing with this topic—studies that had been conducted by researchers working independently in different locations and at different points in time. She then analyzed the results from these 70 investigations using a relatively new statistical procedure called *meta-analysis*. In her article, she reports the following:

> The sign of the summary effect reported above (r_+ = .19) was positive, indicating a general trend that greater exposure to violent video games was associated with greater aggressive behavior displayed by the participant in real life. However, it was possible that the nature of the relationship between these two variables was different for different subgroups. To investigate this possibility, tests for heterogeneity were performed.
>
> A Q statistic was computed to test the null hypothesis that all 70 studies included in the meta-analysis shared a common effect size. With alpha set at α = .05, this statistic was significant, $Q(69)$ = 93.93, p = .025. Related results showed that I^2 = 26.54, indicating that over 26% of the total variability in effect sizes was due to between-studies variability (variability in true effects in the population). Because of this heterogeneity in effects, additional analyses were performed to investigate possible moderator variables.

There's no cause for alarm. All shall be revealed.

The Basics of this Procedure

A ***meta-analysis*** is a quantitative synthesis of the research finding obtained from previous empirical investigations of a specific topic (Hedges & Olkin, 1985; Konstantopoulos, 2008). Its main purpose is to summarize—in quantitative terms—the results obtained in these investigations. But a good meta-analysis will do much more than just summarize the results.

In a typical meta-analysis, investigators attempt to identify all previous empirical studies—published and unpublished—that had focused on a specific research question. These studies are called the ***primary empirical investigations***.

For each study, the researcher computes an ***index of effect size***: a statistic which represents the strength of the relationship between the predictor variable and criterion variable. The index of effect size is typically either the Pearson correlation coefficient (symbol: r), Cohen's d statistic (Cohen, 1988), or the odds ratio (symbol: OR). The researchers then compute a ***summary effect***—a weighted average of the effect indices from the sample of studies. They may perform analyses to determine whether the relationship between the predictor and criterion is moderated by other variables. They may also determine the extent to which the results were affected by ***publication bias*** (the tendency for research journals to publish only statistically significant results).

Why Meta-Analysis Matters

Meta-analysis is the most important of all statistical procedures. This is because meta-analysis allows us to make use of the research findings produced by hundreds—even thousands—of previous empirical investigations. Consumers of research are often frustrated by the fact that different empirical studies sometimes produce conflicting results. But meta-analysis allows us to see through the haze created by seemingly inconsistent findings. When done correctly, it allows us to see the true relationships between the theoretical constructs of interest.

Of equal importance, meta-analysis allows us to identify ***moderator variables***: third variables whose values determine the nature of the relationship between the predictor and criterion variables. For example, a given meta-analysis might find that the relationship between Variable A and Variable B is moderated by subject sex: There is a strong relationship between the variables for male participants, but only a weak relationship for female participants. This ability to test for moderator variables means that meta-analysis can help us understand the specific conditions under which relationships are likely to exist.

Narrative Literature Reviews Versus Meta-Analysis

Years ago, when researchers wanted to summarize the research literature on a given topic, the primary tool was the narrative literature review. This was essentially just a long essay. To write one, the author read the previously-published articles on a given research question, and then summarized (in verbal terms), the general trends of findings. Narrative literature reviews could be comprehensive and unbiased, or they could be subjective and opinionated. Often, the reader was left wondering: "If I had read the same articles this author had read, would I have arrived at the same conclusions?"

Today, meta-analysis provides a much more powerful set of tools for accomplishing the same task. Meta-analysis allows us to quantify the findings from previous studies. It allows us to compute summary effects, establish confidence intervals around those effects, and perform statistical analyses that provide fairly unambiguous answers to questions about moderator variables, publication bias, and related issues.

Readers may not like the conclusions from a meta-analysis on a given topic. But the information that it provides will allow them to track down the original studies, perform their own meta-analysis, and either replicate (or fail to replicate) the results.

And that is good for science.

Psychometric Versus Non-Psychometric Meta-Analysis

There are two major approaches to meta-analysis. This chapter will refer to these as *psychometric meta-analysis* and *non-psychometric meta-analysis*.

With ***psychometric meta-analysis***, an attempt is made to correct each index of effect size for ***artifacts:*** sources of bias arising out of measurement error, research design, and similar problems that may be displayed by the primary empirical investigations. Artifacts typically cause indices of effect size to be smaller (in absolute value) than they otherwise would be. In psychometric meta-analysis, researchers attempt to correct the individual effects for these artifacts prior to performing any other analyses. Correcting the indices in this way allows the meta-analysis to estimate the actual relationship between the constructs of interest, independent of any bias brought about by flaws in the individual, real-world investigations (Hunter & Schmidt, 2004).

This approach is called *psychometric meta-analysis* because it had originally been used to study the psychometric properties of employee selection systems and similar phenomena (e.g., Pearlman, Schmidt, & Hunter, 1980). Today, however, this approach is used with a wide variety of research topics in many different disciplines. It is widely known as the ***Hunter and Schmidt approach***, as it was originally developed by psychologists John Hunter and Frank Schmidt.

In this chapter, the second major approach will be labeled ***non-psychometric meta-analysis***. This approach makes no attempt to correct indices of effect size for artifacts. Instead, it computes summary effects, tests for moderator variables, and performs other statistical analyses on the uncorrected indices. Important references associated with this approach include Borenstein et al. (2009), Glass, McGaw, and Smith (1981), Hedges and Olkin (1985), and Rosenthal (1991).

In this chapter, non-psychometric meta-analysis is discussed first (in fact, the majority of this chapter is devoted to this approach). In part, this is because non-psychometric meta-analysis consists of a somewhat simpler, more straightforward set of procedures. Once the basic components of non-psychometric meta-analysis are understood, it is easy to move on to the slightly more complex procedures used with psychometric meta-analysis. Therefore, a separate section devoted to that topic appears later in the chapter.

Illustrative Investigation

Imagine that our fictitious researcher, Dr. O'Day, receives an e-mail from her sister:

> *Are violent video games bad for kids? My son Waldo wants me to buy a violent game for him, but I'm afraid it will make him more aggressive. Do violent video games have that effect?*

Dr. O'Day takes a quick look at some published research articles on the topic, but the results seem to be mixed: Some articles find that exposure to violent video games causes the players to become more aggressive in real life, and some show that they have no effect at all. What is she to believe?

To get to the bottom of things, Dr. O'Day decides that she will perform a meta-analysis. She will locate the studies that have investigated this topic, will determine the size of the "effects" that were found in the studies, and will compute the mean effect, averaged across the studies.

Variables to be Investigated

In this meta-analysis, she wants to investigate the relationship between *exposure to video game violence* (the predictor variable) and *subject aggressive behavior* (the criterion variable). Based on social learning theories from the field of social psychology, she predicts that playing violent video games will increase the frequency with which the player engages in aggressive behavior in the real world. This hypothesis is illustrated in Figure 19.1.

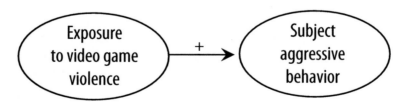

Figure 19.1. Hypothesized relationship between exposure to video game violence and subject aggressive behavior.

It is at this point that this book would usually describe exactly how the predictor variable and criterion variable were measured or manipulated in Dr. O'Day's study. However, in the present situation, Dr. O'Day is not conducting the empirical investigation herself—she is merely collecting results from studies reported by others. She will find that different studies operationalized the two variables in different ways. For example, consider the variable *exposure to video game violence*. In some studies, this was a manipulated independent variable in a true experiment: Some participants were assigned to an experimental condition in which they played violent video games, and others were assigned to a control condition in which they played nonviolent (but still exciting) video games. In other studies, *exposure to video game violence* was a non-manipulated subject variable in a correlational study: The researcher merely used a questionnaire to measure how often participants played violent video games on their own. In other studies, the variable was assessed in still other ways.

The point is this: In the current investigation, Dr. O'Day is not designing an empirical study—she is merely collecting studies that have already been conducted by other researchers, and is classifying these studies according to their characteristics (e.g., experimental study versus nonexperimental study; strong research design versus weak research design, etc.). Later, she may use these characteristics as potential moderator variables in her meta-analysis.

Research Questions Addressed

Because it synthesizes information from multiple empirical investigations, the typical meta-analysis can answer a variety of questions. Here are some of the questions that Dr. O'Day's analysis might address:

- Is there a relationship between exposure to violent video games and subject aggressive behavior in the real world?

- If yes, how strong is the relationship?

- Does *quality of the research methodology* serve as a moderator variable? That is, do we see stronger effects with strong research designs or with weak research designs?

- Does *location of the study* (United States versus Japan) serve as a moderator variable? That is, do we see stronger effects with studies conducted in the United States or with studies conducted in Japan?

- Is there reason to believe that the results have been affected by **publication bias** (the tendency of research journals to publish results only if they are statistically significant)? If yes, what would the summary effect from the meta-analysis probably be if there had been no publication bias?

Details, Details, Details

This section introduces the nuts and bolts of this procedure: the general steps that are followed when conducting a meta-analysis, the indices of effect size that are computed, the tests for heterogeneity that are performed, and related issues.

Steps Performed in a Meta-Analysis

Some readers associate "meta-analysis" with just the statistical analyses that are performed, such as computing the summary effect. But these statistical analyses are just one step, and they come fairly late in the process. The real value of the meta-analysis to the research community will be determined by the steps that precede the statistical analyses. Below is a snapshot of these steps, based largely on Field (2009b).

REFINING THE RESEARCH QUESTION. Did the researcher have a clear, well-defined research question? Did the researcher clearly define the variables to be studied?

STATING THE CRITERIA FOR INCLUDING A STUDY. The researcher should clearly describe the rules that were used for deciding whether to include a given study in the meta-analysis. In her meta-analysis dealing with video games, Dr. O'Day might decide that she will include only true experiments. Or, she might decide to include both true experiments and nonexperimental studies, but she will use *type of research design* as a moderator variable. Either way, she should have clear criteria and should describe these criteria in her article.

CONDUCTING THE SEARCH FOR EMPIRICAL INVESTIGATIONS. Did the researcher go to great lengths to find every study dealing with her research topic? At a minimum, she should use online databases (such as PsychINFO) to find relevant published articles. But she should not stop there—a conscientious researcher attempts to locate unpublished research findings as well. This is important, because studies that produce nonsignificant findings are less likely to be submitted for publication (Dickersin, Min, & Meinert, 1992) and are less likely to be published if they are submitted (Hedges, 1984). There is a danger that, if a meta-analysis is based exclusively on published studies, the sample will probably be biased with an overrepresentation of statistically significant studies. This could mislead readers into believing that there is a substantial relationship between the predictor variable and criterion variable when the actual relationship is weak (or even nonexistent).

CALCULATING INDICES OF EFFECT SIZE FOR INDIVIDUAL STUDIES. Different empirical investigations tend to measure predictor and criterion variables in different ways, and they often analyze their data using different statistics. In order to perform a meta-analysis, the researcher must convert all of these different values to a common metric—a common index of effect size. This common index is typically one of the following three: Cohen's d statistic, the Pearson r correlation coefficient, or the odds ratio (or some index related to one of these three). A later section of this chapter describes these indices in greater detail.

PERFORMING THE STATISTICAL META-ANALYSIS. The *statistical meta-analysis* is the step in which the researcher computes mean effect sizes, performs tests for heterogeneity, and performs other data-analysis tasks. At the very least, the statistical meta-analysis usually involves (a) computing a weighted mean of the indices of effect size found in the literature review, and (b) determining whether the results are influenced by moderator variables.

INVESTIGATING POSSIBLE PUBLICATION BIAS. As a final step, the researcher may perform analyses to determine whether her results are likely to have been influenced by ***publication bias***: the tendency for journals to publish statistically significant results, rather than nonsignificant results. There are procedures that allow the researcher to determine the presence and size of a publication-bias problem, and there are also procedures that allow the researcher to estimate what the size of the summary effect would be if there were no publication bias at all in the meta-analysis. These methods are discussed in later sections.

Indices of Effect Size

One of the main goals of meta-analysis is to compute a ***summary effect***: an average that represents the typical index of effect size found in research reports. This is not as simple as it sounds, because many reports (especially older ones) do not include an index of effect size at all—they merely report a significance test, such as the t statistic, the F statistic, or χ^2. Ultimately, the investigator must convert all of these into the same metric (the same index of effect size) before the statistical meta-analysis may be performed.

Fortunately, formulas and software are available that allow the investigator to convert F statistics, t statistics, and other values into a common index of effect size (e.g., Borenstein et al., 2009; Rosenthal, 1991, 1994). This "common index" is usually Cohen's d statistic, the Pearson r, or the odds ratio.

COHEN'S d AND HEDGE'S g. Cohen's d statistic (Cohen, 1988) represents the difference between two means, as measured in standard deviations. The prototypical study that produces Cohen's d is a study in which the predictor variable is a dichotomous variable and the criterion variable is a multi-value variable assessed on an interval scale or ratio scale. Below is an example of such a study:

- In this investigation, the predictor variable is *video-game status*, and this predictor variable includes two groups of participants: video-game players versus non-game players.

- The criterion variable is *rated aggressiveness*: evaluations of just how aggressive the participants are, as made by friends or relatives.

At the end of the investigation, the researcher computes the mean aggressiveness rating displayed by the video-game players, along with the mean aggressiveness rating displayed by the non-game players. She then computes Cohen's d statistic to determine the size of the difference between the two means.

When a researcher performs a study with two independent samples, d is computed as:

$$d = \frac{\bar{X}_1 - \bar{X}_2}{s_{\text{within}}}$$

Where:

d = Cohen's d statistic

$\bar{X}_1$ = Mean score on the criterion variable for the first condition

$\bar{X}_2$ = Mean score on the criterion variable for the second condition

S_{within} = the within-groups standard deviation, pooled across groups

Cohen's *d* is a *standardized difference statistic*: It estimates what the difference between the two condition means would be if the criterion variable was standardized so that it had a standard deviation equal to 1 (Borenstein et al., 2009). When *d* is equal to zero, it means that there was no difference between the two conditions. This tells us that the *no-effect* value for Cohen's *d* is *d* = 0.00. The larger the absolute value of *d*, the bigger the "effect" (*d* may assume either positive or negative values). For example, when *d* = 1.5, it indicates that one mean was 1.5 standard deviations higher than the other mean.

When evaluating *d* for a specific study, it is best to compare the obtained value of *d* to the values that have been obtained by other researchers investigating the same topic. When this is not possible, Cohen (1988) offers these criteria as a last resort:

$d \geq \pm.20$ is considered a small effect.

$d \geq \pm.50$ is considered a medium effect.

$d \geq \pm.80$ is considered a large effect.

In smaller samples, Cohen's *d* overestimates the actual difference between the conditions in the population. To address this, some researchers multiply *d* by a correction factor, producing Hedge's *g*: an unbiased estimate of the difference between population means (Hedges, 1981). Under typical research conditions, however (i.e., when *df* > 10), the difference between *d* and *g* will be trivial.

THE CORRELATION COEFFICIENT, *r*. You learned about the Pearson *r* in *Chapter 6* of this book. In addition to being a widely used correlation coefficient, the *r* statistic is also used as an index of effect size in meta-analyses.

In the prototypical study producing data that might be analyzed using Pearson's r, both the predictor variable and the criterion variable are quantitative variables assessed on an interval or ratio scale. In most cases, both variables are multi-value variables. Here's an example:

- In this investigation, the predictor variable is *game-playing hours:* the number of hours that the participants spend playing video games each week.
- The criterion variable is once again *rated aggressiveness*: evaluations of just how aggressive the participants are, as rated by friends or relatives.

The Pearson *r* statistic summarizes the direction and strength of the relationship between two variables. Under typical research conditions, its values may range from −1.00 through 0.00 through +1.00. A value of 0.00 indicates that there is no relationship between the two variables; the larger the absolute value of the statistic, the stronger the relationship. A value of −1.00 or +1.00 indicates a perfect relationship.

Positive values of the correlation coefficient (such as $r = +.61$) indicate a positive relationship between the two variables: As values on the predictor variable increase, values on the criterion variable also tend to increase. In contrast, negative values (such as $r = -.61$) indicate a negative relationship: As values on the predictor variable increase, values on the criterion variable tend to decrease.

When evaluating r for a specific study, it is best to compare the obtained value to the values that have been reported by other researchers investigating the same topic. When this is not possible, Cohen (1988) offers these criteria as a last resort:

$r \geq \pm.10$ is considered a small effect.

$r \geq \pm.30$ is considered a medium effect.

$r \geq \pm.50$ is considered a large effect.

Earlier, this section indicated that, when $r = 0.00$, it means that there is no relationship between the two variables. This is just another way of saying that the *no-effect* value for Pearson's r is $r = 0.00$.

Many researchers do not perform the meta-analysis on the r statistics in their natural form. Instead, they use Fisher's r-to-z transformation to convert the r statistics to Fisher's z scale. The meta-analysis is then performed on these transformed values (e.g., means are computed, confidence intervals are established), and the resulting summary statistics are then converted back to Pearson r statistic for presentation in the published article.

THE ODDS RATIO, OR. The *odds ratio* (abbreviation: *OR*) represents the ratio of (a) the odds for one subgroup to (b) the odds for the second subgroup (Warner, 2008). It is an index of effect size that is often reported for studies whose results may be summarized in 2×2 frequency tables. These are cross-tabulation tables in which the predictor variable consists of two groups, the criterion variable consists of two groups, and the researcher tabulates the number of participants who appear in each of four squares that constitute the resulting 2×2 table.

The odds ratio is a relatively difficult statistic for laypeople to understand. It is also a relatively difficult statistic for professional researchers to understand. For example, Holcomb et al. (2001) found that 26% of authors in the most prestigious medical journals misinterpreted the meaning of the odds ratios that they reported.

Given findings like this, it is tempting to just sort of omit any discussion of the odds ratio in this chapter. That, unfortunately, is a luxury that we cannot afford. As a consumer of research, you are often going to encounter the odds ratio, especially in meta-analyses from the fields of medicine and health. You *do* want to read that meta-analysis that tells the real truth about low-carb diets, don't you? So here we go:

The prototypical investigation producing data to be analyzed with an odds ratio is a study in which the predictor variable is a dichotomous variable and the criterion variable is also a dichotomous variable. Here's an illustration:

- In this study, the predictor variable is *video-game status*. This predictor variable includes two groups of participants: video-game players versus non-players.
- The criterion variable is *assault status*. This variable also consists of two groups: participants who have committed assault within the past 12 months, and participants who have not.

Computing the odds ratio for a study such as this consists of three steps: (a) computing the odds of the event for participants in the first subgroup under the predictor variable, (b) computing the odds of the event for participants in the second subgroup, and then (c) using the two resulting odds to compute the odds ratio.

The **odds** can be defined as the ratio of (a) the probability that an event of interest will occur to (b) the probability that the event of interest will not occur (Yaremko et al., 1982). In the current investigation, the event of interest is *committing assault*. In general, we use the following formula to compute the odds that someone in a given subgroup will display the event of interest (Osbourne, 2008):

$$odds = \frac{number\ of\ participants\ in\ subgroup\ who\ display\ the\ event}{number\ of\ participants\ in\ subgroup\ who\ do\ not\ display\ the\ event}$$

To calculate the odds of committing assault for the video-game-player subgroup, we divide the number of people in that subgroup who committed assault by the number of people in that subgroup who did not commit assault. The result is the odds of committing assault among the video-game players.

We then repeat this process for the non-players. We divide the number of people in that subgroup who committed assault by the number of people in that subgroup who did not commit assault. The result is the odds of committing assault among the non-players.

Finally, we compute the odds ratio itself. When two subgroups are being compared, the odds ratio is computed as the ratio of the odds for one subgroup to the odds for the second subgroup:

$$OR = \frac{odds\ for\ subgroup\ 1}{odds\ for\ subgroup\ 2}$$

We are now ready for a definition. An **odds ratio** (*OR*) represents the ratio of (a) the odds for one subgroup to (b) the odds for a second subgroup (Warner, 2008). Its values may range from zero through positive infinity.

This is just a fancy way of saying that an odds ratio cannot be any lower than 0.00, but there is theoretically no limit as to how large it may be—it may assume values such as 0.14, 1.31, 5.62, 12.33, and so forth. Given the formula for the *OR* presented above:

- When the *OR* is *equal to exactly 1.00*, it means that the there was no difference between the two subgroups with respect to the odds of the event.

- When the *OR* is *greater than 1.00* (such as 1.21, 3.48, or 18.90), it indicates that the odds for subgroup 1 were larger than the odds for subgroup 2.

- When the *OR* is *less than 1.00* (such as 0.11, 0.28, or 0.83), it indicates that the odds for subgroup 1 were smaller than the odds for subgroup 2.

The first of the three points listed above is the most important: When the *OR* is equal to 1.00, it means that there is no difference between the two subgroups regarding the odds of the event occurring. This means that *OR* = 1.00 is the *no-effect* value for the odds ratio.

With the two other effect indices presented here (Cohen's *d* and the correlation coefficient, *r*), this chapter summarized Cohen's "last resort" criteria for classifying an effect as being a small, medium, or large effect. Alas, Cohen's (1988) book provides no such criteria for evaluating the odds ratio. There are, however, formulas for converting the *OR* to *d* (e.g., Borenstein et al., 2009, p. 47), so it is theoretically possible for researchers to convert a summary effect odds ratio to *d*, and then apply Cohen's criteria for evaluating the size of the resulting *d* statistic (if they just absolutely *have* to use those criteria). Similarly, Bonett (2007) shows how an *OR* can be converted into an approximate product-moment correlation coefficient.

As an alternative, Haddock et al. (1998, p. 342) offer the following general rules of thumb for interpreting odds ratios:

- Odds ratios *close to 1.00* represent weak relationships.

- Odds ratios *over 3.0* represent strong positive relationships.

- Odds ratios *under 0.33* represent strong negative relationships.

When a meta-analysis is performed on odds ratios from a sample of investigations, a standard convention is to take the natural logarithm of each odds ratio. The natural logarithm is the log base *e*. The resulting index is called the *log-odds ratio* and is represented using the symbols *log(OR)* or *ln(Odds Ratio)*. All statistical analyses in the meta-analysis (such as computing the mean effect size) are then performed on the resulting sample of log-odds ratios. The results are typically then converted back to regular odds ratios prior to reporting them in the published article (Borenstein et al., 2009; Konstantopoulos, 2008).

You must be prepared to interpret the meaning of log-odds ratios for those meta-analyses in which the author does not convert the log-odds ratios back to odds ratios. Possible values of the log-odds ratio range from negative infinity through zero through positive infinity. This is just a fancy way of saying that, theoretically, a log-odds ratio may assume any negative or positive number that you can imagine. The *no-effect* value for the log-odds ratio is zero: *log(OR)* = 0.00 (Haddock et al., 1998).

SUMMARY: THE "NO-EFFECT" VALUE FOR THE INDICES. One of the most important points made in this section is the ***no-effect value*** for each of the indices described above. This is the value that you will see when the meta-analysis indicates that there is no effect for the predictor variable being studied (i.e., no difference between conditions, or no relationship between the variables). Here is a summary of these values:

- If the meta-analysis is based on *Cohen's d* (or its cousin, *Hedges g*), the no-effect value is zero. In other words, if the summary effect is $d = 0.00$ or $g = 0.00$, it means that, in the typical study, there is no difference between conditions. The more the summary effect departs from zero, the greater the "effect."
- If the meta-analysis is based on the *correlation coefficient*, the no-effect value is $r = 0.00$.
- If the meta-analysis is based on the *odds ratio*, the no-effect value is $OR = 1.00$.
- If meta-analysis is based on the odds ratio but the summary effect is instead reported as a log-odds ratio, the no-effect value is $log(OR) = 0.00$.

The Fixed-Effects Versus Random-Effects Models

In the article that reports a meta-analysis, the author should indicate whether they used a fixed-effects model or a random-effects model in their analyses. The difference between these two approaches is a very big deal, as the selection of a model will determine whether the inferences from the findings may be generalized beyond the specific set of studies included in the meta-analysis (Field, 2009b).

It is not terribly difficult to understand the difference between the two models, as long as you understand the concept of *sampling error*. Therefore, this section will begin with a quick refresher on that concept.

SAMPLING ERROR. The concept of sampling error was introduced early in this book, and has been relevant to most of the statistical procedures that the book has covered. In general, **sampling error** can be defined as the difference between some characteristic of a sample versus the corresponding characteristic of the larger population from which the sample was drawn, with the difference being due to the fact that samples typically are not perfectly representative of the populations from which they were drawn.

Think about it this way: Imagine that we conduct a massive investigation in which we assess participants' *exposure to violent video games* (the predictor variable) along with the *aggressive behavior displayed by the participants in real life* (the criterion variable). Imagine that we measure these two variables in the entire population of people on the planet Earth (such a study would be impossible, of course, but bear with me). Upon analyzing the data, we find the Pearson correlation (r) between these two variables in the population is $r = .25$.

Now imagine that we conduct a much less ambitious study: We draw a random sample of just 100 participants from this larger population, measure the same two variables, and find that the correlation between the two variables in this sample is $r = .20$.

Wait a minute—the actual correlation in the population is $r = .25$. Why is the correlation based on this smaller sample a different number: $r = .20$?

It is a different number because of *sampling error*. The sample of just 100 participants was not perfectly representative of the population from which it was drawn, and this caused the sample correlation coefficient to be a $r = .20$, slightly different from the true population value of $r = .25$. If we had drawn a another sample of 100 different participants, the correlation obtained in this sample might have been a bit higher at $r = .29$. If we had drawn yet another sample, the correlation might have been $r = .23$. If we had repeated this process with 10 new samples, most of these samples would have given us slightly different correlation coefficients. And the variability in these obtained sample correlation coefficients would have been due to *sampling error*.

FIXED VERSUS RANDOM EFFECTS MODELS: INTRODUCTION. In any meta-analysis, there will always be variability in the indices of effect size obtained from different studies, with some investigations reporting larger effects than others. Statisticians have developed a variety of models that make different assumptions about why this variability exists. This section will focus on Field's (2009b) discussion of the *fixed-effects model* versus the *random-effects model*.

FIXED-EFFECT MODEL. When researchers use a *fixed-effect model* in meta-analysis, they assume that there is just one true effect size underlying all of the studies included in the analysis. They assume that all of the variability in effect sizes displayed by the various studies is due to *sampling error*, as described above.

This approach assumes that all of the investigations came from a population with a fixed average effect size, hence the named *fixed-effects model*. Its assumes that different studies produced different effects simply because of sampling error—because none of the samples were perfectly representative of the one population from which they were drawn (Borenstein et al., 2009).

RANDOM-EFFECTS MODEL. With a *random-effects model*, the researcher assumes that there are two sources of variability in the effects observed in the different studies: (a) sampling error (as described above) plus (b) additional variability introduced by the fact that the investigations were sampled from multiple populations—populations with different true effect sizes (Field, 2009b).

Consider Dr. O'Day's meta-analysis. If some of the variability in the effect sizes from different studies is due to the fact that there is more than one true effect underlying the various investigations, it may be due to factors such as these:

- Some studies may reflect a true effect that is relatively large because they represent a population of studies in which the researchers used a very powerful "sledge-hammer" independent variable.

- Some studies may reflect a true effect that is relatively small because they represent a population of studies that used a very weak independent variable.

- Some studies may reflect a true effect that is relatively large because they included participants who are very sensitive to video-game violence.

- Some studies may reflect a true effect that is relatively small because they included participants who are not at all sensitive to video-game violence.

Still other studies may reflect different true effects for other reasons. The point is this: with a random-effects model, some of the variability is due to sampling error, and other variability is due to the fact that the various studies have been sampled from populations with different average effect sizes. It assumes that the average effect size in the population varies randomly from investigation to investigation—hence the name *random-effects model*.

IMPLICATIONS. In the published article, the authors should indicate whether their meta-analysis was based on a fixed-effects model or a random-effects model. Field (2009b) argues that the random-effects model is typically most appropriate—and most desirable—for meta-analyses conducted in the field of psychology. He points out that, when researchers use the random-effect model, the inferences that they draw from their results may be generalized beyond the specific set of investigations that were included in their meta-analysis (with the fixed-effects model, such inferences may not be generalized beyond the current set of studies). Since researchers in the social and behavioral sciences almost always hope to generalize their results, the random-effects model will usually be the most desirable approach to meta-analysis.

Forest Plots

Researchers often represent the data points analyzed in a meta-analysis in a special type of figure called a ***forest plot***. Forest plots are useful because they convey a great deal of information in just one figure. This section will use some fictitious results to illustrate the features of a forest plot (the conventions used here are those used by Borenstein et al., 2009).

A forest plot may be used to illustrate any of a number of statistics. The current plot will illustrate Pearson correlation coefficients, r.

Imagine that Dr. O'Day has begun her literature review. She has found 10 studies that investigated the relationship between exposure to violent video games and subject aggressive behavior. For each study, she has computed the Pearson correlation coefficient, r, and the 95% confidence interval around the coefficient. These results are summarized in Figure 19.2.

LABELS FOR THE AXES. The labels on the X axis of Figure 19.2 (the horizontal axis) represent possible values for r—values that range from −1.00 through 0.00 through +1.00. The vertical axis provides a name for each study being displayed. At the top of the axis is the label "Study 1"; below that is "Study 2," and so forth.

POINT ESTIMATES AND CONFIDENCE INTERVALS. Each black square in the forest plot represents the Pearson correlation coefficient, r, which was found in a specific investigation. The location of the square in relation to the X axis tells us the value of r. You can see the correlation coefficient obtained in Study 1 was approximately $r = +.25$, the correlation obtained in Study 2 was approximately $r = +.15$, and so forth.

The error bars that branch out from a given square represent the 95% confidence interval for the correlation coefficient. For Study 1, the 95% CI extends from about .05 to about .45; for Study 2, it extends from about −.10 to about +.40.

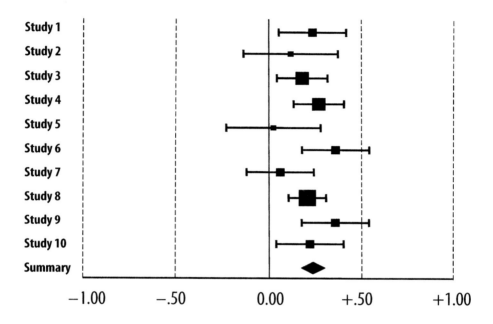

Figure 19.2. Forest plot showing observed correlation coefficients (squares), 95% confidence intervals for each correlation, and the summary effect (diamond) for 10 fictitious studies.

SAMPLE SIZE AND PRECISION. The size of a given square represents the relative number of participants in the study—larger squares represent studies with more participants. The square for Study 1 is somewhat larger than the square for Study 2, indicating that Study 1 had more participants. You can see that studies with small squares tend to have wide confidence intervals (Study 2 is a good example of this). This is what you would expect—a small sample does not result in good precision. Therefore, the confidence interval for r based on a small sample is expected to be wide.

THE SUMMARY EFFECT. Finally, the diamond that appears toward the bottom of Figure 19.2 represents the ***summary effect:*** the weighted mean of the effects obtained in the individual studies. The summary effect is an estimate of the ***true effect size***: the actual strength of the relationship between the predictor variable and criterion variable in the population. The diamond in Figure 19.2 is approximately $r = .25$. This indicates that, if Dr. O'Day's meta-analysis were based on just these 10 studies, she would estimate that the true correlation between exposure to violent video games and real-world aggression is $r = .25$.

Heterogeneity of Effect Sizes

The forest plot in Figure 19.2 shows that that there is variability in the effects observed in the 10 studies: Some studies produced small correlations (such as Study 5, where $r = .05$), other studies produced relatively large correlations (such as Study 6, where $r = .40$), and the remaining studies were somewhere in-between.

When researchers conduct meta-analyses, they know that some of the variability in the indices of effect size is merely due to sampling error. You will recall that ***sampling error*** is variability attributable to the fact that samples are usually not perfectly representative of the populations from which they were drawn. Researchers sometimes refer to this as ***within-study variability*** (Huedo-Medina, Sanchez-Meca, Marin-Martinez, & Botella, 2006). In a meta-analysis, there will always be some variability in the observed effect sizes due to this within-study variability. In a meta-analysis in which *all* of the variability is due merely to sampling error, the effects are said to be ***homogeneous***.

But in some meta-analyses, the effects are not homogeneous. In some meta-analyses, the sample of effects reflects two sources of variability: (a) sampling error plus (b) additional variability due to the fact that different samples represent different populations, with some populations displaying larger true effects than others. Variability of this type is called ***between-studies variability:*** dispersion in observed effect sizes due differences in true effects in the populations that underlie the studies.

This brings us to the word *heterogeneity*. Some authors use this word to refer to any kind of variability in effect sizes—whether due to within-study variability or between-studies variability. However, this chapter will follow the convention of using the term ***heterogeneity*** to refer only to variability in effect sizes attributable to different true effects in the population. In this chapter, *heterogeneity* means *between-studies variability*.

When the indices in a meta-analysis display substantial heterogeneity, it is an important finding. Substantial heterogeneity is usually interpreted as evidence that the results are influenced by one or more *moderator variables*. A ***moderator variable*** is a third variable whose values determine the *nature of the relationship* between the predictor variable and criterion variable being studied. A later section will discuss moderator variables in greater detail. First, however, the current section will describe the statistics and tests that are used to determine whether moderator variables are likely to present. These include tau-squared, the Q statistic, and the I^2 statistic (Borenstein et al., 2009).

VARIANCE OF TRUE EFFECT SIZES. The variance of the true effect sizes is one measure of heterogeneity, and researchers refer to this variance as *tau-squared*. The symbol for the population variance is τ^2 (τ is the lower-case Greek letter *tau*). The symbol for the sample estimate of the population variance is T^2, and the symbol for the sample estimate of the population standard deviation is T. Published meta-analyses often report T^2 in a table; remember that this statistic is an estimate of the between-studies variance.

THE Q STATISTIC. One of the most frequently reported statistics in meta-analyses is the Q statistic. A **Q statistic** may be computed for any of a number of purposes, but it is most frequently used to test the null hypothesis that all of the studies in the meta-analysis share a common effect size.

When it is computed for this purpose, the Q statistic represents a *weighted sum of squares (WSS)*, somewhat analogous to the sums of squares *(SS)* that are computed in a traditional analysis of variance (ANOVA). To understand how the Q statistic is computed, imagine that Dr. O'Day has identified 70 studies that investigate the effects of exposure to video games. Imagine that she has computed an index of effect size for each of these studies (symbol: Y_i), as well as the summary effect (the average effect size for all 70 studies; symbol: M). To compute Q, she:

- Subtracts the summary effect, M, from the effect size for a specific study, Y_i,
- squares the resulting difference,
- weights this squared difference by the inverse variance for that study (this gives greater weight to effects that are more precise estimates), and
- sums these values across the 70 studies included in the meta-analysis.

The result is the Q statistic, a weighted sum of squares. The Q statistic is distributed as chi-square (χ^2), and its degrees of freedom are computed as $k - 1$, where k = the number of studies included in the meta-analysis. Values of Q may range from zero through infinity, with higher values indicating greater between-studies variability.

Earlier, this section indicated that the Q statistic is often used to test the null hypothesis that all of the studies in the meta-analysis share a common effect size. When it is used for this purpose, researchers typically set alpha at $\alpha = .05$ or $\alpha = .10$. For an analysis of this sort, two outcomes are possible:

- *A significant Q statistic.* If the obtained p value for the Q statistic is less than alpha (i.e., if the results are significant), the researcher rejects the null hypothesis and concludes that the studies have been drawn from populations with true effects of different sizes. In other words, a significant Q statistic is consistent with the idea that the effects *are* heterogeneous. In some cases, this heterogeneity may be due to the presence of moderator variables. Therefore, when the Q statistic is significant, researchers often perform subgroup analyses or other analyses to understand the nature of the moderator variables.

- *A nonsignificant Q statistic.* On the other hand, if the p value for the obtained Q statistic is larger than alpha (i.e., if the results are nonsignificant), this is typically interpreted as evidence that all of the studies are *homogeneous*—that they all came from a population with just one true effect size.

Based on the statements made above, it is reasonable to infer that, when the Q statistic is nonsignificant, the researchers will assume that no moderator variables are present. But this is not necessarily the case. If the researchers have *a priori* hypotheses that specific moderator variables should be present, they may still perform subgroup analyses or other analyses to test those hypotheses (an *a priori* hypothesis is a hypothesis based on theory or previous research). So, a nonsignificant Q statistic does not necessarily mean that the meta-analysis comes to a halt.

THE I^2 STATISTIC. The Q statistic by itself is somewhat limited—it only allows the researcher to test the null hypothesis of no between-studies variability. The Q statistic by itself does not indicate how *much* of the total variability in the effect sizes is due to between-studies variability. In response to this shortcoming, Higgins and Thompson (2002) proposed the I^2 statistic to supplement or replace the Q statistic.

The I^2 statistic indicates the percent of total variability in effect sizes that may be attributed to between-studies variability (variability in the true effects in the population). Values of I^2 may range from zero through 100, with higher values indicating greater heterogeneity. For example:

- When $I^2 = 0$, it means that none of the variability in the effects sizes in the meta-analysis is due to between-studies variability. In other words, when $I^2 = 0$, it means that all of the observed variability is merely due to sampling error. The I^2 statistic will be equal to 0 when T^2 (the estimated variance of true effect sizes in the population) is equal to zero.

- When $I^2 = 75$, it means that 75% of the variability in the effects is due to between-studies variability.

Higgins and Thompson (2002) suggested the following tentative criteria for evaluating the size of I^2:

$I^2 = 25$ indicates low heterogeneity

$I^2 = 50$ indicates medium heterogeneity

$I^2 = 75$ indicates high heterogeneity

When $I^2 = 0$, it means that almost all of the variability in observed effect sizes is due to random error, and none of the observed variability is due to "true" heterogeneity. In such a situation, it would not make sense to perform subgroup analyses or other tests for moderator variables, since just about all of the variation is simple random error (Borenstein et al., 2009).

It is possible to create a confidence interval around I^2. When the confidence interval contains the value of zero, it may be interpreted as evidence that the effects are homogeneous. When it does not contain zero, it may be interpreted as evidence that the effects are heterogeneous.

Moderator Variables

In the typical meta-analysis, the researcher tries to determine whether the results are influenced by moderator variables. Before we review the definition for *moderator variable*, let's refresh out memory regarding the basic relationship that is investigated in a meta-analysis.

The typical meta-analysis investigates the relationship between two variables: a predictor and a criterion. In Dr. O'Day's meta-analysis, the predictor was exposure to video game violence, and the criterion was aggressive behavior displayed by participants in real life.

DEFINITION. A ***moderator variable*** is a third variable whose values determine the *nature of the relationship* between the predictor and the criterion. For example, when the moderator assumes one value, the relationship between predictor and criterion may be weak; when the moderator assumes a different value, the relationship may be strong.

EXAMPLE. For example, one of the potential moderator variables in Dr. O'Day's meta-analysis is *location of the study:* United States versus Japan. She describes it in the following research hypothesis:

> Location of study will moderate the relationship between exposure to video game violence and subject aggressive behavior. Specifically, it is predicted that studies conducted in the United States will display a strong positive relationship between exposure to video game violence and subject aggressive behavior, whereas studies conducted in Japan will display a weak positive relationship between exposure to video game violence and subject aggressive behavior.

Maybe Dr. O'Day makes this hypothesis because she believes that people in the United States are particularly prone to violence, and when they play violent video games, it really sets them off. On the other hand, people in Japan are much more restrained, so playing violent video games should have less effect on them (please remember that I am just making all this up).

Do you see how *location of study* qualifies as a moderator variable? Location of study is a third variable whose values (United States versus Japan) could determine the nature of the relationship between exposure to video game violence and subject aggressive behavior. Or so Dr. O'Day predicts.

HOW MODERATOR VARIABLES ARE INVESTIGATED. There are a variety of different approaches for identifying the presence of moderator variables in a meta-analysis. One popular approach involves performing *subgroup analyses—* procedures that are somewhat analogous to analysis of variance (ANOVA).

With the subgroup-analysis approach, the researcher divides the primary empirical investigations into two or more groups, and these groups represent the different values of the moderator variable. For example, one of the potential moderator variables in Dr. O'Day's study is *location of study*. This means that she will divide her studies into two groups: those conducted in the United States, and those conducted in Japan. She will compute a mean effect size for each group; if her research hypothesis is correct, the mean effect size for the studies conducted in the United States will be larger than the mean effect size for the studies conducted in Japan.

In an earlier section, you learned about the Q statistic, which is used to investigate the variability of effect sizes in meta-analysis. When one uses this ANOVA-like approach to performing meta-analysis, it is possible to partition the different sources of variability in the observed effect sizes. In a meta-analysis, each source of variability is represented by a different Q statistic: There is one Q statistic that represents the within-groups variance in effect sizes, another Q statistic that represents the total variance in effect sizes, and so forth.

With this subgroup-analysis approach, one of the statistics that is of great importance is the Q statistic for the *between-groups variance.* In general, this Q statistic tests the null hypothesis that the effect size is equal in the two subgroups being compared. In the current example, it tests the null hypothesis that the mean effect size for studies conducted in the United States is equal to the mean effect size for studies conducted in Japan. This between-groups Q statistic is distributed as chi-square, with degrees of freedom equal to $p - 1$, where p represents the number of subgroups being compared. There are two subgroups in the current meta-analysis, so the $df = p - 1 = 2 - 1 = 1$.

If the Q statistic for the between-groups variance is significant, the researcher concludes that there is a significant difference between the mean effects displayed by the two subgroups. Such a finding would provide support for Dr. O'Day's research hypothesis that location of study is operating as a moderator variable.

Later, this chapter will illustrate this approach to testing moderator variables. For more details regarding this procedure (as well as alternative procedures for investigating moderator variables), see Borenstein et al. (2009).

Fictitious Meta-Analysis on Video Game Violence

This section presents results from Dr. O'Day's meta-analysis. It illustrates how the standard results from a meta-analysis (e.g., the summary effect, the tests for heterogeneity, etc.) are typically presented in the tables and text of a journal article. The study presented here is an example of a non-psychometric meta-analysis. A later section of this chapter presents an example of a psychometric meta-analysis (i.e., the Hunter and Schmidt approach).

Caveat: This Meta-Analysis Is Fictitious

The meta-analysis on video game violence reported in this section is fictitious. I generated and analyzed a fictitious data set in order to illustrate some of the statistical results that you are likely to encounter in articles about actual meta-analyses: summary effects, Q statistics, moderator variables, and so forth.

I created the data so that some of the summary effects would be similar to those reported by Anderson et al. (2010): an actual meta-analysis dealing with the effects of video games. But please do not take any of the results presented here seriously—if you compare the results of Dr. O'Day's current meta-analysis against the results reported in Anderson et al. (2010), you will see plenty of differences.

Besides, if you wish to understand the actual effects of video game violence on real-world behavior, you should consult a real meta-analysis.

Fortunately, you will not have to look far. Research on the effects of violent video games has generated a good deal of interest and even controversy. Some meta-analyses have found that playing violent video games has little or no effect on real-world aggressive behavior (e.g., Ferguson, 2007). Other meta-analyses have concluded that playing violent video games does cause greater aggressive behavior in the real-world, and that the size of the effect is large enough to make it a public policy issue (e.g., Anderson & Bushman, 2001; Anderson et al., 2010). The fact that different meta-analyses have resulted in different conclusions has led to some spirited exchanges between scholars (e.g., Bushman, Rothstein, & Anderson, 2010; Ferguson & Kilburn, 2010; Huesmann, 2010). This means that, if you are new to meta-analysis as a research method, these articles and response papers dealing with the effects of video games are an excellent place to get your feet wet.

The Study

Imagine that Dr. O'Day has now conducted most of the steps required for her meta-analysis. She has refined the research question, clarified the inclusion criteria, searched the databases, located published and unpublished studies, and has computed a common index of effect size (the Pearson correlation coefficient, r), for each study. She is now ready to perform the statistical meta-analysis. This is the step in which she computes a summary effect, performs subgroup analyses, and performs additional analyses to investigate the nature of the relationship between exposure to violent video games and subject aggressive behavior.

Dr. O'Day will use an ordered approach to present the results of the meta-analysis in an article. She will begin with the most general findings (i.e., the summary effect for all studies combined), and will then move to more specific findings (i.e., the summary effects for various subgroups of studies). She begins with the most basic question: Is there any effect at all?

Is There a Relationship Between the Variables?

In the following excerpt from the text of her article, Dr. O'Day presents the aggregate results from the meta-analysis:

> **Results for all studies combined.** The search for published and unpublished research findings resulted in a final set of 70 studies that investigated the relationship between exposure to violent video games and subject aggressive behavior. Combined, 5,460 individuals had participated in these 70 investigations.
>
> The common index of effect size used in all analyses was the Pearson correlation coefficient, r. The Fisher r-to-z transformation was used to convert all indices prior to the statistical analyses. All indices were then converted back to r for presentation in this article. Analyses were performed using the computer application *Comprehensive Meta-Analysis V2*™ (Borenstein et al., 2009).

> For the combined set of 70 investigations, the summary effect (i.e., the weighted average of the effect indices) was $r_+ = .19$, 95% CI [.16, .21]. This summary effect was significantly different from zero, $z = 14.52$, $p < .001$.

The preceding excerpt uses the symbol r_+ to represent the summary effect. Other articles may use other symbols, such as $\bar{r}$ (to reflect the fact that this index is the average of many indices), or the Greek letter rho: ρ (to reflect the fact that this index is their best estimate for the true effect in the population; remember that the Greek letter rho is the standard symbol for the correlation coefficient in the population).

In the preceding excerpt, Dr. O'Day used a z statistic to determine whether the summary effect was significantly different from zero. Some have questioned the use of such significance tests in a meta-analysis, in part because they often have so much power that they can produce statistically significant results even when the effect is relatively small in practical terms (Field, 2009b).

What Is the Nature of the Relationship?

When we ask "what is the nature of the relationship?" in a meta-analysis, we are actually asking a number of related questions. These include "What is the direction of the relationship?" and "Is the relationship affected by moderator variables?" Dr. O'Day begins to address these questions in the following excerpt.

> The sign of the summary effect reported above ($r_+ = .19$) was positive, indicating a general trend that greater exposure to violent video games was associated with greater aggressive behavior in subjects. However, it was possible that the nature of the relationship between these variables was different for different subgroups. To investigate this possibility, tests for heterogeneity were performed.
>
> **Tests for heterogeneity.** A Q statistic was computed to test the null hypothesis that all 70 studies included in the meta-analysis shared a common effect size. With alpha set at $\alpha = .05$, this statistic was significant, $Q(69) = 93.93$, $p = .025$. Related results showed that $I^2 = 26.54$, indicating that over 26% of the total variability in effect sizes was due to between-studies variability (variability in true effects in the population). Because of this heterogeneity in effects, additional analyses were performed to investigate possible moderator variables.
>
> **Quality of methodology as a moderator variable.** Each of the 70 investigations was classified as being either a weak study or a strong study. These classifications were based on the validity of the measures used, the internal validity of the research design, and other factors that reflected the quality of the research methods. Two teams of judges independently classified the studies into these two groups according to written criteria, and there was 100% agreement in their classifications. As a result, 36 studies were classified as having a strong research design, and 34 were classified as having a weak research design.

These classifications were then used in a subgroup analysis to determine whether quality of methodology operated as a moderator variable. Table 19.1 presents the results from this analysis.

Table 19.1

Subject Aggressive Behavior as a Function of Violent Video Game Exposure: Results for the Full Data Set as Well as Subgroup Analyses for Strong Versus Weak Research Design

Design	N	k	Effect size r_+	95% CI	NHST z	p	Heterogeneity Q	df	p	I^2
Strong	2,808	36	.232	[.198, .265]	13.078	<.001	42.073	35	.191	16.81
Weak	2,652	34	.140	[.102, .177]	7.272	<.001	36.698	33	.228	14.725
Total within							80.772	68	.138	
Total between							13.156	1	<.001	
Overall	5,460	70	.189	[.164, .214]	14.518	<.001	93.928	69	.025	26.539

Note. NHST = Null hypothesis significance test (two-tailed) for null hypothesis that summary effect is equal to zero; N = number of participants; k = number of studies; r_+ = weighted mean effect size; 95% CI = 95% confidence interval for r_+; z = obtained z statistic; Q = weighted sum of squares reflecting variability in effect sizes; I^2 = percent of observed variability in effect sizes due to variability in true effects in the population.

Table 19.1 shows that the summary effect for the studies using a strong research design was $r_+ = .23$, whereas the summary effect for studies using a weak research design was lower at $r_+ = .14$. The row headed "Total between" presents the Q statistic for the between-groups variance. This Q statistic tests the null hypothesis that the mean effect size for the two subgroups is equal in the population. This statistic was significant, $Q(1) = 13.156$, $p < .001$. Combined, these results show that "quality of methodology" moderated the relationship between exposure to video game violence and subject aggressive behavior: The relationship was significantly stronger in studies with a strong research design, compared to studies with a weak research design.

Because studies with weak research designs produced effects that were significantly lower, it was not clear that the findings from these studies could be trusted. For this reason, they were excluded from all subsequent analyses to be conducted as part of this meta-analysis.

The concept of a moderator variable is all about the words: *it depends*. If Dr. O'Day's sister asks whether playing violent video games is related to real-world aggressive behavior, Dr. O'Day will have to answer:

It depends. Studies with a weak research design tend to show a very weak relationship between the two variables, but studies with a strong research design tend to show a stronger relationship.

Dr. O'Day is not finished with her tests for moderator variables. Remember her hypothesis about the difference between studies conducted in the United States versus those conducted in Japan? She now turns her attention to that hypothesis:

Table 19.2

Subject Aggressive Behavior as a Function of Violent Video Game Exposure: Subgroup Analyses for United States Versus Japan

Location	N	k	Effect size r_+	95% CI	NHST z	p	Heterogeneity Q	df	p	I^2
Japan	1,170	15	.233	[.184, .281]	9.031	<.001	16.491	14	.284	15.105
United States	1,638	21	.232	[.185, .277]	9.458	<.001	25.581	20	.180	21.818
Total within							42.072	34	.161	
Total between							0.001	1	.973	
Overall	2,808	36	.232	[.198, .265]	13.078	<.001	42.073	35	.191	16.812

Note. NHST = Null hypothesis significance test (two-tailed) for null hypothesis that summary effect is equal to zero; N = number of participants; k = number of studies; r_+ = weighted mean effect size; 95% CI = 95% confidence interval for r_+; z = obtained z statistic; Q = weighted sum of squares reflecting variability in effect sizes; I^2 = percent of observed variability in effect sizes due to variability in true effects in the population.

Location of the study as a moderator variable. A separate set of analyses was performed to determine whether there was a significant difference between the summary effect observed with studies conducted in the United States compared to studies conducted in Japan. All of these analyses were conducted using only the subgroup of 36 studies that had been previously identified as displaying a "strong" research design (for reasons provided earlier).

Table 19.2 shows that the summary effect for the studies conducted in Japan was $r_+ = .233$, whereas the summary effect for studies conducted in the United States was virtually identical at $r_+ = .232$. The Q statistic in the row headed "Total between" tests the null hypothesis that the mean effect size for the two subgroups is equal in the population. With alpha set at $\alpha = .05$, this statistic was nonsignificant, $Q(1) = 0.001$, $p = .973$, indicating that there was no difference between the typical effect found in Japan versus the United States.

How Strong Is the Relationship?

Throughout this book you have been warned that statistical significance does not guarantee that the index of effect size will be large. In the following excerpt, Dr. O'Day evaluates the magnitude of the mean effect in the current meta-analysis.

> Across the 70 studies included in the meta-analysis, the weighted mean correlation was $r_+ = .19$. This indicated that about 4% of the variance in subject aggressive behavior was accounted for by exposure to violent video games. However, the strength of this relationship was moderated by the quality of the research methodology used in the individual studies: For studies with a strong research design, the summary effect was $r_+ = .23$, and for studies with a weak design, the summary effect was $r_+ = .14$. With respect to Cohen's (1988) criteria for evaluating correlation coefficients, each of these indices fell somewhere between a small effect ($r = \pm.10$) and a medium effect ($r = \pm.30$).

Summary of Findings

The summary effect based on all 70 studies included in this meta-analysis was $r_+ = .189$, 95% CI [.164, .214], which was significantly different from zero, $z = 14.518$, $p < .001$. This was a positive correlation, which meant that individuals with greater exposure to violent video games tended to display more real-world aggressive behavior.

Two tests for moderator variables were performed, and alpha was set at $\alpha = .05$ in each test. One test showed that the mean correlation was significantly stronger in studies with a strong research design ($r_+ = .232$) compared to studies with a weak research design ($r_+ = .140$). The second test revealed a nonsignificant difference between studies conducted in Japan ($r_+ = .233$) versus those conducted in the United States ($r_+ = .232$).

None of the mean correlation coefficients reported here were large enough to be labeled a medium effect, according to Cohen's (1988) guidelines for interpreting Pearson correlation coefficients. Each mean correlation fell somewhere between a small effect ($r = \pm.10$) and a medium effect ($r = \pm.30$) according to those guidelines.

Investigating Publication Bias in Meta-Analysis

But wait—*there's more!* Dr. O'Day has not yet investigated the possibility that her meta-analysis may have been affected by publication bias.

In this context, **publication bias** refers to the possibility that the results from a meta-analysis may be misleading due to a bias that journals display regarding the types of articles they are willing to publish. Theoretically, journals might display any conceivable type of bias. In practice, however, scholars are most concerned that journals (and similar entities) are biased in favor of publishing articles that report statistically significant results (Dickersin, Min, & Meinert, 1992).

Some refer to this as the "file-drawer" problem (Rosenthal, 1979). This name connotes the possibility that there may be a large number of completed research reports sitting idly in investigators' file drawers all over the world—studies that were never published because the results were nonsignificant, so the investigators believed that there was no point in submitting them to journals.

If journals are, in fact, biased in favor of publishing articles with statistically significant results, it could mean that the effect sizes in published articles are misleadingly higher than the "true" effect sizes in the population (this is because, with other factors held constant, large effect sizes are more likely to be statistically significant; Hedges, 1984). This, in turn, implies that the summary effects reported in meta-analyses may be misleadingly larger than the true effect sizes in the population. This is because so many of the effects analyzed in a meta-analysis come from published empirical investigations.

Fictitious Example of Publication Bias

All of this is potentially very bad news for meta-analysis as a research method. If journals do, in fact, display a great deal of bias in favor of statistically significant results, and if there is no way to detect this bias (and correct it), it could mean that the results obtained from a meta-analysis could be very misleading.

Consider Dr. O'Day's study. She obtained a summary effect of $r_+ = .19$, and this effect was large enough to be statistically significant and meaningful in practical terms. But *why* did she obtain a summary effect of this size? Was it because the correlation between exposure to video games and real-world aggressive behavior really is equal to .19 in the population?

Or is it possible that the actual correlation between these two variables is actually $r_+ = .00$ in the population? With this latter scenario:

- The actual correlation in the population is $r_+ = .00$: no relationship at all.
- All over the world, hundreds of researchers conduct investigations on this topic, and the vast majority find $r = .00$ (or pretty close to it).
- Most of these researchers never attempted to publish their findings, assuming as they do that journals will not publish nonsignificant findings.

- The few researchers who obtained statistically significant positive correlations (just due to chance) do in fact publish their studies.

- Therefore, when Dr. O'Day eventually performs her meta-analysis, the only studies she can locate are these published studies with significant findings (notice that I am changing her research method here so that she is now using only *published* articles; this is to make a point about publication bias). She computes the summary effect, and finds it to be $r_+ = .19$.

Success! Everywhere, people are talking about the meta-analysis which shows that there is a statistically significant positive relationship between exposure to violent video games and aggressive behavior in the real world. Radio programs want to interview Dr. O'Day. Hollywood wants the movie rights. She is a *star*.

However, unbeknownst to Dr. O'Day (and the rest of us), the actual effect in the population is $r_+ = .00$. Everybody got it wrong because of publication bias.

The good news is that this nightmare scenario does not have to happen in real life (just to be clear—there is no reason to believe that it happened in the real-world study that served as the inspiration for Dr. O'Day's meta-analysis—the study reported by Anderson et al., 2010). Researchers actually have a number of very clever tools that allow them to (a) determine whether it is likely that a meta-analysis has been affected by publication bias, (b) estimate the extent of the bias, and even (c) estimate what their summary effect would have been if there had been no publication bias at all. Some of these tools are briefly described in the following sections.

Rosenthal's Fail-Safe N

One of the earliest tools for addressing the problem of publication bias was **Rosenthal's fail-safe N**, also called the ***classic fail-safe N*** (Rosenthal, 1979). Rosenthal's fail-safe N is the number of unpublished studies with an effect size of zero that would have to exist (hidden away in file drawers) in order to turn a statistically significant summary effect into a nonsignificant summary effect.

Consider Dr. O'Day's meta-analysis (the complete data set, including all 70 studies). Earlier, this chapter indicated that the summary effect for these 70 studies was $r_+ = .189$, and that this summary effect was significantly different from zero ($p < .001$). Rosenthal's fail-safe N for this meta-analysis was 3,544. This means that we would have to find 3,544 unpublished studies (each with $r = .00$) in order to push Dr. O'Day's summary effect low enough so that it becomes statistically nonsignificant (with alpha set at $\alpha = .05$, two-tailed test).

The general rule is that, the larger the value of the fail-safe N, the less we need to worry about possible publication bias. Needless to say, there are probably not 3,544 unpublished studies on this topic hiding in researchers' file drawers. The fact that the fail-safe N is so large (3,544!) would reassure most researchers that we can have faith in Dr. O'Day's findings.

Orwin's Fail-Safe N

Although Rosenthal's fail-safe *N* is often included as part of a meta-analysis, it has its limitations. One common criticism involves the fact that it focuses on the statistical significance of the summary effect, rather than its practical significance.

Orwin's fail-safe N (Orwin & Boruch, 1983) addresses this problem by focusing on the size of the summary effect rather than its statistical significance. **Orwin's fail-safe N** estimates the number of unpublished studies that would have to be found in order to bring the summary effect itself down to some specified value—a value that most researchers would view as being "trivial" in size. For example, with Dr. O'Day's meta-analysis, Orwin's fail-safe *N* estimates that she would have to find 64 studies (each with $r = .00$) in order to bring her summary effect from its current value of $r_+ = .189$ down to $r_+ = .10$. You will recall that, according to Cohen (1988), a correlation of $r = \pm.10$ is typically viewed as being a small effect; for this reason, that value is sometimes targeted when computing Orwin's fail-safe *N*.

Notice that, in this case, Orwin's fail-safe *N* (64) is not a terribly large number. It is not so difficult to imagine that—if we cast a wide net—we might actually be able to find 64 unpublished studies (each with $r = .00$) hidden away in file drawers. This could be bad news for Dr. O'Day's meta-analysis. Maybe she does need to worry about publication bias after all.

The Funnel Plot

One of the most important steps that researchers can take to investigate possible publication bias involves creating a funnel plot (Light & Pillemer, 1984). A *funnel plot* is a type of scatterplot in which the "dots" represent the individual studies that went into the meta-analysis. The horizontal axis represents the effect size for each study, and the vertical axis represents sample size, variance, or some other index of the "precision" of the study.

Figure 19.3 presents the funnel plot for the 70 studies included in Dr. O'Day's fictitious meta-analysis (this plot was created using the computer application *Comprehensive Meta-Analysis V2™*). The horizontal (X) axis plots effect size (converted to Fisher's z). Circles that appear toward the right side of the plot represent studies with larger effects. The vertical (Y) axis plots the standard error for each study (other factors held constant, smaller samples tend to have larger standard errors and therefore poorer precision). The Y axis has been reversed, so that larger samples (with smaller standard errors) appear at the top. As we move down the Y axis, the samples become smaller, and the standard errors become larger.

The vertical line running through the middle of the funnel plot represents the weighted mean effect size obtained for the 70 studies (again, converted to Fisher's z). In general, the studies with a larger *N* appear higher in the plot, and the studies with a smaller *N* appear lower. As you move lower in the plot, the dots should be more spread out, because there is more variability in studies with a smaller *N*.

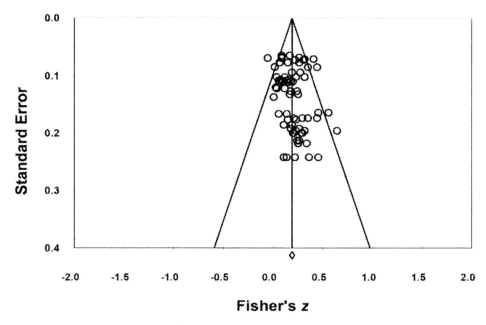

Figure 19.3. Funnel plot for the fictitious data from Dr. O'Day's meta-analysis; each open circle represents one of the 70 studies included in the meta-analysis (Figure created using *Comprehensive Meta-Analysis V2* ™).

When there is no publication bias, the dots in the funnel plot should be symmetrical around the center line. In other words, if there is no publication bias, the dots on the left side of the center line should look something like a mirror image of the dots on the right side. It will not be a perfect mirror image, but the left side of the plot should appear fairly balanced with the right side.

If the meta-analysis suffers from publication bias, the dots will not be symmetrical around the center line—with publication bias, there will be more dots on the right side of the plot. With publication bias, this lack of symmetry should grow worse as you move lower in the plot, due to the fact that studies with a smaller N are least likely to display statistical significance (and thus are least likely to be published). In short, a funnel plot indicates possible publication bias when there is a relative absence of dots on the left side of the plot, especially toward the bottom.

A visual inspection of Figure 19.3 indicates that Dr. O'Day's meta-analysis may, in fact, suffer from publication bias. There appear to be more dots on the right side of the center line, compared to the left side. To obtain a more precise estimate of the extent of possible bias, we will turn to Duval and Tweedie's (2000) trim and fill.

Duval and Tweedie's Trim and Fill

Duval and Tweedie's trim and fill (Duval & Tweedie, 2000) is an algorithm that provides an estimate of what the overall summary effect in a meta-analysis would have been if the funnel plot had been symmetrical. In other words, it provides an estimate of what the summary effect would have been if there had been no publication bias at all.

The "trim" part of trim and fill involves trimming from the funnel plot the small-N studies that appear in the lower right corner. This is done in an iterative fashion, ending in a symmetrical funnel plot and an "adjusted" overall summary effect (this *adjusted effect* is an estimate of what the summary effect would have been if there was no file-drawer problem). The "fill" part of trim and fill involves returning the trimmed dots to the funnel plot, and adding imputed "mirror image" dots to the left side so that the confidence interval for the summary effect is correct.

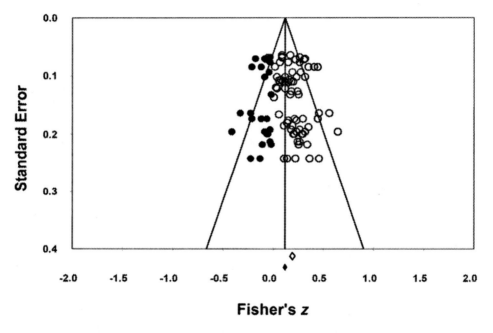

Figure 19.4. Funnel plot for the fictitious data from Dr. O'Day's meta-analysis, following the trim and fill procedure. Each open circle represents one of the 70 observed (although fictitious) studies included in the meta-analysis. Each of the filled-in dots represents a new study added as part of the "fill" process. (Figure created using *Comprehensive Meta-Analysis V2* ™).

For example, Figure 19.4 contains the funnel plot for Dr. O'Day's meta-analysis after the trim and fill procedure has been performed. The 25 filled-in dots that appear on the left side of the plot were imputed by the statistical algorithm as part of the "fill" step. Notice that the resulting funnel plot is now symmetrical.

You will recall that the overall summary effect from Dr. O'Day's meta-analysis (based on all 70 studies) was $r_+ = .189$. After the trim and fill procedure, the adjusted summary effect was somewhat smaller at $r_+ = .118$. This value tells us that, if her meta-analysis had not been affected by publication bias at all, our best estimate is that her summary effect probably would have been $r_+ = .118$.

Duval and Tweedie's trim and fill procedure produces a confidence interval for the adjusted summary effect, and this interval can be used to test the adjusted effect for statistical significance. In the current procedure, the 95% confidence interval for the adjusted summary effect extended from .096 to .140. Since the interval did not contain the value of zero, the adjusted summary effect of $r_+ = .118$ was still statistically significant.

You will also recall that Dr. O'Day's fictitious meta-analysis was very loosely based on the actual meta-analysis conducted by Anderson et al. (2010). The real-world researchers investigated possible publication bias in their study using procedures similar to those discussed here, and found very little evidence that their results had been influenced by the "file-drawer" problem.

Publication Bias in Meta-Analysis: Summary

Theoretically, publication bias could have been the Achilles heel of meta-analysis. If researchers had no way of estimating the extent to which their study suffered from publication bias (and correcting it), meta-analysis would be much less valuable as a research method. Skeptics could dismiss any study they disagreed with as a case of "garbage in—garbage out," and meta-analysis as a tool of research could have just faded away.

But that is not what happened. With Rosenthal's fail-safe N and Orwin's fail-safe N, researchers can determine the number of unpublished studies that a skeptic would have to turn up to bring the current summary effect down to a nonsignificant or trivial value. Creating a funnel plot can help researchers visualize the extent to which their meta-analysis might suffer from the file-drawer problem. And when there is evidence of publication bias, Duval and Tweedie's trim and fill and can even help the researcher estimate what the summary effect would have been if there had been no publication bias at all.

Consumers of research owe a debt of gratitude to the individuals who developed these and similar procedures. Skeptics may not always like the findings from a meta-analysis, but with the tools described here, a carefully-conducted meta-analysis is not so easy to dismiss.

Psychometric Meta-Analysis

An earlier section indicated that there are at least two schools of thought that have contributed to the development of meta-analysis as it is currently performed. One school is represented by Borenstein et al. (2009), Glass et al. (1981), Hedges and Olkin (1985), Rosenthal (1991) and similar texts. This chapter uses the label ***non-psychometric meta-analysis*** to refer to this approach. Non-psychometric meta-analysis is essentially *descriptive* in nature: One of its main objectives is to *describe* the typical effect that was observed in previous investigations on a given topic. This approach typically makes no attempt to correct indices of effect size for ***artifacts:*** imperfections in empirical investigations that can cause their results to be biased or misleading. Until now, this chapter has focused exclusively on non-psychometric meta-analysis.

A second, somewhat different school of thought is called ***psychometric meta-analysis*** or the ***Hunter and Schmidt approach*** (Hunter & Schmidt, 2004). As its name would suggest, psychometric meta-analysis had originally focused on research involving tests, assessments, personnel selection, and related phenomena (although today it is used with all manner of research topics).

CORRECTING INDICES OF EFFECT SIZE FOR ARTIFACTS. One of the defining characteristics of psychometric meta-analysis is its emphasis on *artifacts*. In this context, an ***artifact*** is an imperfection in an individual empirical investigation—an imperfection that may cause the study's results to be biased or misleading. Artifacts are flaws or weaknesses in research design, in the way that variables are measured, or in other aspects of the study.

For example, imagine that, in a specific empirical investigation, the multiple-item summated rating scale used to measure the criterion variable displayed poor reliability: The coefficient alpha reliability estimate for the scale was $\alpha = .59$, which is below the recommended value of $\alpha = .70$ and is *way* below the perfect reliability value of $\alpha = 1.00$. If this occurred, it would indicate that this study suffers from an artifact called "error of measurement in the dependent variable" (Hunter & Schmidt, 2004, p. 35).

The preceding was an example of just one of the artifacts that may be displayed by an investigation. Additional types of artifacts will be listed a bit later in this section.

Artifacts are important because they can cause the index of effect size obtained from an investigation to be biased. In most cases, an artifact causes an index of effect size to be smaller (in absolute value) than it otherwise would be.

To address this issue, a psychometric meta-analysis attempts to correct each index of effect size for as many artifacts as possible. For example, imagine that a given meta-analysis includes 70 indices of effect size (from 70 empirical investigations), and each of these empirical investigations suffers from the error of measurement artifact mentioned above. In this situation, each of the 70 indices would be corrected for this artifact (mathematical formulas are used to make these corrections). If the studies suffered from some additional artifact, each of the 70 studies would be corrected for that artifact as well.

After each individual effect has been corrected in this way, the meta-analysis would proceed in the usual fashion: a summary effect would be computed by taking the weighted-mean of the corrected indices of effect size. Because of the corrections performed, the summary effect obtained in a psychometric meta-analysis tends to be larger (in absolute value), compared to the summary effect obtained in a corresponding non-psychometric meta-analysis.

THE GOALS OF NON-PSYCHOMETRIC VERSUS PSYCHOMETRIC META-ANALYSIS. All of this leads to what may be the most important distinction between the two approaches to meta-analysis. You will recall that, in non-psychometric meta-analysis, no attempt is made to correct for artifacts. Because of this, the results from this approach are *descriptive* in nature: they describe the typical result obtained in a sample of real-world studies, most of which suffer from measurement problems, design flaws, and other imperfections. One of the goals of a non-psychometric meta-analysis is to provide a summary effect that reflects the results that were actually observed in the current sample of real-world investigations, warts and all.

In contrast, a psychometric meta-analysis attempts something more ambitious. In a psychometric meta-analysis, the index of effect size from each investigation is corrected for artifacts before the summary effect is computed. This reflects the goal of psychometric meta-analysis—it attempts to estimate what the relationship between the variables *would have been* if the sample of investigations had not displayed any flaws at all. In this way, psychometric meta-analysis attempts to arrive at a summary effect that represents the actual relationship between the theoretical constructs of interest, independent of the bias and error that can be introduced by flawed, real-world investigations (Hunter & Schmidt, 2004).

GREEK LETTERS AS SYMBOLS FOR TRUE EFFECTS IN POPULATIONS. Much earlier in this book, you learned that Greek letters are typically used to represent various characteristics of populations. These Greek-letter symbols are often used in articles that report a meta-analysis—especially articles that report a psychometric meta-analysis. Their use reflects the fact that this school of meta-analysis strives for the very ambitious goal of estimating the true effect size in the population, independent of artifacts.

In a meta-analysis in which the index of effect size in the individual empirical investigation is the Pearson correlation coefficient (r), the corresponding symbol for the actual correlation between constructs in the population is ρ (the Greek lower-case letter *rho*). When the index of effect size in the empirical investigation is Cohen's d, the symbol for the actual standardized difference in the population is δ (the Greek lower-case letter *delta*). Tables in an article that report a psychometric meta-analysis often include columns headed ρ or δ (or some variation, such as $\bar{\rho}$ or $\bar{\delta}$). These symbols are typically used to represent the estimated value of the population correlation (ρ) or population standardized difference (δ).

Illustrative Investigation

Imagine that Dr. O'Day is interested in an instrument called the *Smith Conscientiousness Inventory*, or *SCI* (a fictitious instrument). This is a multiple-item self-report questionnaire designed to measure the personality trait of *conscientiousness*: the extent to which an individual is reliable, detail-oriented, and organized. With this inventory, possible scores can range from 1-50, with higher scores reflecting higher levels of conscientiousness.

CRITERION VARIABLES. Dr. O'Day believes that scores on this instrument should be positively related to two outcome variables that are often studied by industrial psychologists: job performance and organizational citizenship behavior. When employees score high on *job performance*, it means that they are effective in performing the formal responsibilities of their jobs. When they score high on *organizational citizenship behavior*, it means that they often engage in behaviors that help their organizations, but are not required by formal job descriptions (e.g., making suggestions to improve productivity or helping fellow employees meet a deadline).

RESEARCH QUESTIONS. Over the past 30 years, 77 empirical studies have investigated the validity of the Smith Conscientiousness Inventory. In the typical study, researchers computed the Pearson r correlation between scores on the SCI and scores on one of the two outcome variables described above (job performance or organizational citizenship). Dr. O'Day decides that she will perform a meta-analysis of the results obtained in those studies. She will investigate two research questions:

- What is the true validity of the Smith Conscientiousness Inventory? In other words, what is the actual correlation between scores on the SCI and these two outcome variables in the population, independent of artifacts?

- Do any variables moderate the relationship between SCI scores and these outcome variables? If yes, what are these moderator variables? What is the nature of the moderated relationship?

In reviewing the primary empirical investigations, Dr. O'Day sees that a few of the studies found no relationship at all between SCI scores and the outcome variables (e.g., $r = .00$) and a few of the studies found fairly strong correlations (e.g., $r = .50$), Most studies found correlations that were somewhere between these two extremes.

Correcting Effects for Artifacts

As was stated earlier, in a psychometric meta-analysis, researchers often wish to correct the indices of effect size for sources of error and bias called *artifacts*. For example, imagine that Study 1 (of the 77 studies included in the meta-analysis) found that the Pearson correlation between SCI scores and job performance was $r = .30$. However, when Dr. O'Day looked closely at the investigation, she found that the measure of job performance used in the study was not terribly reliable. Supervisory ratings were used as the measure of job performance, and the coefficient alpha reliability estimate for these supervisory ratings was only $\alpha = .70$ (acceptable, but not great). We know that when we measure a variable with less-than-perfect reliability and then compute the Pearson correlation between that variable and another variable, it typically causes the resulting correlation to be smaller (in absolute value) than it otherwise would be. In this situation, the less-than-perfect reliability of the criterion variable (job performance ratings) is operating as an *artifact*.

Fortunately, formulas are available to correct correlation coefficients for criterion-variable unreliability. These formulas allow us to estimate what a given correlation coefficient would have been if the criterion variable had been measured with perfect reliability. Imagine that Dr. O'Day uses such a formula to correct the correlation coefficient displayed in Study 1. Prior to the correction, the observed correlation between SCI scores and job performance was $r = .30$. After the correction, the correlation was higher at $r = .40$. Dr. O'Day then uses the same formula to correct each of the remaining correlation coefficients for unreliability of the criterion variable.

Having corrected the correlation coefficients for unreliability, Dr. O'Day then corrects each correlation for other artifacts. Hunter and Schmidt (2004) present a fairly long list of artifacts that may influence the size of correlation coefficients. Here are a few examples:

- *Sampling error.* **Sampling error** refers the difference between a sample statistic versus the corresponding population parameter due to the fact that samples are typically not perfectly representative of the populations from which they are drawn. Sampling error becomes worse as sample size decreases.

- *Dichotomization of a continuous criterion variable.* When we say that someone has **dichotomized** a variable, it typically means that they have turned a *continuous quantitative variable*—a variable that consists of many scores—into a *dichotomous variable*—a variable that consists of just two categories. For example: previously you learned that one of the criterion variables in the Dr. O'Day's current meta-analysis is employee job performance. Imagine that, in most of primary empirical investigations, job performance was measured by having supervisors rate each employee using a multiple-item summated rating scale. Scores on this scale could range from 1.00 (which represented poor performance) to 7.00 (which represented excellent performance), with many possible scores in-between—scores such as 1.25, 2.50, 5.75, and so forth. Most researchers would view this as being a **theoretically continuous variable**: a quantitative variable that theoretically could display an infinite number of different scores. But imagine that, in some of the primary investigations, the researchers had dichotomized this variable so that employees with scores below 3.50 were labeled "poor performers" and those with scores equal to or greater than 3.50 were labeled "good performers." The new variable that results from this process is a **dichotomous variable**—a variable that consists of just two categories. Statisticians typically view this practice of dichotomization a continuous variable as a no-no because it often results in biased statistics when the new dichotomous variable is included in any subsequent analysis. For example: If the original researchers were to compute the correlation between SCI scores and this new dichotomous job performance variable, the resulting correlation coefficient would be smaller (in absolute value) than it would have been if the they had simply left the variable in its original continuous form. To make things right again, the meta-analyst who later re-analyses these results must use a formula to any correct correlation coefficient that has been negatively affected in this way. After being corrected, the resulting correlation coefficient will provide an estimate of what the correlation would have been if the original researchers had not dichotomized the job performance variable in the first place.

- *Restriction of range*. When there is a **restriction of range** on a given variable, it means that the participants in the sample do not display the full range of scores that are displayed in the reference population. When there is a restriction of range on either the predictor variable or the criterion variable in a given investigation, it causes the Pearson r to be smaller (in absolute value) than it otherwise would be. For example, imagine that we gave the SCI to the entire population of people who wanted to be employed in a given industry, hired all of them, and six months later obtained job performance scores for all of them. Imagine that, in such a situation, the correlation between SCI scores and job performance scores would be $r = .60$. However, if we had instead hired just half of the participants—the 50% with the highest scores on the SCI—the correlation between SCI scores and job performance scores would be smaller, perhaps $r = .40$. It is smaller in the second scenario because there is now a restriction of range on the predictor variable (the SCI). Fortunately, there are formulas that allow meta-analysts to correct for restriction of range—to estimate what the correlation would have been if the sample had displayed the full range of scores.

This section has reviewed just a few of the artifacts that can introduce bias into indices of effect size. When using psychometric meta-analysis, researchers often correct indices for these artifacts, as well as others described by Hunter and Schmidt (2004). The researchers then compute the weighted mean of the corrected indices, and this mean serves as the estimate of the true effect size in the population (Whitener, 1990).

Confidence Intervals Versus Credibility Intervals

You were first introduced to confidence intervals in *Chapter 5* of this book. In the current section you will see how they are relevant to meta-analysis. You will then learn about a new concept called *credibility intervals*. You will see how credibility intervals are different from confidence intervals, and how they are used to assess something called *validity generalization*.

CONFIDENCE INTERVALS. Meta-analysis often results in a summary effect along with the confidence interval for that summary effect. When a meta-analysis is performed on a sample of correlation coefficients, a ***confidence interval*** (symbol: CI) reflects the amount of variability in the mean correlation coefficient that can be attributed to sampling error. Confidence intervals are computed using the standard error of the mean effect size, and they reflect the accuracy of the estimate (Whitener, 1990). When the confidence interval is relatively narrow, it indicates that the mean effect size is a relatively precise estimate. When the confidence interval is relatively wide, it indicates that the mean effect size is a less precise estimate. Confidence intervals may be used to perform significance tests: If the CI does not contain the value of zero, the summary effect is significantly different from zero.

Bookbyte
The Textbook Way to Save

SHIP TO:
Omar Aziz
7 MONTICELLO DR APT 204
ATHENS, OH 45701-3322
US

Total Items: 1
Black

SHIPPED FROM:
Bookbyte
2800 Pringle Road SE
Salem, OR 97313

YOUR SHIPMENT METHOD:
Standard

Hello Omar Aziz,

Here is your **Order #14163190** from Bookbyte which is also **Order #113-4227698-3681037** from Amazon.com Market.

TITLE: Advanced Statistics in Research: Reading, Understanding, and Writing Up Data Analysis Results, by Hatcher
SKU: 135928692
INVENTORY LOCATION: G-46-4-1
CONDITION: Good Used
DESCRIPTION: Has minor wear and/or markings.

RETURN POLICY: If necessary, you may return any book for any reason within 14 days of delivery. For additional information and instructions, check out our complete Return Policy on our website.

RENTED YOUR BOOKS? HERE'S OUR SIMPLE RETURN PROCESS:

Shipping label goes on the box.

Place your books in any box.

Print your shipping label from your account page.

Drop-off box at nearest FedEx location.

CREDIBILITY INTERVALS. Confidence intervals (as described above) are often reported in both types of meta-analysis: non-psychometric as well as psychometric. However, a psychometric meta-analysis typically produces an additional result called a *credibility interval*. A ***credibility interval*** is computed using the standard deviation of the distribution that results after the correlation coefficients have been corrected for artifacts. Credibility intervals are important, in part, because they may be used to determine whether the results from the current analysis may be generalized across situations—a phenomena known as *validity generalization* (Field, 2009b; Pearlman, Schmidt, & Hunter, 1980). If the lower limit of the credibility interval (called the ***credibility value***, symbol: CV) is larger than zero, this is typically taken as evidence that the presence of validity may be generalized across situations (Pearlman et al., 1980).

For example, imagine that Dr. O'Day performed a meta-analysis of 22 studies that investigated the relationship between SCI scores and job performance scores. Assume that the mean corrected correlation coefficient was $r_c = .46$, which was significantly different from zero. This would be good news—it would indicate that the SCI was valid for predicting job performance ratings among employees. Imagine further that the 90% credibility interval for this correlation was 90% CV [.17, .75]. This would also be good news. Since the lower limit of this interval (.17) was larger than zero, it would mean that the positive validity that she found for the SCI could be generalized across the different situations that were studied in the various investigations included in her meta-analysis. In other words, the SCI should be valid for predicting job performance in all of the situations represented in these 22 studies.

For purposes of contrast, consider a different scenario in which she obtained totally different results: Imagine that the 90% credibility interval had instead been 90% CV [−.10, .90]. This 90% credibility interval now contains the value of zero, and this indicates that the positive validity of the SCI does not generalize across situations. It indicates that the SCI displays a positive correlation with job performance in some situations (perhaps in situations in which the jobs were complex), but a negative correlation with job performance in other situations (perhaps in situations in which the jobs were simple). This would be bad news, as it would mean that the SCI does not display validity generalization. In this second scenario, most researchers would conduct a search for moderator variables in order to discover the specific situations in which the correlations were positive versus negative.

In summary:

- If the credibility interval for a positive mean correlation does not contain zero, it is supportive of validity generalization—it indicates that the positive correlation generalizes across the situations studied in the meta-analysis.

- If the credibility interval for a positive mean correlation contains zero, it is not supportive of validity generalization—it indicates that the positive correlation does not generalize across situations (possibly due to the presence of a moderator variable).

Hunter and Schmidt's 75% Rule

The preceding section indicates that researchers conducting a psychometric meta-analysis sometimes use credibility intervals to determine whether their results are influenced by moderator variables. As an alternative, researchers performing a psychometric meta-analysis sometimes apply a rule of thumb called the *75% rule* for the same purpose.

When researchers plan to use this approach, they first correct the individual indices for artifacts such as dichotomization and restriction of range (as described above). Next, they determine the percent of variance in the indices that is attributable to these artifacts. Once computed, they may then apply the **75% rule** as follows:

- If 75% or more of the variability in indices is due to artifacts, it means that no moderator variables are present.

- If less than 75% of the variability in indices is due to artifacts, it means that moderator variables may be present.

RATIONALE. The 75% rule is based on the assumption that, in a typical meta-analysis, much of the variability in the values of r for the individual studies is due to artifacts such as criterion unreliability, restriction of range, and the other sources described above. Even after correcting for the most common sources of error, it is likely that a certain percent of the variance—say 25%—is still due to artifacts that are difficult to isolate and correct (e.g., clerical errors, typos made when keying data). Therefore, if the known artifacts that have been corrected in the meta-analysis constitute 75% or more of the variance in the individual r statistics, it is safe to assume that *all* of the variance was due to artifacts, and none of it was due to moderator variables. Therefore, if 75% or more of the variability in indices is due to artifacts, there is no need to search for moderator variables.

RESEARCH ON THE 75% RULE. At least one Monte-Carlo investigation has shown that the 75% rule is fairly accurate in identifying the presence of moderator variables (Sackett, Harris, & Orr, 1986). In a *Monte-Carlo study*, a computer application generates large numbers of fictitious data sets to test hypotheses or investigate some statistical procedure. Sackett et al. (1986) found that the 75% rule displayed statistical power that was equal to or greater than competing methods (such as the Q statistic) for correctly identifying the presence of moderator variables. It performed particularly well when the number of studies included in the meta-analysis was small and the sample size of each study was also small.

Summarizing a Psychometric Meta-Analysis in a Table

Table 19.3 shows how the results of Dr. O'Day's meta-analysis of the Smith Conscientiousness Inventory might be summarized in a table. Remember that these results are fictitious.

Table 19.3

Meta-analysis Results: Validity of the Smith Conscientiousness Index for Predicting Job Performance and Organizational Citizenship Behavior

Variable	k	N	r	r_c	95% CI LL, UL	90% CV LL, UL	% Variance due to artifacts
Job performance	22	9,520	.36	.46	.37, .56	.17, .75	81
Citizenship	55	31,360	.44	.57	.50, .64	.23, .91	89

Note. k = number of studies; N = number of participants; r = sample-size weighted mean uncorrected effect size; r_c = sample-size weighted mean corrected effect size; 95% CI = 95% confidence interval; 90% CV = 90% credibility interval; LL = lower limit; UL = upper limit.

In a sense, Dr. O'Day actually performed two meta-analyses. One meta-analysis investigated the validity of the SCI when it is used to predict employee job performance, and the row headed "Job performance" summarizes findings from this analysis. The second meta-analysis investigated the validity of the SCI when it was used to predict organizational citizenship behavior, and the row headed "Citizenship" summarizes those results. The remainder of this section provides a brief description of the information presented in the columns of Table 19.3.

THE COLUMN HEADED "k." This column presents the number of primary empirical investigations included in each meta-analysis. You can see that meta-analysis involving job performance was based on 22 studies, whereas the meta-analysis involving citizenship was based on 55.

THE COLUMN HEADED "N." This column presents of the total number of participants included in the all of the empirical investigations used the meta-analysis. You can see that the meta-analysis for job performance was based on a total of 9,520 participants, whereas the meta-analysis for citizenship was based on 31,360.

THE COLUMN HEADED "r." This column presents the sample-size weighted mean of the uncorrected effect indices from the primary empirical investigations. For example, the value in the row headed *job-performance* is .36. This indicates that the mean correlation between the SCI and job performance (appropriately weighted for sample size) was $r = .36$. Because this correlation was labeled "uncorrected," we know that it was the average of the correlations that had not been corrected for artifacts such as criterion-variable unreliability or restriction of range. Some authors use symbols other than "r" to represent the mean of the uncorrected correlations. Alternative symbols include "Mean r," "M_r," and "$\bar{r}$."

The column headed "r_c." This column presents the single most important result from a psychometric meta-analysis: the sample-size weighted mean of the corrected effect indices from the primary empirical investigations. In other words, this column presents the average of the correlation coefficients after they have been corrected for artifacts. It estimates what the actual correlation between the constructs would have been if the meta-analysis had been performed on a sample of empirical investigations that did not suffer from imperfections such as criterion-variable unreliability or restriction of range. The mean corrected correlation coefficients presented in this column are our best estimates of the true correlation in the population. That's why some authors label this column using the Greek letter rho ("ρ") since rho is the symbol for a correlation coefficient in the population. Alternative symbols include "$\bar{r}_c$," "Mean ρ," "M_ρ," and "$\bar{\rho}$."

The columns headed "95% CI." These columns present the lower limit (symbol: LL) and upper limit (symbol: UL) of the 95% confidence interval for r_c. As was explained earlier, the 95% CI reflects the amount of variability in the mean correlation coefficient that can be attributed to sampling error. When the confidence interval is relatively narrow, it indicates that the mean effect size is a relatively *precise* estimate. When the confidence interval for a summary effect does not contain the value of zero, it indicates that the summary effect is significantly different from zero. For example, Table 19.3 shows that the 95% confidence interval for job performance extends from .37 to .56; since this interval does not contain zero, we know that that the corrected correlation coefficient (r_c = .46) is significantly different from zero.

The columns headed "90% CV." These columns contain the lower limit and upper limit for the 90% credibility interval. The symbol "CV" represents **credibility value**. As was explained earlier, if the credibility interval does not contain the value of zero, this is typically taken as evidence that the validity of the test being investigated (the SCI, in this case) may be generalized across situations (Pearlman et al., 1980). Table 19.3 shows that the credibility interval for job performance does not contain zero: 90% CV [.17, .75]. This indicates that the SCI should be valid for predicting job performance in any of the situations represented by the various empirical investigations included in the meta-analysis. The same conclusion results from inspection of the credibility interval for citizenship, 90% CV [.23, .91].

Some articles report credibility intervals using conventions that differ from the conventions illustrated in Table 19.3. For example, rather than reporting a 90% CV, some articles report an 80% CV (or some other interval size). In addition, some articles present just the lower limit of the credibility interval, rather than both the lower limit and the upper limit. This is because readers really only need to review the lower limit to determine whether the test's validity may be generalized.

The column headed "% Variance due to artifacts." This column reports the percent of variance in the correlation coefficients that can be attributed to artifacts such as criterion unreliability, restriction of range, and other problems of measurement and research design. Researchers apply the 75% rule to the information presented in this column to determine whether moderator variables may be present. According to this rule of thumb, if 75% (or more) of the variance is due to artifacts, it is reasonable to assume that no moderator variables are present. Table 19.3 shows that, for job performance, 81% of the variance displayed by the correlation coefficients was due to artifacts. Therefore, Dr. O'Day concludes that no moderator variables are present in that analysis. For citizenship behavior, 89% of the variance is due to artifacts, so she concludes that no moderator variables are at work in that analysis either.

Conclusion

This section has focused on the differences between non-psychometric meta-analysis versus psychometric meta-analysis. Nevertheless, we should remember that there are more similarities between the two approaches than differences.

Although psychometric meta-analysis had originally been developed to investigate issues related to employee selection, both psychometric and non-psychometric meta-analysis are now used to investigate a wide variety of topics in the social and behavioral sciences, education, business, and medicine. Both approaches allow researchers to synthesize research findings, compute summary effects, and investigate possible moderator variables. And both approaches allow researchers to tap the wealth of information contained in the results produced by previous empirical investigations.

Additional Issues Related to this Procedure

We still have a few loose ends to tie up. This section provides a short introduction to *meta-regression*: the multiple-regression approach to meta-analysis. It ends with some suggestions for learning more about meta-analysis.

Meta-Regression

A number of different approaches are used to investigate potential moderator variables in meta-analysis. Two widely used strategies are the *subgroup-analysis approach* and the *multiple-regression approach*. Most of this chapter has focused on the former. In this section, you will learn about the latter.

THE SUBGROUP-ANALYSIS APPROACH. To refresh your memory: the **subgroup-analysis approach** to meta-analysis is somewhat analogous to analysis of variance (ANOVA) as it is used with traditional empirical investigations. The subgroup-analysis approach is particularly useful when the moderator variable being investigated is categorical in nature (i.e., is a limited-value variable). In these situations, it is possible to classify the individual studies into groups, and then treat these groups as if they were different conditions in a one-way ANOVA. The current chapter's meta-analysis on the effects of video games illustrated this ANOVA-like approach when it investigated *location of the study* as a possible moderator variable. In that analysis, Dr. O'Day determined whether there was a difference between "studies conducted in Japan" versus "studies conducted in the

United States" with respect to the mean effect size obtained. If she had found a significant difference, it would have meant that location of the study was operating as a moderator variable.

META-REGRESSION. But what if the moderator variable of interest is a multi-value quantitative variable rather than a categorical variable? For example, imagine that Dr. O'Day suspects that *sample size* may be operating as a moderator variable. This would be the case if she suspected that larger studies are more likely to report larger effects (i.e., larger r statistics or d statistics) compared to smaller studies. Inspection of her 70 studies showed that the smallest sample includes just 20 participants, the largest sample includes 300 participants, and the sample sizes for the remaining studies are fairly evenly distributed between these two extremes (with sample sizes of $N = 28$, $N = 35$, $N = 49$, and so forth, all the way up to $N = 300$). In this case, the potential moderator variable is *sample size*, and this appears to be a continuous quantitative variable, not a categorical variable. When moderator variables are continuous rather than categorical, it may be appropriate to perform the analysis using a special procedure called ***meta-regression***.

The simplest version of meta-regression includes just one continuous moderator variable, and the meta-analysis is somewhat similar to the bivariate regression procedures covered in *Chapter 7* of this book. To get a rough idea of the analysis, imagine a bivariate regression analysis performed on the 70 empirical investigations that had been included in Dr. O'Day's meta-analysis on the effects of video games. In this regression, the individual empirical investigation is the unit of analysis, which means that the regression will be performed on 70 observations. In this regression, the index of effect size from a given empirical investigation will serve as the criterion variable, and the sample size used with that empirical investigation (N) will serve as the predictor variable. If meta-regression were used to perform this analysis, it would produce a regression coefficient that describes the nature of the relationship between the two variables, a confidence interval for this coefficient, and a significance test for it. All of these results from the meta-regression have analogs in the bivariate regression procedures that you learned about in *Chapter 7*. If the regression coefficient obtained from the meta-regression was a positive value and was statistically significant, this could be seen as evidence that larger studies did, in fact, produce larger effect indices.

None of this is to say that meta-regression is identical to the bivariate regression procedures that you learned about in *Chapter 7*. For example, in meta-regression the analyst must decide whether to use a fixed-effects model versus a random-effects model, must decide how the effect indices from the empirical investigations should be weighted, and must make other decisions that are unique to meta-analysis. For a discussion of these issues, along with a worked-out example of meta-regression, see *Chapter 20* of Borenstein et al. (2009).

THE MULTIPLE-REGRESSION APPROACH. A more complex version of meta-regression allows for multiple moderator variables. This approach is therefore analogous to the traditional multiple regression procedures used with primary empirical investigations as covered in *Chapter 9* and *Chapter 10* of this book.

In the ***multiple-regression approach*** to meta-analysis, indices of effect size serve as the criterion variable and multiple moderator variables serve as the predictors. Multi-value quantitative variables may be used as the predictor (i.e., moderator) variables. If the multiple regression coefficient (i.e., b weight or β weight) for a given moderator variable is statistically significant, it is interpreted as evidence that the variable does, in fact, moderate the relationships being investigated (Viswesvaran & Sanchez, 1998). Early examples of the multiple-regression approach to meta-analysis include the Smith and Glass (1977) meta-analysis dealing with psychotherapy outcomes as well as the Smith and Glass (1980) meta-analysis dealing with class size.

Theoretically, there is much to recommend this multiple-regression approach, as it can bring to a meta-analysis the same strengths and capabilities that multiple regression brings to the analysis of primary empirical investigations. For example, the predictor variables in the analysis may be continuous quantitative variables as well as categorical predictor variables (the latter may be included if they have been appropriately transformed using dummy-coding or some similar transformation). The multiple-regression approach to meta-analysis allows for the investigation of both linear and nonlinear relationships, and also allows predictor variables to be added to the equation in a series of pre-determined steps in order to control for potential confounding variables (Borenstein et al., 2009).

Unfortunately, there are also some disadvantages associated with the multiple-regression approach. First, this procedure requires a relatively large ratio of effect sizes to moderator variables (just as traditional multiple regression requires a large ratio of subjects to predictor variables). This means that the multiple-regression approach may not be appropriate for a meta-analysis based on relatively small number of studies (Borenstein et al., 2009). Similarly, Hunter and Schmidt (2004) warn that the multiple regression approach may suffer from low statistical power, and may capitalize on chance characteristics of the current sample. Finally, when moderator variables are correlated with one another, the beta weights obtained for the moderator variables may be misleading when used as indicators of the relative importance of the moderator variables (Viswesvaran & Sanchez, 1998). This is the same *correlated-predictor variable problem* that was discussed with reference to traditional multiple regression in *Chapter 9* of this book.

To Learn More About Meta-Analysis

But wait. *There's more!*

You can never have too much expertise in meta-analysis. It is, after all, the most important of all statistical procedures. To expand your mastery, this section offers some suggested reading.

Durlak (1995) provides a relatively short, easy-to-grasp introduction to meta-analysis, with an emphasis on non-psychometric approaches. More recent introductions are provided by Field (2009b) and Konstantopoulos (2008).

In-depth guides to performing meta-analysis are provided by Borenstein et al. (2009), Glass et al. (1981), Hedges and Olkin (1985), and Rosenthal (1991). The standard reference for performing psychometric meta-analysis is Hunter and Schmidt (2004), and the classic guide to writing up the results of a meta-analysis is Rosenthal (1995).

Meta-analyses have been published in scholarly journals since the 1970s, and the number of meta-analyses published each year appears to be increasing in a nonlinear fashion (Field, 2009b). It was therefore inevitable that that researchers would begin publishing meta-analyses of meta-analyses (no, I am not making this up).

For instance: Richard, Bond, and Stokes-Zoota (2003) summarize results from 322 meta-analyses from the field of social psychology (the studies represented a combined sample of *8 million subjects!*). Similarly, Meyer et al. (2001) present results from 125 meta-analyses dealing with psychological testing and assessment. These are just a few examples—meta-analyses of meta-analyses have been published on topics relevant to almost all major scholarly disciplines.

And it doesn't stop there. For your next week at the beach, you can bring along a paperback that summarizes meta-analyses on a topic of your choosing. Hunt (1997) describes the historical development of meta-analysis, and offers a potpourri of important findings from various fields (Does marital therapy work? Can juvenile delinquents be rehabilitated?). Preiss et al. (2007) summarizes meta-analyses dealing with the effects of the media (Do advertising campaigns convince people to use seat belts? Does violent television programming cause kids to be more aggressive?). On the best-seller list, Hattie (2009) reports a synthesis of 800+ meta-analyses dealing with the correlates of educational achievement (Is active participation really related to student educational achievement? Do students learn more when their teachers have higher expectations? How important is parental involvement?).

The good news is that many authors are now presenting results from meta-analyses in a way that is more accessible to the general public. The number of primary empirical investigations in the published literature is astronomical, and these studies are no longer just gathering dust. Researchers are synthesizing these findings, computing mean effects, and uncovering moderator variables. And many researchers are now sharing these findings with the people who matter: the parents, voters, teachers, and others who stand to benefit from this knowledge. As the public becomes more familiar with meta-analysis, it will be less confused when conflicting findings are reported by individual investigations. Rather than being flustered or dismissive, people will learn to ask:

    What does the meta-analysis say?

Chapter 20

Learning More About Statistics

I Want *Too* Much

When my niece Maureen was about two years old, my wife scooped a dish of ice cream for her. My wife asked:

> Is that enough?

Maureen shook her head and said:

> I want *too much*.

My wife hypothesized that, on some prior occasion, Maureen must have tried scooping a dish of ice cream on her own. When her mother saw what she had done, she said something like:

> Maureen! That's *too much!*

In this way, Maureen learned the words that describe an appropriate portion of ice cream.

And, now that you have completed this book, *too much* is exactly what I wish for you—too much information from empirical research, that is (not ice cream).

The Middle of the Article

With this book behind you, you can now read empirical research articles in academic journals and understand what you have read. Yes, even the middle part of the articles, where the statistics lurk.

Well, you should be able to understand *most* of what you have read.

The problem, of course, is that no single book can cover everything. Sooner or later, your reading will lead you to some statistical procedure not covered here. What will you do then?

An Embarrassment of Riches

It turns out that this is not the only book that has been published on the topic of statistics. Yes, this is a blessing, not a curse.

This final chapter lists publications that will help you learn more about research, statistics, data analysis, and related topics. It even lists some useful resources (such as online databases) that you will find on the Internet.

It's Raining Statistics Books

There are a *lot* of really good books on statistics out there. In the course of writing this opus, I regularly consulted this body of published work, and I was constantly impressed by its breadth, depth, and quality.

This section recommends a small number of go-to books, divided into categories. I make no claim that these are the very best publications in each category—they are simply the books that I go to when I need help on a specific topic. For those of you who are never happy, the section on the Internet (which follows) explains how you may locate even more books on an even larger assortment of topics.

Publications Related to the Teaching of Statistics

If you are reading this section, there is a good chance that you have either (a) just been asked to teach a statistics course (if you are a tenured faculty member), or (b) just been *told* to teach a statistics course (if you are an untenured faculty member). Either way, you are probably open to suggestions and material for the class.

A good place to start might be Hulsizer and Woolf's (2009) *A Guide to Teaching Statistics*. It is concise, gives tons of useful recommendations, and is actually informed by empirical research.

The Wilkinson et al. (1999) article is the oft-cited report from the *APA Task Force on Statistical Inference*. Many researchers know that this is an important article dealing with the ongoing dust-up over significance testing. But few people realize that it also makes important recommendations for how we should be teaching statistics to our students.

- Hulsizer, M. R., & Woolf, L. M. (2009). *A guide to teaching statistics: Innovations and best practices.* Malden, MA: Wiley-Blackwell.

- Wilkinson, L. & Task Force on Statistical Inference, American Psychological Association (1999). Statistical methods in psychology journals: Guidelines and explanations. *American Psychologist, 54,* 594-604.

Books About Reading Statistics and Research

Since this book does not cover every data analysis procedure under the sun, you should know about some of the other texts that show how to read and understand statistical results. Some of these are oriented toward elementary statistics (e.g., Moore & Notz, 2006), some tackle the more advanced procedures (e.g., Grimm & Yarnold, 1995; 2000), and some are comprehensive, covering elementary through advanced statistics (Huck, 2012).

- Grimm, L. G. & Yarnold, P. R. (Eds.). (1995). *Reading and understanding multivariate statistics.* Washington, DC: American Psychological Association.

- Grimm, L. G. & Yarnold, P. R. (Eds.). (2000). *Reading and understanding more multivariate statistics.* Washington, DC: American Psychological Association.

- Holcomb, Z. C. (2007). *Interpreting basic statistics: A guide and workbook based on excerpts from journal articles* (5th ed.). Glendale, CA: Pyrczak Publishing.

- Huck, S. W. (2012). *Reading statistics and research* (6th ed.). New York: Pearson.

- Huck, S. W. (2009). *Statistical misconceptions.* New York: Routledge, Taylor & Francis Group.

- Jaeger, R. M. (1993). *Statistics: A spectator sport* (2nd ed.). Thousand Oaks, CA: SAGE Publications.

- Locke, L. F., Silverman, S. J., & Spirduso, W. W. (2004). *Reading and understanding research* (2nd ed.). Thousand Oaks, CA: SAGE Publications.

- Moore, D. S., & Notz, W. I. (2006). *Statistics: Concepts and controversies* (6th ed.). New York, NY: W.H. Freeman and Company.

Books That Show How to Write About Statistics

The *Publication Manual of the American Psychological Association* (2010) is the standard reference for preparing research articles in the social and behavioral sciences. Its guidelines have been adopted by many journals in psychology, sociology, communications, political science, education, business, and related fields. If you enjoy excruciatingly detailed instructions regarding the most microscopic minutia of the manuscript-preparation process, you will not be disappointed (e.g., when typing the symbol F_{max}, please remember that although the "*F*" is in italics, the "max" must *not* be in italics).

Also published by the APA, the two short books by Nicol and Pexman (2010a, 2010b) are gold mines for writers who need to prepare tables and figures according to *Publication Manual* guidelines. For a given statistical procedure (such as multiple regression), Nicol and Pexman (2010b) provide one or more "play it safe" examples of how to organize a table, along with additional examples for specific needs. In a separate book, Nicol and Pexman (2010a) show how to prepare figures and graphs for research reports according to APA format.

- American Psychological Association (2010). *Publication manual of the American Psychological Association* (6th ed.). Washington, DC: Author.

- Morgan, S. E., Reichert, T., & Harrison, T. R. (2002). *From numbers to words: Reporting statistical results for the social sciences.* Boston, MA: Allyn and Bacon.

- Nicol, A. A., & Pexman, P. M. (2010a). *Displaying your findings: A practical guide for creating figures, posters, and presentations* (6th ed.). Washington, DC: American Psychological Association.

- Nicol, A. A., & Pexman, P. M. (2010b). *Presenting your findings: A practical guide for creating tables* (6th ed.). Washington, DC: American Psychological Association.

Traditional Textbooks on Elementary Statistics

This category contains texts that are often used to teach introductory statistics courses in the social and behavioral sciences. Most books in this group cover basic concepts through factorial analysis of variance (although some go on to more advanced topics).

- Aron, A., Aron, E. N., & Coups, E. J. (2006). *Statistics for psychology.* (4th ed). Upper Saddle River, NJ: Pearson Prentice Hall.

- Gravetter, F. J., & Wallnau, L. B. (2000). *Statistics for the behavioral sciences* (5th ed.). Belmont, CA: Wadsworth/Thompson Learning.

- Hays, W. L. (1988). *Statistics* (4th ed.). New York: Holt, Rinehart and Winston, Inc.

- Heiman, G. W. (2006). *Basic statistics for the behavioral sciences* (5th ed.). New York, NY: Houghton Mifflin Company.

- Howell, D. C. (2002). *Statistical methods for psychology* (5th ed.). Pacific Grove, CA: Duxbury.

Traditional Textbooks on Advanced Statistics

Most of these books are either textbooks for graduate-level courses on advanced statistics, references for researchers, or both. The very excellent Warner (2008) book is published by SAGE, and it is worth mentioning that SAGE is a publishing powerhouse for texts dealing with research, data analysis, and statistics. Among other things, it publishes the *Sage University Paper Series on Quantitative Application in the Social Sciences*, which many will recognize as the "little green books" dealing with just about every statistical procedure ever devised. You can learn more at the publisher's website: sagepub.com.

- Keith, T. Z. (2006). *Multiple regression and beyond.* New York: Pearson Allyn and Bacon.

- Stevens, J. (1996). *Applied multivariate statistics for the social sciences* (3rd ed.). Mahwah, NJ: Lawrence Erlbaum Associates.

- Tabachnick, B. G., & Fidell, L. S. (2007). *Using multivariate statistics* (5th ed.). Boston: Pearson, Allyn and Bacon.

- Warner, R. M. (2008). *Applied statistics: From bivariate through multivariate techniques.* Los Angeles, CA: SAGE Publications.

Books about the SAS® Application

From the shameless-plug department: Yes, I am author or co-author of several books that show how to use the SAS application to perform data analyses, and most of those books are listed here. Technically, the Lehman et al. (2005) text actually deals with an application called JMP®, but JMP is also published by the SAS Institute, so it is pretty much at home in this list.

- Field, A., & Miles, J. (2010). *Discovering statistics using SAS (and sex and drugs and rock 'n' roll).* Thousand Oaks, CA: SAGE Publications Inc.

- Hatcher, L. (1994). *A step-by-step approach to using the SAS® system for factor analysis and structural equation modeling.* Cary, NC: SAS Institute, Inc.

- Hatcher, L. (2003). *Step-by-step basic statistics using SAS®: Student guide.* Cary, NC: SAS Institute, Inc.

- Hatcher, L. (2003). *Step-by-step basic statistics using SAS®: Exercises.* Cary, NC: SAS Institute, Inc.

- Lehman, A., O'Rourke, N., Hatcher, L., & Stepanski, E. J.(2005). *JMP® for basic univariate and multivariate statistics: A step-by-step guide.* Cary, NC: SAS Institute Inc.

- O'Rourke, N., Hatcher, L., & Stepanski, E. J. (2005). *A step-by-step approach to using SAS® for univariate and multivariate statistics* (2nd ed.). Cary, NC: SAS Institute, Inc.

Books About the SPSS® Application

There are a lot of really good books that show how to use SPSS to perform data analysis, and a few of them are listed here. For those of you who are wondering about the differences between the two Norušis books: (a) there is actually a good deal of overlap between the two in terms of topics covered, (b) the *Guide to Data Analysis* is somewhat more elementary in nature (e.g., it covers basic statistics and ANOVA, but does not cover many advanced statistics), and (c) the *Statistical Procedures Companion* is somewhat more advanced, and covers more multivariate procedures.

- Field, A. (2009a). *Discovering statistics using SPSS (and sex and drugs and rock 'n' roll)*. (3rd ed.). Thousand Oaks, CA: SAGE Publications Inc.

- Green, S. B., & Salkind, N.J. (2010). *Using SPSS for Windows and Macintosh: Analyzing and understanding data* (6th ed.). Upper Saddle River, NJ: Prentice Hall.

- Norušis, M. J. (2009a). *SPSS 17.0 guide to data analysis.* Upper Saddle River, NJ: Prentice Hall.

- Norušis, M. J. (2009b). *SPSS 17.0 statistical procedures companion.* Upper Saddle River, NJ: Prentice Hall.

- Pallant, J. (2005). *SPSS survival manual* (2nd ed.). New York, NY: Open University Press.

Resources for R: The Open-Source Statistics Language

Do you like *open-source*? Do you like *free*? Do you like cutting-edge software that performs elementary procedures, advanced multivariate procedures, and just about everything in-between? Then you will like *R:* an open-source statistics programming language maintained by the *R* Core Development Team. Books, software, and documentation created by the community of *R* users are freely available at the first web site listed below.

Users across the world have developed over 1,000 *R* "packages" that perform a wide variety of statistical analyses, and these packages are constantly critiqued and improved by the *R* community. They are available at the Comprehensive *R* Archival Network (CRAN) at the second link provided below. Kelley (2007) offers *Methods for the Behavioral, Educational, and Social Sciences* (MBESS), a package of *R* software and documentation covering statistical procedures often used by researchers in these disciplines. The book chapter by Kelley, Lai, and Wu (2008) discusses the philosophy of *R* and provides an overview of how *R* can be used to read data, create graphs, perform power analyses and perform several widely used statistical procedures.

- http://r-project.org
- http://cran.r-project.org/
- Kelley, K. (2007). MBESS: Version 0.0.9. [Computer software and manual]. Retrieved from http://www.r-project.org/
- Kelley, K., Lai, K., & Wu, P. (2008). Using R for data analysis: A best practice for research. In J.W. Osborne, (Ed.) *Best practices in quantitative methods* (pp.535-572). Thousand Oaks, CA: SAGE Publications, Inc.

Publications About Meta-Analysis

By now you may have noticed that I am a major fan of meta-analysis. This section lists some important books on the topic, both recent (e.g., Borenstein et al., 2009), and not-so-recent (e.g., Hedges & Olkin, 1985). Cumming (2011) covers meta-analysis along with other statistical procedures (such as confidence intervals), which are increasingly being required by journals. When you are ready to write up your own meta-analysis, be sure to use the Rosenthal (1995) article for guidance.

- Borenstein, M., Hedges, L. V., Higgins, J. P. T., & Rothstein, H. R. (2009). *Introduction to meta-analysis.* West Sussex, UK: John Wiley & Sons.
- Cooper, H., & Hedges, L. V. (Eds.). (1994). *The handbook of research synthesis.* New York: Russell Sage Foundation.
- Cumming, G. (2011). *Understanding the new statistics: Effect sizes, confidence intervals, and meta-analysis.* New York: Routledge Taylor & Francis Group.
- Hedges, L. V., & Olkin, I. (1985). *Statistical methods for meta-analysis.* New York: Academic Press.
- Hunter, J. E., & Schmidt, F. L. (2004). *Methods of meta-analysis: Correcting error and bias in research findings* (2nd ed.). Thousand Oaks, CA: SAGE Publications.
- Rosenthal, R. (1991). *Meta-analytic procedures for social research.* (2nd ed.). Newbury Park, CA: SAGE.
- Rosenthal, R. (1995). Writing meta-analytic reviews. *Psychological Bulletin, 118,* 183-192.

Books About Controversies and Best Practices in Statistics

Even after slugging their way through a graduate-level course on advanced statistics, students often encounter problems and questions that had not been covered in class. "What should I do about the outliers in my sample?" "Is it okay to dichotomize this continuous variable?" "Is it too late to apply to that doctoral program in Literature that I was thinking about?"

The good news is that you are not the first to ask these questions. A sizable cottage industry of after-market books has emerged to satisfy the needs of curious (or floundering) graduate students, college professors, and researchers. I regularly consulted these sources when writing my own books, and often found myself pulled into chapters on interesting topics that were totally unrelated to my original question.

- Lance, C. E., & Vandenberg, R. J. (Eds.) (2009). *Statistical and methodological myths and urban legends: Doctrine, verity, and fable in the organizational and social sciences.* New York: Routledge Taylor & Francis Group.

- Millsap, R. E., & Maydeu-Olivares, A. (Eds.) (2009). *The SAGE handbook of quantitative methods in psychology.* Thousand Oaks, CA: Sage Publications Inc.

- Newton, R. R., & Rudestam, K. E. (1999). *Your statistical consultant: Answers to your data analysis questions.* Thousand Oaks, CA: SAGE Publications.

- Osborne, J. W. (Ed.). (2008). *Best practices in quantitative methods.* Thousand Oaks, CA: SAGE Publications.

There's This Thing Called *the Internet*

Long ago I predicted that the Internet was just a passing fad that everyone would forget about in a couple of years. Although I stand by my original prediction, I guess there is no harm in making a few recommendations about research-related resources that are available there. Until the fad passes, anyway.

Online Databases for Research Journals

Pick any research topic, and there are dozens (perhaps thousands) of empirical research articles addressing it. But what good is all that research if you can't find the relevant articles? That's where *online databases* come in.

Think of an online database as a turbo-charged search engine for finding scholarly articles. Different databases are used by different disciplines, and the same database may change its name over time. This means that you must check with the librarians at your library to identify the best databases for your specific needs. At the time of this writing, I use a database called *CSA Illumina* a good deal, although my university also makes available *ProQuest, PubMed,* and *ScienceDirect,* among others.

As an example: If I wanted to find articles dealing with the effects of cell-phone usage on driving performance, I would type the relevant key terms into the appropriate fields of the database. I press "enter" and *presto*—a zillion hits! Many databases allow me to narrow the search to just "journal articles" or "peer-reviewed articles," and this is just about always a good idea (I don't want to read some article from *Cosmo,* for Pete's sake). In addition to searching by key terms, I can also search by author, title, and many other alternatives.

When students need to get up to speed on some research topic as quickly as possible, I tell them to try to find a meta-analysis on the subject. The best way to do this involves using the "AND" and "OR" operators during the search. For example, the student might type in the key terms for the topic itself (e.g., "cell phone"), AND the key terms "meta-analysis" OR "meta-analyses" OR "meta-analytic." If anyone on the Planet Earth has done a meta-analysis on the subject, it usually appears on the first page of hits.

The Most Important Journal of All Time

No, I do not refer to the *Annalen der Physik*, in which Einstein published his *Annus Mirabilis* papers in 1905. I refer instead to *Psychological Bulletin*, the journal that publishes the most important type of article (meta-analyses) pertaining to the most important of all disciplines (*my* discipline, of course: psychology).

Don't believe me? Consider some recent offerings:

- What really happened to the survivors of the Holocaust? Compared to counterparts who had not gone through the camps, were the Holocaust survivors more likely to display post-traumatic stress disorder? Impairments in cognitive functioning? Problems of physical health? Did they adjust better if they later settled in Israel? For the answers, see the meta-analysis of 71 samples (and over 12,000 participants) reported by Barel, Van IJzendoorn, Sagi-Schwartz, and Bakermans-Kranenburg (2010).

- Is it true that boys are better at math than girls? Do boys display greater variability in math skills compared to girls? All is revealed in the meta-analysis of 242 studies (including 1,286,350 participants!) reported by Lindberg, Hyde, Petersen, and Linn (2010).

- Forget about math skills—what about sex differences in *vocational interests?* Is it true that males generally prefer working with things, whereas females prefer working with people? Are males really more interested in engineering? Are females really more interested in art? The answers may be found in the Su, Rounds, and Armstrong (2009) meta-analysis based on 503,188 respondents.

- Does anyone truly know the *real* me? Can the people in my life evaluate my personality with any kind of validity? Can they evaluate my personality more accurately than I can evaluate myself? Connelly and Ones (2010) answer these and related questions in an article that reports not one but *three* meta-analyses of 263 independent samples.

- Can we really predict whether someone is going to engage in violence in the future? How accurate are the tools that are often used to make these predictions? Are they sufficiently accurate so that their use can be justified when making decisions about preventive detention within the criminal justice system? Don't ask me—see the meta-analysis of 28 studies reported by Yang, Wong, and Coid (2010).

- When mothers work at jobs outside of the home, are their children likely to display lower levels of academic achievement? Are they more likely to display behavior problems? Do these effects depend on how young the children are when the mothers become employed? All is revealed in a meta-analysis of 69 studies (with over 1,000 effect sizes!) reported by Lucas-Thompson, Goldberg, and Prause (2010).

And these articles were found by skimming just *two years* (2009-2010) of issues in *Psychological Bulletin*. I could have spent all day reading abstracts if my winter-semester classes were not starting tomorrow.

These articles carry more weight than the typical article describing a single empirical investigation. Remember that they are *meta-analyses:* They are based on dozens—even hundreds—of individual investigations. This means that they often represent thousands of participants. For example: Did you notice that the Lindberg et al. article on math ability was based on 1.2 *million* subjects? If you want to know what the bulk of the evidence suggests on some important topic in psychology, then *Psychological Bulletin* is a great place to start.

Not into psychology? No worries—researchers are doing meta-analyses in your discipline too. Just find a good database for your area and follow my instructions for finding meta-analyses, as described above.

The Ultimate Tool of Science: Amazon.com

Having published some books that show how to use the SAS application to perform data analysis, I occasionally get an email from some desperate graduate student who needs help with some obscure statistical procedure. The vast majority of the time, I have no idea how to answer their questions. And they are often surprised when I advise them to take their question to the online seller of books (and just about everything else): *Amazon.com*.

Go to Amazon.com and type in the search term "structural equation modeling." The result will be a long list of books on the topic, sorted more or less according to relevance, popularity, and availability. And the best part is that they have often been rated (and commented upon) by customers who have actually purchased and used the book. These evaluations usually provide a fairly good preview of what to expect, as long as the sample includes 10 or so reviews.

With many books listed on Amazon.com, you can also "look inside" to check out the table of contents, the index, and even a sample chapter or two. As a reader, I love this feature, and I often use it to decide whether the book will tell me what I want to know.

There's Also This Thing Called *the Library*

I hope that you are the kind of person who buys books, whether they are the download-to-your-e-reader type, or the hardcopy-on-the-bookshelf-to-impress-your-friends type. But, hey, you can't buy *all* the books that you will need in a lifetime. That's why they invented libraries.

I have always been impressed by the staff at my library. In particular, I have been impressed at their success at locating hard-to-find books on obscure topics and putting that book in my hands in a very short period of time. Most libraries are members of regional or national consortiums, so if that book on *Multiplicative and Curvilinear Relationships in Logistic Regression* is not available at my local library (and believe me, it's not) they can usually get it via interlibrary loan within a week or so.

And it's *free*.

You May Now Begin

There are oceans of research out there—rolling, boiling seas of information that will help you better manage your government, your career, your family, and your day-to-day life. If you are a tax payer, you have already paid for it. And, having slugged your way through *structural equation modeling*, you've paid your academic dues as well.

So why are you still reading? It's time put down this book and *dive in*.

Appendix A

Questionnaire on Eating and Exercising

Introduction

This appendix contains the *Questionnaire on Eating and Exercising*. Many of the statistical procedures described in this book were illustrated using data obtained from this questionnaire. Although the researcher described in this book (Dr. O'Day) is fictitious, the data obtained from the *Questionnaire on Eating and Exercising* is an actual (non-fictitious) dataset.

This questionnaire was administered to about 260 research participants during winter semester of 2010 (January through May). The participants were students enrolled in a variety of psychology courses at a regional state university in the Midwest. The university was Saginaw Valley State University, and that is why some of the questionnaire items refer to "SVSU." Most students completed the survey in exchange for course credit. The students did not sign their completed questionnaires, and the procedure used with the questionnaire was approved by the university's IRB committee.

Questionnaire on Eating and Exercising

Larry Hatcher, Psychology Dept., SVSU
December 30, 2009 Revision

Overview. This questionnaire deals with your exercising habits and your eating habits. Specifically, it assesses your beliefs, attitudes, and behaviors as they pertain to exercising, the consumption of healthy foods, and related issues.

Instructions: In the questionnaire, we ask questions which make use of rating items with seven spaces in which you may mark your responses. You are to **circle a number** in the place that best describes your opinion. For example, imagine you were asked to rate the statement "In will rain in this area today." If you think it is **extremely likely** to rain in this area today, then you would place your mark as follow:

```
Unlikely    1       2        3         4        5       6      (7)    Likely
         extremely quite  slightly  neither  slightly quite extremely
```

If you think it is **quite unlikely** to rain in this area today, then you would place your mark as follows:

```
Unlikely    1      (2)       3         4        5       6       7     Likely
         extremely quite  slightly  neither  slightly quite extremely
```

In making your ratings, please remember the following points:
1. Feel free to mark any number from 1 through 7 to indicate your opinion.
2. Be sure to answer all items. Please do not leave any blank.
3. Only circle one number per item—please do NOT circle two response numbers because you have difficulty choosing between them.
4. Please read all instructions and items carefully.
5. You may notice that the "question numbers" for the questions are not in perfect sequence—this is not a problem.
6. Remember that questions appear on the front and back of each page.
7. If you have any questions, please ask the researcher for assistance.

Definition of "healthy diet." The following section asks questions about eating a healthy diet. For purposes of this questionnaire, a "healthy diet" is defined as a diet that is high is vegetables, fruits, and whole grains, and is low in saturated fats.

Subjective Norms

1. Most of my friends would think that I should eat a healthy diet.

```
Unlikely    1       2        3         4        5       6       7     Likely
         extremely quite  slightly  neither  slightly quite extremely
```

2. Most of my relatives would think that I should follow a healthy diet.

```
Unlikely    1       2        3         4        5       6       7     Likely
         extremely quite  slightly  neither  slightly quite extremely
```

3. Most of the people that I respect would think that I should eat a healthy diet.

Unlikely 1 2 3 4 5 6 7 Likely
          extremely quite slightly neither slightly quite extremely

Attitudes

Instructions: This section will measure your personal attitudes toward being on a healthy diet, and it will use a different type of rating item to do this. Below, we have provided several different "bipolar adjective items." You will use these items to rate how you feel about being on a healthy diet. For example, look at the first item. If you feel that consistently following a healthy diet would be "Extremely Rewarding," circle the "7." If you feel that it would be "Extremely Punishing," circle the "1." Feel free to circle any number from 1 through 7 for each item. (Notice that there is a different set of adjectives for each item).

This is how I feel about consistently following a healthy diet:

5. Punishing   1 2 3 4 5 6 7 Rewarding
                extremely quite slightly neither slightly quite extremely

6. Useless   1 2 3 4 5 6 7 Useful
                extremely quite slightly neither slightly quite extremely

7. Bad   1 2 3 4 5 6 7 Good
                extremely quite slightly neither slightly quite extremely

8. Foolish   1 2 3 4 5 6 7 Wise
                extremely quite slightly neither slightly quite extremely

9. Unattractive 1 2 3 4 5 6 7 Attractive
                extremely quite slightly neither slightly quite extremely

Perceived Behavioral Control

Instructions: In this section, you will rate how easy it would be for you to follow a healthy diet. Notice that different end-points are used for some of the items in this section.

11. For me to consistently eat a healthy diet would be:

Difficult 1 2 3 4 5 6 7 Easy
           extremely quite slightly neither slightly quite extremely

12. How much control do you have over consistently sticking to a healthy diet?

Little control 1 2 3 4 5 6 7 Complete control
                extremely quite slightly neither slightly quite extremely

13. If I wanted to, I could easily follow a healthy diet.

Unlikely 1 2 3 4 5 6 7 Likely
          extremely quite slightly neither slightly quite extremely

14. Consistently sticking to a healthy diet would be an easy task for me.

Unlikely	1	2	3	4	5	6	7	Likely
	extremely	quite	slightly	neither	slightly	quite	extremely	

Behavioral Intentions

Instructions: In this section, you will indicate whether you intend to follow a healthy diet. Notice that different end-points are used for the items in this section.

16. I intend to consistently eat a healthy diet.

Disagree	1	2	3	4	5	6	7	Agree
	extremely	quite	slightly	neither	slightly	quite	extremely	

17. I will do my best to regularly follow a healthy diet.

Disagree	1	2	3	4	5	6	7	Agree
	extremely	quite	slightly	neither	slightly	quite	extremely	

18. I intend to eat only low-fat foods.

Disagree	1	2	3	4	5	6	7	Agree
	extremely	quite	slightly	neither	slightly	quite	extremely	

19. I plan to eat lots of vegetables.

Disagree	1	2	3	4	5	6	7	Agree
	extremely	quite	slightly	neither	slightly	quite	extremely	

Current Healthy-Diet Behavior

Instructions: In this section, you will rate how frequently you currently eat certain foods. Notice that different end-points are used with the items in this section.

21. I currently eat leafy green vegetables:

Infrequently	1	2	3	4	5	6	7	Frequently
	extremely	quite	slightly	neither	slightly	quite	extremely	

22. I eat colorful vegetables such as carrots:

Infrequently	1	2	3	4	5	6	7	Frequently
	extremely	quite	slightly	neither	slightly	quite	extremely	

23. I eat whole-grain foods (such as whole-grain bread):

Infrequently	1	2	3	4	5	6	7	Frequently
	extremely	quite	slightly	neither	slightly	quite	extremely	

24. I **avoid** fried foods:

Infrequently	1	2	3	4	5	6	7	Frequently
	extremely	quite	slightly	neither	slightly	quite	extremely	

25. I **avoid** "junk food" that contains a lot of fat:

    Infrequently   1   2   3   4   5   6   7   Frequently
                   extremely   quite   slightly   neither   slightly   quite   extremely

Health Motivation

Instructions. In this section, you will rate how important it is to you to maintain good health. Notice that different end-points are used with the items in this section.

27. It is important to me to be healthy.

    Disagree   1   2   3   4   5   6   7   Agree
               extremely   quite   slightly   neither   slightly   quite   extremely

28. I am willing to sacrifice in order to stay healthy.

    Disagree   1   2   3   4   5   6   7   Agree
               extremely   quite   slightly   neither   slightly   quite   extremely

29. I am willing to work hard in order to avoid medical problems.

    Disagree   1   2   3   4   5   6   7   Agree
               extremely   quite   slightly   neither   slightly   quite   extremely

Desire to be Thin

Instructions. In this section, you will rate how important it is to you to have a thin body (e.g., a relatively low body weight and a thin physical appearance).

31. It is important to me to be thin.

    Disagree   1   2   3   4   5   6   7   Agree
               extremely   quite   slightly   neither   slightly   quite   extremely

32. I very much want to have a fairly low body weight.

    Disagree   1   2   3   4   5   6   7   Agree
               extremely   quite   slightly   neither   slightly   quite   extremely

33. It is important to me to have a lean physical appearance.

    Disagree   1   2   3   4   5   6   7   Agree
               extremely   quite   slightly   neither   slightly   quite   extremely

Appetite for High-Fat Foods

Instructions. In this section, you will rate how strong your appetite is for fatty foods (e.g., fried foods, meat with fat, potato chips, ice cream, and so forth).

35. I love to eat fatty foods.

Disagree   1   2   3   4   5   6   7   **Agree**
extremely   quite   slightly   neither   slightly   quite   extremely

36. I have a big appetite for fried foods.

Disagree   1   2   3   4   5   6   7   **Agree**
extremely   quite   slightly   neither   slightly   quite   extremely

37. I like high-fat foods a lot more than low-fat foods.

Disagree   1   2   3   4   5   6   7   **Agree**
extremely   quite   slightly   neither   slightly   quite   extremely

Dislike of Vegetables

Instructions. In this section, you will rate how much you **dislike** eating vegetables (e.g., carrots, broccoli, lettuce, cauliflower, and so forth).

39. I **hate** to eat vegetables.

Disagree   1   2   3   4   5   6   7   **Agree**
extremely   quite   slightly   neither   slightly   quite   extremely

40. I **don't like** the taste of vegetables.

Disagree   1   2   3   4   5   6   7   **Agree**
extremely   quite   slightly   neither   slightly   quite   extremely

41. When I see a plate of vegetables, I think "Yuck!"

Disagree   1   2   3   4   5   6   7   **Agree**
extremely   quite   slightly   neither   slightly   quite   extremely

Lack of Time for Food Preparation

Instructions. In this section, you will indicate the extent to which you are too busy to prepare and cook food (e.g., too busy for washing vegetables, chopping food, cooking food).

43. I am too busy to prepare and cook food the way they did years ago.

Disagree   1   2   3   4   5   6   7   **Agree**
extremely   quite   slightly   neither   slightly   quite   extremely

44. I am too busy to cook meals that take a lot of time.

Disagree   1   2   3   4   5   6   7   **Agree**
extremely   quite   slightly   neither   slightly   quite   extremely

45. With my busy schedule, there is very little time for cooking.

    Disagree   1   2   3   4   5   6   7   Agree
            extremely quite slightly neither slightly quite extremely

Dislike of Aerobic Exercise

Instructions. In this section, you will rate how much you **dislike** aerobic exercise (e.g., exercise that raises your heart rate and breathing rate).

47. I **hate** to engage in aerobic exercise.

    Disagree   1   2   3   4   5   6   7   Agree
            extremely quite slightly neither slightly quite extremely

48. I find aerobic exercise to be unpleasant.

    Disagree   1   2   3   4   5   6   7   Agree
            extremely quite slightly neither slightly quite extremely

49. I find aerobic exercise to be punishing.

    Disagree   1   2   3   4   5   6   7   Agree
            extremely quite slightly neither slightly quite extremely

Lack of Time for Aerobic Exercise

Instructions. In this section, you will indicate the extent to which you are too busy to engage in frequent aerobic exercise (here, "frequent aerobic exercise" means at least three times per week participating in activities such as brisk walking, jogging, swimming, using exercise videotapes, etc.).

51. I am too busy to engage in frequent aerobic exercise.

    Disagree   1   2   3   4   5   6   7   Agree
            extremely quite slightly neither slightly quite extremely

52. I am under too much time pressure to engage in frequent aerobic exercise.

    Disagree   1   2   3   4   5   6   7   Agree
            extremely quite slightly neither slightly quite extremely

53. With my busy schedule, there is very little time for frequent aerobic exercise.

    Disagree   1   2   3   4   5   6   7   Agree
            extremely quite slightly neither slightly quite extremely

Current Physical Health

Instructions: In this section, you will rate how good your health is in general.

55. In general, my health is good.

| **Disagree** | 1 | 2 | 3 | 4 | 5 | 6 | 7 | **Agree** |
| | extremely | quite | slightly | neither | slightly | quite | extremely | |

56. Physically, I feel good on most days.

| **Disagree** | 1 | 2 | 3 | 4 | 5 | 6 | 7 | **Agree** |
| | extremely | quite | slightly | neither | slightly | quite | extremely | |

57. I am healthier than most people my age.

| **Disagree** | 1 | 2 | 3 | 4 | 5 | 6 | 7 | **Agree** |
| | extremely | quite | slightly | neither | slightly | quite | extremely | |

58. Compared to most people my age, I am relatively **free** of medical problems.

| **Disagree** | 1 | 2 | 3 | 4 | 5 | 6 | 7 | **Agree** |
| | extremely | quite | slightly | neither | slightly | quite | extremely | |

60. Your **sex** (check one):

   _____ Male (1)

   _____ Female (2)

62-63. Please indicate your **age in years**: _____

65. Your **student classification** (check one):

   _____ Freshman (1)

   _____ Sophomore (2)

   _____ Junior (3)

   _____ Senior (4)

   _____ Graduate student (5)

   _____ Other (6)

67. Do you **live on campus?** (check one):

   _____ Yes. (1)

   _____ No. (2)

69. Which of the following best describes you? (check one):

   _____ I **eat meat**. (1)

   _____ I am a **vegetarian** (I do not consume meat, but I do eat dairy). (2)

   _____ I am a **vegan** (I do not consume meat, dairy, or any other animal products). (3)

71. **How many children** do you have? (Write zero if you do not have any children).

   Number of children:_____

73-74. **How many hours per week are you employed at a job** outside of the home? (Write zero if you are not employed).

   Total job hours per week:_____

76-77. **How many times each week do you engage in a session of aerobic exercise** that lasts at least 15 minutes? (Write zero if you never engage in aerobic exercise).

   Number of exercise sessions per week:_____

79-81. How many **minutes per week (total) do you engage in aerobic exercise**? (Write zero if you never engage in aerobic exercise).

   Total exercise minutes per week:_____

83. Which of the following best describes your **current level of physical fitness?** (check one):

   _____ My current level of physical fitness is **poor**. (1)

   _____ My current level of physical fitness is **pretty good**. (2)

   _____ My current level of physical fitness is **excellent**. (3)

85. Are you currently a **member of an athletic sports team that is officially representing SVSU?** (such as the SVSU Women's Track Team or the SVSU Football Team?) (check one):

   _____ Yes. (1)

   _____ No. (2)

87. Are you currently a **member of any other athletic sports team** NOT described in the previous question? (e.g., an intramural team or a city league team or anything like that?) (check one):

   _____ Yes. (1)

   _____ No. (2)

End of Questionnaire

Appendix B

Basics of APA Format (and How this Book Sometimes Deviates)

Introduction

One purpose of this book is to show how the results of investigations may be summarized according to the guidelines in the *Publication Manual of the American Psychological Association* (2010). For the sake of brevity, this book will refer to these guidelines simply as *APA format*.

To show how the results from advanced statistical analyses can be summarized, this book provides a large number of excerpts from the *Results* sections of fictitious articles. Although most of the conventions used in these excerpts are consistent with APA format, there are some exceptions. So that you will not be led too far astray, this appendix reviews the basics of APA format and then summarizes the ways that these excerpts sometimes deviate from the official format.

The Basics of APA Format

Since this is a book on advanced statistics (as opposed to elementary statistics), I began with the assumption that most readers are at least vaguely familiar with APA format. For those who have gotten rusty, the current section provides a quick review.

Format Used in Typing the Manuscript

We begin with the basics: the font to be used, double-spacing versus single spacing, and related issues.

- Not all journals require that manuscripts comply with APA format. To be sure, check the *instructions for submissions* page in a recent issue or at the journal's website.
- In the *page layout* menu of your word processing application, use the default paper size of 8.5 ×11-inch (i.e., letter size).
- The journal will most likely request that the manuscript be submitted electronically, as an email attachment. If you need to print it on paper for any reason, use a good-quality printer (preferably a laser printer).
- The preferred font (i.e., typeface) is *Times New Roman*, and the preferred font size is 12-point. Do not use a compressed or narrow font, and do not use your word processing application to compress the font.
- When I was an undergrad and papers were typed using a typewriter, we were instructed to set the line spacing on double-space and leave it there. Period. For today's writers there is one exception: It is now acceptable to use single spacing, 1.5 spacing, or double spacing within the body of a table or a figure. But everything else should be double-spaced.
- Each page of the manuscript should have margins of at least 1 inch at the top, bottom, left, and right. This means that no line of text should be more than 6 ½ inches in length.
- Text should be left-justified. This means that the left margin should line up smooth and straight and the right margin should be ragged (i.e., uneven). In other words, do not justify the lines so that both margins are smooth and straight.
- Begin each new paragraph (except the abstract) with an indentation of five to seven spaces.

Order in Which the Manuscript Parts Should Be Arranged

The various parts of the manuscript should be arranged in the order presented below. The page numbers should run consecutively (starting with the title page) through all sections of the manuscript.

- Title page
- Abstract
- Body of the manuscript (this is sometimes called the *text of the manuscript* or *text of the article*; most of the manuscript will consist of this section)
- Reference list
- Tables (each table should begin on a new page; very long tables may extend over more than one page)
- Figures (each figure should begin on a new page; the caption for a given figure should appear on the same page as the figure itself)
- Appendices (each appendix should begin on a new page)

Sections That Constitute the Body of the Manuscript

After the title page and abstract, comes the body of the manuscript. The specific sections that should be included in this part of the manuscript will be determined by the nature of investigation. Most articles describing empirical investigations will follow a format something like this:

- *Introduction.* In this section, you will introduce the general topic and the questions being investigated. Explain why this issue is important and summarize theory and previous research that is most relevant to the current investigation. State the hypotheses to be investigated and briefly explain why the current research design is appropriate for testing these hypotheses.

- *Method.* In the *Method* section, describe how the current investigation was conducted providing adequate detail so that other researchers can replicate it. This section is often divided into subsections such as *Participants* and *Measures*. In the *Participants* subsection, explain how the subjects were selected and describe their basic characteristics (e.g., age and gender). In the *Measures* subsection, provide operational definitions for the study's variables. In the *Research Design* subsection, indicate whether the study was a true experiment, a correlational study, or some other type of investigation. If appropriate, provide details as to how the independent variables were manipulated.

- *Results.* In this section, describe the data analysis procedures used to investigate the study's hypotheses, along with the results of those analyses. Traditionally, researchers have emphasized null-hypothesis significance tests, although the APA's *Publication Manual* (2010, pp. 33-34) now strongly encourages the reporting of confidence intervals and indices of effect size.

- *Discussion.* In this section, interpret the meaning and importance of the current investigation's findings. Draw conclusions as to the extent to which the current findings support (or fail to support) the hypotheses stated earlier in the manuscript. Review the limitations of the study, relate the current findings to previous research, and discuss its implications for future research.

Levels of Headings in the Body of the Manuscript

APA format allows for up to five levels of headings in a manuscript, although most articles require fewer than five levels. This section illustrates the format that would be used in a manuscript that requires just three levels of headings:

Level 1 is Centered, is in Boldface, and Includes Uppercase and Lowercase Letters

Level 2 is Flush Left, is in Boldface, and Has Uppercase and Lower Letters

Level 3 is paragraph-indented, includes lowercase letters, and ends with a period. The text of the paragraph itself appears next.

As an illustration, below is an example of the headings that might be used in the *Method* section, the *Results* section, and the *Discussion* section of an APA-format article that reports an empirical investigation. The actual headings would vary, determined by the specifics of the study, but the following headings are fairly representative of those often used:

Method

Participant Characteristics

The text of the paragraph appears here. The text of the paragraph appears here. The text of the paragraph appears here.

Questionnaire Scales and Other Measures

Summated rating scales on the questionnaire. The text of the paragraph appears here. The text of the paragraph appears here. The text of the paragraph appears here.

Other variables assessed. The text of the paragraph appears here. The text of the paragraph appears here. The text of the paragraph appears here.

Results

Correlations Between Latent Variables

The text of the paragraph appears here. The text of the paragraph appears here. The text of the paragraph appears here.

Results from the Path Analysis

Results from the sample of male participants. The text of the paragraph appears here. The text of the paragraph appears here. The text of the paragraph appears here.

Results from the sample of female participants. The text of the paragraph appears here. The text of the paragraph appears here. The text of the paragraph appears here.

Discussion

Implications for the Theory of Planned Behavior

The text of the paragraph appears here. The text of the paragraph appears here. The text of the paragraph appears here.

Limitations of the Current Investigation

The text of the paragraph appears here. The text of the paragraph appears here. The text of the paragraph appears here.

Recommendations for Future Research

The text of the paragraph appears here. The text of the paragraph appears here.

The text of the paragraph appears here.

Fonts (Typefaces) Used in the Manuscript

The *Publication Manual of the American Psychological Association* (2010) recommends that researchers prepare their manuscript using a serif font, preferably Times New Roman (a recommendation that this book did not follow). This section discusses the differences between serif fonts versus sans serif fonts and summarizes APA recommendations regarding their use.

Serif Fonts Versus Sans Serif Fonts

A *serif font* has very short lines extending at perpendicular angles from the main strokes that constitute a letter. For example, below is the upper-case letter "T" from a serif font (Times New Roman):

T

Notice the very short lines that appear at the left and right side of the horizontal upper stroke of the letter? Those structures are called *serifs*. APA format recommends that researchers use serif fonts for the vast majority of their manuscripts, because serif fonts are generally easier to read in the text of a document. In this context, "vast majority" means everything except the text that appears in figures.

This book did not use a serif font in the fictitious article excerpts that it presented. Instead, it used a *sans serif font:* a font that does not include the short lines at perpendicular angles to the main strokes of letters. For example, below is the upper case letter T from a sans serif font called Phoenica Condensed Standard Office:

T

Notice that the above letter does not have the short lines at the ends of the main strokes. That is how we know it is a sans serif font.

APA-Recommended Fonts Versus this Book's Fonts

As was stated earlier, the *Publication Manual* of the APA recommends that researchers use Times New Roman for typing the vast majority of a manuscript, including tables. The current book did not comply with this guideline. Instead, the current book used the following fonts:

- Phoenica Condensed Standard Office was used for excerpts taken from the main body (text) of the fictitious articles.

- Phoenica Standard Mono was used for tables from the fictitious articles.

The two preceding fonts were used (instead of the font recommended by APA format) so that it would be easier for readers to distinguish between this book's regular text versus the text of the fictitious article excerpts. This book's regular text was set in a font called *Utopia Standard*, which is a serif font. I was concerned that if I set the fictitious article excerpts in *Times New Roman* (which is also a serif font), readers would have a difficult time distinguishing between the two.

CONDENSED FONTS. As long as we are talking about taboos, APA format recommends against the use of narrow or condensed fonts in the text or tables of manuscripts. This would be an additional reason to not use the Phoenica fonts that I used in a real manuscript, as they were both condensed fonts.

SANS SERIF FONTS IN FIGURES. The *Publication Manual* of the APA does not have a problem with sans serif fonts when they are used for the text that appears within the bodies of figures (such as the figures used to illustrate the path models in *Chapter 16*). This book generally used a sans serif font called Myriad Pro Condensed within the body of figures.

SUMMARY: APA RECOMMENDATIONS FOR FONTS. In summary: To be compliant with APA recommendations:

- Use a serif font (such as Times New Roman) for text and tables in your manuscript.

- Use a sans serif font (such as Myriad Pro or Helvetica) within the body of a figure.

Italicizing Symbols for Statistics

When I was a young man, APA format did not require that anything be placed in italic type. That's because, when I was a young man, we prepared our papers using an ancient instrument called a *typewriter*—a device so primitive it could not even connect to Wikipedia.

Today's students are both blessed and cursed. On the positive side, they can now prepare slick-looking research reports complete with tables, figures, and statistical symbols that did not even exist back during the Nixon administration.

That, too, is their curse. Should the symbol for the chi-square statistic be in italic type? Should the "*F*" in the "F_{max}" symbol be in italic? What about the "max" in the subscript?

No, this section will not answer all of your questions (that's what Wikipedia is for). But it will touch on some of the more important guidelines for the use of italic type according to the *Publication Manual of the American Psychological Association 6th ed.* (2010).

What, Exactly, is Italic Type?

In your research report, words and symbols will typically be in one of three possible typefaces:

- Standard typeface is text that has not been set in italics or boldface (or anything else unusual).
- *Italic type is text that is slanted and squiggly, like this.*
- **Boldface is dark and thick like this.**

The majority of your manuscript will be in standard typeface. In the typical paper, just about nothing is in bold. However, many statistical symbols are in italics, as we shall see.

Values to Place in Italics

First, you may recall that **Latin letters** are the letters that constitute the alphabet that most of us learned in elementary school (e.g., A, a, B, b, C, c). Statisticians have traditionally used Latin letters as symbols for statistics (such as *M* for the mean, *n* for the number of participants in a subgroup, and so forth). To increase the likelihood that these Latin letters will stand out (and will be recognized as statistical symbols by the reader), APA format requires that they be placed in italics when used for certain purposes. This section and the following section summarizes some of the dos and don'ts related to the use of italics.

DO ITALICIZE LATIN LETTERS USED AS STATISTICAL SYMBOLS OR ALGEBRAIC VARIABLES. When Latin letters are used as symbols for statistics or algebraic variables in research articles, they are usually placed in italics. For example, the following are statistics that represent various characteristics of samples and are, therefore, placed in italics: *M*, *SD*, *Mdn*, *n*, and S^2. The following symbols do not necessarily represent the characteristics of samples (strictly speaking), but they are statistical symbols represented by Latin letters, so they are also italicized: *t*, *F*, *df*, *d*, *r*, *R*, *p*. Finally, the following are algebraic variables represented by Latin letters, so they are placed in italics as well: *X*, *Y*.

DO ITALICIZE LATIN-LETTER ABBREVIATIONS FOR STATISTICAL SYMBOLS. If an abbreviation is being used as a symbol for a statistic, it should be italicized. For example, "*MSE*" is an abbreviation for "mean square error." Since a mean square error is a statistic (i.e., a characteristic of a sample), the abbreviation "*MSE*" is placed in italics.

Similarly, "*SD*" is an abbreviation for the statistic "standard deviation," so it is placed in italics as well.

Values That Should Not Be Placed in Italics

In general, APA format recommends that symbols should be placed in italics only if they (a) are Latin letters and (b) are being used as symbols for statistics or algebraic variables. Just because a set of Latin letters are being used as an abbreviation does not necessarily mean that they should be italicized (I mean, what cement-head would italicize "ACLU?"). This section gives examples of some of the terms that should not be placed in italics.

DO NOT ITALICIZE LATIN-LETTER ABBREVIATIONS THAT ARE NOT STATISTICAL SYMBOLS. Although it is okay to italicize Latin-letter abbreviations for *statistics*, it is typically not okay to italicize Latin-letter abbreviations for statistical *procedures* (i.e., data-analysis procedures). For example, "EFA" is the abbreviation for "exploratory factor analysis." Since EFA is the abbreviation for a procedure but is not the abbreviation for a specific statistic, it is not placed in italics. In the same way, "ANOVA" is the abbreviation for the procedure "analysis of variance," and it is not the abbreviation for a specific statistic. Therefore, ANOVA is not placed in italics.

DO NOT ITALICIZE SUBSCRIPTS FOR TERMS THAT WOULD NOT OTHERWISE BE ITALICIZED. If you type a statistic such as F_{crit} (the symbol for the critical value of F), you know that you should put the "F" in italics because it is the symbol for a statistical test. However, you should not put the "crit" in italics because it is merely the abbreviation for the words "critical value," and "critical value" is not a statistic. Therefore, the correct form would be F_{crit}, not F_{crit}.

DO NOT ITALICIZE GREEK LETTERS, REGARDLESS OF WHAT THEY REPRESENT. Greek letters include characters such as η, Φ, β, χ, Λ, and λ. Throughout this book, you saw that certain Greek letters are used to represent population parameters, specific results from a structural equation modeling analysis, and other concepts. Regardless of the purpose for which they are used, Greek letters are typically *not* placed in italics. Within the body of an article, Greek letters already stand out because, well, they are *Greek letters* and they already look kind of strange. Therefore there is no need to place them in italics.

Preparing Tables

When researchers have a large quantity of information to present, they often organize it as a table. This section reviews the conventions recommended for an APA-format table and summarizes the ways that the tables presented in this book sometimes diverge from APA format.

An APA-Format Table

In *Chapter 5* of this book, you read about a fictitious study in which a researcher randomly assigned each of 150 participants to be one of two treatment conditions. Each of the 75 participants in the *high-caffeine condition* consumed a relatively large amount of caffeine each day, and each of the 75 participants in the *no-caffeine condition* consumed no caffeine at all. At the end of a period of time, the researchers obtained four separate measures of *subject irritability* (how emotionally upset the participants felt or acted). The results are presented in Table B.1.

Table B.1

Scores on Four Measures of Irritability as a Function of the Amount of Caffeine Consumed

Dependent variable	High caffeine[a]		No caffeine[b]		Difference between means		
	M_1	(SD_1)	M_2	(SD_2)	$M_1 - M_2$ [95% CI]	$t(148)$	r_{pb}
Self-rated irritability	575.00	(35.49)	544.73	(36.13)	30.27 [18.71, 41.82]	5.18***	.39
Other-rated irritability	556.46	(35.86)	546.27	(36.10)	10.19 [-1.42, 21.80]	1.73*	.14
Self-reported outbursts	6.24	(2.67)	5.03	(2.69)	1.21 [0.35, 2.08]	2.77**	.22
Other-reported outbursts	5.24	(2.67)	4.99	(2.65)	0.25 [-0.61, 1.11]	0.58	.05

Note. $N = 150$. CI = Confidence interval; r_{pb} = Point-biserial correlation between the independent variable and dependent variable.
[a]$n = 75$. [b]$n = 75$.
*$p < .05$. **$p < .01$. ***$p < .001$.

Table B.1 is generally compliant with the *Publication Manual* of the APA. Specifically:

- The table was prepared using the Times New Roman font.

- The table was single-spaced, although extra blank lines were inserted at strategic locations to improve its organization and clarity. APA format allows the table itself to be single-spaced or double-spaced (American Psychological Association, 2010, p. 141) For table titles, the line spacing may be set at single-space, 1.5 space, or double-space (Nicol & Pexman, 2010b, p. 9).

- APA format suggests that columns of numbers should be lined up by aligning the decimal points of the numbers in a given column, and this was attempted in the current table. Some word processing applications have tools which allow the writer to automatically line up columns of numbers according to the decimal points. I really must learn how to do this someday.

A Slightly Non-APA Format Table from this Book

For purposes of comparison, the following table (Table B.2) is more representative of the typical table that actually appears in this book. The information that it contains is the same as that presented in Table B.1, but there are a few differences with respect to format.

Table B.2

Scores on Four Measures of Irritability as a Function of the Amount of Caffeine Consumed

Dependent variable	High caffeine[a] M_1 (SD_1)	No caffeine[b] M_2 (SD_2)	Difference between means $M_1 - M_2$ [95% CI]	$t(148)$	r_{pb}
Self-rated irritability	575.00 (35.49)	544.73 (36.13)	30.27 [18.71, 41.82]	5.18***	.39
Other-rated irritability	556.46 (35.86)	546.27 (36.10)	10.19 [-1.42, 21.80]	1.73*	.14
Self-reported outbursts	6.24 (2.67)	5.03 (2.69)	1.21 [0.35, 2.08]	2.77**	.22
Other-reported outbursts	5.24 (2.67)	4.99 (2.65)	0.25 [-0.61, 1.11]	0.58	.05

Note. $N = 150$. CI = Confidence interval; r_{pb} = Point-biserial correlation between the independent variable and dependent variable.
[a]$n = 75$. [b]$n = 75$.
*$p < .05$. **$p < .01$. ***$p < .001$.

Here is a summary of the ways that Table B.2 diverges from strict APA format:

- It is typed using a sans serif font called Phoenica Standard Mono. APA format suggests that researchers should instead use a serif font such as Times New Roman.
- APA format discourages the use of condensed fonts, such as the one used here.
- The Table is numbered as "Table B.2," and APA format requires simple, consecutive numbers such as Table 1, Table 2, and so forth.

Notes at the Bottom of an APA-Format Table

A table in an APA-format article may include one or more notes at the bottom, and these notes are used to explain the contents of the table. There are three types of notes that may be used:

GENERAL NOTES. A *general note* provides an explanation pertaining to the table as a whole. It may also include an explanation for any symbols or abbreviations that appear. Here is the general note that appeared at the bottom of Table B.1:

> *Note.* $N = 150$. CI = Confidence interval; r_{pb} = Point-biserial correlation between the independent variable and dependent variable.

SPECIFIC NOTES. A *specific note* is identified by a superscript lowercase letter such as [a], [b], or [c]. A specific note explains some specific column, row, or other part of a table. Here are the specific notes that appeared at the bottom of Table B.1:

> [a]$n = 75$. [b]$n = 75$.

PROBABILITY NOTES. *Probability notes* are used to identify the significance levels that are used with null hypothesis significance tests. In most cases, one or more asterisk is used for probability notes. Here are the probability notes from Table B.1:

> *$p < .05$. **$p < .01$. ***$p < .001$.

ORDER OF THE NOTES. When more than one type of note appears at the bottom of a table, they should be presented in the following order:

- General notes come first.
- Specific notes come next.
- Probability notes come last.

Conclusion

And that's it. You now know everything there is to know about APA format.

Except for the 67,236 rules and guidelines that this appendix left out.

Seasoned researchers know that there are a lot of things to worry about when preparing a manuscript according to APA format. This appendix has barely touched the tip of the iceberg. If you plan to do much research in the social and behavioral sciences, there is really no substitute for buying a copy of the *Publication Manual of the American Psychological Association* (2010) and spending a lot of time between its covers.

Serious students and researchers should also consider purchasing two short books that focus on APA format as it applies to figures and tables:

- Nicol, A. A., & Pexman, P. M. (2010a). *Displaying your findings: A practical guide for creating figures, posters, and presentations* (6th ed.). Washington, DC: American Psychological Association.

- Nicol, A. A., & Pexman, P. M. (2010b). *Presenting your findings: A practical guide for creating tables* (6th ed.). Washington, DC: American Psychological Association.

Whereas the *Publication Manual* of the APA provides broad coverage of all aspects of the manuscript, the two books by Nicol and Pexman focus exclusively on tables, figures, and charts. They provide a multitude of examples of the ways that tables and figures can be prepared to illustrate results from empirical investigations according to APA format. The books address a wide variety of research designs and statistical procedures, from the most elementary (e.g., means and standard deviations) to the most advanced (e.g., structural equation modeling and meta-analysis). Both texts are concise, user-friendly, authoritative, and highly recommended.

Appendix C

Statistical Tables

Table C.1: Critical values of the χ^2 (chi-square) distribution
Table C.2: Critical values of the t distribution
Table C.3: Critical values of the F distribution (alpha = .05)
Table C.4: Critical values of the F distribution (alpha = .01)

Table C.1
Critical Values of the χ^2 (Chi-square) Distribution

	Alpha Level	
df	$\alpha = .05$	$\alpha = .01$
1	3.84146	6.63490
2	5.99147	9.21034
3	7.81473	11.3449
4	9.48773	13.2767
5	11.0705	15.0863
6	12.5916	16.8119
7	14.0671	18.4753
8	15.5073	20.0902
9	16.9190	21.6660
10	18.3070	23.2093
11	19.6751	24.7250
12	21.0261	26.2170.
13	22.3621	27.6883
14	23.6848	29.1413
15	24.9958	30.5779
16	26.2962	31.9999
17	27.5871	33.4087
18	28.8693	34.8053
19	30.1435	36.1908
20	31.4104	37.5662
21	32.6705	38.9321
22	33.9244	40.2894
23	35.1725	41.6384
24	36.4151	42.9798
25	37.6525	44.3141
26	38.8852	45.6417
27	40.1133	46.9630
28	41.3372	48.2782
29	42.5569	49.5879
30	43.7729	50.8922
40	55.7585	63.6907
50	67.5048	76.1539
60	79.0819	88.3794
70	90.5312	100.425
80	101.879	112.329
90	113.145	124.116
100	124.342	135.807

Note. From Thompson, Catherine M., Table of percentage points of the χ^2 distribution, *Biometrika*, (Oct., 1941), *Vol. 32*, No. 2, pp. 187-191. By permission of Oxford University Press and the Biometrika Trust.

Table C.2

Critical Values of the t Distribution, Alpha = .05

df	One-tailed test[a]	Two-tailed test
1	6.3138	12.706
2	2.9200	4.3027
3	2.3534	3.1825
4	2.1318	2.7764
5	2.0150	2.5706
6	1.9432	2.4469
7	1.8946	2.3646
8	1.8595	2.3060
9	1.8331	2.2622
10	1.8125	2.2281
11	1.7959	2.2010
12	1.7823	2.1788
13	1.7709	2.1604
14	1.7613	2.1448
15	1.7530	2.1315
16	1.7459	2.1199
17	1.7396	2.1098
18	1.7341	2.1009
19	1.7291	2.0930
20	1.7247	2.0860
21	1.7207	2.0796
22	1.7171	2.0739
23	1.7139	2.0687
24	1.7109	2.0639
25	1.7081	2.0595
26	1.7056	2.0555
27	1.7033	2.0518
28	1.7011	2.0484
29	1.6991	2.0452
30	1.6973	2.0423
40	1.6839	2.0211
60	1.6707	2.0003
120	1.6577	1.9799
∞	1.6449	1.9600

Note. From Merrington, Maxine, Table of percentage points of the t-distribution, *Biometrika*, (Apr., 1942), Vol. *32*, No. 3/4, p. 300. By permission of Oxford University Press and the Biometrika Trust.
[a] Depending upon the nature of the hypothesis, the critical value of *t* for a one-tailed (directional) significance test may be either a positive value or a negative value.

Table C.3

Critical Values of the F Distribution, Alpha = .05

Degrees of freedom in denominator (within groups)	Degrees of freedom in numerator (between groups)								
	1	2	3	4	5	6	7	8	9
1	161.45	199.50	215.71	224.58	230.16	233.99	236.77	238.88	240.54
2	18.513	19.000	19.164	19.247	19.296	19.330	19.353	19.371	19.385
3	10.128	9.5521	9.2766	9.1172	9.0135	8.9406	8.8868	8.8452	8.8123
4	7.7086	6.9443	6.5914	6.3883	6.2560	6.1631	6.0942	6.0410	5.9988
5	6.6079	5.7861	5.4095	5.1922	5.0503	4.9503	4.8759	4.8183	4.7725
6	5.9874	5.1433	4.7571	4.5337	4.3874	4.2839	4.2066	4.1468	4.0990
7	5.5914	4.7374	4.3468	4.1203	3.9715	3.8660	3.7870	3.7257	3.6767
8	5.3177	4.4590	4.0662	3.8378	3.6875	3.5806	3.5005	3.4381	3.3881
9	5.1174	4.2565	3.8626	3.6331	3.4817	3.3738	3.2927	3.2296	3.1789
10	4.9646	4.1028	3.7083	3.4780	3.3258	3.2172	3.1355	3.0717	3.0204
11	4.8443	3.9823	3.5874	3.3567	3.2039	3.0946	3.0123	2.9480	2.8962
12	4.7472	3.8853	3.4903	3.2592	3.1059	2.9961	2.9134	2.8486	2.7964
13	4.6672	3.8056	3.4105	3.1791	3.0254	2.9153	2.8321	2.7669	2.7144
14	4.6001	3.7389	3.3439	3.1122	2.9582	2.8477	2.7642	2.6987	2.6458
15	4.5431	3.6823	3.2874	3.0556	2.9013	2.7905	2.7066	2.6408	2.5876
16	4.4940	3.6337	3.2389	3.0069	2.8524	2.7413	2.6572	2.5911	2.5377
17	4.4513	3.5915	3.1968	2.9647	2.8100	2.6987	2.6143	2.5480	2.4943
18	4.4139	3.5546	3.1599	2.9277	2.7729	2.6613	2.5767	2.5102	2.4563
19	4.3808	3.5219	3.1274	2.8951	2.7401	2.6283	2.5435	2.4768	2.4227
20	4.3513	3.4928	3.0984	2.8661	2.7109	2.5990	2.5140	2.4471	2.3928
21	4.3248	3.4668	3.0725	2.8401	2.6848	2.5727	2.4876	2.4205	2.3661
22	4.3009	3.4434	3.0491	2.8167	2.6613	2.5491	2.4638	2.3965	2.3419
23	4.2793	3.4221	3.0280	2.7955	2.6400	2.5277	2.4422	2.3748	2.3201
24	4.2597	3.4028	3.0088	2.7763	2.6207	2.5082	2.4226	2.3551	2.3002
25	4.2417	3.3852	2.9912	2.7587	2.6030	2.4904	2.4047	2.3371	2.2821
26	4.2252	3.3690	2.9751	2.7426	2.5868	2.4741	2.3883	2.3205	2.2655
27	4.2100	3.3541	2.9604	2.7278	2.5719	2.4591	2.3732	2.3053	2.2501
28	4.1960	3.3404	2.9467	2.7141	2.5581	2.4453	2.3593	2.2913	2.2360
29	4.1830	3.3277	2.9340	2.7014	2.5454	2.4324	2.3463	2.2782	2.2229
30	4.1709	3.3158	2.9223	2.6896	2.5336	2.4205	2.3343	2.2662	2.2107
40	4.0848	3.2317	2.8387	2.6060	2.4495	2.3359	2.2490	2.1802	2.1240
60	4.0012	3.1504	2.7581	2.5252	2.3683	2.2540	2.1665	2.0970	2.0401
120	3.9201	3.0718	2.6802	2.4472	2.2900	2.1750	2.0867	2.0164	1.9588
∞	3.8415	2.9957	2.6049	2.3719	2.2141	2.0986	2.0096	1.9384	1.8799

Note. From Merrington, Maxine, and Thompson, Catherine M. Tables of percentage points of the inverted beta (*F*) distribution, *Biometrika*, (Apr., 1943), Vol. *33*, No. 1, pp. 73-88. By permission of Oxford University Press and the Biometrika Trust.

Table C.4

Critical Values of the F Distribution, Alpha = .01

Degrees of freedom in denominator (within groups)	Degrees of freedom in numerator (between groups)								
	1	2	3	4	5	6	7	8	9
1	4052.2	4999.5	5403.3	5624.6	5763.7	5859.0	5928.3	5981.6	6022.5
2	98.503	99.000	99.166	99.249	99.299	99.332	99.356	99.374	99.388
3	34.116	30.817	29.457	28.710	28.237	27.911	27.627	28.489	27.345
4	21.198	18.000	16.694	15.977	15.522	15.207	14.976	14.799	14.659
5	16.258	13.274	12.060	11.392	10.967	10.672	10.456	10.289	10.158
6	13.745	10.925	9.7795	9.1483	8.7459	8.4661	8.2600	8.1016	7.9761
7	12.246	9.5466	8.4513	7.8467	7.4604	7.1914	6.9928	6.8401	6.7188
8	11.259	8.6491	7.5910	7.0060	6.6318	6.3707	6.1776	6.0289	5.9106
9	10.561	8.0215	6.9919	6.4221	6.0569	5.8018	5.6129	5.4671	5.3511
10	10.044	7.5594	6.5523	5.9943	5.6363	5.3858	5.2001	5.0567	4.9424
11	9.6460	7.2057	6.2167	5.6683	5.3160	5.0692	4.8861	4.7445	4.6315
12	9.3302	6.9266	5.9526	5.4119	5.0643	4.8206	4.6395	4.4994	4.3875
13	9.0738	6.7010	5.7394	5.2053	4.8618	4.6204	4.4410	4.3021	4.1911
14	8.8616	6.5149	5.5639	5.0354	4.6950	4.4558	4.2779	4.1399	4.0297
15	8.6831	6.3589	5.4170	4.8932	4.5556	4.3183	4.1415	4.0045	3.8948
16	8.5310	6.2262	5.2922	4.7726	4.4374	4.2016	4.0259	3.8896	3.7804
17	8.3997	6.1121	5.1850	4.6690	4.3359	4.1015	3.9267	3.7910	3.6822
18	8.2854	6.0129	5.0919	4.5790	4.2479	4.0146	3.8406	3.7054	3.5971
19	8.1850	5.9259	5.0103	4.5003	4.1708	3.9386	3.7653	3.6305	3.5225
20	8.0960	5.8489	4.9382	4.4307	4.1027	3.8714	3.6987	3.5644	3.4567
21	8.0166	5.7804	4.8740	4.3688	4.0421	3.8117	3.6396	3.5056	3.3981
22	7.9454	5.7190	4.8166	4.3134	3.9880	3.7583	3.5867	3.4530	3.3458
23	7.8811	5.6637	4.7649	4.2635	3.9392	3.7102	3.5390	3.4057	3.2986
24	7.8229	5.6136	4.7181	4.2184	3.8951	3.6667	3.4959	3.3629	3.2560
25	7.7698	5.5680	4.6755	4.1774	3.8550	3.6272	3.4568	3.3239	3.2172
26	7.7213	5.5263	4.6366	4.1400	3.8183	3.5911	3.4210	3.2884	3.1818
27	7.6767	5.4881	4.6009	4.1056	3.7848	3.5580	3.3882	3.2558	3.1494
28	7.6356	5.4529	4.5681	4.0740	3.7539	3.5276	3.3581	3.2259	3.1195
29	7.5976	5.4205	4.5378	4.0449	3.7254	3.4995	3.3302	3.1982	3.0920
30	7.5625	5.3904	4.5097	4.0179	3.6990	3.4735	3.3045	3.1726	3.0665
40	7.3141	5.1785	4.3126	3.8283	3.5138	3.2910	3.1238	2.9930	2.8876
60	7.0771	4.9774	4.1259	3.6491	3.3389	3.1187	2.9530	2.8233	2.7185
120	6.8510	4.7865	3.9493	3.4796	3.1735	2.9559	2.7918	2.6629	2.5586
∞	6.6349	4.6052	3.7816	3.3192	3.0173	2.8020	2.6393	2.5113	2.4073

Note. From Merrington, Maxine, and Thompson, Catherine M. Tables of percentage points of the inverted beta (*F*) distribution, *Biometrika*, (Apr., 1943), Vol. 33, No. 1, pp. 73-88. By permission of Oxford University Press and the Biometrika Trust.

References

Ajzen, L., & Madden, T. J. (1986). Prediction of goal directed behavior: Attitudes, intentions, and perceived behavioral control. *Journal of Experimental Social Psychology, 22*, 453-474.

Akaike, H. (1987). Factor analysis and AIC. *Psychometrika, 52*, 317-332.

American Psychological Association (2010). *Publication manual of the American Psychological Association* (6th ed.). Washington, DC: Author.

Anderson, C. A., & Bushman, B. J. (2001). Effects of violent video games on aggressive behavior, aggressive cognition, aggressive affect, physiological arousal, and prosocial behavior: A meta-analytic review of the scientific literature. *Psychological Science, 12*, 353-359.

Anderson, C. A., Shibuya, A., Ihori, N., Swing, E. L., Bushman, B. J., Sakamoto, A., et al., Saleem, M. (2010). Violent video game effects on aggression, empathy, and prosocial behavior in eastern and western countries: A meta-analytic review. *Psychological Bulletin, 136*, 151-173.

Anderson, J. C., & Gerbing, D. W. (1988). Structural equation modeling in practice: A review and recommended two-step approach. *Psychological Bulletin, 103*, 411-423.

Aron, A., Aron, E. N., & Coups, E. J. (2006). *Statistics for psychology* (4th ed.). Upper Saddle River, NJ: Pearson Prentice Hall.

Azen, R., & Budescu, D. V. (2003). The dominance analysis approach for comparing predictors in multiple regression. *Psychological Methods, 8*, 129-148.

Azen, R., & Budescu, D. (2009). Applications of multiple regression in psychological research. In R. E. Millsap & A. Maydeu-Olivares (Eds.), *The SAGE handbook of quantitative methods in psychology* (pp. 285-310). Thousand Oaks, CA: Sage Publications Inc.

Azen, R., Budescu, D. V., & Reiser, B. (2001). Criticality of predictors in multiple regression. *British Journal of Mathematical and Statistical Psychology, 54*, 201-225.

Bandalos, D. L., & Boehm-Kaufman, M. R. (2009). Four common misconceptions in exploratory factor analysis. In C. E. Lance & R. J. Vandenberg (Eds.) *Statistical and methodological myths and urban legends: Doctrine, verity, and fable in the organizational and social sciences* (pp. 61-88). New York: Routledge Taylor & Francis Group.

Barel, E., Van IJzendoorn, M. H., Sagi-Schwartz, A., & Bakermans-Kranenburg, M. J. (2010). Surviving the Holocaust: A meta-analysis of the long-term sequelae of a genocide. *Psychological Bulletin, 136*, 677-698.

Bentler, P. M. (1990). Comparative fit indices in structural models. *Psychological Bulletin, 107*, 238-246.

Bentler, P. M. (1995). *EQS structural equations program manual.* Encino, CA: Multivariate Software.

Bentler, P. M., & Bonett, D. G. (1980). Significance tests and goodness of fit in the analysis of covariance structures. *Psychological Bulletin, 88,* 588-606.

Bentler, P. M. & Chou, C. (1988). Practical issues in structural modeling. In J. S. Long, (Ed.) *Common problems/proper solutions* (pp. 161-192). Newbury Park, CA: SAGE Publications.

Billings, R. S., & Wroten, S. P. (1978). Use of path analysis in industrial/organizational psychology: Criticisms and suggestions. *Journal of Applied Psychology, 63,* 677-688.

Bollen, K. A. (1989). *Structural equations with latent variables.* New York: Wiley.

Bonett, D. G. (2007). Transforming odds ratios into correlations for meta-analytic research. *American Psychologist, 62,* 254-255. doi:10.1037/0003-066X.62.3.254

Borenstein, M., Hedges, L. V., Higgins, J. P. T., & Rothstein, H. R. (2009). *Introduction to meta-analysis.* West Sussex, UK: John Wiley & Sons.

Bray, J. H., & Maxwell, S. E. (1982). Analyzing and interpreting significant MANOVAs. *Review of Educational Research, 52,* 340-367.

Browne, M. W. (1984). Asymptotically distribution-free methods for the analysis of covariance structures. *British Journal of Mathematical and Statistical Psychology, 37,* 62-83.

Bryant, F. B., & Yarnold, P. R. (1995). Principal-components analysis and exploratory and confirmatory factor analysis. In L. G. Grimm & P. R. Yarnold (Eds.), *Reading and understanding multivariate statistics* (pp. 99-136). Washington, DC: American Psychological Association.

Bushman, B. J., Rothstein, H. R., & Anderson, C. A. (2010). Much ado about something: Violent video game effects and a school of red herring: Reply to Ferguson and Kilburn (2010). *Psychological Bulletin, 136,* 182-187.

Campbell, D. T., & Fiske, D. W. (1959). Convergent and discriminant validation by the multitrait-multimethod matrix. *Psychological Bulletin, 56,* 81-105.

Cattell, R. B. (1966). The scree test for the number of factors. *Multivariate Behavioral Research, 1,* 245-276.

Cohen, J. (1988). *Statistical power analysis for the behavioral sciences* (2nd ed.). Hillsdale, NJ: Lawrence Erlbaum Associates.

Cohen, J. (1994). The earth is round ($p < .05$). *American Psychologist, 49,* 997-1003.

Cohen, J., & Cohen, P. (1983). *Applied multiple regression/correlation analysis for the behavioral sciences* (2nd ed.). Hillsdale, NJ: Lawrence Erlbaum Associates.

Cohen, B. H. & Lea, R. B. (2004). *Essentials of statistics for the social and behavioral sciences.* Hoboken, NJ: John Wiley & Sons.

Connelly, B. S., & Ones, D. S. (2010). An other perspective on personality: Meta-analytic integration of observers' accuracy and predictive validity. *Psychological Bulletin, 136,* 1092-1122. doi: 10.1037/a0021212

Cook, T. D., & Campbell, D. T. (1979). *Quasi-experimentation: Design and analysis issues for field settings.* Boston: Houghton Mifflin.

Cortina, J. M. (2002). Big things have small beginnings: An assortment of "minor methodological misunderstandings." *Journal of Management, 28,* 339-362.

Cronbach, L. J. (1951). Coefficient alpha and the internal structure of tests. *Psychometrika, 16,* 297-334.

Cudeck, R., & duToit, S. H. C. (2009). General structural equation models. In R. E. Millsap, & A. Maydeu-Olivares (Eds.), *The SAGE handbook of quantitative methods in psychology* (pp. 515-539). Los Angeles, CA: SAGE Publications.

Cumming, G. (2011). *Understanding the new statistics: Effect sizes, confidence intervals, and meta-analysis.* New York: Routledge Taylor & Francis Group.

Cumming, G., & Finch, S. (2005). Inference by eye: Confidence intervals and how to read pictures of data. *American Psychologist, 60,* 170-180.

Darlington, R. B. (1968). Multiple regression in psychological research and practice. *Psychological Bulletin, 69,* 161-182.

Dickersin, K., Min, Y. I., & Meinert, C. L. (1992). Factors influencing publication of research results: Follow-up of applications submitted to two institutional review boards. *Journal of the American Medical Association, 267,* 374-378.

Durlak, J. A. (1995). Understanding meta-analysis. In L. G. Grimm & P. R. Yarnold (Eds.), *Reading and understanding multivariate statistics* (pp. 319-352). Washington, DC: American Psychological Association.

Duval, S., & Tweedie, R. (2000). Trim and fill: A simple funnel-plot-based method of testing and adjusting for publication bias in meta-analysis. *Biometrics, 56,* 455-463.

Fabrigar, L. R., Wegener, D. T., MacCallum, R. C., & Strahan, E. J. (1999). Evaluating the use of exploratory factor analysis in psychological research. *Psychological Methods, 4,* 272-299.

Ferguson, C. J. (2007). The good, the bad and the ugly: A meta-analytic review of positive and negative effects of violent video games. *Psychiatric Quarterly, 78,* 309-316.

Ferguson, C. J., & Kilburn, J. (2010). Much ado about nothing: The misestimation and overinterpretation of violent video game effects in eastern and western nations: Comment on Anderson et al. (2010). *Psychological Bulletin, 136,* 174-178.

Fidler, F., & Cumming, G. (2008). The new stats: Attitudes for the 21st century. In J. W. Osborne (Ed.), *Best practices in quantitative methods* (pp. 1-12). Thousand Oaks, CA: SAGE Publications.

Field, A. (2009a). *Discovering statistics using SPSS (and sex and drugs and rock 'n' roll)* (3rd ed.). Thousand Oaks, CA: SAGE Publications Inc.

Field, A. (2009b). Meta-analysis. In R. E. Millsap, & A. Maydeu-Olivares (Eds.), *The SAGE Handbook of Quantitative Methods in Psychology* (pp. 404-422). Thousand Oaks, CA: Sage Publications Inc.

Field, A., & Miles, J. (2010). *Discovering statistics using SAS (and sex and drugs and rock 'n' roll)*. Thousand Oaks, CA: SAGE Publications Inc.

Fornell, C. & Larcker, D. F. (1981). Evaluating structural equation models with unobservable variables and measurement error. *Journal of Marketing Research, 18*, 39-50.

Gabriel, K. R. (1964). A procedure for testing the homogeneity of all sets of means in analysis of variance. *Biometrics, 20*, 459-477.

Gabriel, K. R. (1969). Simultaneous test procedure—Some theory of multiple comparisons. *Annals of Mathematical Statistics, 40*, 224-250.

Games, P. A., & Howell, J. F. (1976). Pairwise multiple comparison procedures with unequal N's and/or variances: A Monte Carlo study. *Journal of Educational Statistics, 1*, 113-125.

Gilbert, R., Salanti, G., Harden, M., & See, S. (2005). Infant sleeping position and the sudden infant death syndrome: Systematic review of observational studies and historical review of recommendations from 1940 to 2002. *International Journal of Epidemiology, 34*, 874-887. doi: 10.1093/ije/dyi088

Glass, G. V., McGaw, B., & Smith, M. L. (1981). *Meta-analysis in social research*. Beverly Hills, CA: Sage.

Grabe, S., Ward, L. M., & Hyde, J. S. (2008). The role of the media in body image concerns among women: A meta-analysis of experimental and correlational studies. *Psychological Bulletin, 134*, 460-476.

Gravetter, F. J., & Wallnau, L. B. (2000). *Statistics for the behavioral sciences* (5th ed.). Belmont, CA: Wadsworth/Thompson Learning.

Green, S. B. (1991). How many subjects does it take to do a regression analysis? *Multivariate Behavioral Research, 26*, 449-510.

Greenhouse, S. W., & Geisser, S. (1959). On methods in the analysis of profile data. *Psychometrika, 24*, 95-112.

Grimm, L. G., & Yarnold, P. R. (Eds.). (1995). *Reading and understanding multivariate statistics*. Washington, DC: American Psychological Association.

Grimm, L. G., & Yarnold, P. R. (Eds.). (2000). *Reading and understanding more multivariate statistics*. Washington, DC: American Psychological Association.

Grissom, R. J., & Kim, J. J. (2005). *Effect sizes for research: A broad practical approach*. Mahwah, NJ: Lawrence Erlbaum Associates.

Haddock, C. K., Rindskopf, D., & Shadish, W. R. (1998). Using odds ratios as effect sizes for meta-analysis of dichotomous data: A primer on methods and issues. *Psychological Methods, 3,* 339-353.

Harrison, D. A. (1995). Volunteer motivation and attendance decisions: Competitive theory testing in multiple samples from a homeless shelter. *Journal of Applied Psychology, 80,* 371-385.

Hatcher, L. (1994). *A step-by-step approach to using the SAS® system for factor analysis and structural equation modeling.* Cary, NC: SAS Institute, Inc.

Hatcher, L. (2006). *Choosing the correct basic or intermediate statistic: A simple reference for SAS® Users.* Saginaw, MI: Unpublished manuscript.

Hatcher, L. (2003a). *Step-by-step basic statistics using SAS®: Student guide.* Cary, NC: SAS Institute, Inc.

Hatcher, L. (2003b). *Step-by-step basic statistics using SAS®: Exercises.* Cary, NC: SAS Institute, Inc.

Hattie, J. A. C. (2009). *Visible learning: A synthesis of over 800 meta-analyses relating to achievement.* New York: Routledge Taylor & Francis Group.

Hays, W. L. (1988). *Statistics* (4th ed.). New York: Holt, Rinehart and Winston, Inc.

Hedges, L. (1981). Distribution theory for Glass's estimator of effect size and related estimators. *Journal of Educational Statistics, 6,* 107-128.

Hedges, L. V. (1984). Estimation of effect size under nonrandom sampling: the effects of censoring studies yielding statistically insignificant mean differences. *Journal of Educational Statistics, 9,* 61-85.

Hedges, L. V., & Becker, B. J. (1986). Statistical methods in the meta-analysis of research on gender differences. In J. S. Hyde & M. C. Linn (Eds.), *The psychology of gender* (pp. 14-50). Baltimore: Johns Hopkins University Press.

Hedges, L. V., & Olkin, I. (1985). *Statistical methods for meta-analysis.* New York: Academic Press.

Heiman, G. W. (2006). *Basic statistics for the behavioral sciences* (5th ed.). New York, NY: Houghton Mifflin Company.

Hershberger, S. L. (2006). The problem of equivalent structural models. In G.R. Hancock & R.O. Mueller (Eds.), *Structural equation modeling: A second course* (pp. 13-41). Greenwich, CT: Information Age Publishing.

Higgins, J. P. T., & Thompson, S. G. (2002). Quantifying heterogeneity in meta-analysis. *Statistics in Medicine, 21,* 1539-1558.

Holcomb, W. L., Jr., Chaiworaponga, T., Luke, D. A., & Burgdorf, K. D. (2001). An odd measure of risk: Use and misuse of the odds ratio. *Obstetrics and Gynecology, 84,* 685-688.

Holcomb, Z. C. (2007). *Interpreting basic statistics: A guide and workbook based on excerpts from journal articles* (5th ed.). Glendale, CA: Pyrczak Publishing.

Horrey, W. J., & Wickens, C. D. (2006). Examining the impact of cell phone conversations on driving using meta-analytic techniques. *Human Factors, 48*, 196-205.

Hosmer, D. W., & Lemeshow, S. (1989). *Applied logistic regression.* New York: Wiley.

Howell, D. C. (2002). *Statistical methods for psychology* (5th ed.). Pacific Grove, CA: Duxbury.

Howell, D. (2008). Best practices in the analysis of variance. In J. W. Osborne (Ed.), *Best practices in quantitative methods* (pp. 341-357). Thousand Oaks, CA: SAGE Publications.

Hu, L., & Bentler, P. M. (1999). Cutoff criterion for fit indexes in covariance structure analysis: Conventional criteria versus new alternatives. *Structural Equation Modeling: A Multidisciplinary Journal, 6*, 1-55.

Huberty, C. J., & Smith, J. D. (1982). The study of effects in MANOVA. *Multivariate Behavioral Research, 17*, 417-482.

Huck, S.W. (2004). *Reading statistics and research* (4th ed.). Boston: Pearson/Allyn & Bacon.

Huck, S. W. (2009). *Statistical misconceptions.* New York: Routledge, Taylor & Francis Group.

Huck, S. W. (2012). *Reading statistics and research* (6th ed.). New York: Pearson.

Huedo-Medina, T. B., Sanchez-Meca, J., Marin-Martinez, F., & Botella, J. (2006). Assessing heterogeneity in meta-analysis: Q statistic or I^2 index? *Psychological Methods, 11*, 193-206.

Huesmann, L. R. (2010). Nailing the coffin shut on doubts that violent video games stimulate aggression: Comment on Anderson et al. (2010). *Psychological Bulletin, 136*, 179-181.

Huitema, B. (1980). *The analysis of covariance and alternatives.* New York: Wiley.

Hulsizer, M. R., & Woolf, L. M. (2009). *A guide to teaching statistics: Innovations and best practices.* Malden, MA: Wiley-Blackwell.

Hunt, M. (1997). *How science takes stock: The story of meta-analysis.* New York: Russell Sage Foundation.

Hunter, J. E. (1997). Needed: A ban on the significance test. *Psychological Science, 8*, 3-7.

Hunter, J. E., & Schmidt, F. L. (2004). *Methods of meta-analysis: Correcting error and bias in research findings* (2nd ed.). Thousand Oaks, CA: SAGE Publications.

Huynh, H., & Feldt, L. (1976). Estimation of the Box correction for degrees of freedom from sample data in the randomized block and split plot designs. *Journal of Educational Statistics, 1*, 69-82.

Jackson, D. L., Gillaspy, J. A., Jr., & Purc-Stephenson, R. (2009). Reporting practices in confirmatory factor analysis: An overview and some recommendations. *Psychological Methods, 14*, 6-23.

Jaeger, R. M. (1993). *Statistics: A spectator sport* (2nd ed.). Thousand Oaks, CA: SAGE Publications.

James, L. R., Mulaik, S. A., & Brett, J. M. (1982). *Causal analysis: Assumptions, models, and data*. Beverly Hills: SAGE Publications.

Jöreskog, K. G. & Sörbom, D. (1984). *LISREL VI users guide* (3rd ed.). Mooresville, IN: Scientific Software.

Jöreskog, K. G. & Sörbom, D. (1989). *LISREL 7: A guide to the program and applications*. Chicago: SPSS Inc.

Judge, T. A., Hurst, C., & Simon, L. S. (2009). Does it pay to be smart, attractive, or confident (or all three)? Relationships among general mental ability, physical attractiveness, core self-evaluations, and income. *Journal of Applied Psychology, 94*, 742-755.

Kaiser, H. F. (1970). A second generation Little Jiffy. *Psychometrika, 35*, 401-415.

Kanji, G. K. (1999). *100 Statistical tests: New edition*. Thousand Oaks, CA: SAGE Publications.

Keith, T. Z. (2006). *Multiple regression and beyond*. New York: Pearson Allyn and Bacon.

Kelley, K. (2007). MBESS: Version 0.0.9. [Computer software and manual]. Retrieved from http://www.r-project.org/

Kelley, K., Lai, K., & Wu, P. (2008). Using R for data analysis: A best practice for research. In J. W. Osborne (Ed.), *Best practices in quantitative methods* (pp. 535-572). Thousand Oaks, CA: SAGE Publications, Inc.

Kenny, D. A. (1970). *Correlation and causality*. New York: John Wiley & Sons.

Keppel, G. (1982). *Design and analysis: A researcher's handbook* (2nd ed.). Englewood Cliffs, NJ: Prentice-Hall.

Keppel, G., & Zeddeck, S. (1989). *Data analysis for research designs: Analysis of variance and multiple regression/correlation approaches*. New York: W.H. Freeman and Company.

Killeen, P. R. (2005a). An alternative to null hypothesis significance tests. *Psychological Science, 16*, 345-353.

Killeen, P. R. (2005b). Replicability, confidence, and priors. *Psychological Science, 16*, 1009-1012.

Killeen, P. R. (2008). Replication statistics. In J.W. Osborne (Ed.), *Best practices in quantitative methods* (pp. 103-124). Thousand Oaks, CA: SAGE Publications.

King, J. E. (2008). Binary logistic regression. In J.W. Osborne (Ed.), *Best practices in quantitative methods* (pp. 358-384). Thousand Oaks, CA: SAGE Publications Inc.

Kirk, R. E. (1995). *Experimental design: Procedures for the behavioral sciences* (3rd ed.). Pacific Grove, CA: Brooks/Cole Publishing.

Kirk, R. E. (1996). Practical significance: A concept whose time has come. *Educational and Psychological Measurement, 56,* 746-759.

Klem, L. (1995). Path analysis. In L. G. Grimm & P. R. Yarnold (Eds.), *Reading and understanding multivariate statistics* (pp. 65-98). Washington, DC: American Psychological Association.

Klem, L. (2000). Structural equation modeling. In L. G. Grimm & P. R. Yarnold (Eds.), *Reading and understanding more multivariate statistics* (pp. 227-260). Washington, DC: American Psychological Association.

Kline, R. (2004). *Beyond significance testing: Reforming data analysis methods in behavioral research.* Washington, DC: American Psychological Association.

Kline, R. B. (1998). *Principles and practices of structural equation modeling.* New York: Guilford.

Konstantopoulos, S. (2008). An introduction to meta-analysis. In J. W. Osborne (Ed.), *Best practices in quantitative methods* (pp.177-196). Thousand Oaks, CA: SAGE Publications, Inc.

Lachenbruch, P. A. (1967). An almost unbiased method of obtaining confidence intervals for the probability of misclassification in discriminant analysis. *Biometrics, 23,* 639-645.

Lance, C. E., & Vandenberg, R. J. (Eds.) (2009). *Statistical and methodological myths and urban legends: Doctrine, verity, and fable in the organizational and social sciences.* New York: Routledge Taylor & Francis Group.

Lehman, A., O'Rourke, N., Hatcher, L., & Stepanski, E. J. (2005). *JMP® for basic univariate and multivariate statistics: A step-by-step guide.* Cary, NC: SAS Institute Inc.

Light, R. J., & Pillemer, D. B. (1984). *Summing up: The science of reviewing research.* Cambridge, MA: Harvard University Press.

Lindberg, S. M., Hyde, J. S., Petersen, J. L., & Linn, M. C. (2010). New trends in gender and mathematics performance: A meta-analysis. *Psychological Bulletin, 136,* 1123-1135.

Locke, L. F., Silverman, S. J., & Spirduso, W. W. (2004). *Reading and understanding research* (2nd ed.). Thousand Oaks, CA: SAGE Publications.

Long, J. S. (1983a). *Confirmatory factor analysis: A preface to LISREL.* Sage University Paper Series on Quantitative Application in the Social Sciences, 07-033. Beverly Hills: Sage.

Long, J. S. (1983b). *Covariance structure models: An introduction to LISREL*. Sage University Paper Series on Quantitative Application in the Social Sciences, 07-034. Beverly Hills: Sage.

Lucas-Thompson, R. G., Goldberg, W. A., & Prause, J. (2010, October 4). Maternal work early in the lives of children and its distal associations with achievement and behavior problems: A meta-analysis. *Psychological Bulletin, 136*, 915-942. doi: 10.1037/a0020875

MacCallum, R. C. (2009). Factor analysis. In R. E. Millsap & A. Maydeu-Olivares (Eds.), *The SAGE handbook of quantitative methods in psychology* (pp. 123-147). Thousand Oaks, CA: SAGE Inc.

MacCallum, R. C., Browne, M. W., & Cai, L. (2006). Testing differences between nested covariance structure models: Power analysis and null hypotheses. *Psychological Methods, 11*, 19-35.

MacCallum, R. C., Brown, M. W. & Sugawara, H. M. (1996). Power analysis and determination of sample size for covariance structure modeling. *Psychological Methods, 1*, 130-149.

MacCallum, R. C., Roznowski, M., & Necowitz, L. B. (1992). Model modifications in covariance structure analysis: The problem of capitalization on chance. *Psychological Bulletin, 111*, 490-504.

McDonald, R. P., & Ho, M. R. (2002). Principles and practice in reporting structural equation analyses. *Psychological Methods, 7*, 64-82. doi: 10.1037//1082-989X.7.1.64

Menard, S. (1995). *Applied logistic regression analysis*. Sage University Paper Series On Quantitation Application in The Social Sciences, 07-106. Thousand Oaks, CA: Sage.

Meyer, G. J, Finn, S. E., Eyde, L. D., Kay, G. G., Moreland, K. L., Dies, R. R., et al. (2001). Psychological testing and psychological assessment: A review of evidence and issues. *American Psychologist, 56*, 128-165.

Miceli, M. P., & Near, J. P. (1984). The relationships among beliefs, organizational position, and whistle-blowing status: A discriminant analysis. *Academy of Management Journal, 27*, 697-705.

Millsap, R. E., & Maydeu-Olivares, A. (Eds.) (2009). *The SAGE handbook of quantitative methods in psychology*. Thousand Oaks, CA: Sage Publications Inc.

Moore, D. S., & Notz, W. I. (2006). *Statistics: Concepts and controversies* (6th ed.). New York: W.H. Freeman and Company.

Morgan, S. E., Reichert, T., & Harrison, T. R. (2002). *From numbers to words: Reporting statistical results for the social sciences*. Boston, MA: Allyn and Bacon.

Morrison, D. F. (1976). *Multivariate statistical methods*. New York: McGraw-Hill.

Mueller, R. O., & Hancock, G. R. (2008). Best practices in structural equation modeling. In J. W. Osborne (Ed.), *Best practices in quantitative methods* (pp. 488-508). Los Angeles, CA: SAGE Publications.

Mulaik, S. A., James, L. R., Alstine, J. V., Bennett, N., Lind, S., & Stilwell, D. (1989). Evaluation of goodness-of-fit indices for structural equation models. *Psychological Bulletin, 105*, 430-445.

Muthén, B. (1984). A general structural equation model with dichotomous, ordered categorical and continuous latent variable indicators. *Psychometrika, 49*, 115-132.

Muthén, L. K., & Muthén, B. O. (2001). *Mplus User's Guide* (2nd ed.). Los Angeles, CA: Author.

Myers, J. L., & Well, A. D. (1991). *Research design and statistical analysis.* New York: Harper Collins.

Nagelkerke, N. J. D. (1991). A note on the general definition of the coefficient of determination. *Biometrika, 78*, 691-692.

Netemeyer, R. G., Johnston, M. W., & Burton, S. (1990). Analysis of role conflict and role ambiguity in a structural equations framework. *Journal of Applied Psychology, 75*, 148-157.

Newton, R. R., & Rudestam, K. E. (1999). *Your statistical consultant: Answers to your data analysis questions.* Thousand Oaks, CA: SAGE Publications.

Nicol, A. A., & Pexman, P. M. (2010a). *Displaying your findings: A practical guide for creating figures, posters, and presentations* (6th ed.). Washington, DC: American Psychological Association.

Nicol, A. A., & Pexman, P. M. (2010b). *Presenting your findings: A practical guide for creating tables* (6th ed.). Washington, DC: American Psychological Association.

Norušis, M. (2005). *SPSS 14.0 statistical procedures companion.* Upper Saddle River, NJ: Prentice Hall.

Norušis, M. (2006). *SPSS 14.0 guide to data analysis.* Upper Saddle River, NJ: Prentice Hall.

Norušis, M. J. (2009a). *SPSS 17.0 guide to data analysis.* Upper Saddle River, NJ: Prentice Hall.

Norušis, M. J. (2009b). *SPSS 17.0 statistical procedures companion.* Upper Saddle River, NJ: Prentice Hall.

Nunnally, J. (1978). *Psychometric theory.* New York: McGraw-Hill.

O'Boyle, E. H., Jr., & Williams, L. J. (2011). Decomposing model fit: Measurement vs. theory in organizational research using latent variables. *Journal of Applied Psychology, 96*, 1-12. doi:10.1037/a0020539

Olkin, I., & Finn, J. D. (1995). Correlations redux. *Psychological Bulletin, 118*, 155-164.

O'Rourke, N., Hatcher, L., & Stepanski, E. J. (2005). *A step-by-step approach to using SAS® for univariate and multivariate statistics* (2nd ed.). Cary, NC: SAS Institute, Inc.

Orwin, R. G., & Boruch, R. F. (1983). RRT meets RDD: statistical strategies for assuring response privacy in telephone surveys. *Public Opinion Quarterly, 46*, 560-571.

Orwin, R. G., & Cordray, D. S. (1985). Effects of deficient reporting on meta-analysis: A conceptual framework and reanalysis. *Psychological Bulletin, 97*, 134-147.

Osborne, J. W. (Ed.). (2008). *Best practices in quantitative methods*. Thousand Oaks, CA: SAGE Publications.

Osborne, J. W. (2008). Bringing balance and technical accuracy to reporting odds ratios and the results of logistic regression analyses. In J. W. Osborne (Ed.), *Best practices in quantitative methods* (pp. 385-389). Thousand Oaks, CA: SAGE Publications.

Osborne, J. W., Costello, A. B., & Kellow, J. T. (2008). Best practices in exploratory factor analysis. In J. W. Osborne (Ed.), *Best practices in quantitative methods* (pp. 86-102). Thousand Oaks, CA: SAGE Publications, Inc.

Pallant, J. (2005). *SPSS survival manual* (2nd ed.). New York: Open University Press.

Pearlman, K., Schmidt, F. L., & Hunter, J. E. (1980). Validity generalization results for tests used to predict job proficiency and training success in clerical occupations. *Journal of Applied Psychology, 65*, 373-406.

Pedhazur, E. J. (1982*). Multiple regression in behavioral research: Explanation and prediction* (2nd ed.). New York: Holt, Rinehart, and Winston.

Preiss, R. W., Gayle, B. M., Burell, N., Allen, M., & Bryant, J. (Eds.). (2007). *Mass media effects research: Advances through meta-analysis*. New York: Routledge Taylor & Francis Group.

Press, S. J., & Wilson, S. (1978). Choosing between logistic regression and discriminant analysis. *Journal of the American Statistical Association, 73*, 699-705.

Richard, F. D., Bond, C. F., Jr., & Stokes-Zoota, J. J. (2003). One hundred years of social psychology quantitatively described. *Review of General Psychology, 7*, 331-363.

Rosenthal, R. (1979). The file drawer problem and tolerance for null results. *Psychological Bulletin, 86*, 638-641.

Rosenthal, R. (1991). *Meta-analytic procedures for social research* (2nd ed.). Newbury Park, CA: SAGE.

Rosenthal, R. (1994). Parametric measures of effect size. In H. Cooper & L. V. Hedges (Eds.), *The handbook of research synthesis* (pp. 231-244). New York: Russell Sage Foundation.

Rosenthal, R. (1995). Writing meta-analytic reviews. *Psychological Bulletin, 118,* 183-192.

Rosenthal, R., & Rubin, D. B. (1986). Meta-analytic procedures for combining studies with multiple effect sizes. *Psychological Bulletin, 99,* 400-406.

Rosnow, R. L., & Rosenthal, R. (1996). Computing contrasts, effect sizes, and counternulls on other people's published data: General procedures for research consumers. *Psychological Methods, 1,* 331-340.

Rummel, R. J. (1970). *Applied factor analysis.* Evanston, IL: Northwestern University Press.

Sackett, P. R., Harris, M. M., & Orr, J. M. (1986). On seeking moderator variables in the meta-analysis of correlational data: A Monte Carlo investigation of statistical power and resistance to Type I error. *Journal of applied Psychology, 71,* 302-310.

SAS® Institute, Inc. (2004). *SAS/STAT 9.1 User's Guide.* Cary, NC: Author.

Schmidt, F. L. (1996). Statistical significance testing and cumulative knowledge in psychology: Implications for training of researchers. *Psychological Methods, 1,* 115-129.

Schmidt, F. L., & Hunter, J. E. (1977). Development of a general solution to the problem of validity generalization. *Journal of Applied Psychology, 62,* 529-540.

Schmidt, F. L., Hunter, J. E., & Raju, N. S. (1988). Validity generalization and situational specificity: A second look at the 75% rule and Fisher's r to z transformation. *Journal of Applied Psychology, 73,* 665-672.

Sheng, Y. (2008). Testing the assumptions of analysis of variance. In J. W. Osborne (Ed.), *Best practices in quantitative methods* (pp. 324-340). Thousand Oaks, CA: SAGE Publications.

Silva, A. P. D., & Stam, A. (1995). Discriminant analysis. In L. G. Grimm & P. R. Yarnold (Eds.), *Reading and understanding multivariate statistics* (pp. 277-318). Washington, DC: American Psychological Association.

Smith, M.L., & Glass, G.V. (1977). Meta-analysis of psychotherapy outcome studies. *American Psychologist, 32,* 752-760.

Smith, M. L., & Glass, G. V. (1980). Meta-analysis of research on class size and its relationship to attitudes and instruction. *American Educational Research Journal, 17,* 419-433.

Stevens, J. P. (1980). Power of the multivariate analysis of variance. *Psychological Bulletin, 88,* 728-737.

Stevens, J. (1996). *Applied multivariate statistics for the social sciences* (3rd ed.). Mahwah, NJ: Lawrence Erlbaum Associates.

Stevens, J. (2002). *Applied multivariate statistics for the social sciences* (4th ed.). Hillside, NJ: Erlbaum.

Stevens, S. S. (1946). On a theory of scales of measurement. *Science, 103*, 677-680.

Su, R., Rounds, J., & Armstrong, P. I. (2009). Men and things, women and people: A meta-analysis of sex differences in interests. *Psychological Bulletin, 135*, 859-884.

Tabachnick, B. G., & Fidell, L. S. (2007). *Using multivariate statistics* (5th ed.). Boston: Pearson, Allyn and Bacon.

Thompson, B. (2000). Canonical correlation analysis. In L. G. Grimm & P. R. Yarnold (Eds.), *Reading and understanding more multivariate statistics* (pp. 285-316). Washington, DC: American Psychological Association.

Thompson, B. (2000). Ten commandments of structural equation modeling. In L. G. Grimm & P. R. Yarnold (Eds.), *Reading and understanding more multivariate statistics* (pp. 261-283). Washington, DC: American Psychological Association.

Thompson, B. (2008). Computing and interpreting effect sizes, confidence intervals, and confidence intervals for effect sizes. In J. W. Osborne (Ed.), *Best practices in quantitative methods* (pp. 246-262). Thousand Oaks, CA: Sage.

Toothaker, L. E. (1993). *Multiple comparison procedures.* Sage University Paper Series on Quantitative Applications in the Social Sciences, 07-089. Newbury Park, CA: SAGE Publications, Inc.

Tucker, L. R., & Lewis, C. (1973). A reliability coefficient for maximum likelihood factor analysis. *Psychometrika, 38*, 1-10.

Tukey, J. W. (1977). *Exploratory data analysis.* Reading, MA: Addison-Wesley.

Twenge, J. M., Zhang, L., & Im, C. (2004). It's beyond my control: A cross-temporal meta-analysis of increasing externality in locus of control, 1960-2002. *Personality and Social Psychology Review, 8*, 308-319.

Ullman, J. B. (2007). Structural equation modeling. In B. G. Tabachnick & L. S. Fidell (Eds.), *Using multivariate statistics* (5th ed.) (pp. 676-780). Boston: Pearson, Allyn and Bacon.

VandenBos, G. R. (Ed.). (2007). *APA dictionary of psychology.* Washington, DC: American Psychological Association.

Velicer, W. F., Eaton, C. A., & Fava, J. L. (2000). Construct explication through factor or component analysis: A review and evaluation of alternative procedures for determining the number of factors or components. In R. D. Goffin & E. Helmes (Eds.), *Problems and solutions in human assessment* (pp. 41-71). Boston: Kluwer.

Velicer, W. F., & Jackson, D. N. (1990). Component analysis versus common factor-analysis: Some further observations. *Multivariate Behavioral Research, 25*, 97-114.

Viswesvaran, C., & Sanchez, J. I. (1998). Moderator search in meta-analysis: A review and cautionary note on existing approaches. *Educational and Psychological Measurement, 58*, 77-88. Gale Document Number A20533929.

Warner, R. M. (2008). *Applied statistics: From bivariate through multivariate techniques.* Los Angeles, CA: SAGE Publications.

Weinfurt, K. P. (1995). Multivariate analysis of variance. In L. G. Grimm & P. R. Yarnold (Eds.), *Reading and understanding multivariate statistics* (pp. 245-276). Washington DC: American Psychological Association.

Weinfurt, K. P. (2000). Repeated-measures analyses: ANOVA, MANOVA, and HLM. In L. G. Grimm & P. R. Yarnold (Eds.), *Reading and understanding more multivariate statistics* (pp. 317-362). Washington DC: American Psychological Association.

Whitener, E. M. (1990). Confusion of confidence intervals and credibility intervals in meta-analysis. *Journal of Applied Psychology, 75*, 315-321.

Wilcox, R. R. (1987). New designs in analysis of variance. *Annual Review of Psychology, 38*, 29-60.

Wilkinson, L. & the Task Force on Statistical Inference (1999). Statistical methods in psychology journals: Guidelines and explanations. *American Psychologist, 54*, 594-604. doi:10.1037/0003-066X.54.8.594

Wirth, R. J., & Edwards, M. C. (2007). Item factor analysis: Current approaches and future directions. *Psychological Methods, 12*, 58-79.

Wright, R. E. (1995). Logistic regression. In L. G. Grimm & P. R. Yarnold (Eds.), *Reading and understanding multivariate statistics* (pp. 217-244). Washington, DC: American Psychological Association.

Yang, M., Wong, S. C. P., & Coid J. (2010). The efficacy of violence prediction: A meta-analytic comparison of nine risk assessment tools. *Psychological Bulletin, 136*, 740-767.

Yaremko, R. M., Harari, H., Harrison, R. C., & Lynn, E. (Eds.). (1982). *Reference handbook of research and statistical methods in psychology for students and professionals.* New York: Harper & Row.

Zwick, W. R., & Velicer, W. F. (1986). Comparison of five rules for determining the number of components to retain. *Psychological Bulletin, 99*, 432-442.

Index

−2 log likelihood, 323
−2LL, 323
75% rule, 566
a priori hypothesis, 369
absolute value, 74, 178
adjusted group means, 377
adjusted means, 392
adjusted odds ratio, 330
adjusted R^2, 258
alpha, 139
alpha level, 133, 136
alternative explanation, 21
alternative theoretical models, 504
antecedent event, 18
antecedent variable, 445
approximately normal distribution, 58, 102
arithmetic mean, 63
artifacts, 531, 560
attributes, 296
average, 63
beta weight, 215, 264, 266, 286
between-groups design, 24
between-studies variability, 544
between-subjects design, 24
between-subjects factor, 344
biased estimate of population variance, 80
bimodal, 67
binary logistic regression, 341
biserial correlation coefficient, 190
bivariate, 196
bivariate correlations, 172
bivariate normality, 47
Bonferroni adjustment, 140, 359
Bonferroni correction, 140
box length, 91
boxplot, 51, 89
canonical discriminant functions, 299
canonical variables, 299
canonical variates, 299
categorical variables, 15
causal relationship, 18
centroid, 305
chi-square difference statistic, 461
chi-square difference test, 461
classic fail-safe N, 555
classification variable, 15, 32
coefficient alpha, 51, 87, 255
coefficient of determination, 186, 209
coefficient of multiple determination, 257
combined model, 501, 515
common factor, 406, 411
communality, 416
comparison-wise Type I error rate, 139
composite, 489
composite model, 501
confidence interval, 147, 183, 215, 331, 564
confirmatory factor analysis, 483
confounding variable, 228
consequent event, 18
consequent variable, 445
constant, 202, 229, 259
continuous variable, 38
control condition, 23
convergence, 323
convergent validity, 489
correlated predictor variable problem, 265
correlated-samples design, 25
correlation ratio, 189, 192
correlational research, 19

count variable, 37
counterbalancing, 25
covariate, 375
covariate × treatments interaction, 403
credibility interval, 565
credibility value, 565, 568
criterion variable, 19
critical value, 129
cross validation, 311
cumulative percentage, 55
curvilinear relationship, 47, 188
data set, 14
data-driven model modifications, 451
degrees of freedom, 81, 134
degrees of freedom between groups, 366
degrees of freedom within groups, 366
dependent variable, 23
descriptive statistic, 17, 50, 79, 85, 98, 370
deviance, 323
deviation, 72
dichotomize, 563
dichotomous, 251
dichotomous variable, 39, 316, 563
dimension-reduction procedure, 406
direct effect, 445, 449, 467, 470, 521
direct logistic regression, 319
directional significance test, 134
discrete variable, 38
discriminant analysis, 296
discriminant function, 299
discriminant function analysis., 296
discriminant function variate, 349
discriminant function variates, 299
discriminant scores, 299
disturbance term, 500, 520
dominance analysis, 269
dummy coding, 30, 288
Duval and Tweedie's trim and fill, 557
effect coefficient, 472, 521
effect size, 133, 161
eigenvalue, 417
eigenvalue-1 criterion, 419
endogenous latent factor, 500
endogenous variable, 448
equal intervals, 32
estimated factor scores, 434
estimated population standard deviation (the unbiased estimate), 84
Estimation, 147
estimation procedure, 452
estimators, 17
eta, 189, 192
eta squared, 192, 363, 367, 369
eta, the correlation ratio, 192
exogenous variable, 449
experimental condition, 23
Experimental control, 228
experiment-wise Type I error rate, 139
exploratory factor analysis, 406
extreme value, 95
F statistic, 367
factor, 413, 478
factor loading, 423, 481, 489
factor matrix, 423
factor pattern matrix, 425
factor rotation, 423
factor structure, 408, 425, 478
factor structure matrix, 430
factorial ANCOVA, 375

factorial ANOVA, 370
factorial MANOVA, 344
false negative rate, 335
false positive rate, 335
familywise Type I error rate, 139, 353
final communality estimate, 417, 433
final preferred model, 505
first-order partial correlation, 237
first-order semipartial correlation, 242
fit indices, 412
fixed-effect model, 541
forest plot, 542
four scales of measurement, 32
frequency histogram, 56
frequency table, 52
full structural model), 501
funnel plot, 556
Gaussian distribution, 100, 102
general linear model GLM, 29
generic analysis model, 41
global fit index, 454
grouping variable, 296
heterogeneity, 544
heterogeneity of regression, 384, 402
heterogeneous, 45
heteroscedasticity, 46, 221
hierarchical multiple regression, 268, 274
higher-order partial correlation, 237
higher-order semipartial correlations, 242
hit rate, 310
homogeneity of regression, 384, 402
homogeneity of variance, 45, 220
homogeneous, 45, 544
homoscedasticity, 45, 221
Hunter and Schmidt approach, 531, 560
incremental R^2 statistic, 283
Incremental variance, 274
incremental variance accounted for, 242
independent, 46
independent variable, 23
independent-samples design, 24
index of effect size, 185, 396, 530
indicator variables, 488
indirect effect, 471, 521
inferential statistic, 17, 80, 213
intact groups, 398
intercept, 202
inter-factor correlations, 411, 429
internal-consistency reliability, 87
interpretability criterion, 421
Interpreting the R2 Statistic, 257
interquartile, 90
interquartile range, 71
interval-scale variable, 34
inverse correlation, 177
Kuder-Richardson reliability estimates, 387
Kurtosis, 62
latent factor, 410, 443, 489, 496
latent-variable path model, 507
leave-one-out cross-validation, 311
leave-one-out estimator, 311
likelihood ratio test, 324
Likert item, 36
Likert-scale, 36
Likert-type scale, 36
limited-value variable, 40, 251, 274, 296, 317, 374
line of best fit, 200
linear, 196
linear regression line, 200
linear relationship, 383, 401
linearity, 47

logistic regression coefficient, 322, 327
logit, 320
manifest variable, 406, 410, 443, 478, 496, 498
MANOVA, 344
maximum likelihood estimation (**ML**), 323, 452
mean, 63
mean average deviation (**MAD**), 72
mean square, 366
mean square between groups, 366
mean square error (**MSE**), 366
measure of variability, 69
measurement model, 484, 501, 507
measurement portion of the model, 501
measures of central tendency, 51, 63
measures of variability, 51
median, 64, 91
mediating variable, 471
mesokurtic, 62
meta-regression, 570
Methods of statistical control, 227
mode, 66
model χ^2 statistic, 339
model χ^2 test, 325
model R^2 statistic, 281
model-comparison χ^2 tests, 514
model-improvement χ^2 statistic, 339
moderator variable, 530, 544, 547
multicollinearity, 48
multinomial logistic regression, 341
multiple regression coefficient, 259, 260
multiple-comparison procedure, 361, 368
multiple-item summated rating scale, 34
multiple-regression approach, 571
multi-value variable, 41, 251, 316
multivariate analysis of covariance (MANCOVA), 375
multivariate analysis of variance (MANOVA), 344
multivariate normality, 47
multivariate statistics, 47
naturally occurring variable, 19
negative correlation, 177
negative skew, 59
negative slope, 203
nested models, 460
no-effect value, 539
nominal-scale, 274
nominal-scale variable, 15, 296
nomological-validity χ^2 tests, 514
nondirectional significance test, 133
nonexperimental research, 19
nonnormed fit inde, 487
non-psychometric meta-analysis, 531, 559
nonrecursive models:, 474
nonstandardized regression coefficient, 202
normal curve, 100
normal distribution, 45, 100, 102
normed fit index, 487
null hypothesis, 30, 125
null-hypothesis significance test (**NHST**), 125
oblique factors, 424
oblique rotations, 424
observation, 14
observed X score, 203
obtained statistic, 131
odds, 321, 330, 538
odds ratio, 330, 537, 538
omega squared, 370
one-tailed significance test, 134
one-way ANCOVA, 375
one-way MANOVA, 344
Order effects, 25
ordinal-scale variable, 32

orthogonal factors, 424
orthogonal rotations, 424
Orwin's fail-safe N, 556
outlier, 95
overall correct classification rate, 311
p value, 135, 258
paired-samples design, 25
parameter, 16, 44, 77
parameter estimate, 17, 213, 452
parametric statistics, 44
Parsimonious fit indices, 455
parsimonious normed-fit index, 456
parsimony ratio, 455
part correlation, 224, 238, 266
partial correlation, 224, 230, 236, 238
Partial eta squared, 370, 396
path analysis with latent factors, 443, 496
path analysis with manifest variables, 443
path coefficient, 449, 453
path model, 442, 501
path portion of the model, 501, 515
Pearson correlation coefficient, 172
Pearson product-moment correlation, 172
percentage, 54
percentile, 55
per-comparison error rate, 139
perfect normal distribution, 57, 102
phi coefficient, 191
physical measurements, 36
planned contrast, 368
planned contrasts, 395
Planned contrasts, 369
platykurtic distribution, 62
point estimate, 147, 155, 331
point-biserial correlation coefficient, 190
population, 16, 77, 84
population standard deviation, 84
population variance, 77
positive correlation, 176
positive skew, 59
positive slope, 203
positively biased, 258
post-hoc tests, 368, 395
power, 132
predicted model matrix, 453
predicted Y score, 202
Prediction, 200
predictor variable, 20
p_{rep} *statistic*, 146
primary theoretical model, 504
principle of least squares, 207
prior communality estimate, 416
probability, 320
probability value, 135, 215, 258, 367
proportions, 54
pseudo-R2 statistics, 333
psychometric meta-analysis, 531, 560
psychometric properties, 487
publication bias, 530, 533, 535, 554
Q statistic, 545
qualitative variables, 15
quantitative variables, 15
quartile, 56, 71
R^2 *statistic*, 516
random-effects model, 541
range, 70
ranking variables, 33
ratio-scale variable, 36
raw residual matrix, 454
raw score, 103, 112
Raw variables, 213
raw-score regression coefficient, 202

real-world variable, 205
Recursive models, 474
region of nonsignificance, 129
Region of Rejection, 129
regression analysis, 196
regression line, 200
regression weight, 202, 203
related-samples design, 25
relative frequencies, 54
Reliability, 188, 487
reliability estimate, 51
repeated-measures design, 25
replication statistics, 146
research hypothesis, 125
research question, 125
residual of prediction, 206, 233
residual term, 445, 450
residualized variable, 234
restriction of range, 564
restriction of range problem, 188
root mean square error of approximation, 487
Rosenthal's fail-safe N, 555
sample, 16, 85
sample standard deviation (the biased estimate), 84
sample variance, 79
sampling distribution, 127
sampling error, 130, 541, 544, 563
scatterplot, 176, 198
scree plot, 419
scree test, 419, 420
second-order semipartial correlation, 242
semi-interquartile range, 72
semipartial correlation, 224, 238, 266
sensitivity, 335
sequential logistic regression, 319, 336
serif font, 600
significance level, 133
simple frequencies, 54
simple structure, 423, 428
simultaneous multiple regression, 268
single-item rating scale, 33
skewed distribution, 59, 66
slope, 202, 203
Spearman correlation coefficient for ranked data, 191
specificity, 335
spurious correlation, 228
spurious relationship, 278
squared canonical correlation coefficient, 309
squared semipartial correlation coefficient, 242
standard deviation, 82
standard error, 60, 130, 214
standard error of the estimate, 206
standard normal distribution, 103, 105
standardized difference, 162
standardized discriminant function coefficient, 306
standardized multiple regression coefficient, 263
standardized path coefficients, 517
standardized regression coefficient, 215
standardized root mean square residual, 486
standardized variables, 213
standardized-difference indices of effect size, 162
standardizing, 103
statistic, 17
statistical alternative hypothesis, 126
statistical assumptions, 44
statistical null hypothesis, 125
statistical significance, 181
statistically significant, 30
stepwise multiple regression, 268
structural equation modeling (SEM), 269, 270, 478, 497
structure coefficient, 306
subgroup-analysis approach, 569

subject variables, 19
sum of squares between-groups, 365
sum of the squares, 76, 365
summary effect, 530, 535, 543
summative scale, 34
synthetic variable, 205, 233
terra incognita, 10
test of association, 28
test of differences, 28
tetrachoric correlation coefficient, 191
the critical value of the t statistic, 129
theoretical normal distribution, 57, 102
theoretically continuous variable, 563
total effect, 270, 472, 521
treatment × covariate interaction, 384
treatment conditions, 23
true dichotomy, 190
true effect size, 543
true experiment, 22, 386
two-tailed significance test, 133
Type I error, 138, 352
Type II error, 140, 328
unbiased estimate of population variance, 80
unimodal, 67
unique factors, 411
uniqueness index, 266, 269

unstandardized path coefficients, 517
unstandardized regression coefficient, 202, 214
usefulness index, 266
Validity, 487
value, 15
variable, 15
variance, 77
variance accounted for index of effect size, 164
variate, 350, 413
vector, 288
Wald χ^2, 328
Wald statistic, 327
whiskers, 91
Wilks' lambda, 302, 309
within-groups design, 25
within-study variability, 544
within-subjects design, 25
within-subjects factor, 344
X variable, 175
Y variable, 175
z score, 103
zero correlation, 177
zero slope, 203
zero-order correlation, 172, 225, 230
z-score variable, 103